Fundamentals *of* Mathematics

Seventh Edition

James Van Dyke

James Rogers

Hollis Adams

Portland Community College

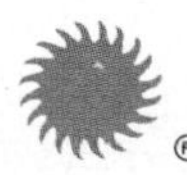

SAUNDERS COLLEGE PUBLISHING

Harcourt Brace College Publishers

Fort Worth Philadelphia San Diego New York Orlando Austin
San Antonio Toronto Montreal London Sydney Tokyo

To our spouses:

Carol Van Dyke
Elinore Rogers
Douglas Adams

Publisher: Emily Barrosse
Acquisitions Editor: Angus McDonald
Developmental Editor: Carol Loyd
Project Management: York Production Services
Art Director: Lisa Adamitis
Text Designers: Rebecca Lemna, Kathleen Flanagan
Production Manager: Alicia Jackson

Cover Credit: Tom Algire/SuperStock

Photo Credits: Chapter 1: Chris Thomaidis/Tony Stone Image©; Chapter 2: Christopher Irion/The Image Bank©; Chapter 3: Garry Adams/ Stock Imagery©; Chapter 4: John Kelly/Tony Stone Images©; Chapter 5: David Madison/Tony Stone Images©; Chapter 6: SuperStock©; Chapter 7: SuperStock©; Chapter 8: James Balog/Tony Stone Images©

Printed in the United States of America

Fundamentals of Mathematics,
Seventh Edition

ISBN: 0-03-022473-X

Library of Congress Catalog Card Number: 98-86825

0123 032 109876543

Contents

Features

Good Advice for Studying: This motivational section, which precedes each chapter, is designed to help students overcome math anxiety through positive reinforcement and to develop the skills necessary for effective learning.

148 CHAPTER 2 MEASUREMENT

Good Advice for Studying

New Habits From Old

If you are in the habit of studying math by only reading the examples to learn how to do the exercises, stop now! Instead, read the assigned section—all of it—before class. It is important that you read more than the examples so that you fully understand the concepts. How to do a problem isn't all that needs to be learned. Where and when to use specific skills are also essential.

When you read, read interactively. This means that you should be both writing and thinking about what you are reading. Write down new vocabulary, perhaps start a list of new terms paraphrased in words that are clear to you. Take notes on the How and Why segments, jotting down questions you may have, for example. As you read examples, work the Warm Up problems in the margin. Begin the exercise set only when you understand what you have read in the section. This process should make your study sessions go much faster and be more effective.

If you have written down questions during your study session, be sure to ask them at the next class session, seek help from a tutor, or discuss them with a classmate. Don't leave these questions unanswered.

Pay particular attention to the objectives at the beginning of each section. Read these at least twice; first, when you do your reading before class and again, after attending class. Ask yourself, "Do I understand what the purpose of this section is?" Read the objectives again before test time to see if you feel that you have met these objectives.

During your study session, if you notice yourself becoming tense and your breathing shallow (light and from your throat or upper part of your lungs), follow this simple coping strategy. Say to yourself: "I'm in control. Relax and take a deep breath." Breathe deeply and properly by relaxing your stomach muscle (that's right, you have permission to let your stomach protrude!) and inhaling so that the air reaches the bottom of your lungs. Hold the air in for a few seconds, then slowly exhale, pulling your stomach muscle in as you exhale. This easy exercise not only strengthens your stomach muscle, but gives your body and brain the oxygen you need to perform free from physical stress and anxiety. This deep breathing relaxation method can be done in one to five minutes. You may want to use it several times a day, especially during an exam.

These techniques can help you to start studying math more effectively and to begin managing your anxiety. Begin today.

Application: Each chapter begins with an illustrated, real-world application, encouraging students to see mathematics as an important part of their everyday lives. Similar problems based on real data from a wide range of subjects also appear in the problem sets and end-of-chapter material.

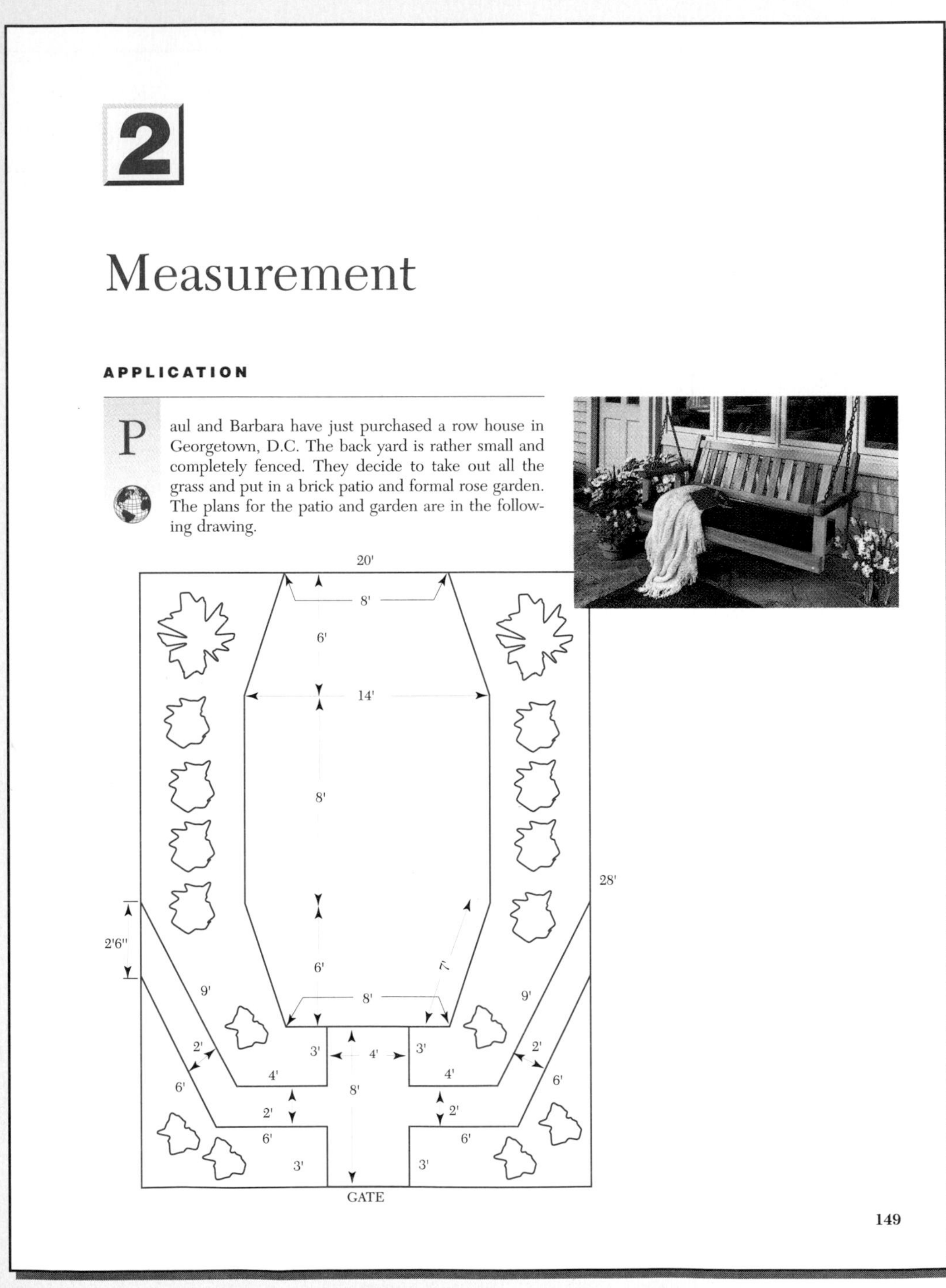

2

Measurement

APPLICATION

Paul and Barbara have just purchased a row house in Georgetown, D.C. The back yard is rather small and completely fenced. They decide to take out all the grass and put in a brick patio and formal rose garden. The plans for the patio and garden are in the following drawing.

149

Learning System: Every section begins with a statement of objectives, followed by a vocabulary list of bolded terms that will familiarize students with the language they need to know in order to master the concepts. Traditional discourse is complemented by the use of graphics and boxed information (e.g., rules, theorems, cautions).

2.4 Volume

OBJECTIVE

Find the volume of common geometric shapes.

VOCABULARY

A **cube** is a three-dimensional geometric solid that has six sides (called **faces**), each of which is a square.

Volume is the name given to the amount of space that is contained inside a three-dimensional object.

HOW AND WHY

Objective

Find the volume of common geometric shapes.

Suppose you have a shoe box that measures 12 in. long by 4 in. wide by 5 in. high that you want to use to store toy blocks that are 1 in. by 1 in. by 1 in. How many blocks will fit in the box?

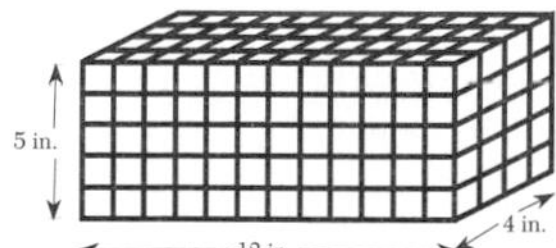

In each layer of blocks there are 12(4) = 48 blocks and there are five layers. Therefore, the box holds 48(5) = 240 blocks.

Volume is a measure of the amount of space that is contained in a three-dimensional object. Often, volume is measured in cubic units. These units are literally *cubes* that measure one unit on each side. For example, pictured below is a cubic inch (1 in^3) and a cubic centimeter (1 cm^3).

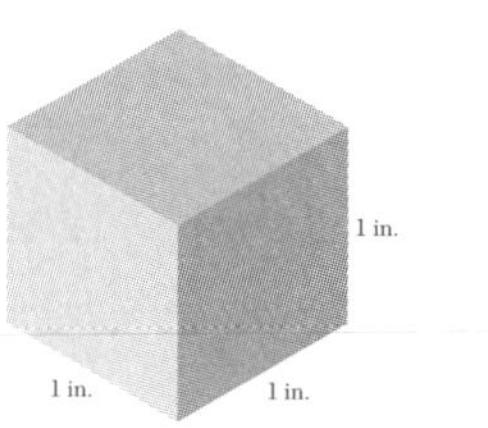

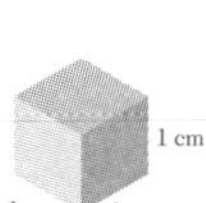

The shoebox discussed previously has a volume of 240 in^3 because exactly 240 blocks, which have volume 1 in^3, can fit inside it and totally fill it up.

In general, volume can be thought of as the number of cubes that fill up a space. If the space is a rectangular solid, like the shoebox, it is a relatively easy matter to determine the volume by making a layer of cubes that covers the bottom, and then deciding how many layers are necessary to fill the box. Note that the number of cubes needed for the bottom layer is the same as the area of the base of the box, ℓw. The number of layers needed is the same as the height of the box, h. So, we come to the following volume formula.

To find the volume of a rectangular solid

Multiply the length by the width by the height.

$V = \ell wh$

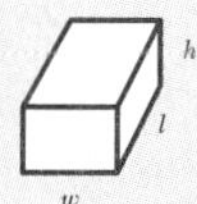

To find the volume of a cube

Cube one of the sides.

$V = s^3$

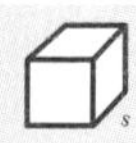

The length, width, and height must all be measured using the same units before the volume formulas may be applied. If the units are different, convert to a common unit before calculating the volume.

The principle used for finding the volume of a box can be extended to any solid with sides that are perpendicular to the base. The area of the base gives the number of cubes necessary to make the bottom layer, and the height gives the number of layers necessary to fill the solid.

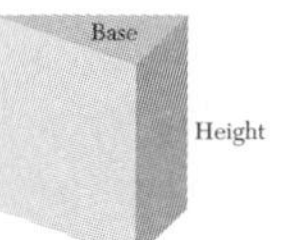

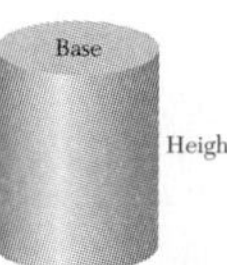

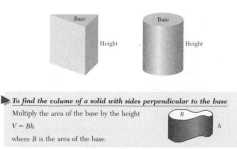

To find the volume of a solid with sides perpendicular to the base

Multiply the area of the base by the height

$V = Bh$

where B is the area of the base.

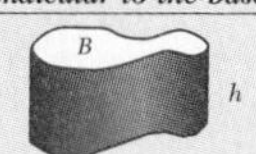

When measuring the capacity of a solid to hold liquid, sometimes special units are used. Recall that in the English system, liquid capacity is measured in ounces, quarts, and gallons. In the metric system, milliliters, liters, and kiloliters are used.

Exercises: A variety of problem types are included at the end of each section, such as enhanced word problems based on real data.

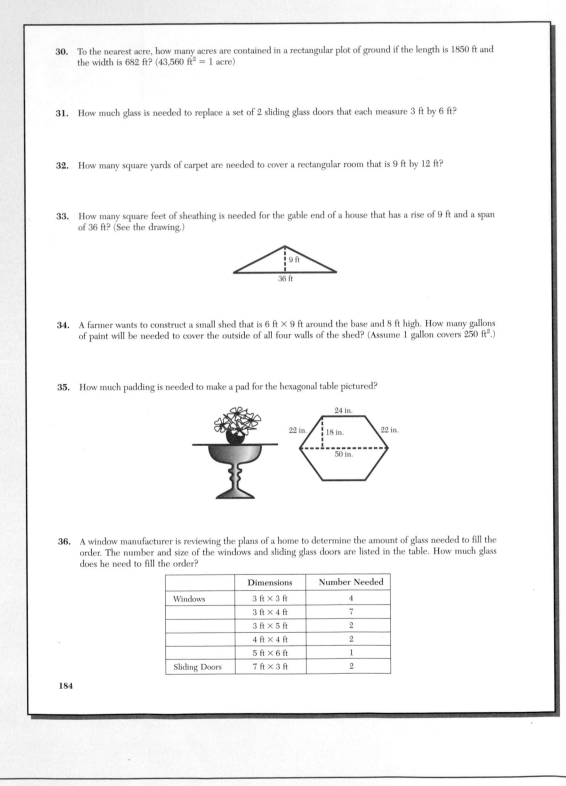

30. To the nearest acre, how many acres are contained in a rectangular plot of ground if the length is 1850 ft and the width is 682 ft? (43,560 ft^2 = 1 acre)

31. How much glass is needed to replace a set of 2 sliding glass doors that each measure 3 ft by 6 ft?

32. How many square yards of carpet are needed to cover a rectangular room that is 9 ft by 12 ft?

33. How many square feet of sheathing is needed for the gable end of a house that has a rise of 9 ft and a span of 36 ft? (See the drawing.)

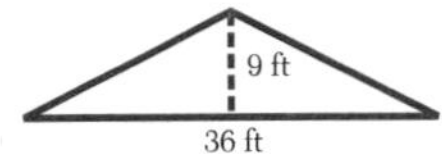

34. A farmer wants to construct a small shed that is 6 ft × 9 ft around the base and 8 ft high. How many gallons of paint will be needed to cover the outside of all four walls of the shed? (Assume 1 gallon covers 250 ft^2.)

35. How much padding is needed to make a pad for the hexagonal table pictured?

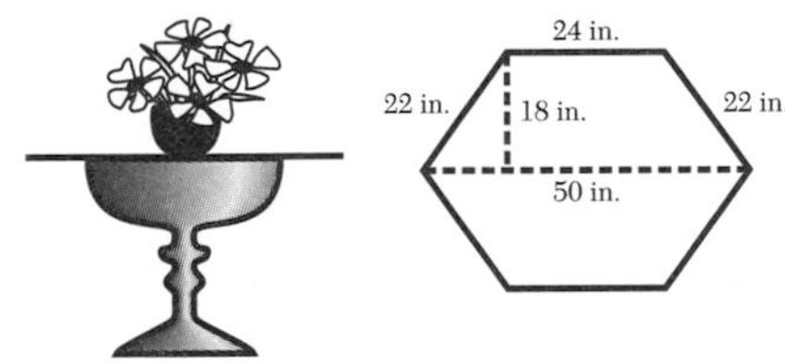

36. A window manufacturer is reviewing the plans of a home to determine the amount of glass needed to fill the order. The number and size of the windows and sliding glass doors are listed in the table. How much glass does he need to fill the order?

	Dimensions	Number Needed
Windows	3 ft × 3 ft	4
	3 ft × 4 ft	7
	3 ft × 5 ft	2
	4 ft × 4 ft	2
	5 ft × 6 ft	1
Sliding Doors	7 ft × 3 ft	2

184

The **"State Your Understanding"** sections require students to analyze and write about the concepts they have learned, thus fostering their ability to communicate about mathematics. **"Challenge"** sets offer several more difficult problems per section, and **"Group Activity"** provides an opportunity for collaborative learning. **"Maintain Your Skills"** reviews material from preceding chapters, so that previous learning is continually reinforced.

NAME ______________________ CLASS ____________ DATE ____________

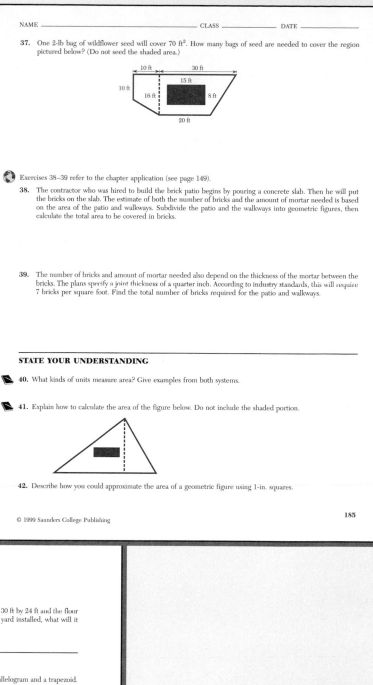

37. One 2-lb bag of wildflower seed will cover 70 ft^2. How many bags of seed are needed to cover the region pictured below? (Do not seed the shaded area.)

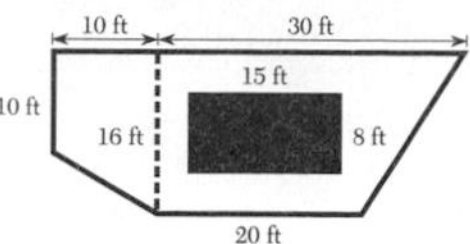

Exercises 38–39 refer to the chapter application (see page 149).

38. The contractor who was hired to build the brick patio begins by pouring a concrete slab. Then he will put the bricks on the slab. The estimate of both the number of bricks and the amount of mortar needed is based on the area of the patio and walkways. Subdivide the patio and the walkways into geometric figures, then calculate the total area to be covered in bricks.

39. The number of bricks and amount of mortar needed also depend on the thickness of the mortar between the bricks. The plans specify a joint thickness of a quarter inch. According to industry standards, this will require 7 bricks per square foot. Find the total number of bricks required for the patio and walkways.

STATE YOUR UNDERSTANDING

40. What kinds of units measure area? Give examples from both systems.

41. Explain how to calculate the area of the figure below. Do not include the shaded portion.

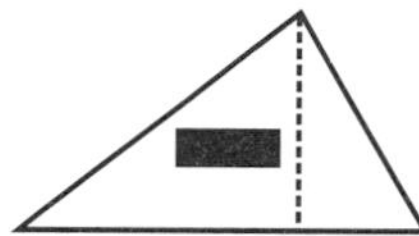

42. Describe how you could approximate the area of a geometric figure using 1-in. squares.

185

CHALLENGE

43. Joe is going to cover his kitchen floor. Along the outside he will put black
The next (inside) row will be white squares that are 6 in. on each si
alternate between black and white squares that are 1 ft on each side. H
he need for the kitchen floor that measures 9 ft by 10 ft?

44. A rectangular plot of ground measuring 120 ft by 200 ft is to have a ce
the inside of the perimeter. How much of the area of the plot will be use
area will remain for the lawn?

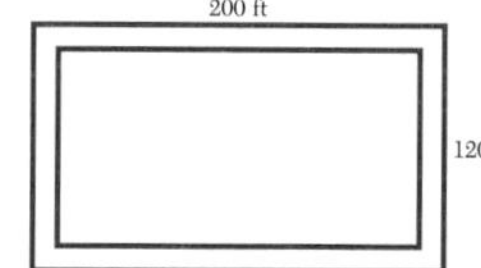

45. Ingrid is going to carpet two rooms in her house. The floor in one room measures 30 ft by 24 ft and the floor in the other room measures 22 ft by 18 ft. If the carpet costs $27.00 per square yard installed, what will it cost Ingrid to have the carpet installed?

GROUP ACTIVITY

46. Use the formula for the area of a rectangle to show how to find the area of a parallelogram and a trapezoid. Draw pictures to illustrate your argument.

47. Determine the coverage of 1 gallon of semi-gloss paint. How much of this paint is needed to paint your classroom, excluding chalkboards, windows, and doors? What would it cost? Compare your results with the other groups in the class. Did all the groups get the same results? Give possible explanations for the differences.

MAINTAIN YOUR SKILLS (SECTIONS 1.1, 1.2, 1.3, 1.4, 1.5, 1.6, 1.7, 1.9, 2.2)

48. Find the product of 47 and 962. Round your answer to the nearest 100.

49. Find the quotient of 295,850 and 97.

50. Simplify: $2(46 - 28) + 50 \div 5$

51. A family of six attends a weaving exhibition. Parking was \$4, adult admission to the exhibition \$6, senior admission \$4, and child admission \$3. How much does it cost for 2 parents, 1 grandmother and 3 children to attend the exhibition?

186

NAME ______________________ CLASS ____________ DATE ____________

CHAPTER 2

Group Project *(2–3 weeks)* OPTIONAL

You are working for a kitchen design firm that has been hired to design a kitchen for the 10-ft-by-12-ft room pictured below.

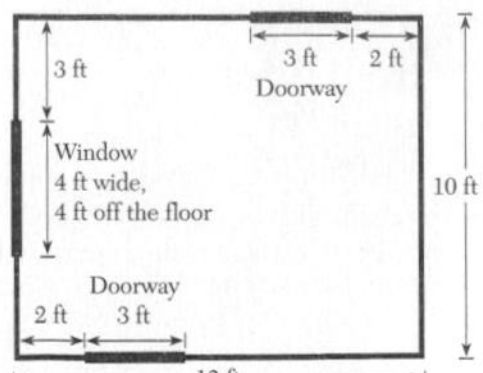

The following table lists appliances and dimensions. Some of the appliances are required and others are optional. All dimensions are in inches.

Appliance	High	Wide	Deep	Required
Refrigerator	68	30 or 33	30	Yes
Range/Oven	30	30	26	Yes
Sink	12	36	22	Yes
Dishwasher	30	24	24	No
Trash compactor	30	15	24	No
Built-in microwave	24	24	24	No

The base cabinets are all 30 in. high and 24 in. deep. The widths can b
tiple of 3 from 12 in. to 36 in. Corner units are 36 in. along the wall
rection. The base cabinets (and the range, dishwasher, and compactor)
installed on 4-in. bases that are 20 in. deep.

The wall (upper) cabinets are all 30 in. high and 12 in. deep. Here too,
can be any multiple of 3 from 12 in. to 36 in. Corner units are 24 in. alo
in each direction.

End-of-Chapter Pedagogy: A **"Group Project"** that takes approximately 2 weeks to complete is provided at the end of each chapter. The **"True-False Concept Review"** requires students to think critically and checks their understanding of the language.

NAME ______________________ CLASS ____________ DATE ____________

CHAPTER 2

True–False Concept Review

Check your understanding of the language of algebra and geometry. Tell whether each of the following statements is true (always true) or false (not always true). For each statement you judge to be false, revise it to make a statement that is true.

ANSWERS

1. English measurements are the most commonly used in the world. **1.** ________

2. Equivalent measures have different units of measurement. **2.** ________

3. A liter is a measure of weight. **3.** ________

4. The perimeter of a square can be found in inches, feet, centimeters, or meters. **4.** ________

5. Area is the measure of the inside of a solid, such as a box or a can. **5.** ________

6. The volume of a square is $V = s^3$. **6.** ________

7. The formula for the area of a trapezoid is $A = \frac{(b_1 + b_2)h}{2}$. **7.** ________

8. It is possible to find equivalent measures without remeasuring the original object. **8.** ________

9. The metric system utilizes the base-ten place value system. **9.** ________

10. Volume is the measure of how much a container will hold. **10.** ________

11. Weight can be measured in pounds, grams, or kilograms. **11.** ________

12. A parallelogram has three sides. **12.** ________

13. Volume can be thought of as the number of squares in an object. **13.** ________

14. One milliliter is equivalent to one cubic centimeter. **14.** ________

 201

Finally, a **"Chapter Test"** gives a comprehensive review of all the material within a chapter, helping students identify areas that might require further review.

NAME ______________________ CLASS __________ DATE __________

CHAPTER 2

Test

ANSWERS

1. 7 m + 454 mm = ? mm — **1.** ______

2. Find the perimeter of a square that is 34 cm on a side. — **2.** ______

3. Find the volume of a drawer that is 4 in. high, 18 in. wide and 24 in. deep. — **3.** ______

4. How much vinyl flooring is needed to cover the room pictured? — **4.** ______

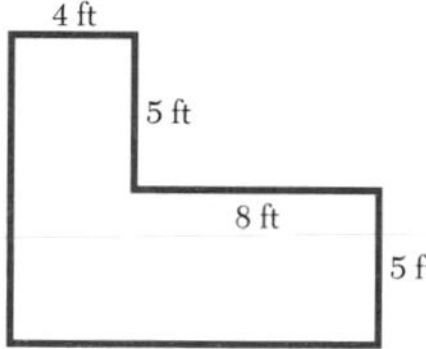

5. Anna has 135 lb of strawberries to divide equally among her 5 children. How many pounds of berries will each one receive? — **5.** ______

6. Find the perimeter of the figure. — **6.** ______

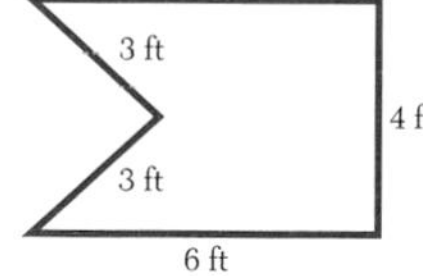

7. Find the area of the triangle. — **7.** ______

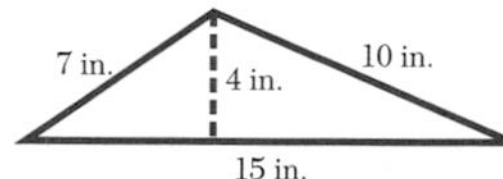

8. Subtract: — **8.** ______

$$\begin{array}{r} 5 \text{ gal } 2 \text{ qt} \\ - \underline{3 \text{ gal } 2 \text{ qt } 1 \text{ pt}} \end{array}$$

To the Student

"It looks so easy when you do it, but when I get home..." is a popular lament of many students studying mathematics.

The process of learning mathematics evolves in stages. For most students, the first stage is listening to and watching others. In the middle stage, students experiment, discover, and practice. In the final stage, students analyze and summarize what they have learned. Many students try to do only the middle stage because they do not realize how important the entire process is.

Here are some steps that will help you to work through all the learning stages:

1. Go to class every day. Be prepared, take notes, and most of all, think actively about what is happening. Ask questions and keep yourself focused. This is prime study time.
2. Begin your homework as soon after class as possible. Start by reviewing your class notes and then read the text. Each section is organized in the same manner to help you find information easily. The objectives tell you what concepts will be covered, and the vocabulary lists all the new technical words. There is a How and Why section for each objective that explains the basic concept, followed by worked sample problems. As you read each example, make sure you understand every step. Then work the corresponding Warm Up problem to reinforce what you have learned. You can check your answer at the bottom of the page. Continue through the whole section in this manner.
3. Now work the exercises at the end of the section. The A group of exercises can usually be done in your head. The B group is harder and will probably require pencil and paper. The C group problems are more difficult, and the objectives are mixed to give you practice at distinguishing the different solving strategies. As a general rule, do not spend more than 15 minutes on any one problem. If you cannot do a problem, mark it and ask someone (your teacher, a tutor, or a study buddy) to help you with it later. Do not skip the Maintain Your Skills problems. They are for review and will help you practice earlier procedures so you do not become "rusty." The answers to the odd exercises are in the back of the text so you can check your progress.
4. In this text, you will find State Your Understanding exercises in every section. You may do these orally or in writing. Their purpose is to encourage you to analyze or summarize a skill and put it into words. Some of these exercises are designated as journal entries. The journal entries are intended to be written. Taken as a whole, the journal entries cover *all* the basic concepts in the text. We recommend that the journal entries be kept together in a special place in your notebook. Then they are readily available as a review for chapter tests and exams.

5. When preparing for a test, work the material at the end of the chapter. The True–False Concept Review and the Chapter Test give you a chance to review the concepts you have learned. You may want to use the Chapter Test as a practice test.

If you have never had to write in a math class, the idea can be intimidating. Write as if you are explaining to a classmate who was absent the day the concept was discussed. Use your own words—*do not copy out of the text.* The goal is that you understand the concept, not that you can quote what the authors have said. Always use complete sentences, correct spelling, and proper punctuation. Like everything else, writing about math is a learned skill. Be patient with yourself and you will catch on.

Since we have many students who do not have a happy history with math, we have included Good Advice for Studying—a series of eight essays that address various problems that are common for students. They include advice on time organization, test taking, and reducing math anxiety. We talk about these things with our own students, and hope that you will find some useful tips.

We really want you to succeed in this course. If you go through each stage of learning and follow all the steps, you will have an excellent chance for success, But remember, you are in control of your learning. The effort that you put into this course is the single biggest factor in determining the outcome. Good luck!

Jim Van Dyke
Jim Rogers
Hollis Adams

CLAST Skills and Their Locations in the Book

Skill		Location in Book
1A1a	Add and subtract rational numbers	Sections 4.7–4.11
1A1b	Multiply and divide rational numbers	Sections 4.3, 4.4, 4.11
1A2a	Add and subtract rational numbers in decimal form	Sections 5.3, 5.9
1A2b	Multiply and divide rational numbers in decimal form	Sections 5.4, 5.5, 5.6, 5.9
1A3	Calculate percent increase and decrease	Section 7.8
2A1	Recognize the meaning of exponents	Sections 1.5, 3.5, 3.6, 5.5
2A2	Recognize the role of the base number in determining place value in the base-ten numeration system and in systems that are patterned after it	Sections 1.1, 5.1
2A3	Identify equivalent forms of positive rational numbers	Sections 5.2, 5.8
2A4	Determine the order relation between magnitudes	Sections 1.1, 4.6, 5.2
4A1	Solve real-world problems which do not require the use of variables and which do not require the use of percent	Chapters 1, 2, 3, 4, 5, 6
4A2	Solve real-world problems that require the use of percent	Section 7.8
4A3	Solve problems that involve the structure and logic of arithmetic	Throughout
1B1	Round measurements to the nearest given unit of the measuring device	Sections 1.1, 5.1
1B2a	Calculate distances	Section 2.2
1B2b	Calculate area	Sections 2.3, 2.4, 4.11
1B2c	Calculate volume	Sections 2.5, 4.11
2B2	Classify simple plane figures by recognizing their properties	Chapter 2
3B1	Infer formulas for measuring geometric figures	Chapter 2, 4.11
3B2	Identify applicable formulas for computing measures of geometric figures	Chapter 2, 4.11
4B1	Solve real-world problems involving perimeter, area, and volume of geometric figures	Sections 2.2, 2.3, 2.4, 2.5
1C1a	Add and subtract real numbers	Sections 8.2, 8.3
1C1b	Multiply and divide real numbers	Sections 8.4, 8.5
1C2	Apply the order of operations agreement to computations involving numbers and variables	Section 8.6
1C4	Solve linear equations and inequalities	Sections 1.2, 1.4, 1.6, 4.4, 4.10, 5.3, 5.6, 5.9, 8.7
1C5	Use given formulas to compute results when geometric measurements are not involved	Throughout
2C3	Recognize statements and conditions of proportionality and variation	Chapter 6
1D1	Identify information contained in a bar, line, and circle graph	Section 1.9
4D1	Interpret real-world data from tables and charts	Section 1.8

ELM Mathematical Skills

The following table lists the California ELM MATHEMATICAL SKILLS and where coverage of these skills can be found in the text. Location of the skills is indicated by chapter section or chapter.

Skill	Location in Text
Cluster A: Algebra	
Real numbers and their operations	Chapters 1, 4, 5
Scientific notation	5.5
Absolute value	8.1
Applications (e.g., estimations, percents, word problems, charts and graphs)	1.2, 1.3, 1.4, 1.8, 1.9, 6.3, 7.8, and throughout the exercise sets
Integer Exponents	1.5, 5.5
Linear equations in one unknown with numerical or literal coefficients	1.2, 1.4, 1.6, 4.4, 4.10, 4.6, 5.3, 5.9, 8.7
Applications and word problems (including ratio and proportion)	6.3 and throughout the exercise sets
Cluster B: Geometry	
Perimeter of triangles	2.2
Area of triangles	2.3
Perimeter of squares, rectangles, and parallelograms	2.2
Areas of squares, rectangles, and parallelograms	2.3
Radius and diameter of circle	4.11
Circumference of a circle	4.11
Area of a circle	4.11
Volume of rectangular solids	2.4
Volume of cylinders	4.11
Volume of spheres	4.11
Interior and exterior angles of triangles: sum of interior angles	Available Geometry Supplement
Angles formed by parallel and perpendicular lines	Available Geometry Supplement
Equilateral and isosceles triangles	Available Geometry Supplement
Right triangles and the Pythagorean Theorem	Available Geometry Supplement
45° - 45° - 90° and 30° - 60° - 90° triangles	Available Geometry Supplement
Congruent and similar triangles	
Cluster C: Data interpretation, Counting, Probability, and Statistics	
Reading data from graphs and charts	1.7
Computation with data	1.7, 1.8, and throughout text
Average (arithmetic mean)	1.6, 4.11, 5.9

TASP Skills and Their Location in the Book

Skill	Location in Book
Use number concepts and computation skills	Throughout
Solve word problems involving integers, fractions, or decimals (including percents, ratios, and proportions)	Chapters 4–8
Solve one- and two-variable equations	Sections 1.2, 1.4, 1.6, 4.4, 4.10, 5.3, 5.6, 5.9, 8.7
Solve problems involving geometric figures	Chapter 2, 4.11

To the Instructor

Fundamentals of Mathematics, seventh edition, is a work text for college students who need to review the basic skills and concepts of arithmetic in order to pass competency or placement exams, or to prepare for courses such as business mathematics or elementary algebra. The text is accompanied by a complete system of ancillaries in a variety of media, affording great flexibility for individual instructors and students.

A Textbook for Adult Students

Though the mathematical content of *Fundamentals of Mathematics* is elementary, students using the text are most often mature adults, bringing with them adult attitudes and experiences and a broad range of abilities. Teaching elementary content to these students, therefore, is effective when it accounts for their distinct and diverse adult needs. As you read about and examine the features of *Fundamentals of Mathematics* and its ancillaries, you will see how they especially meet these three needs of your students:

- Students must establish good study habits and overcome math anxiety.
- Students must see connections between mathematics and the modern, day-to-day world of adult activities.
- Students must be paced and challenged according to their individual level of understanding.

A Textbook for Many Course Formats

Fundamentals of Mathematics is suitable for individual study or for a variety of course formats: lab, both supervised and self-paced; lecture; group; or combined formats. For a lecture-based course, for example, each section is designed to be covered in a standard 50-minute class. The lecture can be interrupted periodically so that students individually can work the warm up exercises or work in small groups on the group activities. In a self-paced lab course, warm up exercises give students a chance to practice while they learn, and get immediate feedback since warm up answers are printed on the bottom of each page. Using the text's ancillaries, instructors and students have even more options available to them. Computer users, for example, can take advantage of complete electronic tutorial and testing systems that are fully coordinated with the text. A detailed description of each print, video, and computer ancillary is found on page xxi.

Teaching Methodology

As you examine the seventh edition of *Fundamentals of Mathematics,* you will see distinctive format and pedagogy that reflect these aspects of teaching methodology:

Teaching by Objective Each section focuses on a short list of objectives, stated at the beginning of the section. The objectives correspond to the sequence of exposition and tie together other pedagogy, including the highlighted content, the examples, and the exercises.

Teaching by Application Applications are included in the examples for most objectives. Other applications appear in exercise sets. These cover a diverse range of fields, demonstrating the utility of the content in business, environment, personal health, sports, and daily life.

Stressing Language New words of each section are explained in the vocabulary segment that precedes the exposition. Exercise sets include questions requiring responses written in the students' own words.

Stressing Skill, Concept, and Problem Solving Each section covers concepts *and* skills that are fully explained and demonstrated in the exposition for each objective. Carefully constructed examples for each objective are connected by a common strategy that reinforces both the skill and the underlying concepts. Skills are not treated as isolated feats of memorization but as the practical result of conceptual understanding: Skills are strategies for solving related problems. Students learn to see the connections between problems and their common solution.

Allowing for Group Work There are short group activities in each section and a concluding chapter group activity. These may be assigned according to student and instructor goals as time permits.

Topics and Sequence **Chapter 1** begins with our numeration system so the student has a thorough understanding of the concept of "number." The basic operations of addition, subtraction, multiplication, and division are reviewed quickly. Concurrently the student is introduced to estimating when performing operations. This ability to estimate increases self-confidence with respect to doing calculations. Following Section 1.2 there begins an ongoing set of introductions to equation solving. The optional *Getting Ready for Algebra* sections use the same skills covered in the preceding section. Exponents are examined so they can be used in the order of operations with whole numbers. Powers of ten are introduced to prepare for mental multiplication and division and to subsequently introduce Scientific Notation (Chapter 5). After covering order of operations and average, the chapter concludes with sections on tables, charts and graphs as ways to organize and display data. Information is compiled from charts using material commonly found in newspapers and magazines.

In **Chapter 2** measurements in both the English and metric systems are studied. Conversions are done in both systems using whole numbers. Measurement is applied to geometric figures covering perimeter, area, and volume where these lend themselves to whole number arithmetic. Compound figures are introduced in area and volume. Formulas for circles, cones, and cylinders are postponed to Sections 4.11 and 5.9.

Chapter 3 starts with divisibility tests that will be helpful in prime factoring and subsequently in building and simplifying fractions. Whole number classifications of multiples, divisors, factors, primes, and composites follow and give the student useful practice with multiplication and division. The chapter concludes with prime factorization and least common multiple computations, two concepts that play an important role in the next chapter.

Chapter 4 begins with a discussion of the meaning of a fraction, using shaded unit regions and rulers to model fractions. This visual presentation backs the mathematical concepts. Operations on fractions and mixed numbers are covered. Building and simplifying fractions utilize the concepts and skills acquired in Chapter 3. In adding fractions, the idea of least common denominator flows from the Chapter 2 presentation of least common multiple. Proficiency in these skills is ensured by concluding the chapter with a presentation of order of operations, average, and geometric formulas involving π, where the approximation $\frac{22}{7}$ is used.

Chapter 5 covers decimals, following an approach parallel to the presentation of whole numbers. The concepts of place value, word names, expanded form, rounding, and inequality are extended to decimals. The basic operations are covered. A special section on multiplying and dividing by powers of ten and Scientific Notation is included that uses the exponent skills developed in Chapter 1. Conversions between fractions and decimals are given to show the relationship between the two ways of writing rational numbers. The fact that not all fractions have an exact decimal representation demonstrates the need for fractions and the practical use of rounding. Conversion of units is extended to include conversions between the English and metric systems and conversion of various rates. The chapter ends with a review of the order of operations, average, and geometric formulas involving π, where the approximation 3.14 is used.

Chapter 6 incorporates discussions of ratios, rates, and proportions. Each concept leads to practical applications of mathematics. The student is introduced to the formal process of translating from a written statement of facts to a mathematical model that can be solved. This skill is reinforced in the next chapter. Solving a proportion is related to solving an algebraic equation.

Chapter 7 presents percent as a useful way to describe numerical comparisons. Students practice change from percents to decimals to fractions so they can see the relationship of percent to the rational numbers. Students become skilled at expressing a number in any one of the forms. Solutions of percent problems are covered using either ratios or the formula $R \times A = B$. The formula $R \times A = B$ is presented in the form of a triangle for quick recall. The chapter ends with a section on applications that incorporates business, sports, environmental issues, and other topics. Circle graphs are introduced in this chapter.

Chapter 8 expands the number system to include signed numbers. Operations on signed numbers include absolute value, opposites, addition, subtraction, multiplication, and division. Order of operations and the solution of equations with signed numbers conclude the chapter. This chapter together with the previous sections on *Getting Ready for Algebra* serve as a bridge for the students' future study of algebra.

Special Content

Special content focuses on study skills and math anxiety, calculators, and simple algebraic equations.

Good Advice for Studying is continued from the sixth edition. Originally written by the team of Dorette Long and Sylvia Thomas, of Rogue Community College, these essays address the unique study problems that students of *Fundamentals of Mathematics* experience. Students learn general study skills and study skills specific to mathematics and to the pedagogy and ancillaries of *Fundamentals of Mathematics*. Special techniques are described to overcome the pervasive problem of math anxiety. Though an essay begins each chapter, students may profit by reading all the essays at once returning to them as the need arises. A description of how to learn from these essays appears in To the Student.

Calculator examples, marked by the symbol 🖩, demonstrate how a calculator may be used, though the use of a calculator is left to the discretion of the instructor. Nowhere is the use of a calculator required. Appendix A reviews the basics of operating a calculator.

Getting Ready for Algebra segments follow sections 1.2, 1.4, 1.6, 4.4, 4.10, 5.3, 5.6, and 5.9. The operations from these sections lend themselves to solving simple algebraic equations. Though entirely optional, each of these segments includes its own exposition, examples with warm ups and exercises. Instructors may cover these segments as part of the normal curriculum or assign them to individual students.

Special Pedagogy

The pedagogical system of *Fundamentals of Mathematics* meets two important criteria: coordinated purpose and consistency of presentation.

Each section begins with numbered **Objectives,** followed by definitions of new **Vocabulary** to be encountered in the section. Following the vocabulary, **How and Why** segments, numbered to correspond to the objectives, explain and demonstrate concepts and skills. Throughout the How and Why segments, **skill boxes** clearly summarize and outline the skills in step-by-step form. Also throughout the segments, **concept boxes** highlight appropriate properties, formulas, and theoretical facts underlying the skills. Following each How and Why segment are **Examples** and **Warm Ups.** Each example of an objective is paired with a warm up, with workspace provided. Solutions to the warm ups are given at the bottom of the page, affording immediate feedback. The examples also include, where suitable, a relevant application of the objective. Examples similar to each other are linked by common **Directions** and a common **Strategy** for solution. Directions and strategies are closely related to the skill boxes. Connecting examples by a common solution method helps students recognize the similarity of problems and their solutions, despite their specific differences. In this way, students may improve their problem solving skills. In both How and Why segments and in the examples, **Caution** remarks help to forestall common mistakes.

Exercises, Reviews, Tests

Thorough, varied, properly paced, and well-chosen exercises are a hallmark of *Fundamentals of Mathematics.* Exercise sets are provided at the end of each section and at the end of each chapter. Necessary workspace is provided for all exercises and each exercise set can be torn out and handed in without disturbing other parts of the book.

Section exercises are paired so that virtually each odd-numbered exercise, in Sections A and B, is paired with an even-numbered exercise that is equivalent in type and difficulty. Since answers for odd-numbered exercises are in the back of the book, students can be assigned odd-numbered exercises for practice, and instructors can assign even numbers exercises for home work.

Section exercises are categorized to satisfy teaching and learning aims. Exercises for estimation, mental computation, pencil and paper computation, applications and calculator skills are provided, as well as opportunities for students to challenge their abilities, master communications skills, and to participate in group problem solving.

- **Category A** exercises, organized by section objective, are those that most students should be able to solve mentally, without pencil, paper, or calculator. Mentally working problems improves students' estimating abilities. These can often be used in class as oral exercises.
- **Category B** exercises, also organized by objective, are similar except for level of difficulty. All students should be able to master Category B.
- **Category C** exercises contain applications and more difficult exercises. Since these are not categorized by objective, the student must decide on the strategy needed to set up and solve the problem. These applications are drawn from business, health and nutrition, environment, consumer, sports, and science fields. Both professional and daily-life uses of mathematics are incorporated.

State Your Understanding exercises require a written response, usually no more than two or three sentences. **Journal-writing** exercises are included here and are identified by a journal icon. These offer the student a chance to explain key concepts in their own words. Maintaining a journal allows students to review concepts as they have written them. These writing opportunities facilitate student writing in accordance with **AMATYC** and **NCTM** endorsed standards.

Challenge exercises stretch the content and are more demanding computationally and conceptually.

Group Activity exercises and **Group Projects** provide opportunities for small groups of students to work together to solve problems and create reports. While the use of these is optional, the authors suggest the assignment of two or three of these per semester or term to furnish students with an environment for exchanging ideas. Group Activity exercises encourage cooperative learning as recommended by **AMATYC** and **NTCM** guidelines.

Maintain Your Skills exercises continually reinforce mastery of skills and concepts from previous sections. These exercises are referenced so students can return to a section for needed reexamination.

Chapter True–False Concept Review exercises require students to judge whether a statement is true or false and, if false, to rewrite the sentence to make it true. Students evaluate their understanding of concepts and also gain experience using the vocabulary of mathematics.

Chapter Test exercises end the chapter. Written to imitate a 50-minute exam, each Chapter Test covers all of the chapter content. Students can use the Chapter Test as a self-test before the classroom test.

Ancillaries

The following supplements to accompany *Fundamentals of Mathematics,* seventh edition, are available to enhance the presentation and understanding of the course:

Student Solutions Manual This guide contains worked-out solutions to one quarter of the problems in the exercise sets (every other odd-numbered problem) to help the student learn and practice the techniques used in solving problems.

Instructor's Manual This supplement features instructor-appropriate solutions to problems in the text. All solutions have been reviewed for accuracy.

Test Bank and Prepared Tests This resource contains written tests, including both open-ended and multiple choice questions, for each chapter of the book. In addition, it provides a printed test bank generated from a computerized test bank.

ExaMaster+™ A flexible, powerful testing system, ExaMaster+™ offers instructors a wide range of integrated testing options and features. For each chapter, test items are provided that can be selected with or without multiple-choice distractors. Teachers can select test items according to a variety of other criteria, including section, objective, focus (skill, concept, or application), and difficulty (easy, medium, or hard). Teachers can scramble the order of test items, administer tests on-line, and print objective-referenced answer keys. ExaMaster+™ can also be used to create extra practice worksheets, and includes a full-function gradebook and graphing features.

MathCue Interactive Software Available in Macintosh and Windows® versions, this interactive software provides additional, self-paced support to students and is free to adopters. The new Windows version combines Tutorial Practices and Solution Finder programs on one disk and includes new features, such as hot links to math summaries, formulas, pop-up definitions, and a searchable index. The following features apply to both platforms:

Tutorial Keyed to topics in *Fundamentals,* this software allows students to test their skills and pinpoint and correct weak areas. Students may choose to see step-by-step solutions or partial solutions to all problems. The software features a Missed Problem and Disk Review to enable students to review problems answered incorrectly and to summarize all topics on the disk.

Practice This algorithm-based software allows students to generate large numbers of practice problems keyed to problem types from each section of the book. Student performance is scored and saved for each session.

Solution Finder This software allows students to input their own questions through use of an expert system, a branch of artificial intelligence. Students can check answers or receive help as if they were working with a tutor. The software will refer the student to the appropriate section of the text and will record the number of problems entered and evaluate a function at a point, graph up to four functions simultaneously, and save and retrieve function setups via disk files.

Videotapes These section-by-section videos use a newscast format to review problem-solving methods and guide the students through practice problems.

Core Concepts Video This four-hour video tutorial covers the core concepts and works through selected examples from each section of the text. Students with access to a VCR can use the video as a take-home tutorial.

Saunders College Publishing may provide complimentary instructional aids and supplements or supplement packages to those adopters qualified under our adoption policy. Please contact your sales representative for more information. If as an adopter or potential user you receive supplements you do not need, please return them to your sales representative or send them to
Attn: Returns Department
Troy Warehouse
465 South Lincoln Drive
Troy, MO 63379

Changes in the Seventh Edition

Instructors who have used a previous edition of *Fundamentals of Mathematics* will see changes and improvements in format, pedagogy, exercises, and the sectioning of content. Many of these changes are in response to comments and suggestions offered by users and reviewers of the manuscript. Others have been made to bring the text into line with math reform standards and to give the instructor the chance to follow educational guidelines recommended by AMATYC and NCTM.

Changes in Content The text has been reduced from 62 sections to 57 sections by combining related sections in Chapters 1, 2 (old chapter 7), and 5 (old Chapter 4).

Sections on drawing and interpreting graphs and reading and interpreting tables have been moved to the end of Chapter 1. This allows the use of graphs and tables throughout the text, giving students more exposure to actual uses of mathematics.

Part of the chapter on measurement has been moved to Chapter 2, following whole numbers. The example and exercises involve geometric formulas that can be restricted to whole numbers. Providing this material earlier in the text allows for the use of geometry throughout, thus providing a richer variety of applications. Successive chapters on fractions and decimals incorporate the same formulas in context. Spiraling these topics furnishes continuous review of basic formulas while incorporating new arithmetic skill. Geometric formulas involving π have been deferred to the chapters on fractions $\left(\frac{22}{7}\right)$ and decimals (3.14).

Conversion of units within and between English and metric measures is now in Sections 4.5 and 5.7. These placements provide students with earlier applications of the multiplication property of one.

Estimations of sums, products, differences, and quotients have been included for both whole numbers and decimals. Students practice making quick estimates to determine whether computed answers (or calculator displays) are reasonable.

Changes in Pedagogy The following changes are pedagogical and enhance the teachability of the text.

- Each chapter begins with a **theme application** that gives the students real-world context for the mathematics of that chapter. Exercises concerning or related to the theme problem are included in almost every section of the chapter.
- Each objective in a given section is treated separately in the **How and Why,** and followed by examples. Students can now begin by focusing on a single topic until they understand it.
- The word "application" has been eliminated from the word problems so that students will view these as "just another exercise or problem" and not as special or difficult exercises.
- To avoid confusion, calculator keystrokes have been removed from the body of the text because of the wide variety of calculators available. The appendix on calculators has been updated to familiarize students with some of the technology now accessible.
- The **Exercises** have been reorganized. The drill exercises (both A and B groups) are separated by objective. The more difficult C exercises and former "application" problems mix objectives, giving the student more practice in applying skills to real situations.
- **Chart, graph, and table readings** are included in most sections so students gain skill in acquiring or displaying information in these formats.
- **Journal-writing** exercises are included and identified by a Journal icon. These may be used at the instructor's discretion as homework, as a mathematical journal reference for the individual student, or as short class projects in writing.
- For those instructors who desire to emphasize writing in mathematics, *all* answers to problems and exercises written in words are given in complete sentences.
- A discretionary **open-ended project** is included at the end of each chapter. Students can have a chance to work in groups and practice applying their new skills in a real setting.

Acknowledgments

The authors appreciate the unfailing and continuous support of their spouses—Carol Van Dyke, Elinore Rogers, and Doug Adams—who made the completion of this work possible. We are grateful to Angus McDonald, Carol Loyd, Terri Ward, Alicia Jackson, Lisa Adamitis, and Kathleen Flanagan of Saunders College Publishing for their suggestions during the preparation of the text. We also want to express our thanks to the following professors and reviewers for their many excellent contributions to the development of the text:

Kinley Alston, *Trident Technical College*
Kristi Bowers, *Mount Aloysius College*
Lawrence Chernoff, *Miami-Dade Community College*
Ann Corbeil, *Massasoit Community College*
Frances Cummins, *Orange County Community College*
Karen Driskell, *Calhoun State Community College*
Nina Gath, *Washington State Community College*
Judith Jones, *Valencia Community College*
Maryanne Kirkpatrick, *University of Wyoming*
Kathryn Lavelle, *Westchester Community College*
Hazel McKenna, *Utah Valley State College*
Ellen Montoya, *Walla Walla Community College*
Lily O'Reilly, *Portland Community College*
Mary Lee Seitz, *Erie Community College*
Jacci Wozniak, *Brevard Community College*
Ed Zanella, *Community College of Rhode Island*
Stephen Zona, *Quinsigamond Community College*

Special thanks to Ann Ostberg and Sudhir Goel for their careful reading of the text and for the accuracy review of all the problems and exercises in the text.

Jim Van Dyke
Jim Rogers
Hollis Adams

Good Advice for Studying

Strategies for Success

Are you afraid of math? Do you panic on tests or "blank out" and forget what you have studied, only to recall the material after the test? Then you are just like many other students. In fact, research studies estimate that as many as 50% of you have some degree of math anxiety.

What is math anxiety? It is a learned fear response to math that causes disruptive, debilitating reactions to tests. It can be so encompassing that it becomes a dread of doing *anything* that involves numbers. Although some anxiety at test time is beneficial—it can motivate and energize you, for example—numerous studies show that too much anxiety results in poorer test scores. Besides performing poorly on tests, you may be distracted by worrisome thoughts, and be unable to concentrate and recall what you've learned. You may also set unrealistic performance standards for yourself and imagine catastrophic consequences for your failure to be successful in math. Your physical signs could be muscle tightness, stomach upset, sweating, headache, shortness of breath, shaking, or rapid heart beat.

The good news is that anxiety is a learned behavior and therefore can be unlearned. If you want to stop feeling anxious, the choice is up to you. You can choose to learn behaviors that are more useful to achieve success in math. You can learn and choose the ways that work best for you.

To achieve success, you can focus on two broad strategies. First, you can study math in ways *proven* to be effective in learning mathematics and taking tests. Second, you can learn to physically and mentally *relax,* to manage your anxious feelings and to think rationally and positively. Make a time commitment to practice relaxation techniques, study math, and record your thought patterns. A commitment of one or two hours a day may be necessary in the beginning. Remember, it took time to learn your present study habits and to be anxious. It will take time to unlearn these behaviors. After you become proficient with these methods, you can devote less time to them.

Begin now to learn your strategies for success. Be sure you have read the Preface to the Student in the beginning of this book. The purpose of this preface is to introduce you to the authors' plan for this text. This will help you to understand the authors' organization or "game plan" for your math experience in this course.

At the beginning of each chapter you will find more Good Advice for Studying that will help you study and take tests more effectively, and manage your anxiety. You may want to read ahead so that you can improve even more quickly. Good Luck!

Whole Numbers

APPLICATION

At the end of the 1997 baseball season, many people observed the unequal payrolls for the various clubs. Specifically, people are worried about the fact that the clubs in postseason play are almost without exception the clubs with the highest payrolls. Many have argued that because of money differences there are actually only 9 or 10 teams that have a realistic chance at making postseason play. Table 1.1 shows a summary of the top eight teams in payroll and the other two who made the 1997 postseason playoffs.

For the 1997 season, the four division winners were the Yankees, Orioles, Indians, and Braves. The Marlins beat the Indians in the World Series that year.

1. Does it appear to you that salary paid to the players goes a long way in determining who has a winning season?
2. What other explanation might there be for the connection between winning and payroll?

TABLE 1.1 PAYROLLS OF MAJOR LEAGUE BASEBALL TEAMS

Team	Payroll	Payroll Rank	Postseason
New York Yankees	$63,700,000	1	Yes
Baltimore Orioles	$58,700,000	2	Yes
Cleveland Indians	$56,700,000	3	Yes
Florida Marlins	$53,300,000	4	Yes
Atlanta Braves	$50,800,000	5	Yes
St. Louis Cardinals	$47,700,000	6	No
Los Angeles Dodgers	$47,000,000	7	No
Seattle Mariners	$44,300,000	8	Yes
San Francisco Giants	$42,400,000	11	Yes
Houston Astros	$33,200,000	19	Yes

1.1

Whole Numbers: Writing, Rounding, and Inequalities

OBJECTIVES

1. Write word names from place value names and place value names from word names.
2. Write an inequality statement about two numbers.
3. Round a whole number.

VOCABULARY

The **digits** are 0, 1, 2, 3, 4, 5, 6, 7, 8, and 9.

The **natural numbers (counting numbers)** are 1, 2, 3, 4, 5, and so on.

The **whole numbers** are 0, 1, 2, 3, 4, 5, and so on. Numbers larger than 9 are written in **place value name** by writing the digits in positions having standard **place value.**

Word names are written words that represent numerals. The word name of 125 is one hundred twenty-five.

The symbols **less than** (<) and **greater than** (>) are used to compare two whole numbers that are not equal. So, $2 < 9$, and $13 > 6$.

To **round** a whole number means to give an approximate value. The symbol $\approx$ means "approximately equal to."

HOW AND WHY

Objective 1

Write word names from place value names and place value names from word names.

In our written whole number system (called the "Hindu-Arabic system"), digits and commas are the only symbols used. This system is a positional base ten (decimal) system. The location of the digit determines its value, from right to left. The first three place value names are one, ten, and hundred. See Figure 1.1.

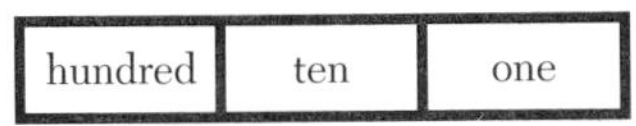

Figure 1.1

For the number 782,

2 is in the ones place, so it contributes 2 ones, or 2, to the value of the number,

8 is in the tens place, so it contributes 8 tens, or 80, to the value of the number,

7 is in the hundreds place, so it contributes 7 hundreds, or 700, to the value of the number.

So, 782 is 7 hundreds + 8 tens + 2 ones, or 700 + 80 + 2. These are called **expanded forms** of the number.

For numbers larger than 999, we use commas to separate groups of three digits. The first four groups are unit, thousand, million, and billion (Figure 1.2). The group on the far left may have one, two, or three digits. All other groups must have three digits. Within each group the names are the same (hundred, ten, and one).

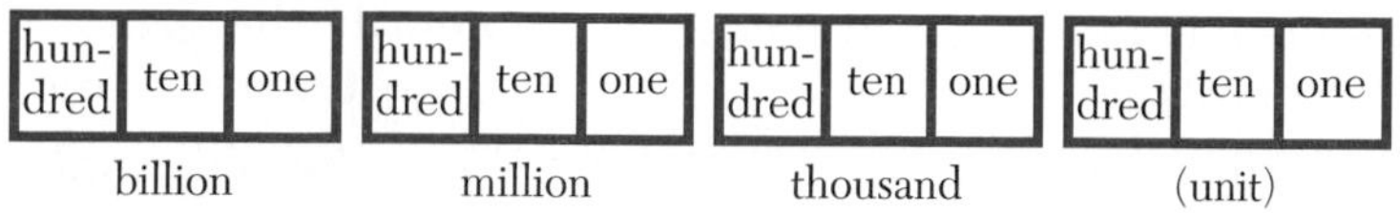

Figure 1.2

For 74,896,314,555 the group names are:

74	896	314	555
billion	million	thousand	unit

The number is read "74 billion, 896 million, 314 thousand, 555." The group "units" is not read.

The word name of a three-digit or smaller number is written by first writing the name of the digit in the hundreds place followed by "hundred" and then the name of the two-digit number in the tens and ones places. The name of the two-digit number in the tens and ones place is the name used to count from one to ninety-nine. So

123 is written one hundred twenty-three,
583 is written five hundred eighty-three,
65 is written sixty-five, and
8 is written eight.

The word name of a number larger than 999 is written by writing the word name for each set of three digits followed by the group name. For example,

TABLE 1.2 WORD NAMES FROM PLACE VALUE NAMES

Place value name	415	609
Word name for each group	four hundred fifteen	six hundred nine
Group name	thousand	(unit)

The word name is written: Four hundred fifteen thousand, six hundred nine.

To write the word name from a place value name

1. From left to right, write the word name for each set of three digits, followed by the group name (except units).
2. Insert a comma after each group name.

CAUTION

The word "and" is not used to write names of whole numbers. So write: three hundred ten, NOT three hundred and ten, also one thousand, two hundred twenty-three, NOT one thousand and two hundred twenty-three.

To write the place value name from the word name of a number, we first identify the group names. Then write each group name in the place value name. Remember to write a 0 for each missing place value. Consider

Two billion, three hundred forty-five million, six thousand, one hundred thirteen

Two *billion,* three hundred forty-five *million,* six *thousand,* one hundred thirteen	**Identify the group names. (*Hint:* Look for the commas.)**
2 billion, 345 million, 6 thousand, 113	**Write the place value name for each group.**
2,345,006,113	**Drop the group names. Keep all commas. Zeros must be inserted to show that there are no hundreds or tens in the thousands group.**

To write the place value name from a word name

1. Identify the group names.
2. Write the three-digit number before each group name followed by a comma. (The first group on the left may have fewer than three digits.)

Numbers like 13,000,000,000, with all zeros following a single group of digits, are often written in a combination of place value name and word name. The first set of digits on the left is written in place value notation followed by the group name. So 13,000,000,000 is written 13 billion.

Warm Ups A–B

A. Write the word name for 6,455,091.

B. Write the word name for 7,597,234.

Examples A–B

Directions: Write the word name.

Strategy: Write the word name of each set of three digits, from left to right, followed by the group name.

A. Write the word name for 17,698,453.

17,	698,	453
seventeen	six hundred ninety-eight	four hundred fifty-three
million	thousand	(unit)

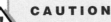

CAUTION
Do not use the word "and" when reading or writing a whole number.

The word name is seventeen million, six hundred ninety-eight thousand, four hundred fifty-three.

B. Write the word name for 3,189,025. Write the solution without the help of a chart.

Three million, one hundred eighty-nine thousand, twenty-five

Answers to Warm Ups A. Six million, four hundred fifty-five thousand, ninety-one B. Seven million, five hundred ninety-seven thousand, two hundred thirty-four

Examples C–E

Directions: Write the place value name.

Strategy: Write the three-digit number for each group followed by a comma.

C. Write the place value name for two million, thirty-seven thousand, five hundred sixty-four.

2,	**Millions group.**
037,	**Thousands group. (Note that a zero is inserted on the left to fill out the three digits in the group.)**
564	**Units group.**

The place value name is 2,037,564.

D. Write the place value name for 346 million.

The place value name is 346,000,000. **Replace the word million with six zeros.**

E. The purchasing agent for the Russet Corporation received a telephone bid of twenty-three thousand eighty-one dollars as the price of a new printing press. What is the place value name of the bid that she will include in her report to her superior?

twenty-three thousand, eighty-one
23, 081

The place value name she reports is $23,081.

Warm Ups C–E

C. Write the place value name for thirty-two million, twenty-seven thousand, nine hundred ten.

D. Write the place value name for five thousand.

E. The purchasing agent for the Russet Corporation also received a bid of seventeen thousand, two hundred eighteen dollars for a supply of paper. What is the place value name of the bid that she will include in her report to her superior?

HOW AND WHY

Objective 2

Write an inequality statement about two numbers.

If two whole numbers are not equal, then the first is either *less than* or *greater than* the second. Look at this number line (or ruler):

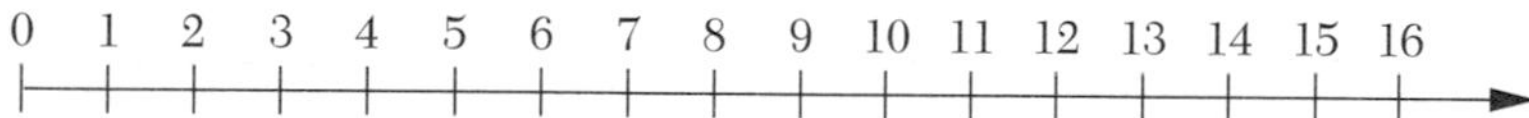

Given two numbers on a number line or ruler, the number on the right is the larger. For example,

$8 > 5$	**8 is to the right of 5, so 8 is greater than 5.**
$7 > 1$	**7 is to the right of 1, so 7 is greater than 1.**
$14 > 12$	**14 is to the right of 12, so 14 is greater than 12.**
$15 > 0$	**15 is to the right of 0, so 15 is greater than 0.**

Given two numbers on a number line or ruler, the number on the left is the smaller. For example,

$2 < 6$	**2 is to the left of 6, so 2 is less than 6.**
$7 < 10$	**7 is to the left of 10, so 7 is less than 10.**
$5 < 9$	**5 is to the left of 9, so 5 is less than 9.**
$11 < 13$	**11 is to the left of 13, so 11 is less than 13.**

Answers to Warm Ups C. 32,027,910 D. 5,000 E. The place value name she reports is $17,218.

For larger numbers, imagine a longer number line. Notice how the symbols < and > point to the smaller of the two numbers. For example,

109 < 405

34 > 25

1009 > 1007

To write an inequality statement about two numbers

1. Insert < between the numbers if the number on the left is smaller.
2. Insert > between the numbers if the number on the left is larger.

Warm Ups F–G

F. Insert the appropriate inequality symbol:
118 134

G. Insert the appropriate inequality symbol:
3,678 3,499

Examples F–G

Directions: Insert < or > to make a true statement.

Strategy: Imagine a number line. The smaller number is on the left. Insert the symbol that points to the smaller number.

F. Insert the appropriate inequality symbol: 67 97

67 < 97

G. Insert the appropriate inequality symbol: 1,314 1,299

1,314 > 1,299

HOW AND WHY

Objective 3

Round a whole number.

Many numbers that we see in daily life are approximations. These are used to indicate the approximate value when it is believed that the exact value does not lend to the discussion. So attendance at a political rally may be stated at 15,000 when it was actually 14,783. The amount of a deficit in the budget may be stated as \$2,000,000 instead of \$2,067,973. In this chapter, we use approximations to estimate the outcome of operations with whole numbers. The symbol ≈, read "approximately equal to," is used to show the approximation. So \$2,067,973 ≈ \$2,000,000.

We approximate numbers by **rounding.** The number line can be used to see how whole numbers are rounded. Suppose we wish to round 27 to the nearest ten.

The arrow under the 27 is closer to 30 than to 20. We say "to the nearest ten, 27 rounds to 30."

We use the same idea to round any number, although we usually make only a mental image of the number line. The key question is: Is this number closer to the

Answers to Warm Ups F. < G. >

smaller rounded number or closer to the larger one?" Practically, we need only to determine if the number is more or less than half the distance between the rounded numbers.

To round 34,568 to the nearest thousand without a number line, draw an arrow under the digit in the thousands place.

34,568
↑

Because 34,568 is between 34,000 and 35,000, we must decide which number it is closer to. Because 34,500 is halfway between 34,000 and 35,000 and because 34,568 > 34,500, we conclude that 34,568 is more than halfway to 35,000.

Whenever the number is halfway or closer to the larger number, we choose the larger number.

34,568 ≈ 35,000 **34,568 is closer to 35,000 than to 34,000.**

To round a whole number to a given place value

1. Draw an arrow under the given place value.
2. If the digit to the right of the arrow is 5, 6, 7, 8, or 9, add one to the digit above the arrow. (Round to the larger number.)
3. If the digit to the right of the arrow is 0, 1, 2, 3, or 4, do not change the digit above the arrow. (Round to the smaller number.)
4. Replace all the digits to right of the arrow with zeros.

Examples H–I

Directions: Round to the indicated place value.

Strategy: Choose the larger number if the digit to the right of the round-off place is five or more, otherwise, choose the smaller number.

H. Round 127,456 to the nearest ten thousand.

127,456 **Draw an arrow under the ten thousands place.**
↑

130,000 **The digit to the right of the arrow is 7. Because 127,456 > 125,000, choose the larger number.**

So 127,456 ≈ 130,000.

I. Round to the indicated place value in the table.

Number	Ten	Hundred	Thousand
365,733	365,730	365,700	366,000
98,327	98,330	98,300	98,000

Warm Ups H–I

H. Round 99,858 to the nearest ten.

I. Round to the indicated place value in the table.

Number	Ten	Thousand
491,356		
480,639		

Answers to Warm Ups H. 99,860 I.

Number	Ten	Thousand
491,356	491,360	491,000
480,639	480,640	481,000

NAME ______________________ CLASS __________ DATE __________

Exercises 1.1

OBJECTIVE 1: *Write word names from place value names and place value names from word names.*

A.

Write the word names of each of these numbers.

1. 542

2. 391

3. 890

4. 500

5. 7015

6. 40,051

Write the place value name.

7. Fifty-seven

8. Seventy-eight

9. Seven thousand, five hundred

10. Seven thousand, five

11. 10 million

12. 123 thousand

B.

Write the word name of each of these numbers.

13. 25,310

14. 25,031

15. 205,310

16. 250,031

17. 45,000,000

18. 750,000

Write the place value name.

19. Two hundred forty-three thousand, seven hundred

20. Two hundred forty-three thousand, seven

21. Twenty-three thousand, four hundred seventy

22. Twenty-three thousand, four hundred seventy-seven

23. Seventeen million

24. Seven hundred thousand seven

OBJECTIVE 2: *Write an inequality statement about two numbers.*

A.

Insert < or > between the numbers to make a true statement.

25. 18 21 **26.** 33 29 **27.** 51 44 **28.** 62 71

B.

29. 145 152 **30.** 212 208 **31.** 348 351 **32.** 275 269

OBJECTIVE 3: *Round a whole number.*

A.

Round to the indicated place value.

33. 694 (ten) **34.** 786 (ten) **35.** 1658 (hundred)

36. 3450 (hundred)

B.

	Number	Ten	Hundred	Thousand	Ten Thousand
37.	102,385				
38.	689,377				
39.	7,250,978				
40.	4,309,498				

C.

Write the place value name.

41. Five hundred sixty million, three hundred fifty-three thousand, seven hundred thirty.

42. Three hundred fifty million, six hundred sixty-three

NAME ______________________ CLASS ____________ DATE ____________

Use the graph for Exercises 43–44. The graph shows the average income of the top 20% of the families and the bottom 20% of the families in Iowa.

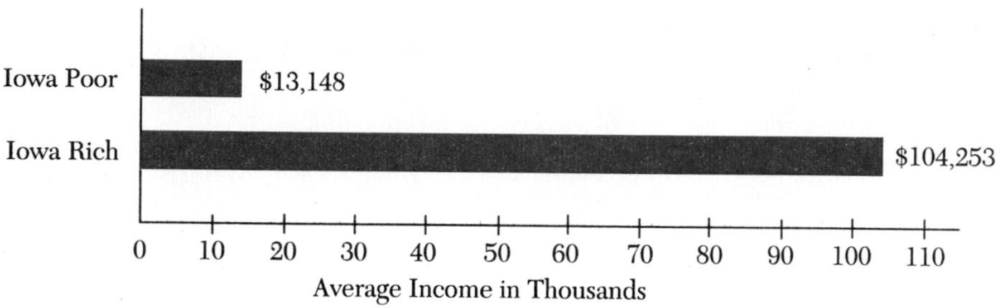

43. Write the word name for the average salary for the poor in Iowa.

44. Write the word name for the average salary of the rich in Iowa.

Insert > or < between the numbers to make a true statement.

45. 5634 5637

46. 10,276 10,199

47. What is the smallest four-digit number?

48. What is the largest five-digit number?

Round to the indicated place value.

49. 43,784,675 (ten thousand)

50. 47,078,665 (hundred thousand)

51. Round 63,749 to the nearest hundred. Round 63,749 to the nearest ten and then round your result to the nearest hundred. Why did you get a different result the second time? Which method is correct?

52. Hazel bought a jet ski boat for $2075. She wrote a check to pay for it. What word name did she write on the check?

53. Kimo bought a used Infinity for $18,465 and wrote a check to pay for it. What word name did he write on the check?

54. The U.S. Fish and Wildlife Department estimates that salmon runs could be as high as 154,320 fish by 2002 on the Rogue River if new management practices are used in logging along the river. Write the word name for the number of fish.

55. Ducks Unlimited estimated that 389,500 ducks spent the winter at a Wisconsin refuge. Write the word name for the number of ducks.

56. The world population during 1990 exceeded 5 billion, 3 hundred thousand. Write the place value name for the world population.

57. The purchasing agent for Print-It-Right received a telephone bid for thirty-six thousand, four hundred seven dollars as the price for a new printing press. What is the place value name for the bid?

58. An office building in downtown Birmingham sold for $2,458,950. Give the purchase price of the building to the nearest hundred thousand dollars.

59. Ten thousand shares of Intel Corporation sold for $846,560. What is the value of the sale, to the nearest thousand dollars?

For Exercises 60–62, the figure shows the quantity of the top three industrial releases of toxic materials a few years ago. These values have been decreasing in recent years.

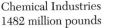

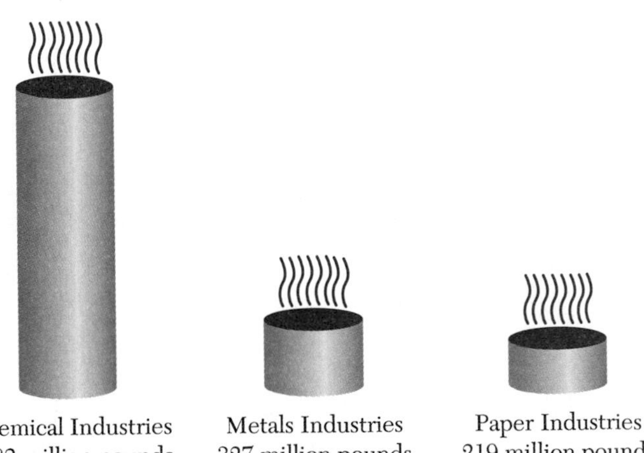

60. Write the place value name for the number of pounds of toxic material released by chemical industries.

NAME ________________ CLASS ________ DATE ________

61. Write the place value name for the number of pounds of toxic material released by metals industries.

62. Write the place value name for the number of pounds of toxic material released by paper industries.

For Exercises 63–66, the per capita personal income in the New England states is given in the following table.

Massachusetts	$26,994
New Hampshire	$25,151
Maine	$20,527
Connecticut	$30,303
Rhode Island	$23,310
Vermont	$20,927

63. Write the word name for the per capita personal income in Maine.

64. Round the per capita personal income in Massachusetts to the nearest thousand.

65. Which state has the smallest per capita personal income?

66. Of Massachusetts and New Hampshire, which has the larger per capita personal income?

67. The distance from the earth to the sun was measured and determined to be 92,875,328 miles. To the nearest million miles, what is the distance?

68. The *National Petroleum News* reported the following number of branded retail gasoline outlets in 1995.

Brand	Number of Outlets
Citgo	14,054
Texaco	13,023
Amoco	9600
Shell	8667
Exxon	8250
Chevron	7988
Mobil	7689
Phillips	7106
BP America	6800
Conoco	5017

Have any of these numbers been rounded? If so, which ones and explain how you know. Revise the table, rounding all the figures to the nearest hundred.

 For Exercises 69–71, refer to the chapter application on page 1.

69. Write the word name for the Yankees' payroll.

70. To what place value do most of the baseball payrolls seem to be rounded? Is it possible to tell for sure that the payrolls have been rounded? Why or why not?

71. Write an inequality for the payrolls of the Marlins and the Indians.

72. The state motor vehicle department estimated the number of licensed automobiles in the state to be 2,376,000, to the nearest thousand. A check of the records indicated that there were actually 2,376,499. Was their estimate correct?

73. The total land area of the earth is approximately 52,425,000 square miles. What is the land area to the nearest million square miles?

NAME ______________________ CLASS ______________ DATE ______________

Exercises 74–76 relate to the chapter application. During the 1996–97 NBA season, the Chicago Bulls had the following payroll and won the NBA championship:

Player	Salary	Player	Salary
Michael Jordan	\$30,140,000	Dennis Rodman	\$9,000,000
Toni Kukoc	\$3,960,000	Ron Harper	\$3,840,000
Luc Longley	\$2,790,000	Scottie Pippen	\$2,250,000
Randy Brown	\$1,300,000	Dickey Simpkins	\$1,040,000
Robert Parrish	\$1,000,000	Bill Wennington	\$1,000,000
Steve Kerr	\$750,000	Jason Caffey	\$700,000
Judd Buechler	\$500,000		

74. Round all the salaries to the nearest million dollars.

75. Write the word name for Scottie Pippen's salary.

76. Write an inequality statement about Toni Kukoc's and Ron Harper's salaries.

Exercises 77–78. The following chart lists some nutritional facts about two brands of peanut butter.

Skippy Super Chunk

Nutrition Facts
Serving Size 2 Tbsp (32g)
Servings Per Container about 15

Amount Per Serving	
Calories 190	**Calories from Fat 140**
	% Daily Values
Total Fat 17g	**26%**
Saturated Fat 3.5g	**17%**
Cholesterol 0mg	**0%**
Sodium 140mg	**6%**
Total Carbohydrate 7g	**2%**
Dietary Fiber 2g	**8%**
Sugars 3g	
Protein 7g	

Jif Creamy
Simply Jif contains 2g sugar per serving.
Regular Jif contains 3g sugar per serving.

Nutrition Facts
Serving Size 2 Tbsp (31g)
Servings Per Container about 16

Amount Per Serving	
Calories 190	**Calories from Fat 130**
	% Daily Values
Total Fat 16g	**25%**
Saturated Fat 3g	**16%**
Cholesterol 0mg	**0%**
Sodium 65mg	**3%**
Total Carbohydrate 6g	**2%**
Dietary Fiber 2g	**9%**
Sugars 2g	
Protein 8g	

77. List the categories of nutrients for which Jif has a smaller amount than Skippy.

78. Round the sodium content in each brand to the nearest hundred. Do the rounded numbers give a fair comparison of the amount of sodium in the brands?

STATE YOUR UNDERSTANDING

79. Explain why base-ten is a good name for our number system.

80. Explain what the digit 7 means in 175,892.

81. What is rounding? Explain how to round 87,452 to the nearest thousand and to the nearest hundred.

CHALLENGE

82. What is the place value for the digit 4 in 3,456,709,230,000?

83. Write the word name for 3,456,709,230,000.

84. Arrange the following numbers from smallest to largest: 1234, 1342, 1432, 1145, 1243, 1324, and 1229.

85. What is the largest value of X that makes $2X56 > 2649$ false?

86. Round 8275 to the nearest ten thousand.

87. Round 37,254 to the nearest hundred thousand.

GROUP ACTIVITY

88. Two other methods of rounding are called the "odd-even method" and "truncating." Find out about these methods and be prepared to explain them in class. (*Hint:* Try the library or talk to science and business instructors.)

1.2

Adding and Subtracting Whole Numbers

OBJECTIVES

1. Find the sum of two or more whole numbers.
2. Estimate the sum of a group of whole numbers.
3. Find the difference of two whole numbers.
4. Estimate the difference of two whole numbers.

VOCABULARY

Addends are numbers that are added. In 9 + 20 + 3 = 32, the addends are 9, 20, and 3.

The result of adding is called the **sum.** In 9 + 20 + 3 = 32, the sum is 32.

The result of subtracting is called the **difference.** So, in 62 − 34 = 28, 28 is the difference.

HOW AND WHY

Objective 1

Find the sum of two or more whole numbers.

When Jose graduated from high school he received cash gifts of $35, $30, and $12. The total number of dollars received is found by adding the individual gifts. The total number of dollars he received is $77.

The addition facts and place value are used to add whole numbers written with more than one digit. Let's use these to find the sum of the cash gifts that Jose received. We need to find the sum of

35 + 30 + 12.

By writing the numbers in expanded form and putting the same place values in columns it is easy to add.

$$\begin{array}{rl} 35 = & 3 \text{ tens} + 5 \text{ ones} \\ 30 = & 3 \text{ tens} + 0 \text{ ones} \\ +\ \underline{12} = & \underline{1 \text{ tens} + 2 \text{ ones}} \\ & 7 \text{ tens} + 7 \text{ ones} = 77 \end{array}$$

So, 35 + 30 + 12 = 77. Jose received $77 in cash gifts.

Because each place can contain only a single digit, it is often necessary to rewrite the sum of a column.

$$\begin{array}{rl} 57 = & 5 \text{ tens} + \ \ 7 \text{ ones} \\ +\ \underline{28} = & \underline{2 \text{ tens} + \ \ 8 \text{ ones}} \\ & 7 \text{ tens} + 15 \text{ ones} \end{array}$$

Because 15 ones is a two-digit number it must be renamed:

$$\begin{aligned} 7 \text{ tens} + 15 \text{ ones} &= 7 \text{ tens} + 1 \text{ ten} + 5 \text{ ones} \\ &= 8 \text{ tens} + 5 \text{ ones} \\ &= 85 \end{aligned}$$

So the sum of 57 and 28 is 85.

The common shortcut is shown in the following sum. To add 567 + 204 + 198, write the numbers in a column.

$$\begin{array}{r} 567 \\ 204 \\ +\ \underline{198} \end{array}$$

Written this way, the digit in the ones, tens, and hundreds places are aligned.

$$\begin{array}{r} {\scriptstyle 1} \\ 567 \\ 204 \\ +\ \underline{198} \\ 9 \end{array}$$

Add the digits in the ones column: 7 + 4 + 8 = 19. Write "9" and carry the "1" (1 ten) to the tens column.

$$\begin{array}{r} {\scriptstyle 11} \\ 567 \\ 204 \\ +\ \underline{198} \\ 69 \end{array}$$

Add the digits in the tens column: 1 + 6 + 0 + 9 = 16. Write "6" and carry the "1" (10 tens = 1 hundred) to the hundreds column.

$$\begin{array}{r} {\scriptstyle 11} \\ 567 \\ 204 \\ +\ \underline{198} \\ 969 \end{array}$$

Add the digits in the hundreds column: 1 + 5 + 2 + 1 = 9

To add whole numbers

1. Write the numbers in a column so that the place values are aligned.
2. Add each column, starting with the ones (or units) column.
3. If the sum of any column is greater than nine, write the ones digit and "carry" the tens digit to the next column.

Warm Ups A–D

A. Add: 864 + 657

Examples A–D

Directions: Add.

Strategy: Write the numbers in a column. Add the digits in the columns starting on the right. If the sum is greater than 9, "carry" the tens digit to the next column.

A. Add: 788 + 643

$$\begin{array}{r} {\scriptstyle 11} \\ 788 \\ +\ \underline{643} \\ 1431 \end{array}$$

Add the numbers in the ones column. 8 + 3 = 11. Because the sum is greater than 9, write 1 in the ones column and carry the 1 to the tens column. Add the numbers in the tens column. 1 + 8 + 4 = 13. Write 3 in the tens column and carry the 1 to the hundreds column. Add the numbers in the hundreds column. 1 + 7 + 6 = 14. Because all columns have been added there is no need to carry.

Answer to Warm Up A. 1521

B. Find the sum: 1773 + 5486 + 3497

```
  121
  1773
  5486
+ 3497
 10756
```

B. Find the sum: 3467 + 8912 + 4569

C. Add 59, 423, 5, and 1607. Round the sum to the nearest ten.

```
  1 2
    59
   423
     5
+ 1607
  2094
```

When writing in a column, make sure the place values are aligned properly.

2094 ≈ 2090 **Round to the nearest ten.**

C. Add 87, 6598, 47, and 3. Round the sum to the nearest ten.

Calculator Example

D. Add: 4509 + 678 + 2345 + 1923 + 6789

Calculators have an internal program that adds numbers just as we have been doing by hand. No special preparation is required on the part of the operator. Simply enter the exercise as it is written horizontally and the calculator will do the rest. The sum is 16,244.

D. Add: 5482 + 9742 + 847 + 1324 + 7321

HOW AND WHY

Objective 2

Estimate the sum of a group of whole numbers.

The sum of a group of numbers can be estimated by rounding each member of the group to the largest place value in the group and then adding the rounded values. For instance,

```
  5392      5000
  4220      4000
  7685      8000
  6615      7000
+  945  +   1000
           25000
```

The largest place value is thousand, so round each number to the nearest thousand.

The estimate of the sum is 25,000. One use of the estimate is to see if the sum of the group of numbers is correct. If the calculated sum is not close to the estimated sum, 25,000, you should check the addition by re-adding. In this case the calculated sum, 24,857, is close to the estimate.

To estimate the sum of a group of numbers

1. Round each number in the group to the largest place in the group.
2. Add the rounded numbers.

Answers to Warm Ups B. 16,948 C. 6740 D. 24,716

Warm Up E

E. 543 + 12 + 792 + 395 + 3 + 87

Example E

Directions: Estimate the sum. Then add and compare.

Strategy: Round each number to the largest place value of all the addends. Then add and compare.

E. 356 + 7895 + 679 + 4567 + 3188 + 12

356	0	**Round each number to the nearest thousand. With practice, this can be done mentally for a quick check.**
7895	8000	
679	1000	
4567	5000	
3188	3000	
+ 12	+ 0	
	17000	

Now add and compare.

$$\begin{array}{r} 356 \\ 7895 \\ 679 \\ 4567 \\ 3188 \\ +\ \underline{12} \\ 16697 \end{array}$$

The sum, 16,697, is close to the estimation, 17,000.

HOW AND WHY

Objective 3

Find the difference of two whole numbers.

Felicia went shopping with $95. She made purchases totaling $53. How much money does she have left? Finding the difference in two quantities is called subtraction. When we subtract $53 from $95 we get $42.

Subtraction can be thought of as finding the missing addend in an addition exercise. For instance, 9 − 5 = ? asks 5 + ? = 9. Because 5 + 4 = 9, we know that 9 − 5 = 4. Similarly, 47 − 15 = ? asks 15 + ? = 47. Because 15 + 32 = 47, we know that 47 − 15 = 32.

For larger numbers, such as 965 − 534, we take advantage of the column form and expanded notation to find the missing addend in each column.

$$\begin{array}{rl} 965 & = 9 \text{ hundreds} + 6 \text{ tens} + 5 \text{ ones} \\ -\ \underline{534} & = \underline{5 \text{ hundreds} + 3 \text{ tens} + 4 \text{ ones}} \\ 431 & = 4 \text{ hundreds} + 3 \text{ tens} + 1 \text{ one} = 431 \end{array}$$

Check by adding:

$$\begin{array}{r} 534 \\ +\ \underline{431} \\ 965 \end{array}$$

So, 965 − 534 = 431.

Answer to Warm Up E. The sum, 1832, is close to the estimation, 1800.

Now consider the difference 784 − 369. Write the numbers in column form.

$$\begin{array}{rl} 784 = & 7 \text{ hundreds} + 8 \text{ tens} + 4 \text{ ones} \\ -\underline{369} = & \underline{3 \text{ hundreds} + 6 \text{ tens} + 9 \text{ ones}} \end{array}$$

Here we cannot subtract 9 ones from 4 ones, so we rename by "borrowing" one of the tens from the 8 tens (1 ten = 10 ones) and adding the 10 ones to the 4 ones.

$$\begin{array}{rl} 784 = & 7 \text{ hundreds} + \not{8} \text{ tens} + \not{4} \text{ ones} \\ -\underline{369} = & \underline{3 \text{ hundreds} + 6 \text{ tens} + 9 \text{ ones}} \\ = & 4 \text{ hundreds} + 1 \text{ ten} + 5 \text{ ones} = 415 \end{array}$$

Check by adding:

$$\begin{array}{r} {\scriptstyle 1} \\ 369 \\ +\underline{415} \\ 784 \end{array}$$

We generally do not bother to write the expanded form when we subtract. We show the shortcut for borrowing in the examples.

To subtract whole numbers

1. Write the numbers in a column so that the place values are aligned.
2. Subtract in each column, starting with the ones (or units) column.
3. When the numbers in a column cannot be subtracted, borrow 1 from the next column and rename by adding 10 to the upper digit in the current column and then subtract.

Examples F–K

Directions: Subtract and check.

Strategy: Write the numbers in columns. Subtract in each column. Rename by borrowing when the numbers in a column cannot be subtracted.

F. Subtract: 85 − 42

$$\begin{array}{r} 85 \\ -\underline{42} \\ 43 \end{array}$$

Subtract the ones column: 5 − 2 = 3.
Subtract the tens column: 8 − 4 = 4.

CHECK:

$$\begin{array}{r} 42 \\ +\underline{43} \\ 85 \end{array}$$

So, 85 − 42 = 43.

Warm Ups F–K

F. Subtract: 97 − 25

Answer to Warm Up F. 72

G. Find the difference:
567 − 398

G. Find the difference: 752 − 295

$$\begin{array}{r} {\scriptstyle 4\,12} \\ 7\not{5}2 \\ -\ \underline{295} \\ 7 \end{array}$$

In order to subtract in the ones column we borrow 1 ten (10 ones) from the tens column and rename the ones (10 + 2 = 12).

$$\begin{array}{r} {\scriptstyle 6\,14} \\ {\scriptstyle \not{4}\,12} \\ \not{7}\not{5}2 \\ -\ \underline{295} \\ 457 \end{array}$$

Now in order to subtract in the tens column we must borrow 1 hundred (10 tens) from the hundreds column and rename the tens (10 + 4 = 14).

CHECK:

295 + 457 = 752

So, 752 − 295 = 457.

H. Subtract: 4500 and 891

H. Subtract: 3700 and 948

$$\begin{array}{r} 3700 \\ -\ \underline{948} \end{array}$$

We cannot subtract in the ones column, and since there are 0 tens, we cannot borrow from the tens column.

$$\begin{array}{r} {\scriptstyle 6\,10} \\ 3\not{7}\not{0}0 \\ -\ \underline{948} \end{array}$$

We borrow 1 hundred (1 hundred = 10 tens) from the hundreds place.

$$\begin{array}{r} {\scriptstyle 9\,10} \\ {\scriptstyle 6\,\not{10}} \\ 3\not{7}\not{0}\not{0} \\ -\ \underline{948} \end{array}$$

Now borrow 1 ten (1 ten = 10 ones). We can now subtract in the ones and tens column, but not in the hundreds column.

$$\begin{array}{r} {\scriptstyle 2\,16\,9\,10} \\ {\scriptstyle \not{6}\,\not{10}} \\ \not{3}\not{7}\not{0}\not{0} \\ -\ \underline{948} \\ 2752 \end{array}$$

Now borrow 1 thousand (1 thousand = 10 hundreds) and we can subtract in every column.

CHECK:

948 + 2752 = 3700

Let's try Example H again using a technique called reverse adding. Just ask yourself "What do I have to add to 948 to get 3700?"

$$\begin{array}{r} 3700 \\ -\ \underline{948} \\ 2 \end{array}$$

Begin with the ones column. 8 is larger than 0, so ask "What do I add to 8 to make 10?"

$$\begin{array}{r} 3700 \\ -\ \underline{948} \\ 52 \end{array}$$

Because 8 + 2 = 10, we write the 2 in the ones column and carry the one over to the 4 to make 5. Now ask "What do I add to 5 to make 10?"

$$\begin{array}{r} 3700 \\ -\ \underline{948} \\ 752 \end{array}$$

Write 5 in the tens column and carry the one to the nine in the hundreds column. Now ask "What do I add to 10 to make 17?"

$$\begin{array}{r} 3700 \\ -\ \underline{948} \\ 2752 \end{array}$$

Finally, ask "What do I add to the carried 1 to make 3?"

Answers to Warm Ups G. 169 H. 3609

The advantage of this method is that 1 is the largest amount carried, so most people can do this process mentally.

So 3700 − 948 = 2752.

I. Find the difference between 7061 and 736 and round to the nearest hundred.

```
 6 10 5 11
 7 0  6 1
−   7 3 6
 6 3  2 5
```

The "borrowing" is all shown at once.

CHECK:

736 + 6325 = 7061

6325 ≈ 6300 **Round to the nearest hundred.**

The difference is about 6300.

I. Find the difference between 6051 and 827 and round to the nearest hundred.

Calculator Example

J. Subtract 14,691 from 33,894.

CAUTION

When a subtraction exercise is worded "Subtract *A* from *B*," it is necessary to reverse the order of the numbers. Write *B* − *A*.

Enter 33,894 − 14,691.

The difference is 19,203.

J. Subtract 17,358 from 44,679.

K. Maxwell Auto is advertising a $742 rebate on all new cars priced above $12,000. What is the cost after rebate of a car originally priced at $13,763?

Strategy: Since the price of the car is over $12,000, we subtract the amount of the rebate to find the cost.

```
  13763        CHECK:    13021
−   742                +   742
  13021                  13763
```

The car costs $13,021.

K. Maxwell Auto is also advertising a $1438 rebate on all new cars priced above $21,000. What is the cost after rebate of a car originally priced at $27,829?

HOW AND WHY

Objective 4

Estimate the difference of two whole numbers.

The difference of two whole numbers can be estimated by rounding each number to the largest place value in the two numbers and then subtracting these rounded numbers. For instance,

```
  7330      7000
− 3282    − 3000
            4000
```

The largest place value is thousand. Round each number to the nearest thousand and subtract.

Answers to Warm Ups I. 5200 J. 27,321 K. The car costs $26,391.

The estimate of the difference is 4000. One use of the estimate is to see if the difference is correct. If the calculated difference is not close to 4000, you should check the subtraction. In this case the difference is 4048, which is close to the estimate.

To estimate the difference of two whole numbers

1. Round each number to the largest place value in either number.
2. Subtract the rounded numbers.

Warm Ups L–M

L. 7555 and 2956

M. Estimate the difference of 32,450 and 9874. Then subtract and compare.

Examples L–M

Directions: Estimate the difference, then subtract and compare.

Strategy: Round each number to the largest place value in either number and then subtract.

L. 6490 and 1350

$$\begin{array}{r} 6000 \\ -\ 1000 \\ \hline 5000 \end{array}$$

Round to the nearest thousand.
Subtract.

$$\begin{array}{r} 6490 \\ -\ 1350 \\ \hline 5140 \end{array}$$

The difference is estimated to be 5000; it is 5140.

M. Estimate the difference of 73,425 and 48,240. Then subtract and compare.

$$\begin{array}{r} 70,000 \\ -\ 50,000 \\ \hline 20,000 \end{array}$$

Round each number to the nearest ten thousand.

Now subtract.

$$\begin{array}{r} 73,425 \\ -\ 48,240 \\ \hline 35,185 \end{array}$$

The estimated difference and the calculated difference are not close. Check to see if the subtraction was done correctly.

$$\begin{array}{r} 73,425 \\ -\ 48,240 \\ \hline 25,185 \end{array}$$

In the first subtraction, the fact that 1 was borrowed from the 7 was ignored.

Now the estimated difference, 20,000, is close to the actual difference, 25,185.

Answers to Warm Ups L. The estimated difference, 5000, is close to the actual difference, 4599. M. The estimated difference, 20,000, is close to the actual difference, 22,576.

NAME ______________________ CLASS ____________ DATE ____________

Exercises 1.2

OBJECTIVE 1: *Find the sum of two or more whole numbers.*

A.

Add.

1. 65 + 32 **2.** 87 + 12 **3.** 748 + 231 **4.** 533 + 254

5. $\begin{array}{r} 146 \\ +\ \underline{363} \end{array}$ **6.** $\begin{array}{r} 756 \\ +\ \underline{236} \end{array}$

7. When you add 36 and 48, the sum of the ones column is 14. You must carry the ________ to the tens column.

8. In 563 + 275 the sum is *X*38. The value of *X* is ________.

B.

Add.

9. 586 + 3492 + 321 **10.** 783 + 5703 + 529

11. 8 + 90 + 403 + 6070 **12.** 6 + 80 + 608 + 4030

13. 2795 + 3643 + 7055 + 4004 (Round sum to the nearest hundred.)

14. 6832 + 8712 + 9032 + 5111 (Round sum to the nearest hundred.)

OBJECTIVE 2: *Estimate the sum of a group of whole numbers.*

A.

Estimate the sum.

15. 345 + 782

16. 595 + 812

17.
$$\begin{array}{r} 3411 \\ 2001 \\ +\ \underline{4561} \end{array}$$

18.
$$\begin{array}{r} 4567 \\ 3611 \\ +\ \underline{2399} \end{array}$$

B.

Estimate the sum and then add.

19.
$$\begin{array}{r} 3209 \\ 7095 \\ 4444 \\ 2004 \\ +\ \underline{3166} \end{array}$$

20.
$$\begin{array}{r} 6073 \\ 3284 \\ 1212 \\ 3593 \\ +\ \underline{5606} \end{array}$$

21.
$$\begin{array}{r} 45{,}902 \\ 33{,}333 \\ 57{,}700 \\ +\ \underline{23{,}653} \end{array}$$

22.
$$\begin{array}{r} 11{,}923 \\ 30{,}871 \\ 21{,}211 \\ +\ \underline{74{,}486} \end{array}$$

OBJECTIVE 3: *Find the difference of two whole numbers.*

A.

Subtract.

23.
$$\begin{array}{r} 4 \text{ hundreds} + 9 \text{ tens} + 3 \text{ ones} \\ -\ \underline{2 \text{ hundreds} + 9 \text{ tens} + 2 \text{ ones}} \end{array}$$

24.
$$\begin{array}{r} 6 \text{ hundreds} + 3 \text{ tens} + 5 \text{ ones} \\ -\ \underline{4 \text{ hundreds} + 2 \text{ tens} + 4 \text{ ones}} \end{array}$$

25. 608 − 82

26. 642 − 80

27. 689 − 238

28. 848 − 611

29. When subtracting 34 − 19, you must "borrow" one from the three. The value of the "borrowed 1" is __________ ones.

30. To subtract 526 − 462, you must borrow from the __________ column to subtract in the __________ column.

B.

Subtract.

31. 741 − 583 **32.** 932 − 857 **33.** 800 − 378 **34.** 600 − 338

35. 8743 − 4078 (Round difference to the nearest hundred.)

36. 9045 − 5786 (Round difference to the nearest hundred.)

OBJECTIVE 4: *Estimate the difference of two whole numbers.*

A.

Estimate the difference.

37. 673 − 423 **38.** 854 − 392 **39.** 45,678 − 34,722 **40.** 67,235 − 58,991

B.

Estimate the difference and then subtract.

41. 875 − 406 **42.** 455 − 207 **43.** 6580 − 3217 **44.** 8732 − 5569

45. Estimate the sum of 23,706, 34, 7561, 9346, and 236, then find the sum.

46. Estimate the sum of 96, 783, 3678, 12,555, and 8999, then find the sum.

47. Estimate the difference of 203,855 and 195,622, then find the difference.

48. Estimate the difference of 423,876 and 298,788, then find the difference.

C.

For Exercises 49–56, a survey of car sales in Wisconsin shows the following distribution of sales among these dealers.

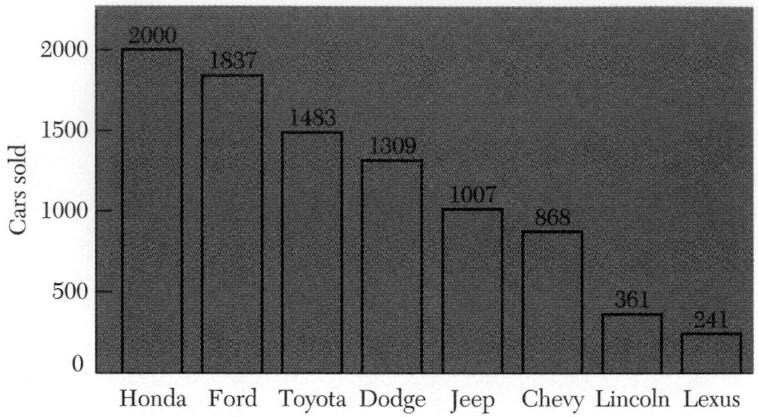

49. What is the total number of Fords, Toyotas, and Lexuses sold?

50. What is the total number of Chevys, Lincolns, Dodges, and Hondas sold?

51. How many more Hondas are sold than Fords?

52. How many more Toyotas are sold than Jeeps?

53. What is the total number sold of the three best-selling cars?

54. What is the total number sold of the three worst-selling cars?

55. Estimate the total number of cars sold and then find the exact total.

56. What is the difference in cars sold between the best-selling car and the worst-selling car?

Exercises 57–58 relate to the chapter application (see page 1).

57. What is the sum of the payrolls of the four division winners for the 1997 season?

58. How much more is the payroll of the Yankees than the payroll of the Astros?

59. The biologist at the Bonneville fish ladder counted the following number of coho salmon during a one-week period: Monday, 895; Tuesday, 675; Wednesday, 124; Thursday, 1056; Friday, 308; Saturday, 312; and Sunday, 219. How many salmon went through the ladder last week? How many more salmon went through the ladder on Tuesday than on Saturday?

For Exercises 60–62, a new car dealership has the following sales in a given week:

Monday	$36,750
Tuesday	$46,780
Wednesday	$21,995
Thursday	$35,900
Friday	$67,950
Saturday	$212,752
Sunday	$345,720

60. What is the difference between the largest and the smallest daily gross sales?

61. What is the gross sales for the week?

62. How much less was the gross sales on Wednesday than on Friday?

For Exercises 63–65, according to the FBI's *Uniform Crime Reports,* the estimated arrests in 1993 in the United States were as follows:

Murder/manslaughter	18,856
Forcible rape	29,432
Robbery	143,877
Aggravated assault	408,148

Burglary	308,849
Larceny-theft	1,131,768
Motor vehicle theft	156,711
Arson	14,504

63. Find the total number of estimated arrests for violent crimes (murder/manslaughter, forcible rape, robbery, and aggravated assault).

64. Find the total number of estimated arrests for property crimes (burglary, larceny-theft, motor vehicle theft, and arson).

65. How many more aggravated assaults than robberies were there?

Exercises 66–68. A home furnace uses natural gas, oil, or electricity for the energy needed to heat the house. We humans get our energy for body heat and physical activity from calories in our food. Even when resting we use energy for muscle actions such as breathing, heartbeat, digestion, and other functions. If we consume more calories than we use up, we gain weight. If we consume less than we use, we lose weight. Some nutritionists recommend about 2270 calories per day for women aged 18–30 who are reasonably active.

Pasta

Nutrition Facts

Serving Size 2 oz (56g) dry
(1/8 of the package)
Servings Per Container 8

Amount Per Serving	
Calories 200	Calories from Fat 10
	% **Daily Value***
Total Fat 1 g	2%
Saturated Fat 0g	0%
Cholesterol 0mg	0%
Sodium 0mg	0%
Total Carbohydrate 41g	14%
Dietary Fiber 2g	8%
Sugars 2g	
Protein 7g	
Vitamin A 0% •	Vitamin C 0%
Calcium 0% •	Iron 10%
Thiamin 35% •	Riboflavin 15%
Niacin 15% •	

* Percent Daily Values are based on a 2,000 calorie diet. Your daily values may be higher or lower depending on your calorie needs.

Marinara Sauce

Nutrition Facts

Serving Size 1/2 cup (125g)
Servings per Container approx 6

Amount Per Serving	
Calories 60	Calories from Fat 20
	% **Daily Value***
Total Fat 2g	3%
Saturated Fat 0g	0%
Cholesterol 0mg	0%
Sodium 370mg	15%
Total Carbohydrate 7g	2%
Dietary Fiber 2g	8%
Sugars 4g	
Protein 3g	
Vitamin A 15% •	Vitamin C 40%
Calcium 0% •	Iron 4%

66. Sasha, who is 22 years old, sets 2250 calories per day as her goal. She plans to have pasta with marinara sauce for dinner. The product labels show the number of calories in each food. If she eats two servings each of pasta and sauce, how many calories does she consume?

67. If Sasha has 550 more calories in bread, butter, salad, drink, and desert for dinner, how many total calories did she consume at dinner?

68. If Sasha keeps to her goal, how many calories could she have eaten at breakfast and lunch?

69. The attendance at three consecutive Super Bowls is 78,943, 85,782, and 103,456. What is the total attendance at the three games? How many more fans were at the game with largest attendance as opposed to the one with the smallest attendance?

70. A forester counted 23,679 trees ready for harvest on a certain acreage. If Forestry Service rules require that 8543 mature trees must be left on the acreage, how many trees can be harvested?

71. The new sewer line being installed in downtown Kearney will handle 345,760 gallons of refuse per minute. The old line handled 178,550 gallons per minute. How many more gallons per minute will the new line handle?

72. Fong's Grocery owes a supplier \$25,875. During the month Fong's makes payments of \$460, \$983, \$565, and \$10,730. How much does Fong's still owe, to the nearest hundred dollars?

73. In the spring of 1989, an oil tanker hit a reef and spilled 10,100,000 gallons of oil off the coast of Alaska. The tanker carried a total of 45,700,000 gallons of oil. The oil that did not spill was pumped into another tanker. How many gallons of oil were pumped into the second tanker? Round to the nearest million gallons.

74. The *National Petroleum News* reports that there were 226,000 gasoline stations in the U.S. in the 1970s. This figure dropped to 188,000 stations in 1997. Have these figures been rounded? If so, to what place value? How many gas stations have been lost during this time period?

75. The median family income of a region is a way of estimating the middle income. Half the families in the region make more than the median income and the other half of the families make less than it. In 1997, the median family income for San Francisco was $64,400 and median for Seattle was $55,100. What place value were these figures rounded to and how much higher was San Francisco's median income than Seattle's?

76. The Grand Canyon, Zion, and Bryce Canyon parks are found in the southwestern United States. Geologic changes over a billion years have created these formations and canyons. The table shows the highest and lowest elevations in each of these parks. Find the change in elevation in each park. In which park is the change greatest and by how much?

	Highest Elevation	**Lowest Elevation**
Bryce Canyon	8500 ft	6600 ft
Grand Canyon	8300 ft	2500 ft
Zion	7500 ft	4000 ft

Exercises 77–78 relate to the chapter application. See table page 15, Exercises 74–76.

77. How much more did Dennis Rodman earn than Scottie Pippen?

78. What was the total payroll, paid the players, for the Chicago Bulls during the 1996–1997 season?

Exercises 79–80. The table below gives the World Health Organization's estimates of the HIV infections in adults in 1994.

Region	**HIV Infections**
Sub-Saharan Africa	10 million
South and Southeast Asia	3 million
Latin America/Caribbean	2 million
North America	1 million
Western Europe	500 thousand
North Africa/Middle East	100 thousand
Eastern Europe/Central Asia	50 thousand
East Asia/Pacific	50 thousand
Australia	25 thousand

79. What is the total estimated number of HIV infections for Asia?

NAME ______________________ CLASS ____________ DATE ____________

80. How many more estimated cases of HIV infections are there in Western Europe than in Eastern Europe/Central Asia?

STATE YOUR UNDERSTANDING

81. Explain to an 8-year-old child why 15 − 9 = 6.

82. Explain to an 8-year-old child why 8 + 7 = 15.

83. Define and give an example of a sum.

84. Define and give an example of a difference.

CHALLENGE

85. Add the following numbers, round the sum to the nearest hundred, and write the word name for the rounded sum: one hundred sixty; eighty thousand, three hundred twelve; four hundred seventy-two thousand, nine hundred fifty-two; and one hundred forty-seven thousand, five hundred twenty-three.

86. How much greater is seven million, two hundred forty-seven thousand, one hundred ninety-five than two million, eight hundred four thousand, fifty-three? Write the word name for the difference.

87. Peter sells three Honda Civics for $14,385 each, four Accords for $17,435 each, and two Acuras for $26,548 each. What is the total dollar sales for the nine cars? How many more dollars were paid for the four Accords than the three Civics?

Complete the sum or difference by writing in the correct digit whenever you see a letter.

88.
$$\begin{array}{r} 5A68 \\ 241 \\ +\ \underline{10A9} \\ B64C \end{array}$$

89.
$$\begin{array}{r} 4A6B \\ -\ \underline{C251} \\ 15D1 \end{array}$$

GROUP ACTIVITY

90. Add and round to the nearest hundred.

$$\begin{array}{r} 14{,}657 \\ 3{,}766 \\ 123{,}900 \\ 569 \\ 54{,}861 \\ +\ \underline{346{,}780} \end{array}$$

Now round each addend to the nearest hundred and then add. Discuss why the answers are different. Be prepared to explain why this happens.

91. If Ramon delivers 112 loaves of bread to each store on his delivery route, how many stores are on the route if he delivers a total of 4368 loaves? (*Hint:* Subtract 112 loaves for each stop from the total number of loaves.) What operation does this perform? Make up three more examples and be prepared to demonstrate them in class.

Getting Ready for Algebra

OBJECTIVE

Solve an equation of the form $x + a = b$ or $x - a = b$, where a, b, and x are whole numbers.

VOCABULARY

An **equation** is a statement about numbers that says that two expressions are equal. Letters, called **variables** or **unknowns,** are often used to represent numbers.

HOW AND WHY

Objective

Solve an equation of the form $x + a = b$ or $x - a = b$, where a, b, and x are whole numbers.

Examples of equations are

$8 = 8$ $\quad 12 = 12$ $\quad 100 = 100$ $\quad 20 + 5 = 25$ $\quad 49 - 9 = 40$

When variables are used, an equation can look like this:

$x = 2$ $\quad x = 5$ $\quad y = 12$ $\quad x + 3 = 10$ $\quad y - 7 = 13$

An equation containing a variable can be true only when the variable is replaced by a specific number. For example,

$x = 2$ is true only when x is replaced by 2.

$x = 5$ is true only when x is replaced by 5.

$y = 12$ is true only when y is replaced by 12.

$x + 3 = 10$ is true only when x is replaced by 7, so that $7 + 3 = 10$.

$y - 7 = 13$ is true only when y is replaced by 20, so that $20 - 7 = 13$.

The numbers that make equations true are called *solutions*. Solutions of equations, such as $x - 4 = 7$, can be found by trial and error, but let's develop a more practical way.

Addition and subtraction are inverse, or opposite, operations. For example, if 12 is added to a number and then 12 is subtracted from that sum, the difference is the original number.

As a specific example, add 11 to 15: $15 + 11 = 26$; the sum is 26. If 11 is subtracted from that sum, $26 - 11$, the result is 15, which was the original number.

We will use this idea to solve the following equation:

$x + 15 = 19$	**15 is added to the number represented by x.**
$x + 15 - 15 = 19 - 15$	**To remove the addition and have only x on the left side of the equal sign, we subtract 15. To keep a true equation, we must subtract 15 from both sides.**
$x = 4$	**This equation will be true when x is replaced by 4.**

To check, replace x in the original equation with 4 and see if the result is a true statement:

$x + 15 = 19$

$4 + 15 = 19$

$19 = 19$ **The statement is true, so the solution is 4.**

We can also use the idea of inverses to solve an equation in which a number is subtracted from a variable (letter):

$x - 4 = 7$	**Since 4 is subtracted from the variable, we eliminate**
$x - 4 + 4 = 7 + 4$	**the subtraction by adding 4 to both sides of the equation. Recall that addition is the inverse of subtraction.**
$x = 11$	**This equation will be true when x is replaced by 11.**

To solve an equation using addition or subtraction

1. Add the same number to each side of the equation to isolate the variable, or
2. Subtract the same number from each side of the equation to isolate the variable.
3. Check the solution by substituting it for the variable in the original equation.

Warm Ups A–E

A. $x + 7 = 11$

B. $x - 11 = 8$

Examples A–E

Directions: Solve and check.

Strategy: Isolate the variable by adding or subtracting the same number to or from each side.

A. $x + 4 = 9$

$x + 4 = 9$	
$x + 4 - 4 = 9 - 4$	**Because 4 is added to the variable, eliminate the addition by subtracting 4 from both sides of the equation.**
$x = 5$	**Simplify.**

CHECK:

$x + 4 = 9$	**Check by substituting 5 for x in the original equation.**
$5 + 4 = 9$	
$9 = 9$	**The statement is true.**

The solution is $x = 5$.

B. $x - 9 = 7$

$x - 9 = 7$	
$x - 9 + 9 = 7 + 9$	**Because 9 is subtracted from the variable, eliminate the subtraction by adding 9 to both sides of the equation.**
$x = 16$	**Simplify.**

CHECK:

$x - 9 = 7$	**Check by substituting 16 for x in the original equation.**
$16 - 9 = 7$	
$7 = 7$	**The statement is true.**

The solution is $x = 16$.

Answers to Warm Ups A. $x = 4$ B. $x = 19$

C. $22 = y + 15$

In this example we do the subtraction vertically.

$$\begin{array}{rl} 22 = y + 15 \\ \underline{-15} \quad \underline{-15} \\ 7 = y \end{array}$$

Subtract 15 from both sides to eliminate the addition of 15.

Simplify.

CHECK:

$22 = y + 15$

$22 = 7 + 15$ **Substitute 7 for y.**

$22 = 22$ **The statement is true.**

The solution is $y = 7$.

C. $37 = z + 18$

D. $z - 27 = 35$

$z - 27 = 35$

$z - 27 + 27 = 35 + 27$ **Add 27 to both sides.**

$z = 62$ **Simplify.**

CHECK:

$z - 27 = 35$

$62 - 27 = 35$ **Substitute 62 for z.**

$35 = 35$ **The statement is true.**

The solution is $z = 62$.

D. $b - 47 = 45$

E. The selling price of a pair of shoes is \$67. If the markup on the shoes is \$23, what is the cost to the store? Cost + markup = selling price.

Strategy: To find the cost of the pair of shoes, first translate the English sentence to an algebraic form. Let C represent cost, M represent markup, and S represent selling price.

$C + M = S$ **Since cost + markup = selling price.**

$C + 23 = 67$ **Substitute 23 for markup and 67 for selling price.**

$C + 23 - 23 = 67 - 23$ **Subtract 23 from each side.**

$C = 44$

CHECK: Does the cost + the markup equal \$67?

\$44	Cost
+ 23	Markup
\$67	Selling price

So the cost of the shoes to the store is \$44.

E. The selling price of a set of dishes is \$145. If the markup is \$39, what is the cost to the store? Cost + markup = selling price.

Answers to Warm Ups C. $z = 19$ D. $b = 92$ E. The cost to the store is \$106.

NAME ______________________ CLASS ____________ DATE ____________

Exercises

OBJECTIVE: *Solve an equation of the form $x + a = b$ or $x - a = b$, where a, b, and x are whole numbers.*

Solve and check.

1. $x + 3 = 12$

2. $x - 7 = 8$

3. $x - 8 = 3$

4. $x + 9 = 43$

5. $z + 12 = 19$

6. $b - 15 = 17$

7. $c + 24 = 53$

8. $y - 22 = 37$

9. $a - 60 = 123$

10. $x + 68 = 96$

11. $x + 89 = 123$

12. $x - 85 = 23$

13. $y + 98 = 145$

14. $z - 56 = 89$

15. $k - 87 = 159$

16. $c + 75 = 75$

17. $60 = x + 23$

18. $712 = a + 688$

19. $52 = w - 78$

20. $271 = d - 175$

21. The selling price for a computer is $1458. If the cost to the store is $1065, what is the markup?

22. The selling price of a trombone is \$675. If the markup is \$235, what is the cost to the store?

23. The length of a rectangular garage is 2 meters more than the width. If the width is 7 meters, what is the length?

24. The width of a rectangular fish pond is 6 feet shorter than the length. If the length is 34 feet, what is the width?

25. A Saturn with manual transmission has an EPA city rating of 4 miles per gallon more than the EPA rating of a Subaru Impreza. Write an equation that describes this relationship. Be sure to define all variables in your equation. If the Saturn has an EPA city rating of 28 mpg, find the rating of the Impreza.

26. In 1993 in the United States, the number of deaths by drowning was 1700 less than the number of deaths by fire. Write an equation that describes this relationship. Be sure to define all variables in your equation. If there were approximately 4800 deaths by drowning that year, how many deaths by fire were there?

Exercises 27–28. A city treasurer made the following report to the city council regarding monies allotted to and dispersed from a City parks bond.

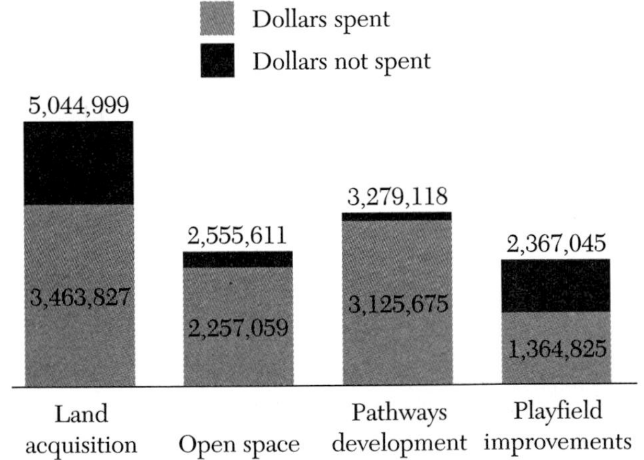

27. Write an equation that relates the total money budgeted per category to the amount of money spent and the amount of money not yet spent. Define all the variables.

28. Use your equation (Exercise 27) to calculate the amount of money not yet spent in each of the four categories.

1.3

Multiplying Whole Numbers

OBJECTIVES

1. Multiply whole numbers.
2. Estimate the product of whole numbers.

VOCABULARY

There are several ways to show a multiplication exercise. Here are examples of most of them, using 18 and 29.

18×29 $\quad$ $18 \cdot 29$ $\quad$ $\begin{array}{r} 29 \\ \times \underline{18} \end{array}$

$(18)(29)$ $\quad$ $18(29)$ $\quad$ $(18)29$

The **factors** of a multiplication exercise are the numbers being multiplied. In $5(8) = 40$, 5 and 8 are the factors.

The **product** is the answer to a multiplication exercise. In $5(8) = 40$, the product is 40.

HOW AND WHY

Objective 1

Multiply whole numbers.

Multiplying whole numbers is a shortcut for repeated addition:

$$\underbrace{8 + 8 + 8 + 8 + 8 + 8}_{\text{6 eights}} = 48 \qquad \text{or} \qquad 6 \cdot 8 = 48$$

As numbers get larger, the shortcut saves time. Imagine adding 146 eights.

$$\underbrace{8 + 8 + 8 + 8 + 8 + \cdots + 8}_{\text{146 eights}} = ?$$

We multiply 8 times 146 using the expanded form of 146.

$$\begin{array}{rl} 146 & = 100 + 40 + 6 \quad \textbf{Write 146 in expanded form.} \\ \times \underline{8} & = \underline{8} \\ & = 800 + 320 + 48 \quad \textbf{Multiply 8 times each addend.} \\ & = 1168 \quad \textbf{Add.} \end{array}$$

The exercise can also be performed in column form without expanding the factors.

$$\begin{array}{rl} 146 & \\ \times \underline{8} & \\ 48 & 8(6) = 48 \\ 320 & 8(40) = 320 \\ \underline{800} & 8(100) = 800 \\ 1168 & \end{array} \qquad \begin{array}{r} {\scriptstyle 3\,4} \\ 146 \\ \times \underline{8} \\ 1168 \end{array}$$

The form on the right shows the usual shortcut. The carried digit is added to the product of each column. Study this example.

```
  629
× 46
```

First multiply 629 by 6.

```
  15
  629    
×  46
 3774
```

6(9) = 54. Carry the 5 to the tens column. 6(2 tens) = 12 tens. Add the 5 tens that were carried: (12 + 5) tens = 17 tens. Carry the 1 to the hundreds column. 6(6 hundreds) = 36 hundreds. Add the 1 hundred that was carried: (36 + 1) hundreds = 37 hundreds.

Now multiply 629 by 40.

```
  13
  15
  629
×  46
 3774
25160
```

40(9) = 360 or 36 tens. Carry the 3 to the hundreds column. 40(20) = 800 or 8 hundreds. Add the 3 hundreds that were carried. (8 + 3) hundreds = 11 hundreds. Carry the 1 to the thousands column. 40(600) = 24,000 or 24 thousands. Add the 1 thousand that was carried: (24 + 1) thousands = 25 thousands. Write the 5 in the thousands column and the 2 in the ten thousands column.

```
  13
  15
  629
×  46
 3774
25160
28934
```

Add the products.

MULTIPLICATION PROPERTY

Multiplication property of zero

$a \cdot 0 = 0 \cdot a = 0$

Any number times zero is zero.

MULTIPLICATION PROPERTY

Multiplication property of one

$a \cdot 1 = 1 \cdot a = a$

Any number times one is that number.

Two important properties of arithmetic and higher mathematics are the *multiplication property of zero* and the *multiplication property of one*.

As a result of the multiplication property of zero, we know that

$0 \cdot 19 = 19 \cdot 0 = 0$ and $0(165) = 165(0) = 0$

As a result of the multiplication property of one, we know that

$1 \cdot 27 = 27 \cdot 1 = 27$ and $1(354) = 354(1) = 354$

Warm Ups A–F

A. (210)(0)

Examples A–F

Directions: Multiply.

Strategy: Write the factors in columns. Start multiplying with the ones digit. If the product is ten or more, carry the tens digit to the next column and add it to the product in that column. Repeat the process for every digit in the second factor. When the multiplication is complete, add to find the product.

A. 0(145)

0(145) = 0 **Multiplication property of zero.**

Answer to Warm Up A. 0

B. Find the product: 6(4592)

```
  351
  4592
×    6
 27552
```

Multiply 6 times each digit, carry when necessary, and add the number carried to the next product.

B. Find the product: 8(3456)

C. Multiply: 38 · 74

```
   1
   3
   74
×  38
  592
 2220
 2812
```

When multiplying by the 3 in the tens place, write a 0 in the ones column to keep the places lined up.

C. Multiply: 59 · 84

D. Find the product of 513 and 205.

Strategy: When multiplying by zero in the tens place, rather than showing a row of zeros, just put a zero in the tens column. Then multiply by the 2 in the hundreds place.

```
    513
×   205
   2565
 102600
 105165
```

D. Find the product of 326 and 707.

Calculator Example

E. 346(76)

Enter the multiplication.

The product is 26,296.

E. 398(148)

F. The Sweet & Sour Company ships 62 cartons of packaged candy to a convenience store. Each carton contains 48 packages of candy. What is the total number of packages of candy shipped?

Strategy: To find the total number of packages, multiply the number of cartons by the number of packages per carton.

```
    62
×   48
   496
  2480
  2976
```

The company shipped 2976 packages of candy.

F. The Sweet & Sour Company ships 68 cartons of chewing gum to the Sweet Shop. If each carton contains 72 packages of gum, how many packages of gum are shipped to the Sweet Shop?

Answers to Warm Ups B. 27,648 C. 4956 D. 230,482 E. 58,904 F. The company shipped 4896 packages of gum.

HOW AND WHY

Objective 2

Estimate the product of whole numbers.

The product of two numbers can be estimated by rounding each factor to the largest place value in each factor and then multiplying the rounded factors. For instance,

$$\begin{array}{r} 8130 \\ \times \underline{\quad 58} \end{array} \qquad \begin{array}{r} 8000 \\ \times \underline{\quad 60} \\ 480000 \end{array} \qquad \begin{array}{l} \textbf{Round to the nearest thousand.} \\ \textbf{Round to the nearest ten.} \\ \textbf{Multiply.} \end{array}$$

The estimate of the product is 480,000. One use of the estimate is to see if the product is correct. If the calculated product is not close to 480,000, you should check the multiplication. In this case the actual product is 471,540, which is close to the estimate.

To estimate the product of two whole numbers

1. Round each number to its largest place value.
2. Multiply the rounded numbers.

Warm Up G

Example G

Directions: Estimate the product. Then find the product and compare.

Strategy: Round each number to its largest place value. Multiply the rounded numbers. Multiply the original numbers.

G. 634 and 578

G. 687 and 347

$$\begin{array}{r} 700 \\ \times \underline{\quad 300} \\ 210000 \end{array} \qquad \begin{array}{l} \textbf{Round to the nearest hundred.} \\ \textbf{Round to the nearest hundred.} \\ \end{array}$$

$$\begin{array}{r} 687 \\ \times \underline{\quad 347} \\ 4809 \\ 27480 \\ \underline{206100} \\ 238389 \end{array}$$

The product is 238,389 and is close to the estimated product of 210,000.

Answer to Warm Up G. The product is 366,452 and is close to the estimated product of 360,000.

NAME ____________________ CLASS ____________ DATE ____________

Exercises 1.3

OBJECTIVE 1: *Multiply whole numbers.*

A.

Multiply.

1. 42×5

2. 54×3

3. 33×4

4. 23×6

5. 72×4

6. 46×6

7. 69×7

8. 94×8

9. $8 \cdot 41$

10. $6 \cdot 55$

11. (75)(0)

12. (0)(167)

13. 66×40

14. 93×70

15. In 326 × 52 the place value of the product of "5" and "3" is __________.

16. In 326 × 52 the product of "5" and "6" is "30" and you must carry the 3 to the __________ column.

B.

Multiply.

17. 242×9

18. 652×6

19. 806×5

20. 505×8

21. (42)(38)

22. (24)(65)

23. (89)(32)

24. (98)(23)

25. 513×300

26. 682×600

27. 703×67

28. 509×58

29. 646×45

30. 328×52

31. (92)(145) Round product to the nearest hundred.

32. (78)(246) Round product to the nearest thousand.

OBJECTIVE 2: *Estimate the product of whole numbers.*

A.

Estimate the product.

33. 38(42) **34.** 71(67) **35.** 213(81) **36.** 365(24)

37. 362×19 **38.** 756×37 **39.** 544×71 **40.** 277×42

B.

Estimate the product and multiply.

41. 312×50 **42.** 675×40 **43.** 412×84

44. 277×53 **45.** 684×47 **46.** 823×65

47. (1234)(54) **48.** (3510)(83)

C.

49. Find the product of 606 and 415.

50. Find the product of 707 and 526.

NAME ______________________ CLASS ____________ DATE ____________

51. Estimate the product and multiply: (633)(2361)

52. Estimate the product and multiply: (6004)(405)

53. During the first week of the Rotary Rose sale 235 dozen roses are sold. It is estimated that a total of 18 times that number will be sold during the sale. What is the estimated number of dozens of roses that will be sold?

54. The Good Food Grocery Store orders 735 cases of fruit cocktail. If each case costs $19, what is the total cost of the fruit cocktail?

Exercises 55–58 refer to the monthly sales at Dick's Country Cars as given in the chart.

Car model	Number of cars sold	Average price per sale
Cougar XR7	27	$17,845
Sable GS	42	$19,340
Villager GS	17	$21,795

55. Estimate the gross receipts from the sale of the Cougars.

56. What are gross receipts from the sale of the Sables?

57. Estimate the gross receipts from the sale of Villagers and find the actual gross receipts.

58. Estimate the gross receipts for the month (the sum of the estimates for each model) and then find the actual gross receipts rounded to the nearest thousand dollars.

59. Estimate the product and multiply: (24)(45)(36)

60. Estimate the product and multiply: (32)(71)(82)

61. An average of 134 salmon per day are counted at the Bonneville fish ladder during a 17-day period. How many total salmon are counted during the 17-day period?

62. During 1997, the population of Washington County grew at a pace of 2043 people per month. What was the total growth in population during 1997?

63. The CEO of Apex Corporation exercised his option to purchase 2355 shares of Apex stock at $13 per share. He immediately sold the shares for $47 per share. If broker fees came to $3000, how much money did he realize from the purchase and sale of the shares?

64. The comptroller of Apex Corporation exercised her option to purchase 895 shares of Apex stock at $18 per share. She immediately sold the shares for $51 per share. If broker fees came to $950, how much money did she realize from the purchase and sale of the shares?

65. A certain bacteria culture triples its count every hour. If the culture has a count of 375 at 10 A.M., what will the count be at 2 P.M. the same day?

Exercises 66–67 relate to the chapter application (see table page 15, Exercises 74–76).

66. If Scottie Pippen would sign a new contract for three times his 1996–1997 salary, what would his new salary be?

67. Is eight times Ron Harper's salary larger or smaller than Michael Jordan's salary?

68. The starling population of the United States doubles every three years. If the current population of these birds is estimated to be 3,575,000, what will the population be in nine years?

69. The water consumption in Hebo averages 320,450 gallons per day. How many gallons of water are consumed in a 31-day month, rounded to the nearest thousand gallons?

70. The property tax in Mt. Pedro averages $1536 per home. What is the total tax collected from 8347 homes, rounded to the nearest ten thousand dollars?

71. Ms. Muzos orders 450 radios for sale in her discount store. If she pays $78 per radio and sells them for $112, how much do the radios cost her and what is the net income from their sale? How much are her profits from the sale of the radios?

72. Mr. Garcia orders 325 snow shovels for his hardware store. He pays $12 per shovel and plans to sell them for $21 each. What do the shovels cost Mr. Garcia and what will be his gross income from their sale? What net income does he receive from the sale of the shovels?

73. In 1997, Bill Gates of Microsoft was the richest person in the United States, with an estimated net worth of $36 billion. Write the place value name for this number. A financial analyst made the observation that the average person has a hard time understanding such large amounts. She gave the example that in order to spend $1 billion, one would have to spend $40,000 per day for 69 years. How much money would you spend if you did this?

74. The median purchase price of a home is a way of estimating the middle price. Half the homes in an area cost more than the median price whereas the other half of the homes are less expensive than the median. In 1997, the median price of a home in Portland was $147,000 and the median price of a home in San Francisco was $288,000. Choose a word to make the statement true. The median price of a home in San Francisco was (almost/more than) twice the median price of a home in Portland.

Exercises 75–76 relate to the chapter application (see page 1).

75. If the Astros doubled their 1997 payroll, where would they rank?

76. If the Yankees doubled their payroll and the Giants tripled their payroll, who would have the largest payroll? By how much?

77. When walking for 1 mile Carol burns 112 calories. If she walks 4 miles each day, how many calories does she burn in a week? In a 30 day month? In a year?

STATE YOUR UNDERSTANDING

78. Explain to an 8-year-old child that $3(8) = 24$.

79. When 74 is multiplied by 8 we carry 3 to the tens column. Explain why this is necessary.

80. Define and give an example of a product.

CHALLENGE

81. Find the product of twenty-four thousand, fifty-five and two hundred thirteen thousand, two hundred seventy-six. Write the word name for the product.

82. Tesfay harvests 75 bushels of grain per acre from his 11,575 acres of grain. If Tesfay can sell the grain for \$27 a bushel, what is the crop worth, to the nearest thousand dollars?

Complete the problems by writing in the correct digit whenever you see a letter.

83.
$$\begin{array}{r} 51A \\ \times \underline{B2} \\ 10B2 \\ \underline{154C} \\ 1A5E2 \end{array}$$

84.
$$\begin{array}{r} 1A57 \\ \times \underline{42} \\ B71C \\ \underline{D428} \\ 569E4 \end{array}$$

GROUP ACTIVITY

85. Multiply 23, 56, 789, 214, and 1345 by 10, 100, and 1000. What do you observe? Can you devise a rule for multiplying by 10, 100, and 1000?

1.4

Dividing Whole Numbers

OBJECTIVES

1. Divide whole numbers.
2. Estimate the quotient of whole numbers.

VOCABULARY

There are a variety of ways to show a division exercise. These are the most commonly used:

$51 \div 3$ $\quad 3\overline{)51}$ $\quad \dfrac{51}{3}$ $\quad 51/3$

The **dividend** is the number being divided, so in $36 \div 4 = 9$, the dividend is 36.

The **divisor** is the number that we are dividing by, so in $36 \div 4 = 9$, the divisor is 4.

The **quotient** is the answer to a division exercise, so in $36 \div 4 = 9$, the quotient is 9.

When a division exercise does not come out even, as in $53 \div 3$, the quotient is not a whole number.

$$\begin{array}{r} 17 \\ 3\overline{)53} \\ \underline{51} \\ 2 \end{array}$$

We call 17 the **partial quotient** and 2 the **remainder.** The quotient is written 17 R 2.

HOW AND WHY

Objective 1

Divide whole numbers.

The division exercise $128 \div 32 = ?$ (read "128 divided by 32") can be interpreted in one of two ways.

How many times can 32 be subtracted from 128?	**This is called the "repeated subtraction" version.**
What number times 32 is equal to 128?	**This is called the "missing factor" version.**

All division problems can be done using repeated subtraction. In $128 \div 32 = ?$, we can find the missing factor by repeatedly subtracting 32 from 128:

$$\begin{array}{r} 128 \\ -\underline{\ 32} \\ 96 \\ -\underline{\ 32} \\ 64 \\ -\underline{\ 32} \\ 32 \\ -\underline{\ 32} \\ 0 \end{array}$$

Four subtractions, so $128 \div 32 = 4$.

The process can be shortened using the traditional method of guessing the number of 32s and subtracting from 128:

```
32)128|  2 thirty-twos            32)128|  3 thirty-twos
    64|                               96|
    64|  2 thirty-twos     or         32|  1 thirty-two
    64|  4                            32|  4
     0|                                0|
```

```
or   32)128|
        128|  4 thirty-twos
          0|  4
```

In each case, 128 ÷ 32 = 4.

We see that the missing factor in (32)(?) = 128 is 4. So 32(4) = 128 or 128 ÷ 32 = 4.

This leads to a method for checking division. If we multiply the divisor times the quotient we will get the dividend. To check 128 ÷ 32 = 4 we multiply 32 and 4.

(32)(4) = 128

So 4 is correct.

This process works regardless of the size of the numbers. If the divisor is considerably smaller than the dividend, you will want to guess a rather large number.

```
44)13,420
    4 400|   100
    9 020|
    4 400|   100
    4 620|
    4 400|   100
      220|
      220|     5
        0|   305
```

So, 13,420 ÷ 44 = 305

All divisions can be done by this method. However, the process can be shortened by finding the number of groups, starting with the largest place value on the left, in the dividend, and then working toward the right. Study the following example. The answer is written above the problem for convenience.

```
23)17135
```

Working from left to right, we note that 23 does not divide 1, and it does not divide 17. However, 23 does divide 171 seven times. Write the 7 above the second 1 in the dividend.

```
     745
23)17135
   161
    103
     92
     115
     115
       0
```

7(23) = 161. Subtract 161 from 171. Because the difference is less than the divisor, no adjustment is necessary. Bring down the next digit, which is 3. Next, 23 divides 103 four times. The 4 is placed above the 3 in the dividend. 4(23) = 92. Subtract 92 from 103. Again, no adjustment is necessary since 11 < 23. Bring down the next digit, which is 5. Finally, 23 divides 115 five times. Place the 5 above the 5 in the dividend. 5(23) = 115. Subtract 115 from 115, the remainder is zero. The division is complete.

CHECK:

$$\begin{array}{r} 745 \\ \times\ \underline{23} \\ 2235 \\ \underline{14900} \\ 17{,}135 \end{array}$$

Check by multiplying the quotient by the divisor.

So 17,135 ÷ 23 = 745.

Not all division problems come out even (have a zero remainder). In

$$\begin{array}{r} 3 \\ 15\overline{)50} \\ \underline{45} \\ 5 \end{array}$$

we see that 50 contains 3 fifteens and 5 toward the next group of fifteen. The answer is written as 3 remainder 5. The word "remainder" is abbreviated "R" and the result is 3 R 5.

A check can be made by finding (15)(3) and adding the remainder.

(15)(3) = 45

45 + 5 = 50

So 50 ÷ 15 = 3 R 5.

Recall that 45 ÷ 0 = ? asks what number times 0 is 45: 0 × ? = 45. According to the multiplication property of zero we know that 0 × ? = 0, so it cannot equal 45.

CAUTION

Division by zero is not defined. It is an operation that cannot be performed.

When dividing by a single-digit number the division can be done mentally using "short division."

$$\begin{array}{r} 423 \\ 3\overline{)1269} \end{array}$$

Divide 3 into 12. Write the answer, 4, above the 2 in the dividend. Now divide the 6 by 3 and write the answer, 2, above the 6. Finally divide the 9 by 3 and write the answer, 3, above the 9.

The quotient is 423.

If the "mental" division does not come out even, each remainder is used in the next division.

$$\begin{array}{l} \quad 4\ 5\ 2\ \text{R}\ 2 \\ 3\overline{)1\ 3\ {}^{1}5\ 8} \end{array}$$

13 ÷ 3 = 4 R 1. Write the four above the 3 in the dividend. Now form a new number, 15, using the remainder 1 and the next digit, 5. Divide 3 into 15. Write the answer, 5, above 5 in the dividend. Because there is no remainder, divide the next digit, 8, by 3. The result is 2 R 2. Write this above the 8.

The quotient is 452 R 2.

Warm Ups A–E

A. Divide: $9\overline{)5436}$

B. Divide: $18\overline{)2970}$

C. Find the quotient: $\dfrac{196008}{365}$

Examples A–E

Directions: Divide.

Strategy: Divide from left to right. Use short division for single-digit divisors.

A. Divide: $7\overline{)4207}$

Strategy: Because there is a single-digit divisor we will use short division.

$$\begin{array}{r} 601 \\ 7\overline{)4207} \end{array}$$

7 divides 42 six times.
7 divides 0 zero times.
7 divides 7 one time.

CAUTION
A zero must be placed in the quotient so that the 6 and the 1 have the correct place values.

The quotient is 601.

B. Divide: $16\overline{)2160}$

Strategy: Write the partial quotients above the dividend with the place values aligned.

$$\begin{array}{rl} 135 & \\ 16\overline{)2160} & \\ \underline{16} & \quad 16(1) = 16 \\ 56 & \\ \underline{48} & \quad 16(3) = 48 \\ 80 & \\ \underline{80} & \quad 16(5) = 80 \\ 0 & \end{array}$$

CHECK:

$$\begin{array}{r} 135 \\ \times \underline{16} \\ 810 \\ \underline{1350} \\ 2160 \end{array}$$

The quotient is 135.

C. Find the quotient: $\dfrac{103297}{365}$

Strategy: When a division is written as a fraction, the dividend is above the fraction bar and the divisor is below.

$$\begin{array}{rl} 283 & \\ 365\overline{)103297} & \quad \textbf{365 does not divide 1.} \\ \underline{730} & \quad \textbf{365 does not divide 10.} \\ 3029 & \quad \textbf{365 does not divide 103.} \\ \underline{2920} & \quad \textbf{365 divides 1032 two times.} \\ 1097 & \quad \textbf{365 divides 3029 eight times.} \\ \underline{1095} & \quad \textbf{365 divides 1097 three times.} \\ 2 & \quad \textbf{The remainder is 2.} \end{array}$$

Answers to Warm Ups A. 604 B. 165 C. 537 R 3

CHECK: Multiply the divisor by the partial quotient and add the remainder.

$$283(365) + 2 = 103295 + 2$$
$$= 103,297$$

The answer is 283 with a remainder of 2. This can also be written 283 R 2.

You may recall other ways to write a remainder using fractions or decimals. These are covered in a later chapter.

Calculator Example

D. Divide 2756 by 143.
Enter the division: 2756 ÷ 143

$2756 \div 143 \approx 19.27272727$

The quotient is not a whole number. This means that 19 is the partial quotient and there is a remainder. To find the remainder, multiply 19 times 143. Subtract the product from 2756. The result is the remainder.

$2756 - 19(143) = 39$

So $2756 \div 143 = 19$ R 39.

D. Divide 4337 by 123.

E. When planting Christmas trees, the Greenfir Tree Farm allows 64 square feet per tree. If there are 43,520 square feet in 1 acre, how many trees will they plant per acre?

Strategy: Because each tree is allowed 64 square feet, we divide the number of square feet in one acre by 64 to find out how many trees will be planted in one acre.

```
    680
64)43520
   384
    512
    512
     00
      0
      0
```

There will be a total of 680 trees planted per acre.

E. The Greenfir Tree Farm allows 256 square feet per large spruce tree. If there are 43,520 square feet in 1 acre, how many trees will they plant per acre?

HOW AND WHY

Objective 2

Estimate the quotient of whole numbers.

The quotient of two numbers can be estimated by rounding the divisor and the dividend to their largest place value and then dividing the rounded numbers. For instance,

27)6345 30)6000 **Round each to its largest place value.**

```
   200
30)6000
   6000
      0
```

Divide.

Answers to Warm Ups D. 35 R 32 E. They will plant 170 trees per acre.

The estimate of the quotient is 200. One use of the estimate is to see if the quotient is correct. If the calculated quotient is not close to 200, you should check the division. In this case the actual quotient is 235, which is close to the estimate.

The estimated quotient will not always come out even. For instance,

$628\overline{)241678}$

$600\overline{)200000}$ **Round each number to its largest place value.**

$\begin{array}{r} 300 \\ 600\overline{)200000} \end{array}$ **Because 3(600) = 1800 is closer to 2000 than is 4(600) = 2400, we choose 3 to use in the estimate, and complete the problem.**

The estimated quotient is 300.

As you get more proficient at estimating you might want to round the dividend to two nonzero digits. This will give you a better estimation.

$628\overline{)241678}$

$600\overline{)240000}$ **Round the dividend to two nonzero digits.**

$\begin{array}{r} 400 \\ 600\overline{)240000} \end{array}$ **Because 600 divides into 2400 exactly 4 times, we conclude that 400 is a better estimate than 300.**

The actual quotient is 384 R 526, so we see that 400 is indeed a better estimate than 300.

To estimate the quotient of two whole numbers

1. Round each number to its largest place value.
2. Divide the rounded numbers.
3. If the first partial quotient has a remainder, multiply the divisor by the partial quotient and one more than the partial quotient. For the estimate choose the number whose product is closer to the dividend. Write zeros to complete the estimated quotient.

Warm Up F

F. 78,624 ÷ 416

Example F

Directions: Estimate the quotient and divide.

Strategy: Round each number to its largest place value and then divide the rounded numbers. If the first partial quotient has a remainder, choose the digit that will give the closer value when multiplied by the divisor. Write zeros to complete the estimated quotient.

F. 58,590 ÷ 158

$158\overline{)58590}$ $200\overline{)60000}$

$\begin{array}{r} 300 \\ 200\overline{)60000} \end{array}$ **Divide.**

$$\begin{array}{r} 370 \\ 158\overline{)58590} \\ \underline{474} \\ 1119 \\ \underline{1106} \\ 130 \\ \underline{0} \\ 130 \end{array}$$

The quotient is estimated to be 300 and is 370 R 130.

Answer to Warm Up F. The quotient is estimated to be 200 and is 189.

NAME ______________________ CLASS ____________ DATE ____________

Exercises 1.4

OBJECTIVE 1: *Divide whole numbers.*

A.

Divide.

1. $8\overline{)96}$ **2.** $7\overline{)56}$ **3.** $9\overline{)45}$ **4.** $5\overline{)60}$

5. $7\overline{)749}$ **6.** $4\overline{)168}$ **7.** $5\overline{)305}$ **8.** $9\overline{)189}$

9. $135 \div 9$ **10.** $192 \div 6$ **11.** $680 \div 17$ **12.** $900 \div 18$

13. $497 \div 7$ **14.** $639 \div 3$ **15.** $19 \div 8$ **16.** $16 \div 5$

17. The division has a remainder when the last difference in the division is smaller than the ________ and is not zero.

18. For $2600 \div 13$, in the partial division $26 \div 13 = 2$, 2 has place value ________.

B.

Divide.

19. $12{,}208 \div 4$ **20.** $12{,}324 \div 6$ **21.** $\frac{768}{32}$ **22.** $\frac{632}{79}$

23. $34\overline{)884}$ **24.** $54\overline{)4212}$ **25.** $24\overline{)2304}$ **26.** $35\overline{)2520}$

27. $234\overline{)11{,}934}$ **28.** $322\overline{)24{,}794}$ **29.** $45\overline{)7862}$ **30.** $24\overline{)9822}$

31. $43\overline{)675}$ **32.** $28\overline{)456}$ **33.** (62)(?) = 3596 **34.** (?)(73) = 2555

35. 46,113 ÷ 57. Round quotient to the nearest ten.

36. 20,000 ÷ 65. Round quotient to the nearest hundred.

OBJECTIVE 2: *Estimate the quotient of whole numbers.*

A.

Estimate the quotient.

37. 625 ÷ 57 **38.** 789 ÷ 29 **39.** 3500 ÷ 43 **40.** 2356 ÷ 33

41. $610\overline{)34{,}560}$ **42.** $459\overline{)55{,}923}$ **43.** $\frac{34{,}976}{712}$ **44.** $\frac{81{,}782}{198}$

B.

Estimate the quotient and divide.

45. $73\overline{)19{,}783}$ **46.** $28\overline{)5460}$ **47.** 17,121 ÷ 39 **48.** 52,812 ÷ 81

49. $103\overline{)59{,}602}$ **50.** $108\overline{)67{,}891}$ **51.** 780,854 ÷ 436 **52.** 560,999 ÷ 356

C.

53. Estimate the quotient and divide: 6,784,821 ÷ 423

54. Estimate the quotient and divide: 4,378,921 ÷ 241

For Exercises 55–58, the revenue department of a central state reported the following data for the first three weeks of March:

Number of Returns	Total Taxes Paid
Week 1—4563	$24,986,988
Week 2—3981	$19,315,812
Week 3—11,765	$48,660,040

55. Estimate the taxes paid per return during week 2.

56. Find the actual taxes paid per return during week 1.

57. Find the actual taxes paid per return during week 3. Round to the nearest hundred dollars.

58. Find the actual taxes paid per return during the 3 weeks. Round to the nearest hundred dollars.

59. Estimate the quotient and divide: $\dfrac{229{,}367}{1216}$

60. Estimate the quotient and divide: $\dfrac{932{,}486}{3722}$

61. A forestry survey finds that 1664 trees are ready to harvest on a 13-acre plot. On the average, how many trees are ready to harvest per acre?

62. Rosebud Lumber Company replants 5696 seedling fir trees on a 16-acre plot of logged-over land. What is the average number of seedlings planted per acre?

63. The estate of Ken Barker totals $347,875. It is to be shared equally by his five nephews. How much will each nephew receive?

64. Eight co-owners of the Alley Cat dress shop share equally in the proceeds when the business is sold. If the business sells for $229,928, how much will each receive?

65. The Nippon Electronics firm assembles radios for export to the United States. Each radio is constructed using 14 resistors. How many radios can be assembled using 32,278 resistors in stock? How many resistors are left over?

66. The Nippon Electronics firm in Exercise 65 assembles a second radio containing 17 resistors. Using the 32,278 resistors in stock, how many of these radios can be assembled? How many resistors are left over?

67. It takes the Morris Packing Plant 8 hours to process 12 tons of lima beans. How long does it take the plant to process 780 tons of lima beans?

68. It takes the Pacific Packing Plant 6 hours to process 4 tons of Dungeness crab. How long does it take the plant to process 392 tons of Dungeness crab?

69. In 1997, Bill Gates of Microsoft was the richest person in the United States, with an estimated net worth of $36 billion. How much would you have to spend per day in order to spend all of Bill Gates' $36 billion in 69 years? Round to the nearest ten thousand dollars.

70. How much money would you have to spend per day in order to spend Bill Gates' $36 billion in 50 years? In 20 years? Round to the nearest one hundred dollars.

Exercises 71–72 relate to the chapter application (see page 1). A major league season consists of 162 games. Assume each game is 9 innings long.

71. To the nearest hundred thousand dollars, what is the Braves payroll per game?

72. To the nearest ten thousand dollars, what is the Mariners payroll per inning?

73. A new subway system is being built in Los Angeles at an estimated cost of $290 million per mile. It is designed to accommodate a peak-hour load of 36,000 passengers. What is the estimated cost per mile per passenger rounded to the nearest ten dollars?

74. Portland, Oregon, is building a surface light-rail system at an estimated cost of $96 million per mile. It is designed to accommodate a peak-hour load of 6000 passengers. What is the estimated cost per mile per passenger?

75. Using the results of Exercises 73 and 74, is Portland or Los Angeles getting the better deal for their transit system? Justify your answer.

Exercises 76–77 relate to the chapter application (see table on page 15, Exercises 74–76).

76. How many players could the Chicago Bulls hire at Steve Kerr's salary before their combined salaries would be more than Michael Jordan's?

77. The regular NBA season runs 82 games. How much is Michael Jordan paid per game? Round to the nearest ten thousand.

78. Juan is advised by his doctor not to exceed 2700 mg of aspirin per day for his arthritic pain. If he takes capsules containing 325 mg of aspirin, how many capsules can he take without exceeding the doctor's orders?

NAME ______________________ CLASS ____________ DATE ____________

STATE YOUR UNDERSTANDING

79. Explain to an 8-year-old child that $45 \div 9 = 5$.

80. Explain the concept of remainder.

81. Define and give an example of a quotient.

CHALLENGE

82. The Belgium Bulb Company has 171,000 tulip bulbs to market. Eight bulbs are put in a package when shipping to the United States and sold for $3 per package. Twelve bulbs are put in a package when shipping to France and sold for $5 per package. In which country will the Belgium Bulb Company get the greatest gross return? What is the difference in gross receipts?

Complete the problems by writing in the correct digit whenever you see a letter.

83. $3\overline{)1653C}$ with quotient $5AB2$

84. $A3\overline{)4CC1}$ with quotient $21B$

GROUP ACTIVITY

85. Divide 23,000,000 and 140,000,000 by 10, 100, 1000, 10,000, and 100,000. What do you observe? Can you devise a rule for dividing by 10, 100, 1000, 10,000, and 100,000?

Getting Ready for Algebra

OBJECTIVE

Solve an equation of the form $ax = b$ or $\frac{x}{a} = b$, where x, a, and b are whole numbers.

HOW AND WHY

In Section 1.4 the equations involved the inverse operations addition and subtraction. Multiplication and division are also inverse operations. We can use this idea to solve equations containing those operations.

For example, if 4 is multiplied by 2, $4 \cdot 2 = 8$, the product is 8. If the product is divided by 2, $8 \div 2$, the result is 4, the original number. In the same manner, if 12 is divided by 3, $12 \div 3 = 4$, the quotient is 4. If the quotient is multiplied by 3, $4 \cdot 3$, the result is 12, the original number. We use this idea to solve equations in which the variable is either multiplied or divided by a number.

When a variable is multiplied or divided by a number, the multiplication symbols ($\cdot$ or $\times$) and the division symbol ($\div$) normally are not written. We write $3x$ for three times x and $\frac{x}{3}$ for x divided by 3.

Consider the following:

$$3x = 9$$

$$\frac{3x}{3} = \frac{9}{3}$$ **Division will eliminate multiplication.**

or $$x = 3$$

If x in the original equation is replaced by 3, we have

$$3x = 9$$

$$3 \cdot 3 = 9$$

$9 = 9$, **which is a true statement.**

Therefore, the solution is 3.

If the variable is divided by a number

$$\frac{x}{5} = 20$$

$$5 \cdot \frac{x}{5} = 5 \cdot 20$$ **Multiplication will eliminate division.**

Thus, $x = 100$

If x in the original equation is replaced by 100, we have

$$\frac{100}{5} = 20$$

$20 = 20$, **which is a true statement.**

Therefore, the solution is 100.

To solve an equation using multiplication or division

1. Divide both sides by the same number to isolate the variable, or
2. Multiply both sides by the same number to isolate the variable.
3. Check the solution by substituting it for the variable in the original equation.

Warm Ups A–E

A. $3y = 15$

B. $\frac{a}{6} = 7$

C. $\frac{c}{3} = 12$

Examples A–E

Directions: Solve and check.

Strategy: Isolate the variable by multiplying or dividing both sides of the equation by the same number. Check the solution by substituting it for the variable in the original equation.

A. $2x = 12$

$2x = 12$	
$\frac{2x}{2} = \frac{12}{2}$	**Isolate the variable by dividing both sides of the equation by 2.**
$x = 6$	**Simplify.**

CHECK:

$2x = 12$	
$2(6) = 12$	**Substitute 6 for *x* in the original equation.**
$12 = 12$	**The statement is true.**

The solution is $x = 6$.

B. $\frac{x}{5} = 4$

$\frac{x}{5} = 4$	
$5 \cdot \frac{x}{5} = 5(4)$	**Isolate the variable by multiplying both sides by 5.**
$x = 20$	**Simplify.**

CHECK:

$\frac{x}{5} = 4$	
$\frac{20}{5} = 4$	**Substitute 20 for *x* in the original equation.**
$4 = 4$	**The statement is true.**

The solution is $x = 20$.

C. $\frac{b}{2} = 9$

$\frac{b}{2} = 9$	
$2 \cdot \frac{b}{2} = 2(9)$	**Isolate the variable by multiplying both sides of the equation by 2.**
$b = 18$	**Simplify.**

Answers to Warm Ups A. $y = 5$ B. $a = 42$ C. $c = 36$

CHECK:

$\frac{b}{2} = 9$

$\frac{18}{2} = 9$ **Substitute 18 for b in the original equation.**

$9 = 9$ **The statement is true.**

The solution is $b = 18$.

D. $3y = 12$

$3y = 12$

$\frac{3y}{3} = \frac{12}{3}$ **Isolate the variable by dividing both sides of the equation by 3.**

$y = 4$ **Simplify.**

CHECK:

$3y = 12$

$3(4) = 12$ **Substitute 4 for y in the original equation.**

$12 = 12$ **The statement is true.**

The solution is $y = 4$.

D. $5z = 35$

E. What is the width (w) of a rectangular lot in a subdivision if the length (ℓ) is 125 feet and the area (A) is 9375 square feet? Use the formula $A = \ell w$.

Strategy: To find the width of the lot, substitute the area, $A = 9375$, and the length, $\ell = 125$, into the formula and solve.

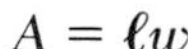

$A = \ell w$

$9375 = 125w$ **$A = 9375$, $\ell = 125$.**

$\frac{9375}{125} = \frac{125w}{125}$ **Divide both sides by 125.**

$75 = w$

CHECK: If the width is 75 feet and the length is 125 feet, is the area 9375 square feet?

$A = (125 \text{ ft})(75 \text{ ft}) = 9375 \text{ sq ft}$ **True**

The width of the lot is 75 feet.

E. What is the length (ℓ) of a second lot in the subdivision if the width (w) is 90 feet and the area (A) is 10,350 square feet? Use the formula $A = \ell w$.

Answers to Warm Ups D. $z = 7$ E. The length of the lot is 115 ft.

NAME ______________________ CLASS ____________ DATE ____________

Exercises

OBJECTIVE: *Solve an equation of the form $ax = b$ or $\frac{x}{a} = b$, where x, a, and b are whole numbers.*

Solve and check.

1. $3x = 15$

2. $\frac{z}{4} = 5$

3. $\frac{c}{3} = 6$

4. $5x = 30$

5. $12x = 48$

6. $\frac{y}{8} = 12$

7. $\frac{b}{8} = 15$

8. $15a = 135$

9. $12x = 144$

10. $\frac{x}{14} = 12$

11. $\frac{y}{13} = 24$

12. $23c = 184$

13. $27x = 648$

14. $\frac{a}{32} = 1536$

15. $\frac{b}{29} = 1566$

16. $63z = 2457$

17. $80 = 16x$

18. $288 = 9y$

19. $71 = \frac{w}{18}$

20. $57 = \frac{c}{23}$

21. Find the width of a rectangular garden plot that has a length of 35 feet and an area of 595 square feet. Use the formula $A = \ell w$.

22. Find the length of a room that has an area of 391 square feet and a width of 17 feet.

23. Crab sells at the dock for \$2 per pound. A fisherman sells his catch and receives \$4680. How many pounds of crab does he sell?

24. Felicia earns \$7 an hour. Last week she earned \$231. How many hours did she work last week?

25. If the wholesale cost of 18 stereo sets is \$5580, what is the wholesale cost of one set? Use the formula $C = np$, where C is the total cost, n is the number of units purchased, and p is the price per unit.

26. Using the formula in Exercise 25, if the wholesale cost of 24 personal computers is \$18,864, what is the wholesale cost of one computer?

27. The average daily low temperature in Toronto in July is twice the average high temperature in January. Write an equation that describes this relationship. Be sure to define all variables in your equation. If the average daily low temperature in July is 60°F, what is the average daily high temperature in January?

28. Car manufacturers recommend that the fuel filter in a car be replaced when the mileage is ten times the recommended mileage for an oil change. Write an equation that describes this relationship. Be sure to define all variables in your equation. If a fuel filter should be replaced every 30,000 miles, how often should the oil be changed?

1.5

Whole Number Exponents and Powers of Ten

OBJECTIVES

1. Find the value of an expression written in exponential form.
2. Multiply or divide a whole number by a power of 10.

VOCABULARY

A **base** is a number used as a repeated factor. An **exponent** indicates the number of times the base is used as a factor and is always written as a superscript to the base. In 2^3, 2 is the base and 3 is the exponent.

The **value** of 2^3 is 8.

An exponent of 2 is often read **"squared"** and an exponent of 3 is often read **"cubed."**

A **power of 10** is the value obtained when 10 is written with an exponent.

HOW AND WHY

Objective 1

Find the value of an expression written in exponential form.

Just as multiplication is repeated addition, exponents show repeated multiplication. Whole number *exponents* greater than 1 are used to write repeated multiplications in shorter form. For example,

3^4 means $3 \cdot 3 \cdot 3 \cdot 3$

and because $3 \cdot 3 \cdot 3 \cdot 3 = 81$ we write $3^4 = 81$. The number 81 is sometimes called the "fourth power of three" or "the *value* of 3^4."

EXPONENT
↓
BASE $\rightarrow 3^4 = 81 \leftarrow$ VALUE

Similarly, the value of 8^5 is

$8^5 = 8 \cdot 8 \cdot 8 \cdot 8 \cdot 8 = 32{,}768.$

The base, the repeated factor, is 8. The exponent, which indicates the number of times the base is used as a factor, is 5.

The exponent 1 is a special case.

In general, $x^1 = x$. So $2^1 = 2$, $11^1 = 11$, $3^1 = 3$, and $(123)^1 = 123$.

EXPONENTIAL PROPERTY

One as an exponent

If 1 is used as an exponent, the value is equal to the base.

We can see a reason for the meaning of 6^1 ($6^1 = 6$) by studying the following pattern.

$6^4 = 6 \cdot 6 \cdot 6 \cdot 6$

$6^3 = 6 \cdot 6 \cdot 6$

$6^2 = 6 \cdot 6$

$6^1 = 6$

To find the value of an expression with a whole number exponent

1. If the exponent is 1, the value is the same as the base.
2. If the exponent is greater than 1, use the base number as a factor as many times as shown by the exponent. Multiply.

Warm Ups A–E

A. Find the value of 11^2.

B. Simplify: 104^1

C. Find the value of 10^6.

D. Evaluate: 7^3

E. Find the value of 5^{11}.

Examples A–E

Directions: Find the value.

Strategy: Identify the exponent. If it is one, the value is the base number. If it is greater than one, use it to tell how many times the base is used as a factor and then multiply.

A. Find the value of 6^3.

$6^3 = 6 \cdot 6 \cdot 6$ **Use 6 as a factor three times.**

$= 216$

The value is 216.

B. Simplify: 29^1

$29^1 = 29$ **If the exponent is one, the value is the base number.**

The value is 29.

C. Find the value of 10^7.

$10^7 = (10)(10)(10)(10)(10)(10)(10)$

$= 10{,}000{,}000$ **Ten million. Note that the value has seven zeros.**

The value is 10,000,000.

D. Evaluate: 5^4

$5^4 = 5(5)(5)(5) = 625$

The value is 625.

Calculator Example

E. Find the value of 7^9.

Calculators have an exponent key marked $\boxed{y^x}$ or $\boxed{\wedge}$. Of course, you can always multiply the repeated factors.

The value is 40,353,607.

Answers to Warm Ups A. 121 B. 104 C. 1,000,000 D. 343 E. 48,828,125

HOW AND WHY

Objective 2

Multiply or divide a whole number by a power of 10.

It is particularly easy to multiply or divide a whole number by a power of 10. Consider the following and their products when multiplied by 10.

$5 \times 10 = 50$ $7 \times 10 = 70$ $13 \times 10 = 130$ $24 \times 10 = 240$

The place value of every digit becomes ten times larger when the number is multiplied by 10. So to multiply by 10, we need to merely write a zero on the right of the whole number. If a whole number is multiplied by 10 more than once, a zero is written on the right for each 10. So,

$24 \times 10^4 = 240{,}000$ **Four zeros are written on the right, one for each 10.**

Because division is the inverse of multiplication, dividing by 10 will eliminate the last zero on the right of a whole number. So,

$240{,}000 \div 10 = 24{,}000$ **Eliminate the final zero on the right.**

If we divide by 10 more than once, one zero is eliminated for each 10. So,

$240{,}000 \div 10^3 = 240$ **Eliminate three zeros.**

To multiply a whole number by a power of 10

1. Identify the exponent of 10.
2. Write as many zeros to the right of the whole number as the exponent of 10.

To divide a whole number by a power of 10

1. Identify the exponent of 10.
2. Eliminate the same number of zeros on the right of the whole number as the exponent of 10.

Using powers of 10, we have a third way of writing a whole number in expanded form.

$$
\begin{aligned}
2345 &= 2000 + 300 + 40 + 5, \text{ or} \\
&= 2 \text{ thousands} + 3 \text{ hundreds} + 4 \text{ tens} + 5 \text{ ones, or} \\
&= (2 \times 10^3) + (3 \times 10^2) + (4 \times 10^1) + (5 \times 1)
\end{aligned}
$$

Examples F–J

Warm Ups F–J

Directions: Multiply or divide.

Strategy: Identify the exponent of 10. For multiplication, write the same number of zeros on the right of the whole number as the exponent of 10. For division, eliminate the same number of zeros on the right of the whole number as the exponent of 10.

F. Multiply: 1699×10^8

F. Multiply: $12{,}748 \times 10^5$

$12{,}748 \times 10^5 = 1{,}274{,}800{,}000$ **The exponent of 10 is 5. To multiply, write 5 zeros on the right of the whole number.**

The product is 1,274,800,000.

G. Simplify: 57×10^4

G. Simplify: 346×10^2

$346 \times 10^2 = 34{,}600$

The product is 34,600.

H. Divide: $\dfrac{1{,}860{,}000}{10^4}$

H. Divide: $\dfrac{975{,}000}{10^2}$

$\dfrac{975{,}000}{10^2} = 9750$ **The exponent of 10 is 2. To divide, eliminate 2 zeros on the right of the whole number.**

The quotient is 9750.

I. Simplify: $281{,}000 \div 10^2$

I. Simplify: $496{,}230{,}000 \div 10^4$

$496{,}230{,}000 \div 10^4 = 49{,}623$ **Eliminate 4 zeros on the right.**

The quotient is 49,623.

J. A survey of 100,000 (10^5) people indicated that they pay an average of $4186 in federal taxes. What was the total paid in taxes?

J. A recent fund-raising campaign raised an average of $146 per donor. How much was raised if there were 10,000 (10^4) donors?

Strategy: To find the total raised, multiply the average donation by the number of donors.

$146 \times 10^4 = 1{,}460{,}000$ **The exponent is 4. To multiply, write 4 zeros on the right of the whole number.**

The campaign raised $1,460,000.

Answers to Warm Ups F. 169,900,000,000 G. 570,000 H. 186 I. 2810 J. The total paid in taxes was $418,600,000.

NAME ______________________ CLASS ____________ DATE ____________

Exercises 1.5

OBJECTIVE 1: *Find the value of an expression written in exponential form.*

A.

Write in exponential form.

1. 12(12)(12)(12)(12)(12)

2. $73 \times 73 \times 73 \times 73 \times 73 \times 73 \times 73 \times 73$

Find the value.

3. 9^2

4. 8^2

5. 2^3

6. 3^3

7. 1^{18}

8. 17^1

9. In $7^3 = 343$, 7 is the ________, 3 is the ________, and 343 is the ________.

10. In $5^4 = 625$, 625 is the ________, 5 is the ________, and 4 is the ________.

B.

Find the value.

11. 6^3

12. 2^6

13. 19^2

14. 21^2

15. 10^4

16. 10^6

17. 8^3

18. 7^4

19. 3^8

20. 5^6

OBJECTIVE 2: *Multiply or divide a whole number by a power of 10.*

A.

Multiply or divide.

21. 45×10^2

22. 56×10^1

23. 7×10^4

24. 13×10^3

25. $1200 \div 10^2$

26. $1600 \div 10^2$

27. $340{,}000 \div 10^3$

28. $4500 \div 10^1$

29. To multiply a number by a power of 10, write as many zeros to the right of the number as the ________ of 10.

30. To divide a number by a power of 10, eliminate as many ________ on the right of the number as the exponent of 10.

B.

Multiply or divide.

31. 435×10^4

32. 276×10^3

33. $1{,}200{,}000 \div 10^3$

34. $35{,}000{,}000 \div 10^4$

35. 3591×10^4

36. 6711×10^3

37. $\frac{30{,}200}{100}$

38. $\frac{95{,}500}{10^2}$

39. 705×10^8

40. 300×10^6

41. $970{,}000{,}000 \div 10^5$

42. $3{,}506{,}000{,}000 \div 10^6$

C.

43. Write in exponent form: 10(10)(10)(10)(10)(10)(10)(10)(10)(10)(10)

44. Write in exponent form: 4(4)(4)(4)(4)(4)(4)(4)(4)(4)(4)(4)(4)(4)(4)(4)(4)

Find the value.

45. 14^4

46. 16^4

47. 9^9

48. 8^8

Multiply or divide.

49. 3350×10^9

50. 420×10^{11}

51. $\frac{438{,}000{,}000{,}000}{10^8}$

52. $\frac{1{,}460{,}000{,}000{,}000{,}000}{10^7}$

53. The operating budget of a community college is approximately 73×10^6 dollars. Write this amount in place value form.

54. A congressional committee proposes to decrease the national debt by 125×10^5 dollars. Write this amount in place value form.

55. During one week last year approximately 32×10^6 shares of Microsoft were traded on the New York Stock Exchange. Write this amount in place value form.

56. The distance from the Earth to the nearest star outside our solar system (Alpha Centauri) is approximately 255×10^{11} miles. Write this distance in place value form.

57. In 1994 the World Health Organization estimated the number of HIV infections in sub-Saharan Africa to be 10 million and in North Africa/Middle East to be 100 thousand. Express each of these estimations as a power of 10.

58. The estimated number of HIV infections in Latin America/Caribbean in 1994 was 2 million and in Australia it was 25 thousand. Express these estimations as the product of a number and a power of ten.

59. The estimated number of HIV infections in Western Europe in 1994 was 500 thousand. Express the number of infections as the product of a number and a power of 10 in three different ways.

60. The number of bacteria in a certain culture doubles every hour. If there are two bacteria at the start, zero hour, how many bacteria will be in the culture at the end of 14 hours? Express the answer as a power of 2 and as a whole number.

61. A high roller in Atlantic City places nine consecutive bets at the "Twenty-one" table. The first bet is $5 and each succeeding bet is five times the one before. How much does she wager on the ninth bet? Express the answer as a power of five and as a whole number.

62. The world population in 1900 was 1625 million. Write this as a product of a number and a power of ten.

63. The distance that light travels in a year is called a light-year. This distance is almost 6 trillion miles. Write the place value name for this number. Write the number as 6 times a power of 10.

64. The average distance from the Earth to the Sun is approximately 93 million miles. Write the place value name for this distance and write it as 93 times a power of 10.

Exercises 65–69 refer to the chapter application (see page 1).

65. Express the payroll of the Cardinals as a product of a number and a power of 10.

66. Express the sum of the payrolls of the Indians and the Marlins as a number times a power of 10.

67. What is the smallest power of 25 that is larger than the payroll of the Astros?

68. What power of 13 is approximately the payroll of the Yankees?

69. If you were offered Michael Jordan's 1997 salary of \$30,140,000 or a salary of $\$5^{11}$, which would pay you the most money? By how much?

STATE YOUR UNDERSTANDING

70. Explain what is meant by 4^{10}.

71. Explain how to multiply a whole number by a power of 10. Give at least two examples.

CHALLENGE

72. Mitchell's grandparents deposit \$3 on his first birthday and triple that amount on each succeeding birthday until he is 12. What amount did Mitchell's grandparents deposit on his twelfth birthday? What is the total amount they have deposited in the account?

73. Find the sum of the cubes of the digits.

74. Find the difference between the sum of the cubes of 4, 8, 11, and 23 and the sum of the fourth powers of 2, 5, 6, and 7.

75. The Sun is estimated to weigh 2 octillion tons. Write the place value for this number. Write it as 2 times a power of ten.

GROUP ACTIVITY

76. Research your local newspaper and find at least five numbers that could be written using a power of 10. Write these as a product using the power of 10.

77. Have each member of your group work the following problem independently.

$19 - 2 \cdot 6 - 5 + 7$

Compare your answers. If they are different, be prepared to discuss in class why they are different.

1.6

Order of Operations

OBJECTIVE

Perform any combination of operations on whole numbers.

VOCABULARY

Parentheses ( ) and **brackets** [ ] are used in mathematics as **grouping** symbols. These symbols indicate that the operations inside are to be performed first. Other grouping symbols that are often used are **braces** { } and the **fraction bar** —.

HOW AND WHY

Objective

Perform any combination of operations on whole numbers.

Without a rule, it is possible to interpret $5 + 3 \cdot 12$ in two ways:

$$\begin{aligned} 5 + 3 \cdot 12 &= 5 + 36 \\ &= 41 \end{aligned}$$

or

$$\begin{aligned} 5 + 3 \cdot 12 &= 8 \cdot 12 \\ &= 96 \end{aligned}$$

In order to decide which answer to use, we agree to use a standard set of rules. Among these is the rule that we multiply before adding. So

$$5 + 3 \cdot 12 = 41$$

The order in which the operations are performed is important because the order often determines the answer. Therefore, there is an established *order of operations*. This established order was agreed upon many years ago, and it is built into most of today's calculators and computers.

> **Order of Operations**
>
> ***To evaluate an expression with more than one operation***
>
> **1.** Parentheses—Do the operations within grouping symbols first (parentheses, fraction bar, etc.), in the order given in steps 2, 3, and 4.
> **2.** Exponents—Do the operations indicated by exponents.
> **3.** Multiply and divide—Do only multiplication and division as they appear from left to right.
> **4.** Add and subtract—Do addition and subtraction as they appear from left to right.

So we see that

$8 - 10 \div 2 = 8 - 5$	**Divide first.**
$= 3$	**Then subtract.**
$(6 - 4)(6) = 2(6)$	**Subtract in parentheses first.**
$= 12$	**Then multiply.**
$54 \div 6 \cdot 3 = 9 \cdot 3$	**Neither multiplication nor division takes preference**
$= 27$	**over the other, so do them from left to right.**

As you can see, the rules for the order of operations are fairly complicated and it is important that you learn them all. A standard memory trick is to use the first letters to make an easy-to-remember phrase.

Parentheses
Exponents
Multiplication/**D**ivision
Addition/**S**ubtraction

Consider the phrase: **P**lease **E**xcuse **M**y **D**ear **A**unt **S**ally. Note that the first letters or the words in this phrase are exactly the same (and in the same order) as the first letters for the order of operations.

Many students use "Please excuse my dear Aunt Sally" to help them remember the order of operations. Why not give it a try?

Exercises involving all of the operations are shown in the examples below.

Warm Ups A–G

A. Simplify: $4 \cdot 3 + 6 \cdot 5$

B. Simplify: $4 \cdot 14 - 9 \div 3 + 6 \cdot 2$

C. Simplify: $24 \div 6 + 6 - 3(5 - 3)$

D. Simplify: $5 \cdot 2^3 - 2 \cdot 4^2 + 25 - 7 \cdot 3$

Examples A–G

Directions: Simplify.

Strategy: The operations are done in this order: operations in parentheses first, exponents next, then multiplication and division, and finally, addition and subtraction.

A. Simplify: $3 \cdot 9 + 7 \cdot 2$

$3 \cdot 9 + 7 \cdot 2 = 27 + 14$	**Multiply first.**
$= 41$	**Add.**

B. Simplify: $29 - 6 \div 2 + 7 \cdot 4$

$29 - 6 \div 2 + 7 \cdot 4 = 29 - 3 + 28$	**Divide and multiply.**
$= 26 + 28$	**Subtract.**
$= 54$	**Add.**

C. Simplify: $5 \cdot 9 + 9 - 6(7 + 1)$

$5 \cdot 9 + 9 - 6(7 + 1) = 5 \cdot 9 + 9 - 6(8)$	**Add in parentheses first.**
$= 45 + 9 - 48$	**Multiply.**
$= 54 - 48$	**Add.**
$= 6$	**Subtract.**

D. Simplify: $3 \cdot 4^3 - 8 \cdot 3^2 + 11$

$3 \cdot 4^3 - 8 \cdot 3^2 + 11 = 3 \cdot 64 - 8 \cdot 9 + 11$	**Do exponents first.**
$= 192 - 72 + 11$	**Multiply.**
$= 120 + 11$	**Subtract.**
$= 131$	**Add.**

Answers to Warm Ups A. 42 B. 65 C. 4 D. 12

E. Simplify: $(2^2 + 2 \cdot 3)^2 + 3^2$

Strategy: First do the operations in the parentheses following the proper order.

$(2^2 + 2 \cdot 3)^2 + 3^2 = (4 + 2 \cdot 3)^2 + 3^2$ **Do the exponent in the parentheses first.**

$= (4 + 6)^2 + 3^2$ **Multiply.**

$= (10)^2 + 3^2$ **Add.**

Now that the operations inside the parentheses are complete, continue using the order of operations.

$= 100 + 9$ **Do the exponents.**

$= 109$ **Add.**

E. Simplify: $(3^3 - 12 \div 4)^2 + 5^2$

Calculator Example

F. Simplify: $1845 + 165 \cdot 18 - 3798$

Enter the numbers and operations as they appear from left to right. The calculator has the order of operations built in.

The answer is 1017.

F. Simplify: $1366 + 19 \cdot 372 \div 12$

G. The Lend A Helping Hand Association prepares two types of food baskets for distribution to the needy. The family pack contains nine cans of vegetables and the elderly pack contains four cans of vegetables. How many cans of vegetables are needed for 125 family packs and 50 elderly packs?

Strategy: To find the number of cans of vegetables needed for the packs, multiply the number of packs by the number of cans per pack. Then add the two amounts.

$125(9) + 50(4) = 1125 + 200$ **Multiply.**

$= 1325$ **Add.**

The Lend A Helping Hand Association needs 1325 cans of vegetables.

G. The Fruit-of-the-Month Club prepares two types of boxes for shipment. Box A contains six apples and Box B contains ten apples. How many apples are needed for 96 orders of Box A and 82 orders of Box B?

Answers to Warm Ups E. 601 F. 1955 G. The orders require 1396 apples.

NAME ____________ CLASS ____________ DATE ____________

Exercises 1.6

OBJECTIVE: *Perform any combination of operations on whole numbers.*

A.

Simplify.

1. $5 \cdot 8 + 13$

2. $17 + 5 \cdot 6$

3. $15 - 5 \cdot 3$

4. $28 \cdot 4 - 7$

5. $24 + 6 \div 2$

6. $36 \div 6 - 3$

7. $30 - (13 + 2)$

8. $(19 - 3) - 8$

9. $30 \div 6 \times 5$

10. $45 \div 5 \times 3$

11. $21 + 5 \cdot 4 - 2$

12. $25 - 3 \cdot 7 + 4$

13. $2^2 - 3 + 3^2$

14. $4^2 \div 8 + 2^3$

15. $4 \cdot 7 + 3 \cdot 5$

16. $6 \cdot 7 - 5 \cdot 4$

B.

Simplify.

17. $3^2 - 4 \cdot 2 + 5 \cdot 6$

18. $5^2 + 12 \div 3 + 3 \cdot 3$

19. $36 \div 9 + 8 - 5$

20. $56 \cdot 3 \div 14 + 4 - 6$

21. $(14 + 28) - (34 - 27)$

22. $(56 - 8) - (17 + 7)$

23. $49 \div 7 \cdot 3^3 + 7 \cdot 4$

24. $75 \div 15 \cdot 2^4 + 3 \cdot 8$

25. $96 \div 12 \cdot 3$

26. $100 \div 4 \cdot 5$

27. $72 - 4(19 - 10) + 11 - 19$

28. $45(18 - 12) \div 3 - 8 + 12$

29. $3^4 + 4^3$

30. $3^5 - 2^4 + 7^2$

31. $7 \cdot 4^2 - 56 \div 4 + 4$

32. $5 \cdot 3^2 - 35 \div 5 + 2$

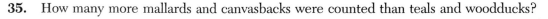

C.

Simplify.

33. $50 - 12 \div 6 - 36 \div 6 + 3$

34. $80 - 24 \div 4 + 30 \div 6 + 4$

Use the following graph to answer the questions in Exercises 35–38. A chapter of Ducks Unlimited counts the following species of ducks at a lake in northern Idaho:

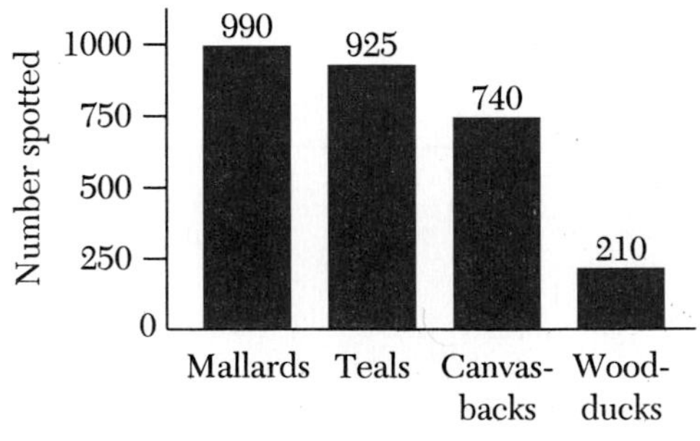

35. How many more mallards and canvasbacks were counted than teals and woodducks?

36. If twice as many canvasbacks had been counted, how many more canvasbacks would there have been than teals?

37. If four times the number of woodducks had been counted, how many more woodducks and mallards would there have been than teals and canvasbacks?

38. If twice the number of teals had been counted, how many more teals and woodducks would there have been compared with mallards and canvasbacks?

Simplify.

39. $7(3^2 \cdot 2 - 8) \div 5 + 4$

40. $11(3^3 \cdot 4 - 98) \div 5 - 21$

41. $4(8 - 3)^3 - 4^3$

42. $3(7 - 3)^3 - 8(3 - 1)^2$

43. Last week, the Sing-Along Music Company advertised guitars for \$635 each and pianos for \$5125 each. They sold 15 guitars and 6 pianos. What were the total sales from the two items?

44. During a year-end sale the Neat-n-Trim clothing store hired two extra clerks. One was paid \$8 per hour and the other \$10 per hour. What was the additional payroll if the first clerk worked 35 hours and the second clerk worked 62 hours?

45. Pete agrees to reward his son for grades earned. He pays \$18 for an A, \$12 for a B, and \$5 for a C or a D. How much does his son earn for two A's, two B's, four C's and one D?

46. Nanette agrees to reward her daughter for grades earned. She decides to give her \$25 for the number of A's she earns in excess of the number of B's, and \$18 for the number of B's she earns in excess of the number of C's. How much does her daughter earn if she has four A's, three B's, and one C?

47. To begin a month, Mario's Hair Barn has an inventory of 76 bottles of shampoo at $5 each, 65 bottles of conditioner at $4 each, and 35 perm kits at $22 each. At the end of the month they have 18 bottles of shampoo, 9 bottles of conditioner, and 7 perm kits left. What is the cost of the supplies used for the month?

48. Juanita's Camera Shop ordered 40 cameras that cost $346 each and 32 lenses at $212 each. She sells the cameras for $575 each and the lenses for $316 each. If they sell 27 cameras and 19 lenses, how much net income is realized on the sale of the cameras and lenses?

49. A long-haul trucker is paid $25 for every 100 miles driven and $15 per stop. If he averages 2500 miles and 40 stops per week, what is his average weekly income? Assuming he takes 2 weeks off each year for vacation, what is his average yearly income?

Exercises 50–51. In an international golf event, players earn points based on how they play each hole. The points are earned as follows: bogie—1, par—2, birdie—3, eagle—4, double eagle—5. All other scores earn no points.

50. What is the score of a golfer who has 1 double bogie, 3 bogies, 7 pars, and 7 birdies for 18 holes?

51. Two golfers had the following scores on the first round: Golfer 1: 2 double bogies, 1 bogie, 8 pars, 5 birdies, 1 eagle, and 1 double eagle. Golfer 2: 4 bogies, 8 pars, 5 birdies, and 1 eagle. Which golfer scored the most points.

Exercises 52–53. The following labels show the nutrition facts for cereal with milk, orange juice, jam, and bread.

Grape Nuts

Nutrition Facts
Serving Size 1/2 cup (58g)
Servings Per Container about 12

Amount Per Serving	Cereal	Cereal with 1/2 cup Skim Milk
Calories	200	240
Calories from Fat	10	10
	% Daily Values**	
Total Fat 1g*	2%	2%
Saturated Fat 0g	0%	0%
Cholesterol 0mg	0%	0%
Sodium 350mg	15%	17%
Potassium 160mg	5%	10%
Total Carbohydrate 47g	16%	18%
Dietary Fiber 5g	21%	21%
Sugars 7g		
Other Carbohydrates 35g		
Protein 6g		

Orange Juice

Nutrition Facts
Serving Size 8 fl oz (240 mL)
Servings Per Container 8

Amount Per Serving	
Calories 110	Calories from Fat 0
	% Daily Value*
Total Fat 0g	0%
Sodium 0mg	0%
Potassium 450mg	13%
Total Carbohydrate 26g	9%
Sugars 22g	
Protein 2g	
Vitamin C 120% • Calcium 2%	
Thiamin 10% • Niacin 4%	
Vitamin B6 6% • Folate 15%	

Not a significant source of saturated fat, cholestrol, dietary fiber, vitamin A and iron.

* Percent Daily Values are based on a 2,000 calorie diet.

Jam

Nutrition Facts
Serving Size 1 Tbsp. (20g)
Servings Per Container about 25

Amount Per Serving	
Calories 50	
	% Daily Value*
Total Fat 0g	0%
Sodium 15mg	1%
Total Carbohydrate 13g	4%
Sugars 13g	
Protein 0g	

* Percent Daily Values are based on a 2,000 calorie diet.

Bread

Nutrition Facts
Serving Size 1 slice (49g)
Servings Per Container 14

Amount Per Serving	
Calories 120	**Calories from Fat 10**
	% Daily Value*
Total Fat 1g	2%
Saturated Fat 0g	0%
Cholesterol 0mg	0%
Sodium 360mg	15%
Total Carbohydrate 24g	8%
Dietary Fiber 1g	5%
Sugars 1g	
Protein 4g	
Vitamin A 0% • Vitamin C 0%	
Thiamin 10% • Riboflavin 8%	
Calcium 2% • Iron 10%	
Niacin 8% • Folic Acid 10%	

52. How many milligrams (mg) of sodium are consumed if Marla has 3 servings of orange juice, 2 servings of cereal, and 2 slices of bread with jam for breakfast? Milk contains 62 mg of sodium per one half-cup serving.

53. How many calories does Marla consume when she eats the breakfast listed in Exercise 52?

STATE YOUR UNDERSTANDING

54. Which of the following is correct? Explain.

$$3 + 5 \cdot 4 = 8 \cdot 4 = 32 \qquad \text{or} \qquad 3 + 5 \cdot 4 = 3 + 20 = 23$$

55. Explain how to simplify $2(1 + 36 \div 3^2) - 3$ using order of operations.

CHALLENGE

56. Simplify: $(6 \cdot 3 - 8)^2 - 50 + 2 \cdot 3^2 + 2(9 - 5)^3$

57. USA Video buys three first-run movies: #1 for \$185, #2 for \$143, and #3 for \$198. During the first 2 months #1 is rented 10 times at the weekend rate of \$5 and 26 times at the weekday rate of \$3; #2 is rented 12 times at the weekend rate and 30 times at the weekday rate; and #3 is rented 8 times at the weekend rate and 18 times at the weekday rate. How much money must still be raised to pay for the cost of the three videos?

GROUP ACTIVITY

58. It is estimated that hot water heaters need to be big enough to accommodate the water usage for an entire hour. To figure what size heater you need, first identify the single hour of the day in which water usage is highest. Next, identify the types of water usage during this hour. The following chart lists estimates of water usage for various activities.

Activity	Gallons Used
Shower	20
Bath	20
Washing hair	4

Activity	Gallons Used
Shaving	2
Washing hands/face	4
Dishwater	14

Now calculate the total number of gallons of water used in your designated hour. Your water heater must have this capacity.

Getting Ready for Algebra

OBJECTIVE

Solve equations of the form $ax + b = c$, $ax - b = c$, $\frac{x}{a} + b = c$, and $\frac{x}{a} - b = c$, in which x, a, b, and c are whole numbers.

HOW AND WHY

Recall that we have solved equations involving only one operation. Let's look at some equations that involve two operations.

To solve $x - 4 = 2$, we added 4 to both sides of the equation. To solve $3x = 12$, we divided both sides of the equation by 3. The following equation requires both steps.

Solve.

$3x - 4 = 2$

$3x - 4 + 4 = 2 + 4$ **First, eliminate the subtraction by adding 4 to both sides.**

$3x = 6$ **Simplify both sides.**

$\frac{3x}{3} = \frac{6}{3}$ **Eliminate the multiplication. Divide both sides by 3.**

$x = 2$ **Simplify.**

CHECK:

$3x - 4 = 2$

$3 \cdot 2 - 4 = 2$ **Replace x by 2.**

$6 - 4 = 2$ **Multiply.**

$2 = 2$ **Subtract.**

Thus, if x is replaced by 2 in the original equation, the statement is true. So the solution is 2. Now solve

$2x + 3 = 9$

$2x + 3 - 3 = 9 - 3$ **First, eliminate the addition by subtracting 3 from both sides.**

$2x = 6$ **Simplify.**

$\frac{2x}{2} = \frac{6}{2}$ **Eliminate the multiplication by dividing both sides by 2.**

$x = 3$ **Simplify.**

CHECK:

$2x + 3 = 9$

$2 \cdot 3 + 3 = 9$ **Replace x by 3.**

$6 + 3 = 9$ **Multiply.**

$9 = 9$ **Add.**

Thus, if x is replaced by 3 in the original equation, the statement is true. So the solution is 3.

Note that in each of the previous examples the operations are eliminated in the opposite order in which they are performed. That is, the addition and subtraction were eliminated first and then the multiplication and division.

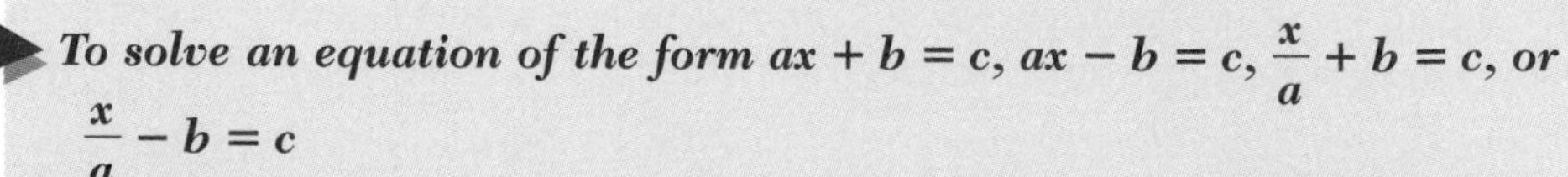

To solve an equation of the form $ax + b = c$, $ax - b = c$, $\frac{x}{a} + b = c$, or $\frac{x}{a} - b = c$

1. Eliminate the addition or subtraction by subtracting or adding the same number to both sides.
2. Eliminate the multiplication or division by dividing or multiplying on both sides.
3. Check the solution by substituting it in the original equation.

Warm Ups A–E

A. $5x - 8 = 17$

B. $\frac{a}{3} + 8 = 12$

Examples A–E

Directions: Solve and check.

Strategy: Isolate the variable by first adding or subtracting the same number from both sides. Second, multiply or divide both sides by the same number.

A. $2x - 7 = 9$

$2x - 7 = 9$

$2x - 7 + 7 = 9 + 7$ **Add 7 to both sides to eliminate the subtraction.**

$2x = 16$ **Simplify.**

$\frac{2x}{2} = \frac{16}{2}$ **Divide both sides by 2 to eliminate the multiplication.**

$x = 8$ **Simplify.**

CHECK:

$2x - 7 = 9$

$2(8) - 7 = 9$ **Substitute 8 for x in the original equation.**

$16 - 7 = 9$ **Simplify.**

$9 = 9$ **The statement is true.**

The solution is $x = 8$.

B. $\frac{y}{5} + 4 = 5$

$\frac{y}{5} + 4 = 5$

$\frac{y}{5} + 4 - 4 = 5 - 4$ **Subtract 4 from both sides to eliminate the addition.**

$\frac{y}{5} = 1$ **Simplify.**

$5\left(\frac{y}{5}\right) = 5(1)$ **Multiply both sides by 5 to eliminate the division.**

$y = 5$ **Simplify.**

Answers to Warm Ups A. $x = 5$ B. $a = 12$

CHECK:

$\frac{y}{5} + 4 = 5$

$\frac{5}{5} + 4 = 5$ **Substitute 5 for y in the original equation.**

$1 + 4 = 5$ **Simplify.**

$5 = 5$ **The statement is true.**

The solution is $y = 5$.

C. $\frac{z}{2} - 6 = 4$

$\frac{z}{2} - 6 = 4$

$\frac{z}{2} - 6 + 6 = 4 + 6$ **Add 6 to both sides to eliminate the subtraction.**

$\frac{z}{2} = 10$ **Simplify.**

$2\left(\frac{z}{2}\right) = 2(10)$ **Multiply both sides by 2 to eliminate the multiplication.**

$z = 20$ **Simplify.**

CHECK:

$\frac{z}{2} - 6 = 4$

$\frac{20}{2} - 6 = 4$ **Substitute 20 for z in the original equation.**

$10 - 6 = 4$ **Simplify.**

$4 = 4$ **The statement is true.**

The solution is $z = 20$.

C. $\frac{x}{5} - 8 = 7$

D. $3b + 4 = 22$

$3b + 4 = 22$

$3b + 4 - 4 = 22 - 4$ **Subtract 4 from both sides to eliminate the addition.**

$3b = 18$ **Simplify.**

$\frac{3b}{3} = \frac{18}{3}$ **Divide both sides by 3 to eliminate the multiplication.**

$b = 6$ **Simplify.**

CHECK:

$3b + 4 = 22$

$3(6) + 4 = 22$ **Substitute 6 for b in the original equation.**

$18 + 4 = 22$ **Simplify.**

$22 = 22$ **The statement is true.**

The solution is $b = 6$.

D. $4c + 9 = 25$

Answers to Warm Ups C. $x = 75$ D. $c = 4$

E. Find the monthly payment on an original loan of \$975 if the balance after 11 payments is \$722.

E. The formula for the balance of a loan (D) is $D + NP = B$, where P represents the monthly payment, N represents the number of payments, and B represents the amount of money borrowed. Find the number of payments that have been made on an original loan of \$630 with a current balance of \$375 if the payment is \$15 per month.

Strategy: Substitute the given values in the formula and solve.

$D + NP = B$	**Formula.**
$375 + N(15) = 630$	**Substitute 375 for D, 630 for B, and 15 for P.**
$15N = 630 - 375$	**Subtract 375 from each side.**
$15N = 255$	
$N = \frac{255}{15}$	**Divide each side by 15.**
$N = 17$	

CHECK: If 17 payments have been made, is the balance \$375?

\$630	
− 255	**17 payments of \$15 is \$255.**
\$375	**True.**

Seventeen payments have been made.

Answer to Warm Up E. The monthly payment is \$23.

NAME ________________ CLASS ________ DATE ________

Exercises

OBJECTIVE: *Solve equations of the form $ax + b = c$, or $ax - b = c$, $\frac{x}{a} + b = c$, and $\frac{x}{a} - b = c$, in which x, a, b, and c are whole numbers*

Solve and check.

1. $4x - 12 = 8$

2. $\frac{a}{3} + 7 = 12$

3. $\frac{y}{2} - 9 = 3$

4. $31 = 5x + 6$

5. $25 = 4x + 9$

6. $\frac{a}{6} + 7 = 12$

7. $\frac{c}{8} + 14 = 25$

8. $12x - 8 = 28$

9. $9x + 24 = 78$

10. $5y + 36 = 151$

11. $12c - 56 = 88$

12. $2 = \frac{w}{15} - 45$

13. $54 = \frac{a}{32} + 29$

14. $\frac{x}{41} + 79 = 187$

15. $429 = 23b - 77$

16. $556 = 36c + 124$

17. Fast-Tix charges \$12 per ticket for a rock concert plus a \$5 service charge. How many tickets did Remy buy if he was charged \$89? Use the formula $C = PN + S$, in which C is the total cost, P is the price per ticket, N is the number of tickets purchased, and S is the service charge.

18. Ticket-Master charges José \$157 for nine tickets to the Festival of Jazz. If the service charge is \$4, what is the price per ticket? Use the formula in Exercise 17.

19. Rana is paid \$40 per day plus \$8 per artificial flower arrangement she designs and completes. How many arrangements did she complete if she earned \$88 for the day? Use the formula $S = B + PN$, in which S is the total salary earned, B is the base pay for the day, P is the pay per unit, and N is the number of units completed.

20. Rana's sister works at a drapery firm where the pay is \$50 per day plus \$12 per unit completed. How many units did she complete if she earned \$122 for the day?

For Exercises 21–24, the following table summarizes several different long distance calling plans.

Company	Monthly Fee	Charge per Minute
AT&T	None	15¢
Tone	\$4.90	10¢
Pace	\$6.96	9¢

21. Jessica has \$30 budgeted for long distance calls each month. Write an equation for the number of minutes she gets from AT&T. Let m be the number of minutes of long distance calls per month. Let C represent the monthly bill in cents. Find the number of minutes that Jessica can purchase from AT&T each month.

22. Jessica has \$30 budgeted for long distance calls each month. Write an equation for the number of minutes she gets from Tone. Let m be the number of minutes of long distance calls per month. Let C represent the monthly bill in cents. Find the number of minutes that Jessica can purchase from Tone each month.

23. Jessica has \$30 budgeted for long distance calls each month. Write an equation for the number of minutes she gets from Pace. Let m be the number of minutes of long distance calls per month. Let C represent the monthly bill in cents. Find the number of minutes that Jessica can purchase from Pace each month.

24. Using the results of Exercises 21–23, which company will give Jessica the most minutes for her \$30?

1.7 Average

OBJECTIVE

Find the average of a set of whole numbers.

VOCABULARY

The **average** or **mean** of a set of numbers is the sum of the set of numbers divided by the total number of numbers in the set.

HOW AND WHY

Objective

Find the average of a set of whole numbers.

The *average,* or *mean,* of a set of numbers is used in statistics. It is one of the ways to find the middle of a set of numbers (like the average of a set of test grades). Mathematicians call the average or mean a "measure of central tendency." The average of a set of numbers is found by adding the numbers in the set and dividing the sum by the number of numbers in the set. For example, to find the average of 11, 17, and 23

$11 + 17 + 23 = 51$ **Find the sum of the numbers in the set.**

$51 \div 3 = 17$ **Divide the sum by the number of numbers.**

The average is 17.

The "central" number or average does not need to be one of the members of the set. For instance, find the average of 27, 36, 49, and 60.

$27 + 36 + 49 + 60 = 172$ **Find the sum of the numbers.**

$172 \div 4 = 43$ **Divide by the number of numbers.**

The average is 43, which is not a member of the set.

To find the average of a set of whole numbers

1. Add the numbers.
2. Divide the sum by the number of numbers in the set.

Examples A–F

Directions: Find the average.

Strategy: Add the numbers in the set. Divide the sum by the number of numbers in the set.

A. Find the average of 103, 98, and 123.

$103 + 98 + 123 = 324$ **Add the numbers in the set.**

$324 \div 3 = 108$ **Divide the sum by the number of numbers.**

The average is 108.

Warm Ups A–F

A. Find the average of 313, 129, and 500.

Answer to Warm Up A. 314

B. Find the average of 9, 27, 46, 58, and 65.

B. Find the average of 7, 40, 122, and 211.

$7 + 40 + 122 + 211 = 380$	**Add the numbers in the set.**
$380 \div 4 = 95$	**Divide the sum by the number of numbers.**

The average is 95.

Calculator Example

C. Find the average of 917, 855, 1014, and 622.

C. Find the average of 345, 567, 824, and 960.
Enter the sum divided by 4.

$(345 + 567 + 824 + 960) \div 4$

The average is 674.

D. The average of 13, 15, 6, 7, and ? is 9. Find the missing number.

D. The average of 12, 8, 20, and ? is 12. Find the missing number.

Strategy: Since the average of the four numbers is 12 we know that the sum of the four numbers is 4(12) or 48. To find the missing number, subtract the sum of the three given numbers from 4 times the average.

$$\begin{aligned} 4(12) - (12 + 8 + 20) &= 4(12) - (40) \\ &= 48 - 40 \\ &= 8 \end{aligned}$$

So the missing number is 8.

E. The local dog food company ships the following cases of dog food: Monday, 3059; Tuesday, 2175; Wednesday, 3755; Thursday, 1851; and Friday, 2875. What is the average number of cases shipped each day?

E. In order to help Pete lose weight, the dietician has him record his caloric intake for a week. He records the following: Monday, 3165; Tuesday, 1795; Wednesday, 1500; Thursday, 2615; Friday, 1407; Saturday, 1850; and Sunday, 1913. What is Pete's average caloric intake per day?

Strategy: Add the calories for each day and then divide by 7, the number of days.

$$\begin{array}{r} 3165 \\ 1795 \\ 1500 \\ 2615 \\ 1407 \\ 1850 \\ +\ \underline{1913} \\ 14245 \end{array} \qquad \begin{array}{r} 2035 \\ 7\overline{)14245} \\ \underline{14} \\ 2 \\ \underline{0} \\ 24 \\ \underline{21} \\ 35 \\ \underline{35} \\ 0 \end{array}$$

Pete's average caloric intake per day is 2035.

Answers to Warm Ups B. 41 C. 852 D. 4 E. The company shipped an average of 2743 cases of dog food each day.

F. During the annual Fishing Derby, 35 fish are entered. The weights of the fish are recorded as shown:

Number of Fish	Weight per Fish
2	6 lb
4	7 lb
8	10 lb
10	12 lb
5	15 lb
4	21 lb
1	22 lb
1	34 lb

What is the average weight of a fish entered in the Derby?

Strategy: First find the total weight of all 35 fish.

$$\begin{aligned} 2(6) &= 12 \\ 4(7) &= 28 \\ 8(10) &= 80 \\ 10(12) &= 120 \\ 5(15) &= 75 \\ 4(21) &= 84 \\ 1(22) &= 22 \\ 1(34) &= +\ \underline{34} \\ & \quad 455 \end{aligned}$$

Multiply 2 times 6 because there are 2 fish that weigh 6 pounds, for a total of 12 pounds, and so forth.

Now divide the total weight by the number of fish, 35.

$455 \div 35 = 13$

The average weight per fish is 13 pounds.

F. In a class of 30 seniors the following weights are recorded on Health Day:

Number of Students	Weight per Student
1	120 lb
3	128 lb
7	153 lb
4	175 lb
5	182 lb
5	195 lb
3	200 lb
2	215 lb

What is the average weight of a student in the class?

Answer to Warm Up F. The average weight of a student is 173 pounds.

NAME ____________________ CLASS ____________ DATE ____________

Exercises 1.7

OBJECTIVE: *Find the average of a set of whole numbers.*

A.

Find the average.

1. 3, 7

2. 7, 11

3. 8, 10

4. 9, 13

5. 8, 14, 17

6. 5, 9, 13

7. 3, 5, 7, 9

8. 5, 5, 9, 9

9. 4, 6, 3, 3

10. 8, 3, 4, 5

11. 3, 5, 7, 8, 2

12. 4, 6, 3, 8, 9

13. 10, 20, 30, 40

14. 20, 20, 25, 25, 10

Find the missing number to make the average correct.

15. The average of 4, 7, 9, and ? is 8.

16. The average of 6, 13, 11, and ? is 9.

B.

Find the average.

17. 14, 18, 30, 42

18. 11, 34, 41, 62

19. 25, 35, 45, 55

20. 18, 36, 41, 25

21. 7, 14, 16, 23, 30

22. 15, 8, 27, 51, 39

23. 8, 11, 19, 28, 44, 76

24. 88, 139, 216, 133

25. 101, 105, 108, 126

26. 281, 781, 513, 413

27. 89, 140, 217, 134

28. 33, 132, 240, 279

29. 45, 67, 42, 145, 215, 92

30. 124, 55, 78, 54, 234, 175

Find the missing number.

31. The average of 34, 81, 52, 74, and ? is 59.

32. The average of 21, 29, 46, 95, 33, and ? is 47.

C.

33. Find the average: 183, 526, 682, 589, 720

34. Find the average: 364, 384, 196, 736, 685

35. A chapter of Ducks Unlimited counts the following species of ducks at a lake in northern Idaho: mallards, 990; teal, 924; canvasbacks, 740; woodducks, 210; and widgeons, 511. Find the average number of ducks per species.

36. The following number of fish were counted at the Bonneville fish ladder during one week in July: coho, 147; shad, 356; silver salmon, 214; and sturgeon, 95. Find the average number of fish per species.

Find the average.

37. 1156, 2347, 5587, 354, 2355, 825

38. 3232, 4343, 5454, 6565, 7676, 8790

39. 23,458, 45,891, 34,652, 17,305, 15,984

40. 112,315, 236,700, 156,865, 103,674, 300,071

41. A golfer shoots the following scores for 11 rounds of golf: 84, 90, 103, 78, 91, 87, 75, 80, 78, 81, 77. What is the average score per round?

42. A bowler has the following scores for nine games: 255, 198, 210, 300, 193, 211, 271, 200, 205. What is the average score per game?

43. Mr. Adams counts his caloric intake for one week prior to starting a diet. He reports the following intake: Monday, 4910; Tuesday, 3780; Wednesday, 3575; Thursday, 4200; Friday, 3400; Saturday, 4350; and Sunday, 3960. What is his average caloric intake?

44. A service station sells the following numbers of gallons of gasoline during a given week: Monday, 2850; Tuesday, 2185; Wednesday, 3760; Thursday, 3264; Friday, 4650; Saturday, 6480; and Sunday, 1885. What is the average number of gallons of gasoline sold per day?

Exercises 45–46 refer to the chapter application problem (see page 1).

45. To the nearest hundred thousand dollars, calculate the average payroll for the four division winners in 1997.

46. To the nearest hundred thousand dollars, calculate the average payroll for the teams that made the playoffs.

47. In 1994 the World Health Organization estimated the number of HIV cases in sub-Saharan Africa at 10 million and the number of cases in North Africa at 100 thousand. Calculate the average estimated number of HIV infections for Africa. Do you think that using the figure you calculated gives an accurate picture of the HIV infections in Africa? Explain.

48. A consumer magazine tests 18 makes of cars for gas mileage. The results are shown in the table:

Number of Makes	Gas Mileage Based on 200 Miles
2	14 mpg
3	20 mpg
2	25 mpg
4	31 mpg
3	36 mpg
2	40 mpg
2	45 mpg

What is the average gas mileage of the cars?

49. A home economist lists the costs of 15 brands of canned fruit drinks. The results are shown in the table:

Number of Brands	Price per Can
3	84¢
1	82¢
4	78¢
3	74¢
2	71¢
2	65¢

What is the average price per can?

50. Twenty-five football players are weighed in on the first day of practice. The weights are recorded in the table:

Number of Players	Weight
1	167 lb
4	172 lb
3	187 lb
5	195 lb
4	206 lb
3	215 lb
3	225 lb
2	245 lb

What is the average weight of the players?

51. At the Rock Creek Country Club ladies championship tournament, the following scores are recorded for the 82 participants:

Number of Golfers	Score
1	66
3	68
7	69
8	70
12	71
16	72
15	74
10	76
6	78
3	82
1	85

What is the average score for the tournament?

52. A west coast city is expanding its mass transit system. It is building a 15-mile east-west light rail line for $780 million and an 11-mile north-south light rail line for $850 million. What is the average cost per mile, to the nearest million dollars, of the new lines?

53. A school district in Iowa has a budget of $7,880,000 and serves 1600 students. A second school district has a budget of $13,850,000 and serves 2500 students. What is the average cost per student in each district? What is the average cost per student in the combined districts?

54. The population of Malaysia in 1997 was 19,700,000. What place value was this statistic rounded to? The national goal is to increase the population to 70,000,000 by the year 2020. What place value is this statistic rounded to? How many additional people are needed in order to meet this goal? True or False: the current population will have to more than triple by 2020 to meet this goal. Explain. Rounded to the nearest hundred thousand, what is the average yearly increase in population necessary to meet the goal?

Exercise 55 relates to the chapter application (see table on page 15), Exercises 74–76.

55. What was the average salary for the Bulls players? Round to the nearest ten thousand dollars. Does this average fairly represent what the players earn?

STATE YOUR UNDERSTANDING

56. Explain what is meant by the average of two or more numbers.

57. Explain how to find the average (mean) of 2, 4, 5, 5, and 9. What does the average of a set of numbers tell you about the set?

CHALLENGE

58. A patron of the arts estimates that the average donation to the fund drive will be $72. She will donate $150 for each dollar by which she misses the average. The 150 donors made the following contributions:

Number of Donors	Donation
5	$153
13	$125
24	$110
30	$100
30	$75
24	$50
14	$25
10	$17

How much does the patron donate to the fund drive?

GROUP ACTIVITY

59. Divide 35, 68, 120, 44, 56, 75, 82, 170, and 92 by 2 and 5. Which ones are divisible by 2 (the division has no remainder)? Which ones are divisible by 5? See if your group can find simple rules for looking at a number and telling whether or not it is divisible by 2 and/or 5.

60. Using the new car ads in the newspaper, find four advertised prices for the same model of a car. What is the average price, to the nearest 10 dollars?

1.8

Reading and Interpreting Tables

OBJECTIVE

Read and interpret information given in a table.

VOCABULARY

A **table** is a method of displaying data in an array using a horizontal and vertical arrangement to distinguish the type of data. A **row** of a table is a horizontal line of a table and reads left to right across the page. A **column** of a table is a vertical line of a table and reads up or down the page. For example, in the table

		Column 2		
	134	**56**	89	102
	14	**116**	7	98
Row 3	**65**	45	**12**	**67**
	23	**32**	7	213

the number 45 is in row 3 and column 2.

HOW AND WHY

Objective

Read and interpret information given in a table.

Data are often displayed in the form of a *table*. We see tables in the print media, in advertisements, and in business presentations. Reading a table involves finding the correct *column* and *row* that describes the needed information and then reading the data at the intersection of that column and that row; for example,

TABLE 1.3 STUDENT COURSE ENROLLMENT

Class	Mathematics	English	Science	Humanities
Freshman	950	1500	500	1200
Sophomore	600	700	650	1000
Junior	450	200	950	1550
Senior	400	250	700	950

To find the number of sophomores who take English, find the column headed English and the row headed Sophomore and read the number at the intersection. The number of sophomores taking English is 700.

We can use the table to find the difference in enrollments by class. To find how many more freshmen take mathematics than juniors, we subtract the entries in the corresponding columns and rows. There are 950 freshmen taking math, as compared to 450 juniors. Because $950 - 450 = 500$, 500 more freshmen take mathematics than juniors.

The total enrollment in science for all four classes can be found by adding all entries in the column marked Science. Because 500 + 650 + 950 + 700 = 2800, there are 2800 enrollments in science classes.

The table can also be used for predicting by scaling the values in the table upward or downward. If the number of seniors doubles next year, we can assume that the number of seniors taking humanities will also double. So the number of seniors taking humanities next year will be 2(950), or 1900.

Other ways to interpret data from a table are shown in the examples.

Examples A–C

Direction: Answer the questions associated with the table.

Strategy: Examine the rows and columns of the table to determine the values that are related.

A. The table below shows the decline in the number of railroad workers in four Western States.

Railroad Workers

State	1980	1988
Oregon	2991	1338
Idaho	3368	1748
Wyoming	3416	1486
Utah	3046	1717

1. Which state had the most railroad workers in 1980?
2. Which state had the fewest railroad workers in 1988?
3. Which state suffered the greatest loss in the number of railroad workers?
4. How many more railroad jobs did Utah have than Oregon in 1988?

1. Wyoming **Read down the column headed "1980" to locate the largest number of workers, 3416. Now read across the row to find the state, Wyoming.**
2. Oregon **Read down the column headed "1988" to find the least number of workers, 1338. Then read across the row and find Oregon.**
3. Oregon: 2991 − 1338 = 1653 **Find the difference in the number of workers for each state.**
   Idaho: 3368 − 1748 = 1620
   Wyoming: 3416 − 1486 = 1930
   Utah: 3046 − 1717 = 1329

   Wyoming suffered the greatest loss, 1930 workers.
4. 1717 − 1338 = 379 **Find the difference between Utah's number of jobs in 1988 and Oregon's in 1988.**

   Utah has 379 more railroad jobs.

Warm Ups A–C

A. Use the table in Example A to answer the questions.

1. What was the total number of railroad workers in 1980 in the four states?
2. How many more railroad workers did Idaho have than Oregon in 1988?
3. What was the total number of railroad workers in Wyoming and Utah in 1988?
4. What was the combined loss in the number of railroad workers in Idaho and Utah?

Answer to Warm Up A. 1. The total number of railroad workers in 1980 was 12,821. 2. In 1988, Idaho had 410 more railroad workers than Oregon. 3. The total number of railroad workers in Wyoming and Utah in 1988 was 3203. 4. The loss of railroad workers in Idaho and Utah was 2949.

B. The table below shows the value of homes sold in the Portland metropolitan area for a given month in 1993.

Values of Houses Sold

Location	Lowest	Highest	Average
N. Portland	$16,000	$ 58,500	$ 34,833
N.E. Portland	$18,000	$120,000	$ 47,091
S.E. Portland	$18,000	$114,000	$ 51,490
Lake Oswego	$40,000	$339,000	$121,080
West Portland	$29,500	$399,000	$112,994
Beaverton	$20,940	$165,000	$ 78,737

1. In which location was the highest-priced home sold?

2. What was the price difference between the average cost of a house and the lowest cost of a house in Lake Oswego?

3. What is the average lowest cost for houses in the region?

4. If 31 houses were sold in Beaverton during the month, what were the total sales in the area?

1. West Portland **Read down the "highest' column and find the largest price, $399,000.**

2. **Subtract the lowest cost from the average cost for Lake Oswego.**

$$\begin{array}{r} \$121080 \\ -\underline{\quad 40000} \\ \$81080 \end{array}$$

The difference is $81,080.

3. **Find the sum of the lowest prices and divide by 6, the number of areas.**

$$\begin{array}{r} \$16000 \\ \$18000 \\ \$18000 \\ \$40000 \\ \$29500 \\ +\underline{\$20940} \\ \$142440 \end{array}$$

$\$142,440 \div 6 = \$23,740$

The average of the low-priced houses is $23,740.

4. **Multiply the average sale price times the number of houses sold.**

$$\begin{array}{r} \$78737 \\ \times\underline{\quad 31} \\ 78737 \\ \underline{236211} \\ \$2440847 \end{array}$$

The total sales in Beaverton for the month is $2,440,847.

B. Use the table in Example B to answer.

1. Which location has the highest average sale price?

2. What is the difference between the highest- and lowest-priced house in Beaverton?

3. What is the difference in the lowest priced house in Lake Oswego and N. Portland?

4. If 67 houses were sold in Lake Oswego, what were the total sales for the month in the area?

Answer to Warm Up B. 1. Lake Oswego has the highest average sale price. 2. The difference between the highest- and lowest-priced house in Beaverton is $144,060. 3. The difference in lowest priced houses in Lake Oswego and N. Portland is $24,000. 4. The total sales for the month is $8,112,360.

C. Use the table in Example C to answer.

1. Which cereal has the most sodium per 1-oz serving?
2. How many milligrams (mg) of potassium are there in 6 oz of wheat flakes?
3. Mary's doctor has counseled her to eat 18 grams (g) of protein for breakfast. How many servings of raisin bran will she need to eat to meet the recommendation?
4. How many ounces of wheat flakes can be eaten before consuming the same amount of potassium as in 1 oz of oat bran?

C. The following table displays nutritional information about four breakfast cereals:

Nutritional Value per 1-oz Serving

Ingredient	Oat Bran	Rice Puffs	Raisin Bran	Wheat Flakes
Calories	90	110	120	110
Protein	6 g	2 g	3 g	3 g
Carbohydrate	17 g	25 g	31 g	23 g
Fat	0 g	0 g	1 g	1 g
Sodium	5 mg	290 mg	230 mg	270 mg
Potassium	180 mg	35 mg	260 mg	4 mg

1. Which cereal has the most calories per serving?
2. How many grams (g) of carbohydrates are in 5 oz of rice puffs?
3. How many more milligrams (mg) of sodium are in a 1-oz serving of wheat flakes as compared with raisin bran?
4. How many ounces of oat bran can be eaten before consuming the same amount of sodium as in 1 oz of rice puffs?

1. Raisin bran **Find the largest value in the calorie row, 120, then read the cereal heading at the top of that column.**

2. $5(25 \text{ g}) = 125 \text{ g}$ **Multiply the grams of carbohydrates in rice puffs by 5.**

   There are 125 g of carbohydrates in 5 oz of rice puffs.

3. $(270 - 230) \text{ mg} = 40 \text{ mg}$ **Subtract the mg of sodium in raisin bran from that of wheat flakes.**

   There are 40 mg more sodium in wheat flakes.

4. $\dfrac{290 \text{ mg}}{5 \text{ mg}} = 58$ **Divide the number of mg of sodium in rice puffs by the number in oat bran.**

   You can eat 58 oz of oat bran.

Answer to Warm Up C. 1. Rice Puffs has the most sodium per 1 oz serving. 2. There are 24 mg of potassium in 6 oz of wheat flakes. 3. She will need 6 servings of raisin bran. 4. It will take 45 oz of wheat flakes.

NAME ______________________ CLASS ____________ DATE ____________

Exercises 1.8

OBJECTIVE: *Read and interpret information given in a table.*

Use the table for Exercises 1–8.

Cost of U.S. Television Rights for the Olympics in Millions

	1984	1988	1992	1994	1996	1998	2000	2002	2004	2006	2008
Winter	ABC $91	ABC $309	CBS $243	CBS $295		CBS $375		NBC $545		NBC $613	
Summer	ABC $25	NBC $300	NBC $401		NBC $456		NBC $705		NBC $793		NBC $894

1. What is the least expensive year for broadcast rights for the Summer Olympics in the 1990s?

2. What is the difference in the cost for broadcast rights for the winter games and the summer games in 1988?

3. What is the total cost of the broadcast rights for the Summer Olympics in the 1990s?

4. What is the predicted increase in cost for broadcast rights for the summer games in 2008 as compared to the 1996 games?

5. What is the difference between the projected costs for broadcast rights to the 2004 and the 2002 games?

6. What is the total cost of broadcast rights for the winter games from 1984 to 2006?

7. Estimate how much NBC will pay for the rights to broadcast the games from 1984 to 2008. Find the actual cost.

8. Estimate how much CBS will pay for the rights to broadcast the games from 1984 to 2008. Find the actual cost.

Use the table for Exercises 9–18.

Nutritional Information Per Serving of Entrée

Ingredient	Fish Cakes	Veal Chops	Chicken Dijon	Pepper Steak
Calories	259	421	247	240
Protein	28 g	34 g	31 g	28 g
Fat	12 g	24 g	12 g	10 g
Carbohydrate	7 g	15 g	2 g	9 g
Sodium	783 mg	687 mg	649 mg	820 mg
Cholesterol	147 mg	115 mg	99 mg	76 mg

9. Which entrée has the highest level of cholesterol per serving?

10. Which entrée has the least number of calories per serving?

11. How much less cholesterol is consumed when ordering chicken dijon as opposed to fish cakes?

12. How much more sodium is consumed when eating pepper steak as opposed to veal chops?

13. How much fat is contained in three servings of veal chops?

14. How many grams of carbohydrate are there in four servings of chicken dijon and two servings of fish cakes?

15. At a buffet, Dan eats one serving each of veal chops, chicken dijon, and pepper steak. How many calories does he consume?

16. At a buffet, Susan eats two servings of fish cakes and one serving of veal chops. How many milligrams (mg) of cholesterol does she consume?

17. Jerry's doctor puts him on a 900-calorie diet. What three individual entrées can he eat and stay within the 900-calorie limit?

18. Jessica is restricted to 30 grams of fat per day. Is there any combination of three entrées that she can eat and remain within the restriction?

Use the table for Exercises 19–26.

Flexible Life Insurance Policy

Age	Death Benefit	Account Value	Cash Surrender Value
57	$400,000	$144,276	$128,677
59	400,000	165,219	150,928
61	400,000	189,522	178,494
63	400,000	217,865	217,865
65	400,000	251,206	251,206

19. What is the gain in the account value from age 57 to age 65?

20. What is the gain in the cash surrender value from age 57 to age 63?

21. If the policy is cashed in at age 59, what is the loss from the account value?

22. If the policy is cashed in at age 65, what is the loss from the account value?

23. What is the difference between the death benefit and the account value at age 61?

24. What is the difference between the death benefit and the surrender value at age 63?

25. Between what two consecutive ages in the table did the largest increase in the account values occur?

26. Between what two consecutive ages in the table did the least increase in the cash surrender value take place?

Use the table for Exercises 27–34.

Visitors at Lizard Lake State Park

	May	June	July	August	September
Overnight camping	231	378	1104	1219	861
Picnics	57	265	2371	2873	1329
Boat rental	29	45	147	183	109
Hiking/climbing	48	72	178	192	56
Horse rental	22	29	43	58	27

27. During which month are the most picnics held?

28. Which month has the fewest horse rentals?

29. How many overnight campers use these facilities during these months?

30. How many boat rentals are there during these months?

31. If it costs $5 to hold a picnic in the park, how much income is realized from picnics in July?

32. If a horse rental costs $8, how much income is realized from horse rentals in June?

33. How many more hikers/climbers are there in August than in May?

34. How many more boat rentals than horse rentals are there in August?

Use the table for Exercises 35–42.

Zoo Attendance

	1992	1993	1994	1995	1996
Fisher Zoo	2,367,246	2,356,890	2,713,455	2,745,111	2,720,567
Delaney Zoo	1,067,893	1,119,875	1,317,992	1,350,675	1,398,745
Shefford Garden	2,198,560	2,250,700	2,277,300	2,278,345	2,311,321
Utaki Park	359,541	390,876	476,200	527,893	654,345

35. Which zoo had the greatest increase in attendance from 1993 to 1996?

36. Which zoo has the smallest increase in attendance from 1993 to 1994?

37. Estimate the attendance for the 5 years at the Fisher Zoo.

38. Estimate the attendance at Utaki Park for the 5 years.

39. Find the average attendance for the 5 years at the Delaney Zoo.

40. Find the average attendance for the 5 years at Utaki Park.

41. If the average attendee at the Fisher Zoo spends $25 during a day, including admission, find the revenue for 1995. Round to the nearest thousand dollars.

42. If the average attendee at Utaki Park spends $32 during a day, including admission, find the revenue for 1996. Round to the nearest thousand dollars.

Use the table for Exercises 43–44. The table gives the number of endangered species in various categories (*Source:* U.S. Fish and Wildlife Service, 1994).

Group	U.S. Only	U.S. and Foreign	Foreign Only
Mammals	36	20	252
Birds	57	16	153
Reptiles	8	8	63
Amphibians	6	0	8
Fishes	60	4	11
Snails	14	0	1
Clams	50	0	2
Crustaceans	11	0	0
Insects	16	3	4
Arachnids	4	0	0
Plants	378	10	1
Total Animals			
Total			

43. Fill in the two bottom rows of the table.

44. Why do you think there are more "foreign only" endangered species of mammals, birds, and reptiles than in the United States? Does this pattern hold for the rest of the species categories? Explain.

STATE YOUR UNDERSTANDING

45. Cite three or four examples of tables that you have used. Include at least two examples from your own experience outside of your class. Are there tables on your list that do not include numbers?

CHALLENGE

Exercises 46–48. Use the table on page 118, Exercises 35–42.

46. Find the estimated total attendance at the four parks for the five years listed in the table.

47. What must the attendance be in 1997 for Utaki Park to average 501,200 in attendance for the years 1992 to 1997?

48. Estimate the income for 1992 to 1996 at Shefford Garden if the average attendee in 1992 spent $21, 1993 spent $24, 1994 spent $28, 1995 spent $31, and in 1996 spent $35.

GROUP ACTIVITY

49. Have your group select 6 grocery items. Then, have each member go to a different market and price the items. Compile the items and prices in a table. Does the table help you see which market has the lower prices?

1.9

Drawing and Interpreting Graphs

OBJECTIVES

1. Read data from bar, pictorial, and line graphs.

2. Construct a bar, pictorial, or line graph.

VOCABULARY

Graphs are used to illustrate sets of numerical information.

A **bar graph** uses solid lines or heavy bars of fixed length to represent numbers from a set. Bar graphs contain two **scales,** a vertical scale and a horizontal scale. The **vertical scale** represents one set of values and the **horizontal scale** represents a second set of values. These values depend on the information to be presented. The following bar graph illustrates four types of cars (first set of values) and the number of each type of car sold (second set of values).

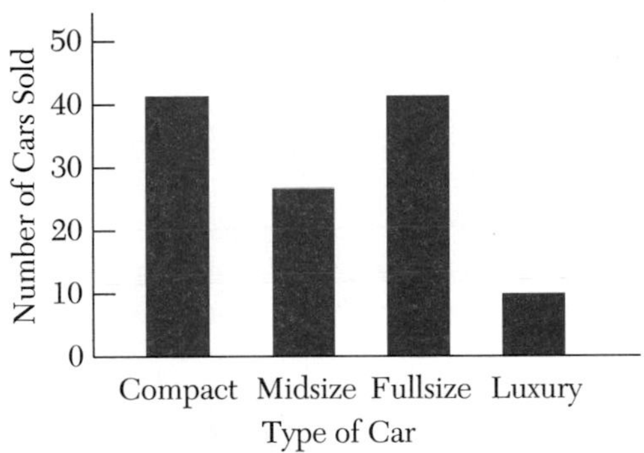

A **line graph** uses lines connecting points to represent numbers from a set. A line graph has a vertical and a horizontal scale, like a bar graph.

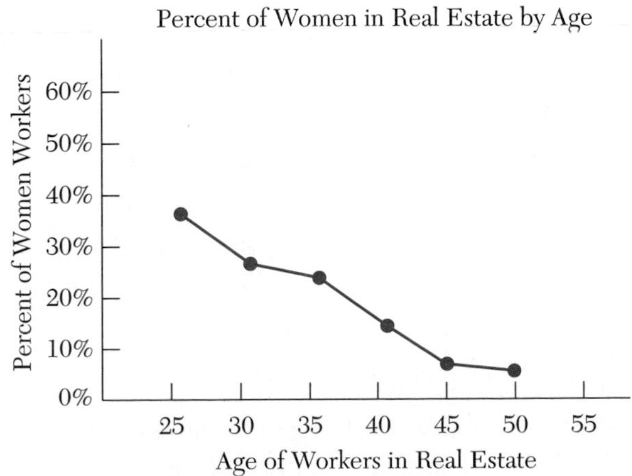

A **pictograph** uses symbols or simple drawings to represent numbers from a set.

DISTRIBUTION OF MATHEMATICS STUDENTS

Mathematics Class	🚶 = 20 students
Prealgebra	🚶 🚶 🚶 🚶 🚶
Algebra	🚶 🚶 🚶 🚶 🚶 🚶 🚶
Calculus	🚶 🚶 🚶 🚶

HOW AND WHY

Objective 1

Read data from bar, pictorial, and line graphs.

A graph or chart is a picture used for presenting data for the purpose of comparison. To "read a graph" means to find values from the graph.

Examine the following bar graph.

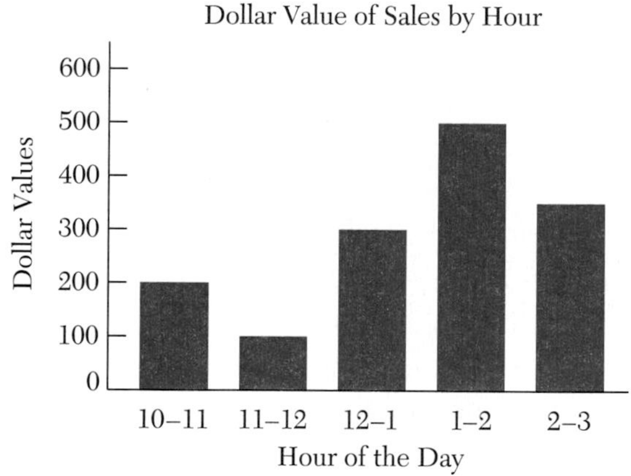

The vertical scale shows dollar values and is divided into units of \$100. The horizontal scale shows time values and is divided into units of one-hour periods from 10:00 A.M. to 3:00 P.M. From the graph we can see that

1. The hour of greatest sales (highest bar) was 1 P.M. to 2 P.M. with \$500 in sales.
2. The hour of least sales (lowest bar) was 11 A.M. to 12 noon was sales of \$100.
3. The total sales during the 2 P.M. to 3 P.M. time period is estimated to be \$350, because the top of the bar falls between the scale divisions.
4. The total of the morning sales was \$300.
5. The total of the afternoon sales was \$1150.

Other observations may be made by studying the graph.

Some advantages of displaying data with a graph:

1. Each person can easily find the data most useful to him or her.
2. The visual display is easy for most people to read.
3. Some questions can be answered by a quick look at the graph. For example, "What time does the store need the most sales clerks?"

Examples A–B

Directions: Answer the questions associated with the graph.

Strategy: Examine the graph to determine the values that are related.

A. The following bar graph shows the number of cars that used Highway 37 during a 1-week period.

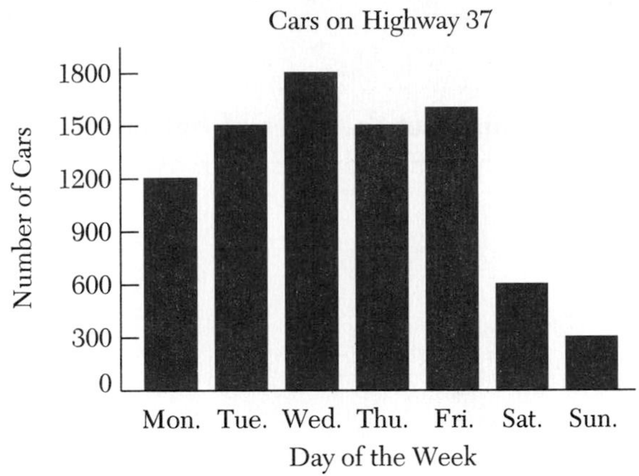

1. What day had the most traffic?
2. What day had the least traffic?
3. How many cars used the highway on Tuesday?
4. How many cars used the highway on the weekend?

1. Wednesday	**The tallest bar shows the largest number of cars.**
2. Sunday	**The shortest bar shows least number of cars.**
3. 1500	**Read the vertical scale at the top of the bar for Tuesday. The value is estimated because the top of the bar is not on a scale line.**
4. 900	**Add the number of cars for Saturday and Sunday.**

Warm Ups A–B

A. The number of bus riders for each day of the week is shown in the following bar graph.

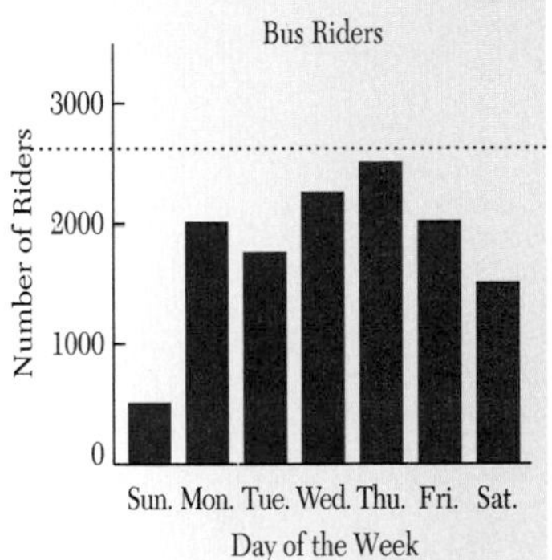

1. What day had the greatest ridership?
2. What day had the least?
3. How many people rode the bus on Friday?
4. How many people rode the bus during the week?

B. The total sales from hot dogs, soda, T-shirts, and buttons during an air show are in the pictograph.

SALES AT AIR SHOW

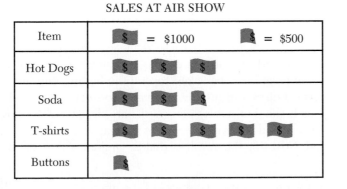

Item	$ = $1000 $ = $500
Hot Dogs	$ $ $
Soda	$ $ $
T-shirts	$ $ $ $ $
Buttons	$

B. The number of birds spotted during a recent expedition of the Huntsville Bird Society is shown in the pictograph.

Birds Spotted

Species	= 50 = 25
Crows	
Woodpeckers	
Wrens	
Canaries	

Answer to Warm Up A. 1. Thursday 2. Sunday 3. 2000 4. 12,500

1. Which species was spotted most often?
2. How many woodpeckers and wrens were spotted?
3. How many more canaries were spotted than crows?

1. What item has the largest dollar sales?
2. What were the total sales from hot dogs and buttons?
3. How many more dollars were realized from the sale of T-shirts than from buttons?

1. T-shirts

There are more bills representing dollar sales in the T-shirt row than any other row.

2. Hot dogs: $3000
   Buttons: $ 500
   Total: $3500

3. T-shirts: $5000
   Buttons: $ 500

   The sale from T-shirts was $4500 more than from buttons.

Subtract the sales of buttons from the sales of T-shirts.

HOW AND WHY

Objective 2

Construct a bar, pictorial, or line graph.

Let us construct a bar graph to show the variation in used car sales at Oldies but Goldies used car lot. The data are shown in the following table:

Month	Cars Sold
January	30
February	15
March	25
April	20
May	15
June	35

To draw and label the bar graph for these data, we show the number of cars sold on the vertical scale and the months on the horizontal scale. This is a logical display because we will most likely be asked to find the highest and lowest months of car sales and a vertical display of numbers is easier to read than a horizontal display of numbers. This is the typical way bar graphs are displayed. Be sure to write the names on the vertical and horizontal scales as soon as you have chosen how the data will be displayed. Now title the graph so that the reader will recognize the data it contains.

The next step is to construct the two scales of the graph. Because each monthly total is divisible by five, we choose multiples of five for the vertical scale. We could have chosen one for the vertical scale, but the bars would be very long and the graph would take up a lot of space. If we had chosen a larger scale, for instance ten, then the graph might be too compact and we would need to find fractional

Answer to Warm Up B. 1. Wrens 2. 350 3. 50

values on the scale. It is easier to draw the graph if we use a scale that divides each unit of data. The months are displayed on the horizontal scale. Be sure to draw the bars with uniform width, because each of them represents a month of sales. A vertical display of between 5 and 12 units is typical. The vertical display should start with zero.

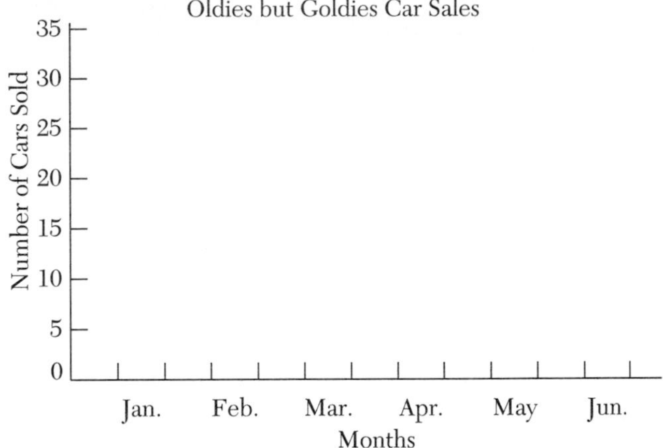

We stop the vertical scale at 35, because that is the maximum number of cars to be displayed. The next step is to draw the bars. Start by finding the number of cars for January. Thirty cars were sold in January, so we draw the bar for January until the height of 30 is reached. This is the top of the bar. Now draw the solid bar for January.

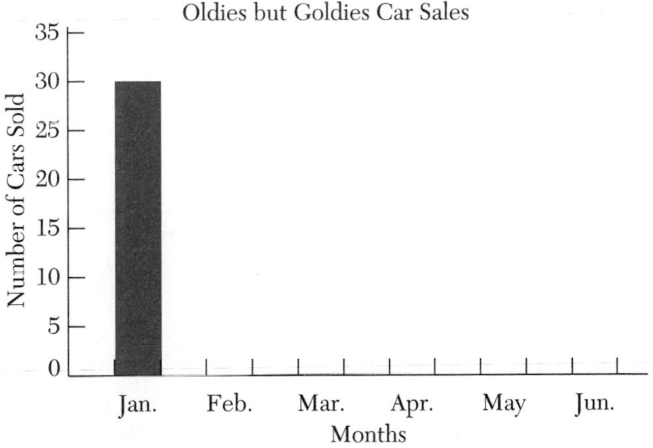

Complete the graph by drawing the bars for the other months.

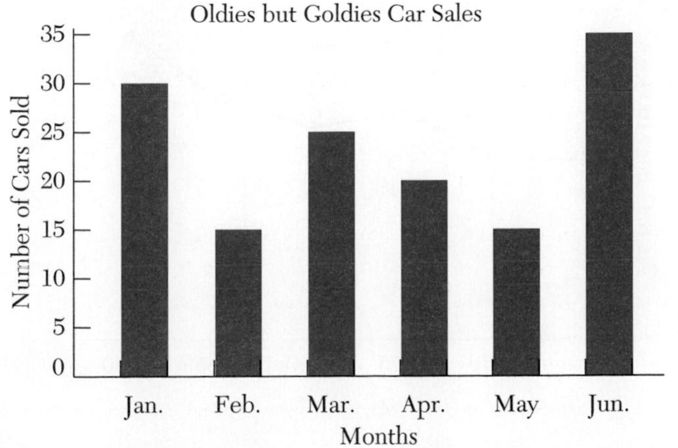

A line graph is similar to a bar graph in that it has vertical and horizontal scales. The data are represented by points rather than bars and the points are connected by line segments. We use a line graph to display the following data:

PROPERTY TAXES

Year	Tax Rate (per $1000)
1970	$12.00
1975	$15.00
1980	$14.00
1985	$16.00
1990	$20.00

The vertical scale represents the tax rate and each unit represents $2. This requires using a half space for the $15.00 rate.

Another possibility is to use a vertical scale in which each unit represents $1, but this would require 20 units on the vertical scale and would make the graph much taller. We opt to save space by using $2 units on the vertical scale.

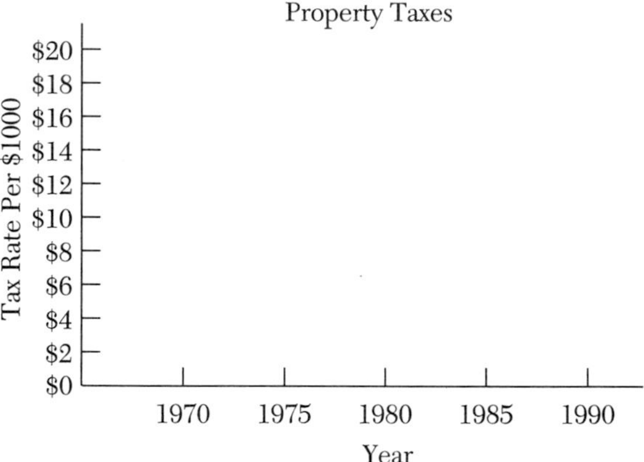

To find the points that represent the data, locate the points that are at the intersection of the horizontal line through the tax rate and the vertical line through the corresponding year. Once all the points have been located, connect them with line segments.

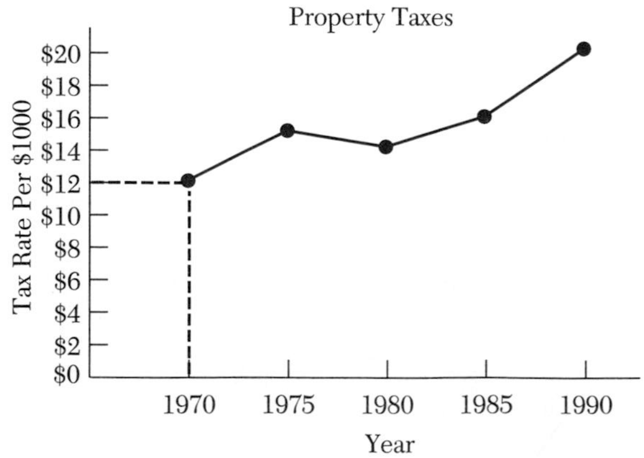

From the graph, we can conclude the following:

1. Only during one five-year period (1975–1980) did the tax rate decline.
2. The largest increase in the tax rate took place from 1985 to 1990.
3. The tax rate has increased \$8 per thousand from 1970 to 1990.

Construct a pictorial graph to show the number of cars sold, by model, at the Western Car Corral. The data are shown in the following table:

Make	Car Sales at Western Car Corral
Ford	50
Chrysler	35
Honda	60
Toyota	25
Pontiac	20

First, select a symbol and the number of cars it represents. Here we use a picture of a car with each symbol representing 10 cars. By letting each symbol represent 10 cars and half a symbol representing 5, we can save space. We could have chosen 5 cars per symbol, but then we would need to display 12 symbols to represent the number of Hondas, and 12 symbols would make the graph quite large. Next, determine the number of symbols we need for each model. The number of symbols can be found by dividing each number of cars by 10:

Make	Number	Symbols Needed
Ford	50	5 tens
Chrysler	35	3 tens and 1 five
Honda	60	6 tens
Toyota	25	2 tens and 1 five
Pontiac	20	2 tens

Now, draw the graph using the symbols (pictures) to represent the data:

CAR SALES AT WESTERN CAR CORRAL

Model	= 10 cars = 5 cars
Ford	
Chrysler	
Honda	
Toyota	
Pontiac	

Warm Up C

C. The number of plants sold at the Pick-a-Posey Nursery: geraniums, 45; fuchsias, 60; marigolds, 150; impatiens, 90; daisies, 120.

Example C

Directions: Construct a bar graph.

Strategy: List the related values in pairs and draw two scales to show the pairs of values.

C. The number of babies born during the first six months at Tuality Hospital: January, 15; February, 9; March, 7; April, 18; May, 10; June, 6.

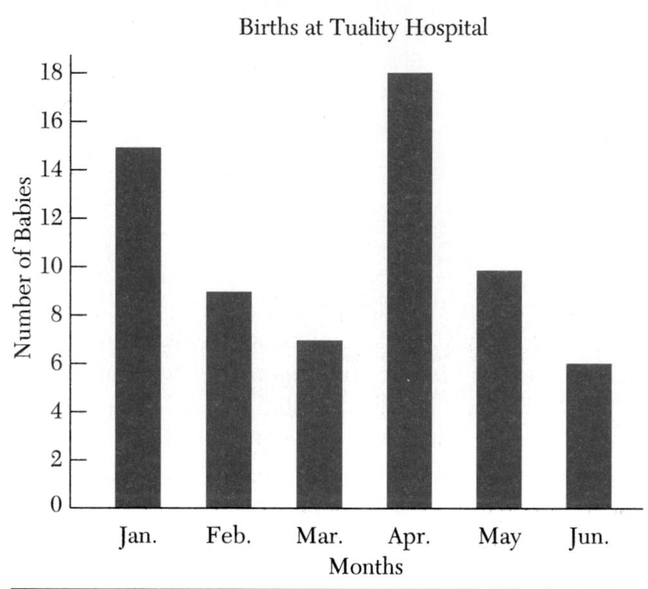

Choose a scale of 1 unit = 2 babies for the vertical scale.

Divide the horizontal scale so that it will accommodate six months with a common space between them.

Construct the graph, label the scales, and give the graph a title.

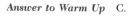

Answer to Warm Up C.

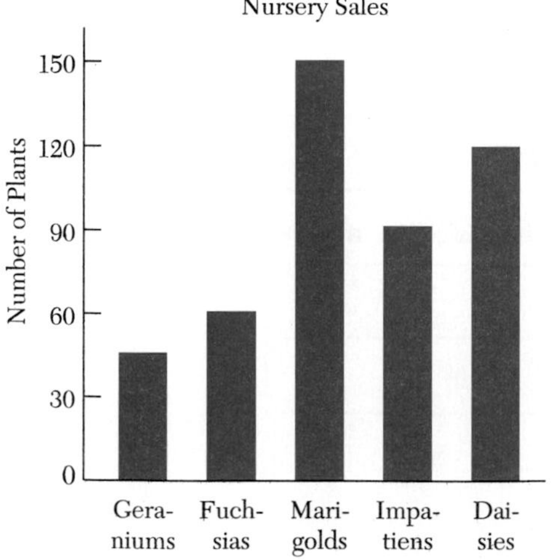

NAME ______________________ CLASS ____________ DATE ____________

Exercises 1.9

OBJECTIVE 1: *Read data from bar, pictorial, and line graphs.*

A.

The graph shows the variation in the number of phone calls during normal business hours:

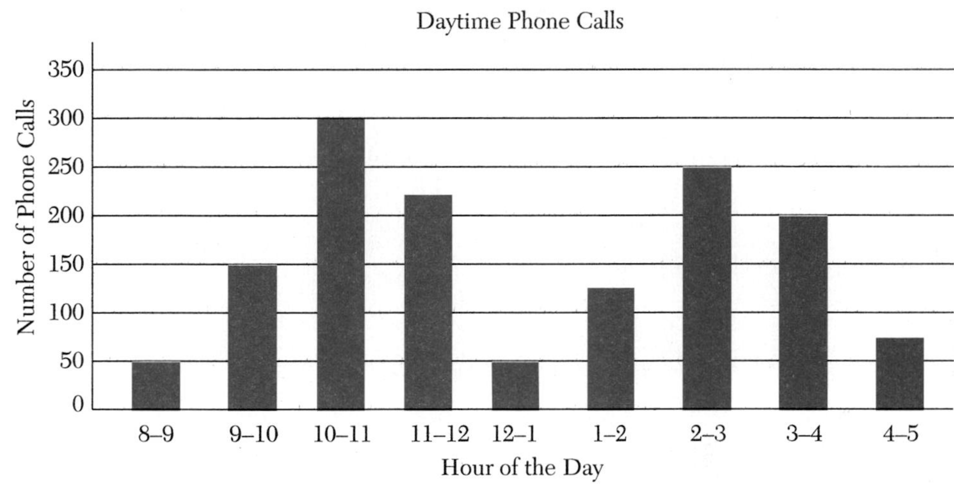

1. At what hour of the day is the number of phone calls greatest?

2. At what hour of the day is the number of phone calls least?

3. What is the number of phone calls made between 2 and 3?

4. What is the number of phone calls made between 8 and 12?

5. What is the total number of phone calls made during the times listed?

6. Are there more phone calls in the morning (8–12) or the afternoon (12–5)?

The graph shows the number of cars in the shop for repair during a given year:

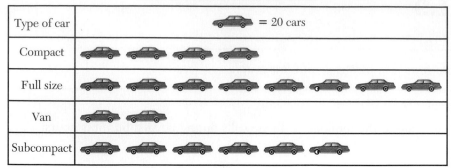

1988 REPAIR INTAKE RECORD

Type of car	= 20 cars
Compact	
Full size	
Van	
Subcompact	

7. How many vans are in for repair during the year?

8. How many compacts and subcompacts are in for repair during the year?

9. What type of car has the most cars in for repair?

10. Are more subcompacts or compacts in for repair during the year?

11. How many cars are in for repair during the year?

12. If the average repair costs for compacts is $210, what is the gross income on compact repairs for the year?

B.

The graph shows the number of production units at NERCO during the period 1992–1996.

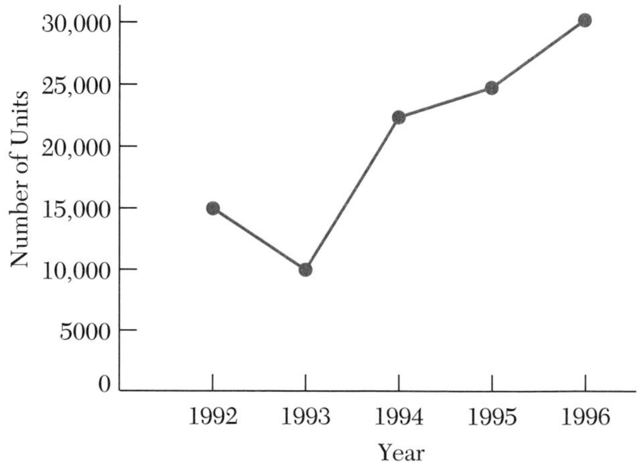

13. What is the greatest production year?

14. What is the year of least production?

15. What is the increase in production between 1993 and 1994?

16. What is the decrease in production between 1992 and 1993?

17. What is the average production per year?

18. If the cost of producing a unit in 1995 is $2750 and the unit is sold for $4560, what is the net income for the year?

The graph shows the amounts paid for raw materials at Southern Corporation during a production period.

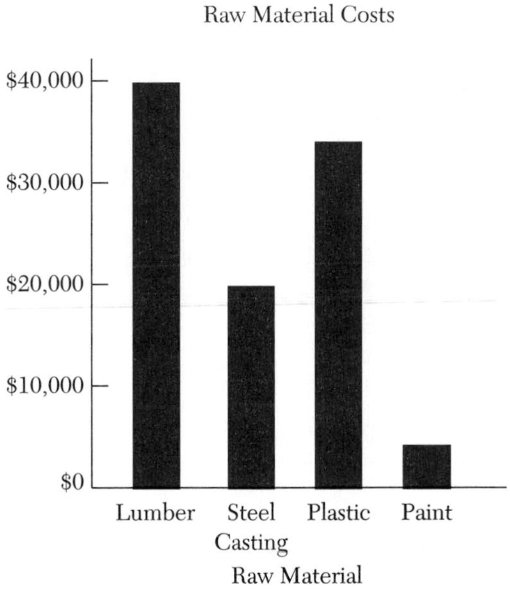

Raw Material Costs

$40,000
$30,000
$20,000
$10,000
$0
Lumber
Steel Casting
Plastic
Paint
Raw Material

19. What is the total paid for paint and lumber?

20. What is the total paid for raw materials?

21. How much less is paid for steel castings than for plastics?

22. How much more is paid for plastic than for paint?

23. If Southern Corporation decides to double its production during the next period, what will it pay for steel casting?

24. If Southern Corporation decides to double its production during the next period, how much more will it pay for lumber and steel castings than for plastic and paint?

OBJECTIVE 2: *Construct a bar, pictorial, or line graph.*

A.

In Exercises 25–36 draw the graph. Be sure to title the graph and label the parts. In Exercises 25–28 draw bar graphs to display the data.

25. Distribution of grades in an algebra class: A, 8; B, 6; C, 15; D, 8; F, 4.

26. Distribution of monthly income: rent, $450; automobile, $300; taxes, $250; clothes, $50; food, $300; miscellaneous, $100.

27. Career preferences as expressed by a senior class: business, 120; law, 20; medicine, 40; science, 100; engineering, 50; public service, 80; armed service, 10.

28. Diners' choices for dinner at the La Plane restaurant in one week: steak, 45; salmon, 80; chicken, 60; lamb, 10; others, 25.

In Exercises 29–32 draw line graphs to display the data.

29. Daily sales at the local men's store: Monday, $1500; Tuesday, $2500; Wednesday, $1500; Thursday, $3500; Friday, $4000; Saturday, $6000; Sunday, $4500.

30. The gallons of water used each quarter of the year by a small city in New Mexico:
January–March 20,000,000
April–June 30,000,000
July–September 45,000,000
October–December 25,000,000

31. Income from various sources for a given year for the Smith family: wages, $36,000; interest, $2000; dividends, $4000; sale of property, $24,000.

NAME ______________________ CLASS ______________ DATE ______________

32. Jobs in the electronics industry in a western state: 1990, 15,000; 1991, 21,000; 1992, 18,000; 1993, 21,000; 1994, 27,000; 1995, 30,000.

B.

In Exercises 33–36 draw pictorial graphs to display the data:

33. The cost of an average three-bedroom house in a rural city:

Year	Cost
1970	$60,000
1975	$70,000
1980	$75,000
1985	$70,000
1990	$65,000
1995	$80,000

34. The population of Wilsonville over 20 years of age: 1975, 22,500; 1980, 30,000; 1985, 32,500; 1990, 37,500; 1995, 40,000.

35. The oil production from a local well over a 5-year period:

Year	Barrels Produced
1991	15,000
1992	22,500
1993	35,000
1994	32,500
1995	40,000

36. The estimated per capita income in the Great Lakes region in 1995:

State	Income
Illinois	$25,000
Michigan	$22,500
Ohio	$22,500
Wisconsin	$20,000
Indiana	$20,000

C.

Use the following graph for Exercises 37–40. The graph shows 1996 U.S. and worldwide employment in the auto industry, according to the American Automobile Manufacturers Association.

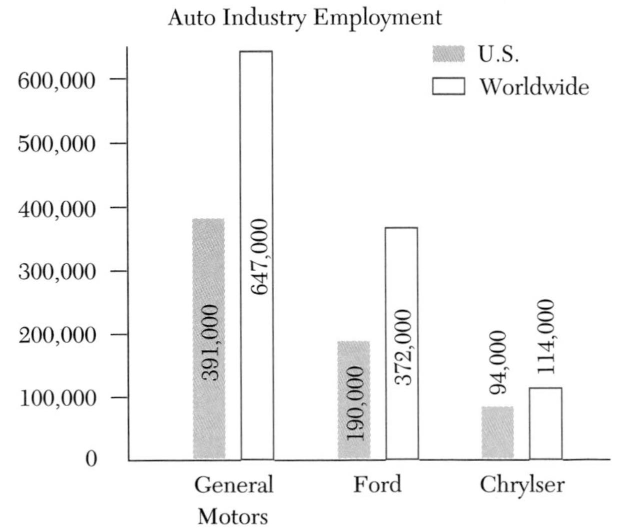

37. Which company has the largest number of employees worldwide? Which has the smallest number of employees worldwide?

38. How many employees does Ford have who work outside the United States?

39. What is the total number of car workers in the United States employed by the three car makers?

40. Do the numbers in the graph appear to be rounded? Why or why not? If so, to what place value were they rounded?

41. Using the information from the table in Exercises 35–42 of Section 1.8, make a bar graph to display the data on attendance at the zoos in 1996. Label carefully.

42. Make a pictograph of the number of gas stations of the various brands in the United States in 1995. The raw data is contained in Section 1.1, Exercise 68. You may want to round the data to the nearest hundred before graphing.

STATE YOUR UNDERSTANDING

43. Explain the advantages of each type of graph. Which do you prefer? Why?

CHALLENGE

44. The figures for U.S. casualties in four wars of the 20th century are: World War I, 321,000; World War II, 1,076,000; Korean War, 158,000; Vietnam War, 211,000. Draw a bar graph, line graph, and a pictorial graph to illustrate the information. Which of your graphs do you think does the best job of displaying the data? (For the pictorial graph you may want to round to the nearest hundred thousand.)

GROUP ACTIVITY

45. Have each member select a country and find its most recent population statistics. Put the numbers together and have each member draw a different kind of graph of the populations.

NAME ______________________ CLASS ______________ DATE ______________

CHAPTER 1

Group Project *(1–2 weeks)*

OPTIONAL

All tables, graphs, and charts should be clearly labeled and computer generated if possible. Written responses should be typed and checked for spelling and grammar.

a. Go to the library and find the population and area for each state in the United States. Organize your information by geographic region. Record your information in a table.

b. Calculate the total population and the total area for each region. Calculate the population density (number of people per square mile, rounded to the nearest whole person) for each region, and put this and the other regional totals in a regional summary table. Then make three separate graphs, one for regional population, one for regional area, and the third for regional population density.

c. Calculate the average population per state for each region, rounding as necessary. Put this information in a bar graph. What does this information tell you about the regions? How is it different from the population density of the region?

d. How did your group decide on the makeup of the regions? Explain your reasoning.

e. Are your results what you expected? Explain. What surprised you?

NAME ______________________ CLASS __________ DATE __________

CHAPTER 1

True–False Concept Review

Check your understanding of the language of basic mathematics. Tell whether each of the following statements is True (always true) or False (not always true). For each statement you judge to be false, revise it to make a statement that is true.

ANSWERS

1. All whole numbers can be written using nine digits. **1.** ________

2. In the number 6731, the digit 7 represents 700. **2.** ________

3. The word "and" is not used when writing the word names of whole numbers. **3.** ________

4. The symbols, $5 < 18$, can be read "five is greater than eighteen." **4.** ________

5. $1345 < 1344$ **5.** ________

6. To the nearest thousand, 8498 rounds to 8000. **6.** ________

7. It is possible for the rounded value of a number to be equal to the original number. **7.** ________

8. The expanded form of a whole number shows the plus signs that are usually not written. **8.** ________

9. The sum of 60 and 4 is 604. **9.** ________

10. The process of "carrying" when doing an addition problem with pencil and paper is based on the place values of the numbers. **10.** ________

11. A line graph has at least two scales. **11.** ________

12. The product of 7 and 4 is 11. **12.** ________

13. It is possible to subtract 37 from 55 without "borrowing." **13.** ________

14. The number 4 is a factor of 56. **14.** ________

15. The multiplication sign is sometimes omitted when writing a multiplication problem. **15.** ______

16. Whenever a number is multiplied by zero, the value remains unchanged. **16.** ______

17. There is more than one method for doing division problems. **17.** ______

18. If ☺ represents 40 people then ☺☺☺☺ represents 80 people. **18.** ______

19. In $92 \div 4 = 23$, the quotient is 92. **19.** ______

20. If a division exercise has a remainder then we know that there is no whole-number quotient. **20.** ______

21. When zero is divided by any whole number from 33 to 78, the result is 0. **21.** ______

22. The result of zero divided by zero can either be 1 or 0. **22.** ______

23. The value of 8^2 is 16. **23.** ______

24. The value of 111^1 is 111. **24.** ______

25. One trillion is a power of ten. **25.** ______

26. The product 340×10^2 is equal to 3400. **26.** ______

27. The quotient of 7000 and 100 is 70. **27.** ______

28. In a pictograph, simple drawings are used as a unit of measure. **28.** ______

29. In the order of operations, exponents always take precedence over addition. **29.** ______

30. In the order of operations, multiplication always takes precedence over subtraction. **30.** ______

31. The value of $2^3 + 2^3$ is the same as the value of 2^4. **31.** ______

32. The average of three different numbers is smaller than the largest of the three numbers. **32.** ______

33. The word "mean" sometimes has the same meaning as "average." **33.** ______

34. A table is a method of displaying data in an array using a horizontal and vertical arrangement to distinguish the type of data. **34.** ______

NAME ______________________ CLASS ____________ DATE ____________

CHAPTER 1

Test

ANSWERS

1. Divide: $54\overline{)5886}$ — 1. ________

2. Subtract: $9123 - 6844$ — 2. ________

3. Simplify: $36 \div 9 + 4 \cdot 5 - 5$ — 3. ________

4. Multiply: 53(768) — 4. ________

5. Insert < or > to make the statement true: 278 ⬜ 201 — 5. ________

6. Multiply: 76×10^4 — 6. ________

7. Multiply: 709(386) — 7. ________

8. Write the place value name for four hundred fifty thousand, eighty-two. — 8. ________

9. Find the average of 1294, 361, 1924, 274, and 682. — 9. ________

16

10. Multiply: 35(2095) Round the product to the nearest hundred. — 10. ________

11. Round 17,852 to the nearest hundred. — 11 ________

12. Estimate the sum of 83,914, 17,348, 47,699, and 10,341. — 12. ________

13. Find the value of 9^3. — 13. ________

14. Add: $39 + 953 + 4 + 4886$ — 14. ________

15. Estimate the product: 478(17) — 15. ________

16. Subtract:

$$\begin{array}{r} 7040 \\ -\ 587 \\ \hline \end{array}$$

16. ________

17. Write the word name for 6007. **17.** ______

18. Simplify: $25 + 2^3 - 24 \div 8$ **18.** ______

19. Add:

$$\begin{array}{r} 45{,}974 \\ 31{,}900 \\ 78{,}211 \\ 12{,}099 \\ +\ 67{,}863 \\ \hline \end{array}$$

19. ______

20. Divide: $6{,}050{,}000{,}000 \div 10^5$ **20.** ______

21. Round 524,942,664 to the nearest ten thousand. **21.** ______

22. Estimate the quotient and then divide: $47{,}125 \div 76$ **22.** ______

23. Simplify: $72 - 8^2 + 27 \div 3$ **23.** ______

24. Simplify: $(4 \cdot 2)^3 + (3^2)^2 + 9 \cdot 2$ **24.** ______

25. Find the average of 582, 678, 425, and 979. **25.** ______

26. A secretary can type an average of 80 words per minute. If there are approximately 700 words per page, how long will it take the secretary to type 12 pages? **26.** ______

27. Twelve people share in the Nationwide lottery jackpot. If the jackpot is worth \$9,456,000, how much will each person receive? If each person's share is to be distributed evenly over a 20-year period, how much will each person receive per year? **27.** ______

NAME ______________________ CLASS ____________ DATE ____________

28. Refer to the graph showing auto sales distribution for a local dealer to answer the following questions. **28.** ____________

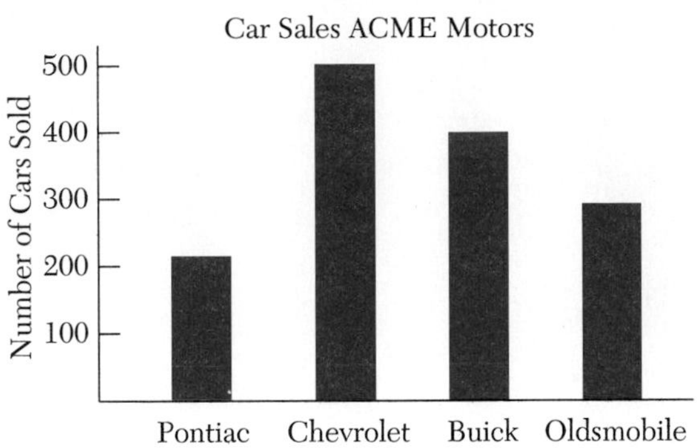

a. What make of auto had the greatest sales?

b. What was the total number of Pontiacs and Chevrolets sold?

c. How many more Buicks were sold than Oldsmobiles?

29. Refer to the table to answer the following questions. **29.** ____________

Employees by Division
Exacto Electronics

Division	**Day Shift**	**Swing Shift**
A	350	175
B	400	125
C	125	25

a. Which division has the greatest number of employees?

b. How many more employees are in the day shift in Division A as compared with the day shift in Division C?

c. How many employees are in the three divisions?

30. Construct a bar graph to display the number of lunches purchased at the local fast food bar over one week: Hamburger, 1100; Fishburger, 300; Chef salad, 500; Roll and soup, 300; Omelet, 400.

30. ______________________

Good Advice for Studying

New Habits From Old

If you are in the habit of studying math by only reading the examples to learn how to do the exercises, stop now! Instead, read the assigned section—all of it—before class. It is important that you read more than the examples so that you fully understand the concepts. How to do a problem isn't all that needs to be learned. Where and when to use specific skills are also essential.

When you read, read interactively. This means that you should be both writing and thinking about what you are reading. Write down new vocabulary, perhaps start a list of new terms paraphrased in words that are clear to you. Take notes on the How and Why segments, jotting down questions you may have, for example. As you read examples, work the Warm Up problems in the margin. Begin the exercise set only when you understand what you have read in the section. This process should make your study sessions go much faster and be more effective.

If you have written down questions during your study session, be sure to ask them at the next class session, seek help from a tutor, or discuss them with a classmate. Don't leave these questions unanswered.

Pay particular attention to the objectives at the beginning of each section. Read these at least twice; first, when you do your reading before class and again, after attending class. Ask yourself, "Do I understand what the purpose of this section is?" Read the objectives again before test time to see if you feel that you have met these objectives.

During your study session, if you notice yourself becoming tense and your breathing shallow (light and from your throat or upper part of your lungs), follow this simple coping strategy. Say to yourself: "I'm in control. Relax and take a deep breath." Breathe deeply and properly by relaxing your stomach muscle (that's right, you have permission to let your stomach protrude!) and inhaling so that the air reaches the bottom of your lungs. Hold the air in for a few seconds, then slowly exhale, pulling your stomach muscle in as you exhale. This easy exercise not only strengthens your stomach muscle, but gives your body and brain the oxygen you need to perform free from physical stress and anxiety. This deep breathing relaxation method can be done in one to five minutes. You may want to use it several times a day, especially during an exam.

These techniques can help you to start studying math more effectively and to begin managing your anxiety. Begin today.

2

Measurement

APPLICATION

Paul and Barbara have just purchased a row house in Georgetown, D.C. The back yard is rather small and completely fenced. They decide to take out all the grass and put in a brick patio and formal rose garden. The plans for the patio and garden are in the following drawing.

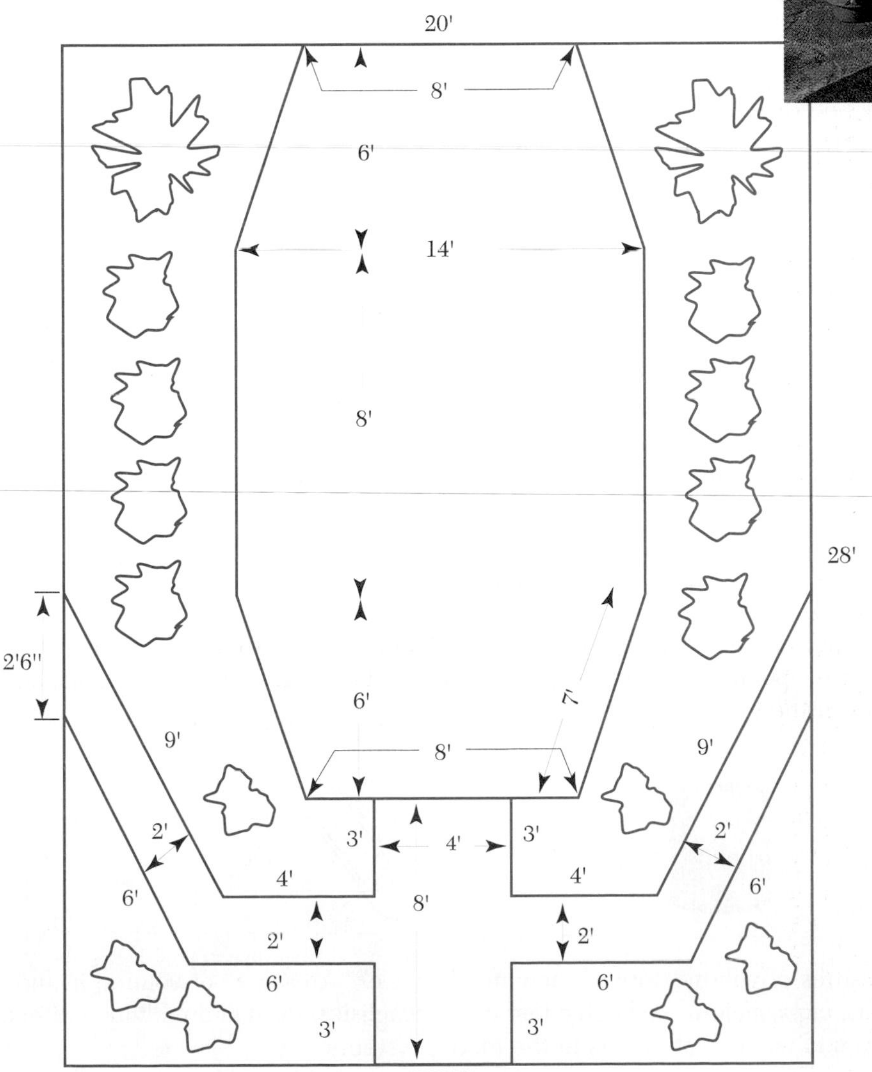

2.1

English and Metric Measurement

OBJECTIVES

1. Multiply and divide a measurement by a number.
2. Add and subtract measurements.

VOCABULARY

A **unit of measure** is the name of a fixed quantity that is used as a standard.

A **measurement** is a number together with a unit of measure.

Equivalent measurements are measures of the same amount but using different units.

The **English system** is the measurement system commonly used in the United States.

The **metric system** is the measurement system used by most of the world.

HOW AND WHY

Objective 1

Multiply and divide a measurement by a number.

One of the primary ways of describing an object is to give its measurements. We measure how long an object is, how much it weighs, how much space it occupies, how long it has existed, how hot it is, and so forth. The units of measure must be universally defined so that we all mean the same thing when we use a measurement. There are two major systems of measurement in use in the United States. One is the English system, so named because we adopted what was used in England at the time. The other is the metric system, which is currently used by almost the entire world.

Measures of length answer questions like "how long?" or "how tall?" or "how deep?" Measures of length include inches, feet, yards, and miles in the English system and millimeters, centimeters, meters, and kilometers in the metric system.

Measures of weight answer the question "how heavy?" Measures of weight include ounces, pounds, and tons in the English system and milligrams, grams, and kilograms in the metric system.

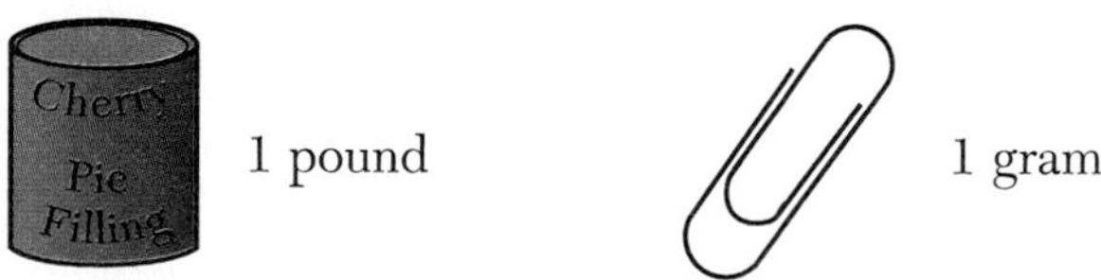

Measures of volume answer "how much space?" Measures of volume include teaspoons, cups, gallons, and cubic feet in the English system and milliliters, liters, kiloliters, and cubic centimeters in the metric system.

To measure objects bigger than a single unit of measure, we count how many of the units are needed. For example, to measure the length of this line,

we count how many inch units are needed for the entire length.

There are four 1-inch units in this length, so we say that it is $4 \cdot (1 \text{ inch}) = 4$ inches.

Similarly, we write

8 centimeters = $8 \cdot (1 \text{ centimeter})$,
45 pounds = $45 \cdot (1 \text{ pound})$, and
327 liters = $327 \cdot (1 \text{ liter})$.

This way of interpreting measurements makes it easy to find multiples of measurements. Consider 3 boards, each 5 feet long. The total length of the boards is

$$\begin{aligned} 3 \cdot (5 \text{ feet}) &= 3 \cdot 5 \cdot (1 \text{ foot}) \\ &= 15 \cdot (1 \text{ foot}) \\ &= 15 \text{ feet} \end{aligned}$$

Similarly, a case of soda holds a total of 12 liters of soda. How much does each bottle hold if there are six bottles per case?

$$\begin{aligned} \frac{12 \text{ liters}}{6} &= \frac{12 \ (1 \text{ liter})}{6} \\ &= \frac{12}{6} (1 \text{ liter}) \\ &= 2 \ (1 \text{ liter}) \\ &= 2 \text{ liters} \end{aligned}$$

To multiply or divide a measurement by a number

Multiply or divide the two numbers and write the unit of measure.

Warm Ups A–B

A. (use centimeters)

B. (use inches)

Examples A–B

Directions: Measure the length of the following lines.

Strategy: Use a ruler and count the number of units.

A. (use centimeters) **Mark off units of centimeters and count.**

The length is 6 centimeters.

B. (use inches) **Mark off units of inches and count.**

The length is 3 inches.

Warm Ups C–D

C. A package of microwave popcorn weighs 101 grams. What is the weight of five packages?

D. A carpenter has a 16-foot board that he must cut into four equal pieces. How long is each piece?

Examples C–D

Directions: Solve.

Strategy: Describe each situation with a statement involving measurements and simplify.

C. What is the total weight of four pieces of cheese that each weigh 30 grams?

Strategy: To find the weight of four pieces, multiply the weight of one piece by 4.

$$\begin{aligned}\text{Total weight} &= 4 \cdot (30 \text{ grams}) \quad \textbf{Multiply.}\\ &= 4 \cdot 30 \cdot (1 \text{ gram})\\ &= 120 \cdot (1 \text{ gram})\\ &= 120 \text{ grams}\end{aligned}$$

The four pieces weigh 120 grams.

D. If 140 ounces of peanut brittle are divided equally among five sacks, how much goes into each sack?

Strategy: To find how much goes in each sack, divide the total weight by 5.

$$\begin{aligned}1 \text{ sack} &= 140 \text{ ounces} \div 5 \quad \textbf{Divide.}\\ &= 140 \cdot (1 \text{ ounce}) \div 5\\ &= (140 \div 5) \cdot (1 \text{ ounce})\\ &= 28 \cdot (1 \text{ ounce})\\ &= 28 \text{ ounces}\end{aligned}$$

Each sack contains 28 ounces.

Answers to Warm Ups A. The length is 3 centimeters. B. The length is 2 inches. C. The weight of five packages is 505 grams. D. Each piece is 4 feet long.

HOW AND WHY

Objective 2

Add and subtract measurements.

The expression "You can't add apples and oranges" applies to adding (and subtracting) measurements. Only measurements with the same unit of measure may be added or subtracted.

10 gallons + 3 gallons = (10 + 3) gallons

= 13 gallons

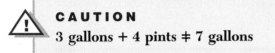

CAUTION

3 gallons + 4 pints ≠ 7 gallons

To add or subtract measurements with the same units of measure

Add or subtract the numbers and write the unit of measure.

If the units of measure do not match, the measurements must first be converted to equivalent measures that do match.

Table 2.1 lists common English measurements, their abbreviations, and their equivalents.

TABLE 2.1 ENGLISH MEASURES AND EQUIVALENTS

Length	Time
12 inches (in.) = 1 foot (ft)	60 seconds (sec) = 1 minute (min)
3 feet (ft) = 1 yard (yd)	60 minutes (min) = 1 hour (hr)
5280 feet (ft) = 1 mile (mi)	24 hours (hr) = 1 day
	7 days = 1 week
Liquid Volume	**Weight**
3 teaspoons (tsp) = 1 tablespoon (tbs)	16 ounces (oz) = 1 pound (lb)
2 cups (c) = 1 pint (pt)	2000 pounds (lb) = 1 ton
2 pints (pt) = 1 quart (qt)	
4 quarts (qt) = 1 gallon (gal)	

Use the table to convert units before adding or subtracting. For instance, if a can weighs 1 lb 3 oz and another can weighs 14 oz, what is the total weight of the two cans?

Total weight = (1 lb 3 oz) + (14 oz)

= (16 oz + 3 oz) + (14 oz) **Convert 1 lb to 16 oz.**

= (19 oz) + (14 oz) **Add.**

= 33 oz

= 2 lb 1 oz **Convert 33 oz to pounds**
33 ÷ 16 = 2, remainder 1.

The two cans weigh 33 oz or 2 lb 1 oz.

The metric system was invented by French scientists in 1799. Their goal was to make a system that was easy to learn and would be used worldwide. They based the system for length on the meter and related it to the earth by defining it as 1/10,000,000 of the distance between the north pole and the equator. (A meter is currently defined by international treaty in terms of the wavelength of the orange-red radiation of the element krypton 86.) Units of measure of volume and weight are related to water.

To make the system easy to use, the scientists based all conversions on powers of 10 and gave the same suffix to all units of measure for the same characteristic. So, all measures of length end in "-meter," all measures of volume end in "-liter," and all measures of weight end in "-gram." A kilometer is 1000 meters, and a kilogram is 1000 grams. Table 2.2 shows the basic units, abbreviations, and conversions for the metric system.

Table 2.2 Metric Measures and Equivalents

Length (Basic Unit Is 1 Meter)	Weight (Basic Unit Is 1 Gram)
1000 millimeters (mm) = 1 meter (m)	1000 milligrams (mg) = 1 gram (g)
100 centimeters (cm) = 1 meter (m)	100 centigrams (cg) = 1 gram (g)
1000 meters (m) = 1 kilometer (km)	1000 grams (g) = 1 kilogram (kg)
Liquid and Dry Measure (Basic Unit Is 1 Liter)	
1000 milliliters (mℓ) = 1 liter (ℓ)	
100 centimeters (cℓ) = 1 liter (ℓ)	
1000 liters (ℓ) = 1 kiloliter (kℓ)	

Warm Ups E–H

E. A set of mixing bowls has capacity of 2 liters, 5 liters, and 8 liters. What is the total capacity of the set?

F. Change 22 qt to gallons and quarts.

Examples E–H

Directions: Solve.

Strategy: Describe each situation with a statement involving measurements and simplify.

E. Ben weighs 86 kg, Chris weighs 75 kg, and Scott weighs 91 kg. What is the total weight of the three boys?

Strategy: To find the total weight, add the weight of the three boys.

Total weight = 86 kg + 75 kg + 91 kg **Add.**
= (86 + 75 + 91) kg
= 252 kg

So the total weight of the three boys is 252 kg.

F. Change 154 sec to minutes and seconds.

Strategy: Divide by the number of seconds in a minute.

154 ÷ 60 = 2, R 34 **Since 60 sec = 1 min, divide by 60.**

So, 154 sec = 2 min 34 sec.

Answers to Warm Ups E. The capacity of the set is 15ℓ. F. 5 gal 2qt

G. If a carpenter cuts a piece of board that is 2 ft 5 in. from a board that is 8 ft 3 in. long, how much board is left? (Disregard the width of the cut.)

Strategy: Subtract the length cut off from the length of the board.

  8 ft 3 in.
− 2 ft 5 in.
remaining board

  7 ft 1 ft 3 in.   **Borrow 1 ft from the 8 ft (1 ft = 12 in.).**
− 2 ft   5 in.

  7 ft 15 in.   **1 ft 3 in. = 15 in.**
− 2 ft 5 in.   **Subtract.**
  5 ft 10 in.

There is 5 ft 10 in. of board remaining.

G. A wine maker draws off 5 gal 3 qt of wine from a full 15-gal keg. How much wine is left in the keg?

H. To run a mile race on the indoor track at the YMCA, the runners have to go around the track eight times. If Abbey runs around the track once, how many feet has she traveled?

Strategy: Divide 1 mile by 8 to find the length of a lap.

Distance = 1 mile ÷ 8
= 5280 ft ÷ 8   **Convert to feet.**
= 660 ft   **Divide.**

Abbey has run 660 ft.

H. A package of Kool-Aid makes 2 quarts of drink. How many cups of drink is this?

Answers to Warm Ups G. There are 9 gal 1 qt of wine left. H. There are 8 cups of Kool-Aid.

NAME ______________________ CLASS ____________ DATE ____________

Exercises 2.1

OBJECTIVE 1: *Multiply and divide a measurement by a number.*

A.

Multiply or divide the following.

1. (4 ft) · 6

2. (4 cups) · (5)

3. (200 mℓ) ÷ 25

4. (28 days) ÷ 4

5. (80 gal) ÷ 20

6. (55 mg) ÷ 5

B.

7. 3 · (317 oz)

8. 23 · (18 mℓ)

9. (400 hours) ÷ 8

10. (357 cm) ÷ 3

11. (2912 lbs) ÷ 14

12. (9105 gal) ÷ 15

13. (56 seconds) · (20)

14. (23 in.) · 174

OBJECTIVE 2: *Add and subtract measurements.*

A.

Add or subtract the following.

15. 6 lb + 14 lb

16. 5 m + 24 m

17. 5 yd + 8 yd + 4 yd

18. 6 hr + 7 hr + 8 hr

19. 32 g − 12 g

20. 20 mi − 13 mi

B.

21. 360 kℓ − 155 kℓ

22. 121 min − 72 min

23. 48 mm + 32 mm + 10 mm

24. 35 mℓ + 14 mℓ + 23 mℓ + 4 mℓ

25. 321 yd − 217 yd

26. 170 kg − 89 kg

27. 624 gal − 209 gal + 138 gal

28. 35 qt − 27 qt − 8 qt

29. 210 cm − 45 cm + 24 cm − 165 cm

30. 190 mi − 78 mi + 25 mi − 64 mi

C.

31. Estimate the length of your shoe in both inches and centimeters. Measure your shoe in both units.

32. Estimate the length of your middle finger in both inches and centimeters. Measure your middle finger in both units.

Do the indicated operations and simplify.

33. (6 ft 5 in.) + (2 ft 7 in.) + (10 ft 7 in.)

34. (7 lb 8 oz) + (2 lb 13 oz) + (11 lb 1 oz)

35. (21 min 39 sec) − (14 min 47 sec)

36. (4 yd 1 ft 5 in.) − (2 yd 2 ft 8 in.)

37. (35 min 12 sec) · 6

38. (2 yd 2 ft 1 in.) · 3

39. 2 yd 2 ft 6 in.
+ 3 yd 1 ft 8 in.

40. 2 gal 3 qt 1 pt
+ 4 gal 2 qt 1 pt

41. During one round of golf, Rick made birdie putts of 3 ft 8 in., 12 ft 10 in., 20 ft 8 in., and 7 ft 4 in. What was the total length of all the birdie putts?

42. The Corner Grocery sold 20 lb 6 oz of hamburger on Wednesday, 13 lb 8 oz on Thursday, and 21 lb 9 oz on Friday. How much hamburger was sold during the 3 days?

43. If a bag contains 298 grams of potato chips, how many grams are contained in 7 bags?

44. Lewis, a lab assistant, has 312 mℓ of acid that is to be divided equally among 24 students. How many milliliters will each student receive?

45. The local newspaper in Green Bay, Wisconsin, charted the overnight low temperatures for the past 5 nights. What was the average low temperature?

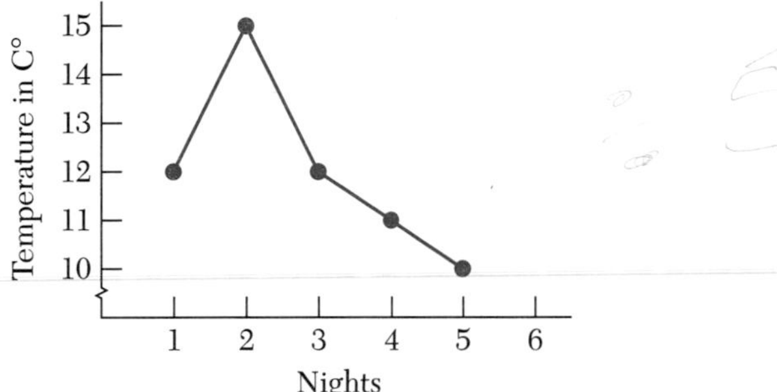

46. If a can of vegetable soup containing 48 oz is split evenly among 6 people, how large a serving will each person receive?

47. A doctor prescribes allergy medication of two tablets, 20 mg each, to be taken three times per day for a full week. How many milligrams of medication will the patient receive in a week?

48. An elevator has a maximum capacity of 2500 lb. A singing group of 8 men and 8 women get on. The average weight of the women is 125 lb, and the average weight of the men is 190 lb. Can they ride safely together?

49. The swimming pool at Tualatin Hills is 50 m long. How many meters of lane dividers should be purchased in order to separate the pool into 9 lanes?

50. A decorator is wallpapering. Each length of wallpaper is 7 ft 4 in. and 8 lengths are needed to cover a wall. If 3 walls are to be covered, how much total wallpaper is needed for the project?

51. The following table lists the longest rivers in the world.

River	Length (mi)
Nile (Africa)	4160
Amazon (South America)	4000
Chang Jiang (Asia)	3964
Ob-Irtysh (Asia)	3362
Huang (Asia)	2903
Congo (Africa)	2900

a. Which, if any, of these figures appear to be estimates? Why?
b. What is the total length of the five longest rivers in the world?
c. The São Francisco River in South America is the twentieth longest river in the world, with a length of 1988 miles. Write a sentence relating its length to that of the Amazon, using multiplication.
d. Write a sentence relating the lengths of the Nile River and the Congo River using addition or subtraction.

Exercises 52–54 refer to the chapter application (see page 149).

52. What are the dimensions of Paul and Barbara's back yard?

53. How wide is the patio? How long is it?

54. How wide are the walkways?

STATE YOUR UNDERSTANDING

55. Give two examples of equivalent measures.

56. Can you add 4 g to 5 in.? Explain how, or explain why you cannot do it.

57. If 8 in. + 10 in. = 1 ft 6 in., why isn't it true that 8 oz + 10 oz = 1 lb 6 oz?

NAME ______________________ CLASS __________ DATE __________

CHALLENGE

Precious metals and gems are measured in Troy Weight according to the following.

1 pennyweight (dwt) = 24 grains
1 ounce troy (oz t) = 20 pennyweights
1 pound troy (lb t) = 12 ounces troy

58. How many grains are in 1 ounce troy? How many grains are in 1 pound troy?

59. Suppose you have a silver bracelet that you want a jeweler to melt down and combine with the silver of two old rings to create a medallion that weighs 5 oz t 14 dwt. The bracelet weighs 3 oz t 18 dwt and one ring weighs 1 oz t 4 dwt. What does the second ring weigh?

60. Kayla has an ingot of platinum which weighs 2 lb t 8 oz t 15 dwt. She wants to divide it equally among her 6 grandchildren. How much will each piece weigh?

GROUP ACTIVITY

61. Measure, as accurately as you can, at least 5 parts of a typical desk in the classroom. Give measurements in both the English system and the metric system. Compare your results with the other groups. Do you all agree? Give some possible reasons for the variations in measurements.

MAINTAIN YOUR SKILLS (SECTIONS 1.1, 1.2, 1.5, 1.6, 1.7)

62. Round 4566 to the nearest ten.

63. Find the sum of 8, 56, 129, 35, and 604.

64. Replace the ? with $<$ or $>$.
a. 14 ? 89 **b.** 4,287,984 ? 4,287,884

65. The auditorium at Lake Community College holds 850. If all but 66 tickets for a concert have been sold, and each ticket costs $5.00, what is the total amount of money taken in?

66. Multiply: 784×10^4

67. Divide: $98{,}000{,}000 \div 10^3$

68. Simplify: $3^2\,(2 + 14 \div 2)$

69. Simplify: $(35 - 17) \div (1 + 5)$

70. Find the average of 5, 8, 12, and 15.

71. Find the average of 15, 18, 112, and 115.

2.2

Perimeter

OBJECTIVE

Find the perimeter of a polygon.

VOCABULARY

A **polygon** is any closed figure whose sides are line segments.

Polygons are named according to the number of sides they have. Table 2.3 lists some common polygons.

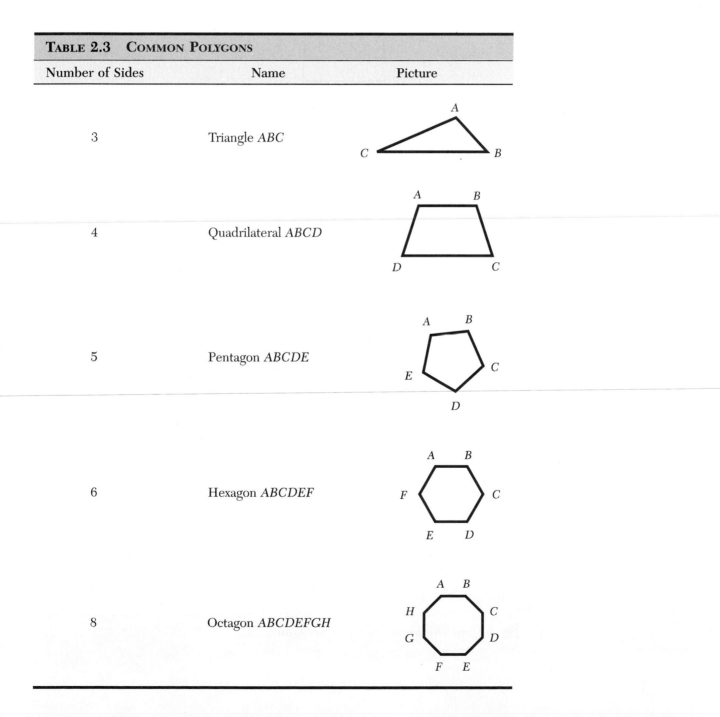

TABLE 2.3 COMMON POLYGONS

Number of Sides	Name	Picture
3	Triangle *ABC*	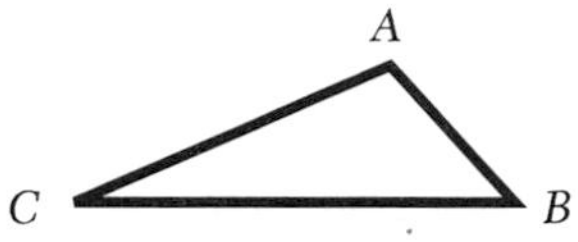 A, B, C
4	Quadrilateral *ABCD*	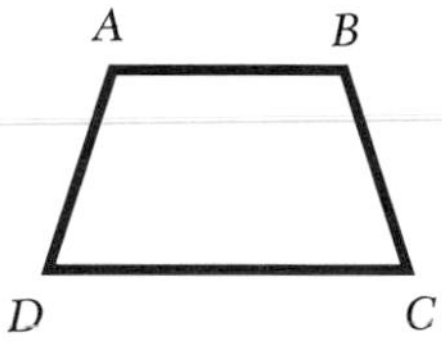 A, B, C, D
5	Pentagon *ABCDE*	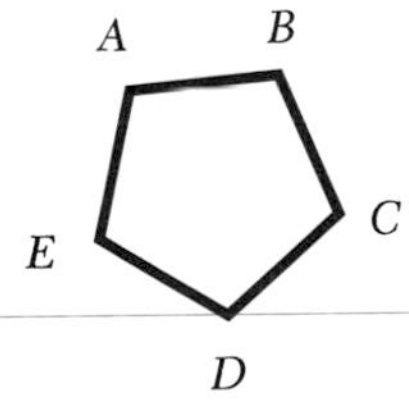 A, B, C, D, E
6	Hexagon *ABCDEF*	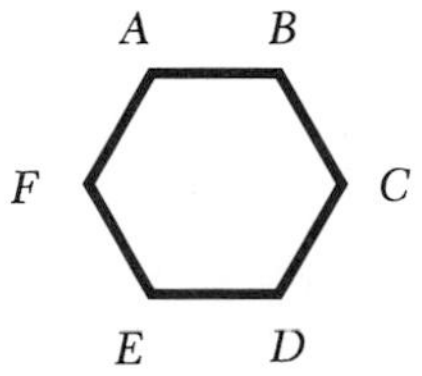 A, B, C, D, E, F
8	Octagon *ABCDEFGH*	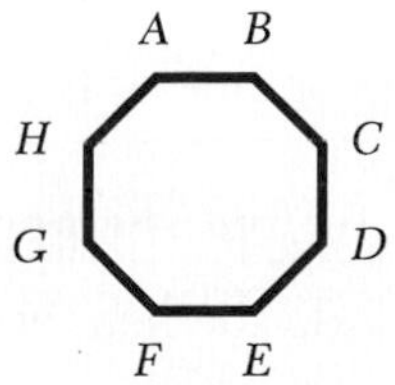 A, B, C, D, E, F, G, H

Quadrilaterals are polygons with four sides. Table 2.4 lists the characteristics of common quadrilaterals.

TABLE 2.4 COMMON QUADRILATERALS

Trapezoid		One pair of parallel sides
Parallelogram		Two pairs of equal parallel sides
Rectangle		A parallelogram with four right angles
Square		A rectangle with all sides equal

The **perimeter** of a polygon is the distance around the outside of the polygon.

HOW AND WHY

Objective

Find the perimeter of a polygon.

The *perimeter* of a figure can be thought of in terms of the distance traveled by walking around the outside of it or by the length of a fence around the figure. The units of measure used for perimeters are length measures (such as inches, feet, meters). Perimeter is calculated by adding the length of all the individual sides.

For example, to calculate the perimeter of this figure, we add the lengths of the sides.

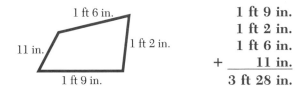

1 ft 6 in.
11 in.
1 ft 2 in.
1 ft 9 in.

1 ft 9 in.
1 ft 2 in.
1 ft 6 in.
+ 11 in.
3 ft 28 in.

28 in. = 2 ft 4 in., so
3 ft 28 in. = 3 ft + 2 ft 4 in.
= 5 ft 4 in.

The perimeter is 5 ft 4 in.

To find the perimeter of a polygon

Add the lengths of the sides.

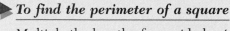

To find the perimeter of a square

Multiply the length of one side by 4.

$P = 4s$

To find the perimeter of a rectangle

Add twice the length and twice the width.

$P = 2\ell + 2w$

Examples A–D

Directions: Find the perimeters of the given polygons.

Strategy: Add the lengths of the sides.

A. Find the perimeter of the trapezoid.

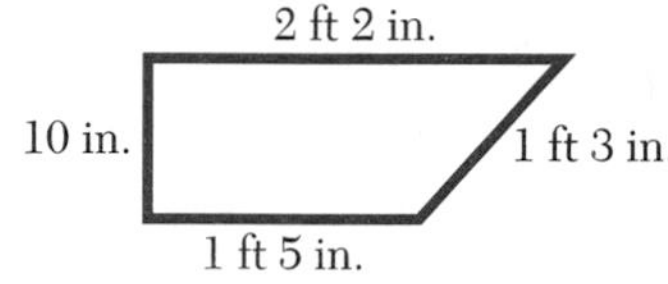

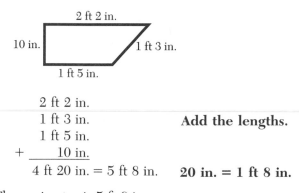

2 ft 2 in.	
1 ft 3 in.	**Add the lengths.**
1 ft 5 in.	
+ 10 in.	
4 ft 20 in. = 5 ft 8 in.	**20 in. = 1 ft 8 in.**

The perimeter is 5 ft 8 in.

B. Find the perimeter of the rectangle.

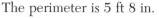

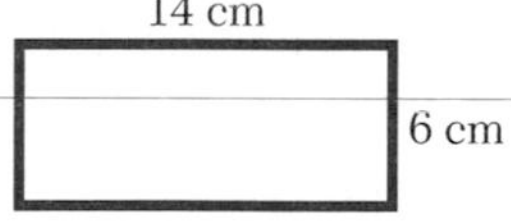

$P = 2\ell + 2w$	**Perimeter formula for rectangles.**
$= 2(14 \text{ cm}) + 2(6 \text{ cm})$	**Substitute.**
$= 28 \text{ cm} + 12 \text{ cm}$	**Multiply.**
$= 40 \text{ cm}$	**Add.**

The perimeter of the rectangle is 40 cm.

Warm Ups A–D

A. Find the perimeter of the polygon.

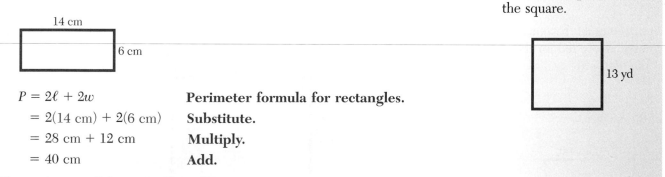

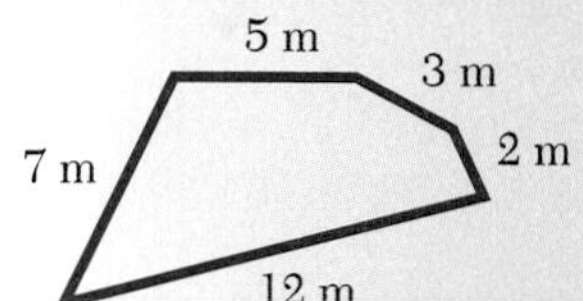

B. Find the perimeter of the square.

Answers to Warm Ups A. The perimeter is 29 m. B. The perimeter is 52 yd.

C. Find the perimeter of the polygon.

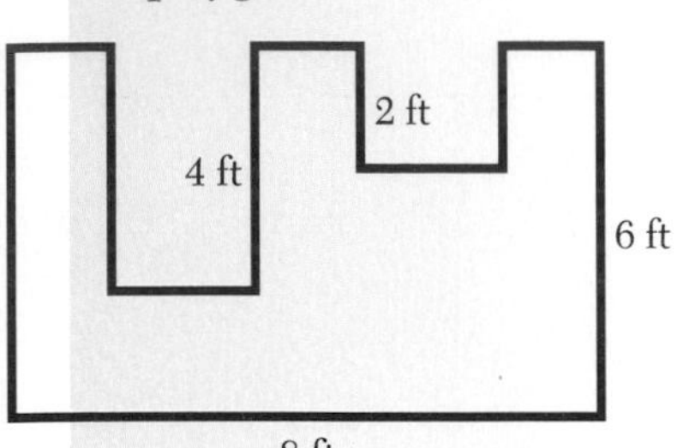

C. Find the perimeter of the polygon.

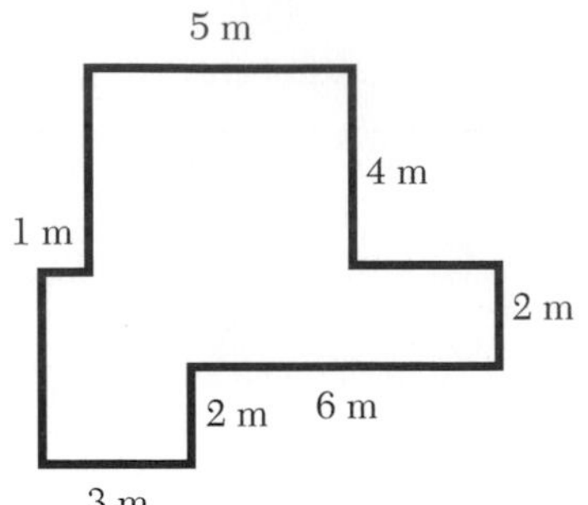

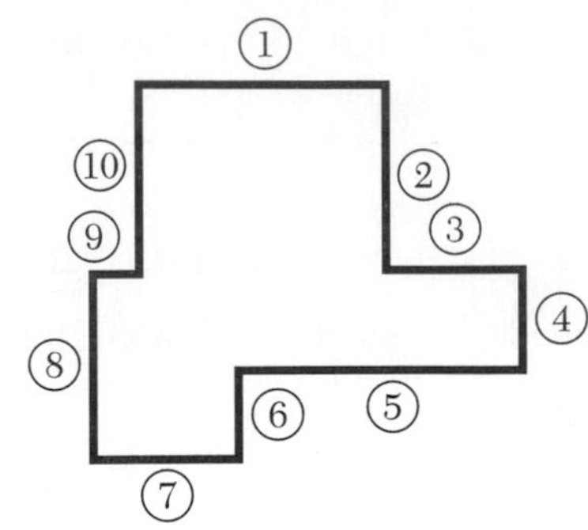

Side	Length
1	5 m
2	4 m
3	?
4	2 m
5	6 m
6	2 m
7	3 m
8	?
9	1 m
10	?
3	3 m
8 + 10	8 m

To find the perimeter, number the sides (there are 10) and write down their lengths.

To find the length of side 3, we determine that

side 3 = side 5 + side 7 − side 9 − side 1
= 6 m + 3 m − 1 m − 5 m
= 3 m

The lengths of sides 8 and 10 are not given but their sum can be found because

side 8 + side 10 = side 2 + side 4 + side 6
= 4 m + 2 m + 2 m
= 8 m

So,

$P = 5\text{ m} + 4\text{ m} + 3\text{ m} + 2\text{ m} + 6\text{ m} + 2\text{ m} + 3\text{ m} + 1\text{ m} + 8\text{ m}$
$= 34\text{ m}$

The perimeter of the polygon is 34 meters.

D. How much baseboard lumber is needed for the room pictured?

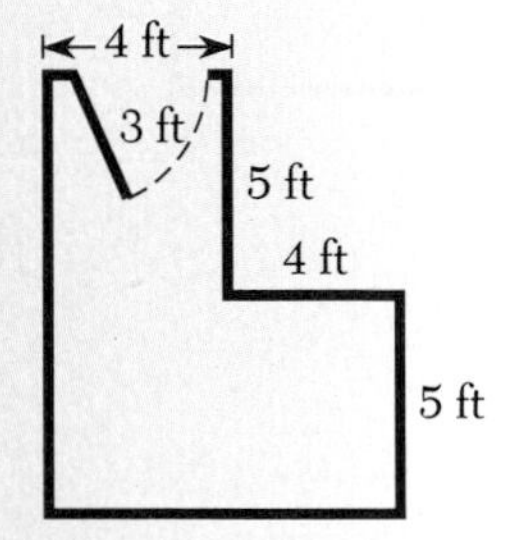

D. A carpenter is replacing the baseboards in a room. The floor of the room is pictured. How many feet of baseboard are needed?

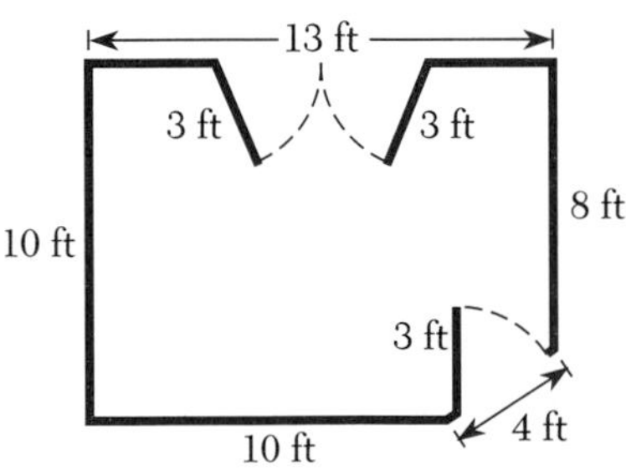

$P = 13\text{ ft} + 10\text{ ft} + 10\text{ ft} + 4\text{ ft} + 8\text{ ft}$
$P = 45\text{ ft}$

Find the perimeter of the room, including all the doors.

Baseboard $= 45\text{ ft} - (6\text{ ft} + 3\text{ ft})$
$= 45\text{ ft} - 9\text{ ft}$
$= 36\text{ ft}$

Subtract the combined width of the doors. Simplify.

Answers to Warm Ups C. The perimeter is 40 ft. D. The carpenter will need 33 ft of baseboard.

NAME ______________________ CLASS __________ DATE __________

Exercises 2.2

OBJECTIVE: *Find the perimeter of a polygon.*

A.

Find the perimeter of the following polygons:

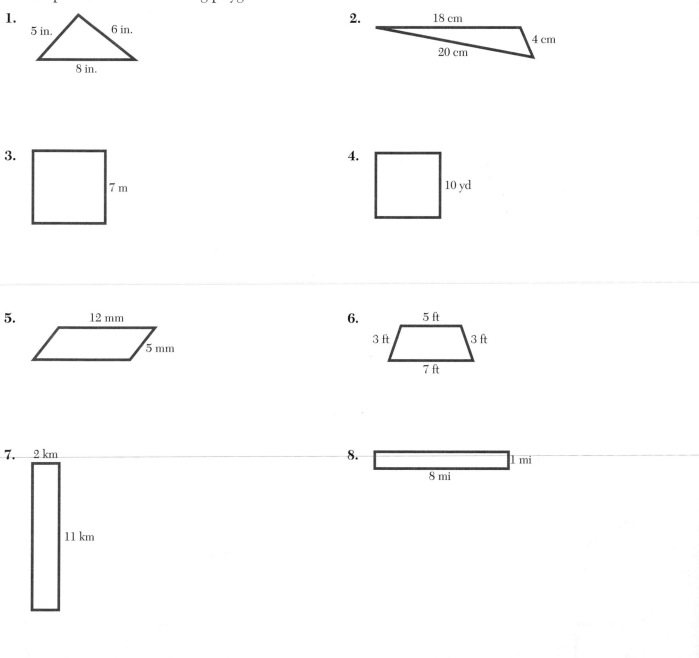

1.

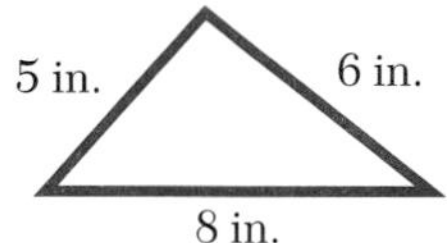

2.

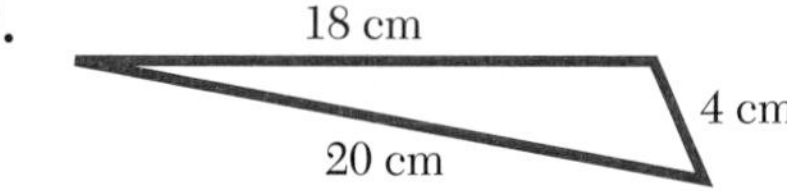

3.

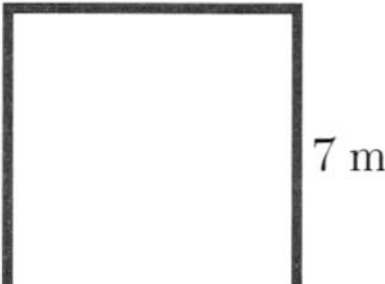

4.

5.

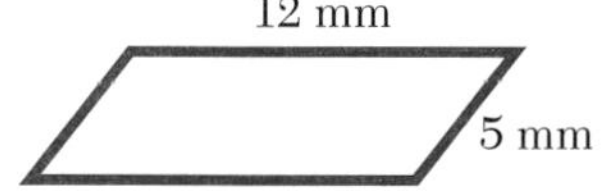

6.

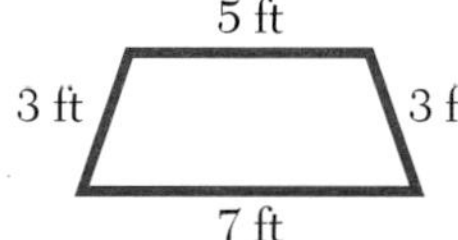

7.

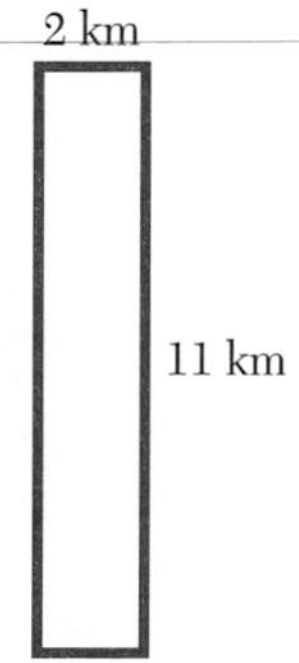

8.

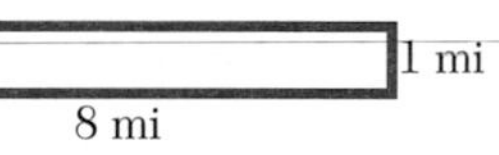

B.

9. Find the perimeter of a triangle with sides 16 mm, 27 mm, and 40 mm.

10. Find the perimeter of a square with sides 230 ft.

11. Find the distance around a rectangular field with length 45 m and width 35 m.

12. Find the distance around a rectangular swimming pool with width 20 yd and length 25 yd.

Find the perimeter of the following polygons:

13.

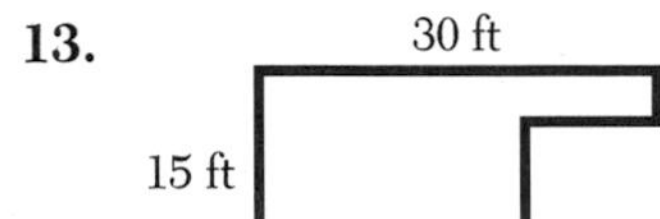

14.

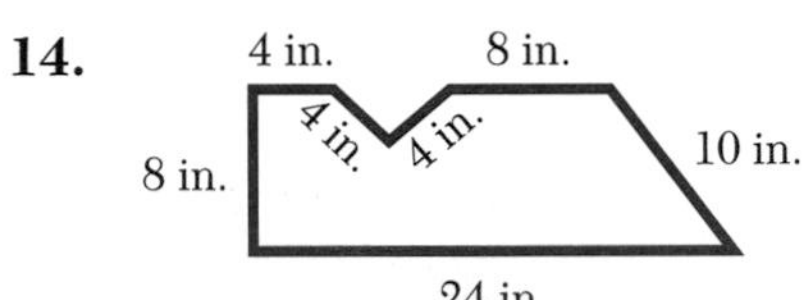

15.

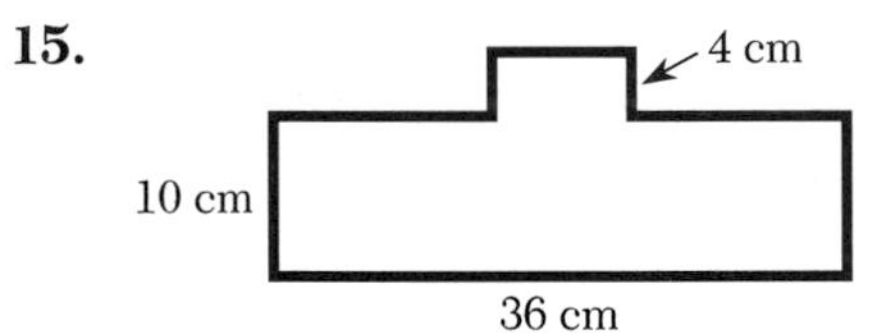

16.

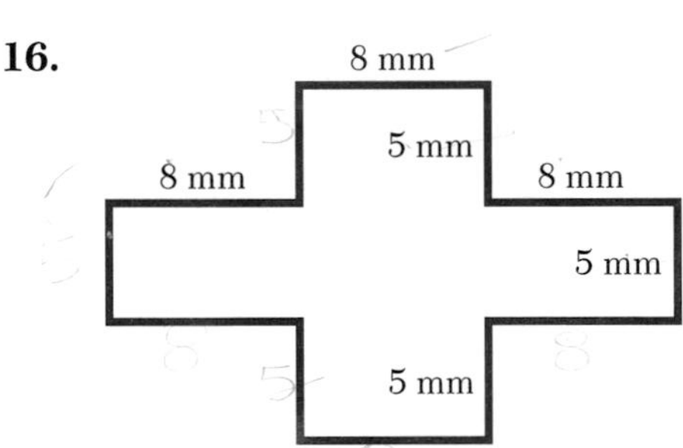

17.

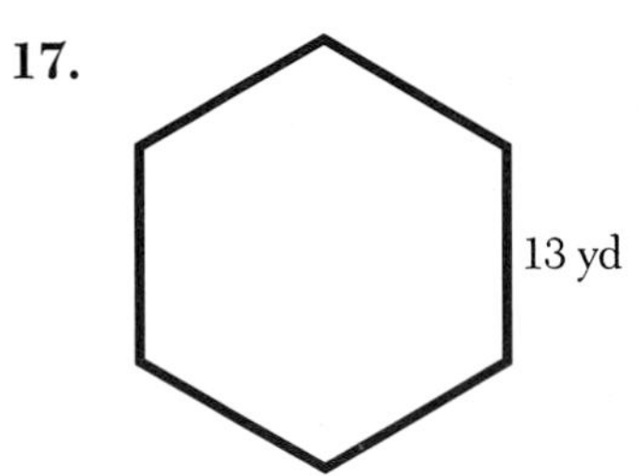

18.

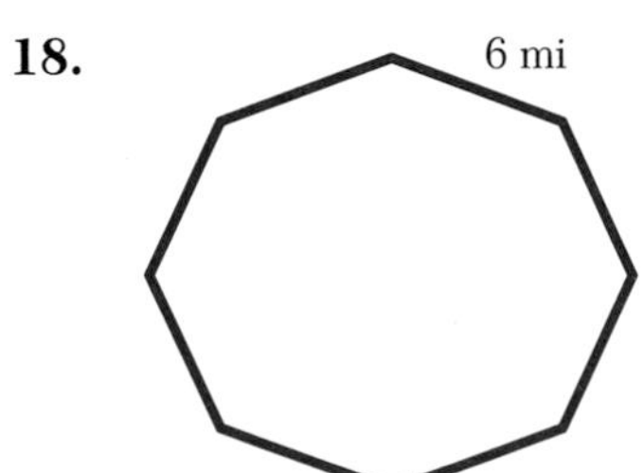

19.

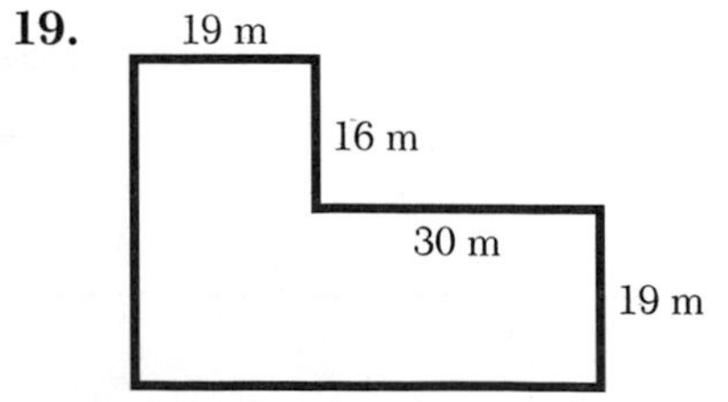

20.

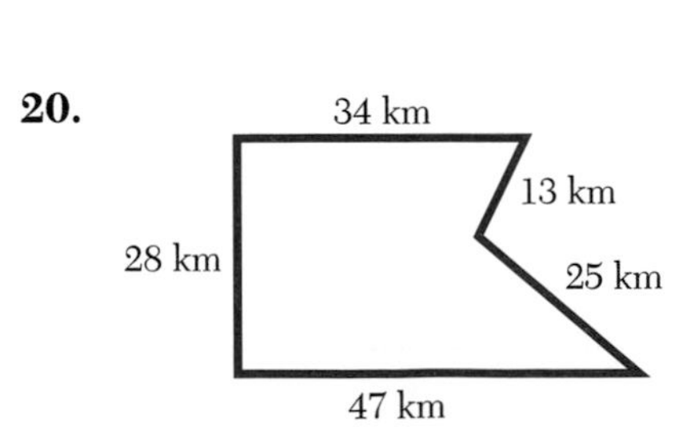

NAME ______________________ CLASS ____________ DATE ____________

C.

21.

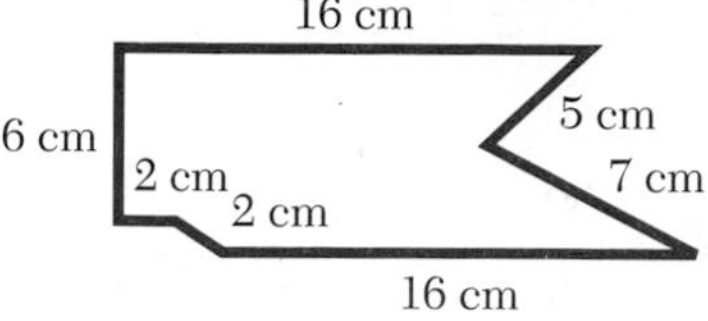

22.

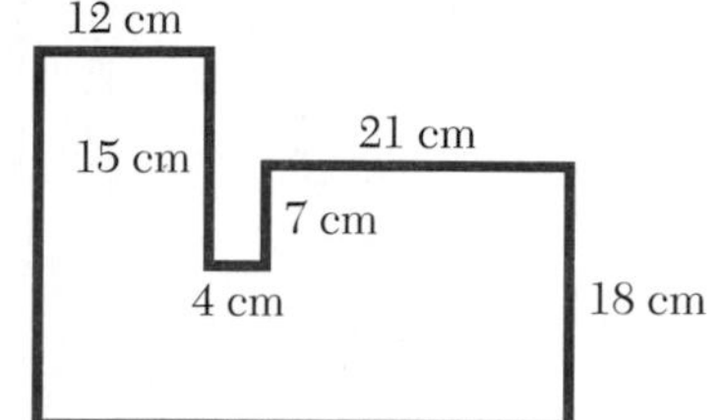

23.

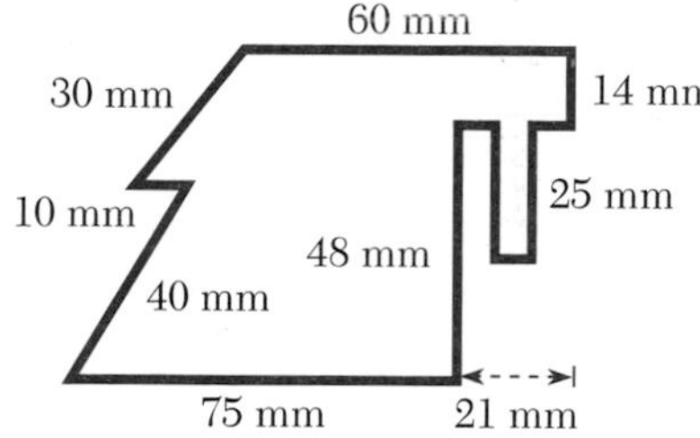

24.

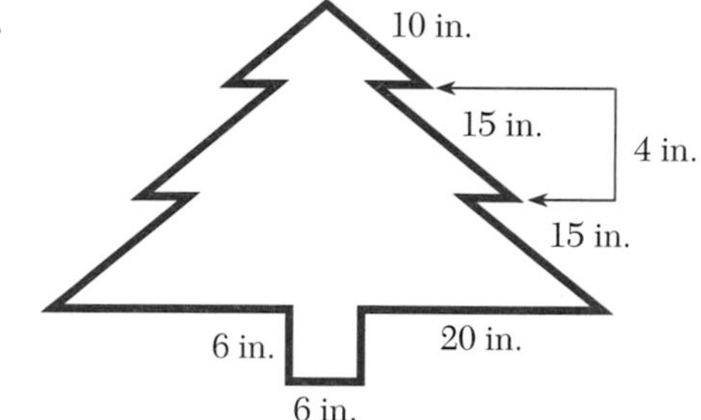

25. How many feet of picture molding is needed to frame four pictures, each measuring 8 in. by 10 in.? The molding is wide enough to require an extra inch added to each dimension to allow for the corners to be mitred.

26. If fencing costs $12 per meter, what will be the cost of fencing a rectangular lot that is 120 km long and 24 km wide?

27. If Hazel needs 2 minutes to put 1 ft of binding on a rug, how long will it take her to put the binding on a rug that is 15 ft by 12 ft?

28. How much fencing is needed to fence a lot that is rectangular in shape if it is 92 ft wide and 35 yd long?

29. Holli has a watercolor picture that is 14 in. by 20 in. She puts it in a mat that is 3 in. wide on all sides. What is the inside perimeter of the frame she needs to buy?

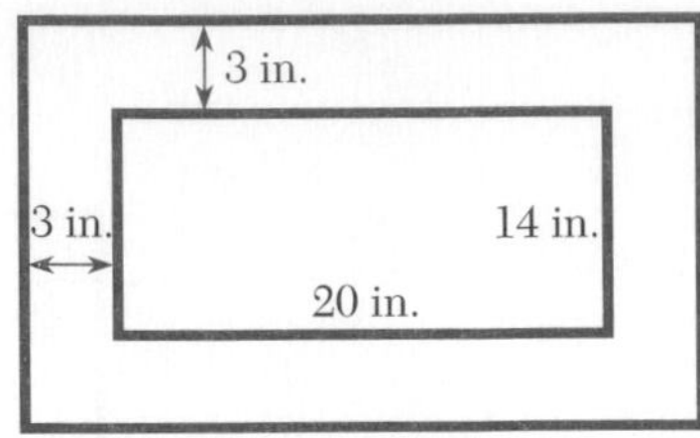

30. Jorge is lining the windows in his living room with Christmas lights. He has one picture window that is 5 ft 8 in. by 4 ft. On each side there is a smaller window that is 2 ft 6 in. by 4 ft. What length of Christmas lights does Jorge need for the three windows?

31. Jenna and Scott just bought a puppy and need to fence their back yard. How much fence should they order?

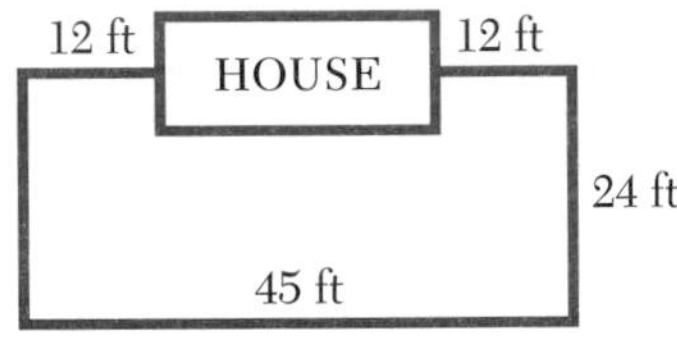

32. As a conditioning exercise, a soccer coach has his team run around the outside of the field three times. If the field measures 60 yd × 100 yd, how far did the team run?

33. A high school football player charted the number of laps he ran around the football field during the first 10 days of practice. If the field measures 120 yd by 53 yd, how far did he run during the 10 days? Convert your answer to the nearest whole mile.

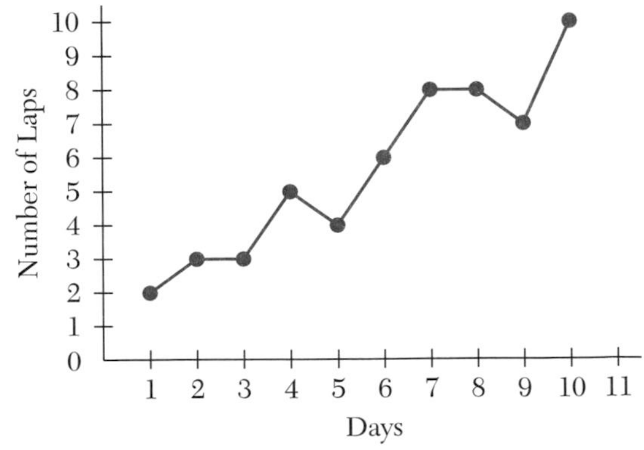

34. A carpenter is putting baseboard in the family room/dining room pictured below. How many feet of baseboard molding are needed?

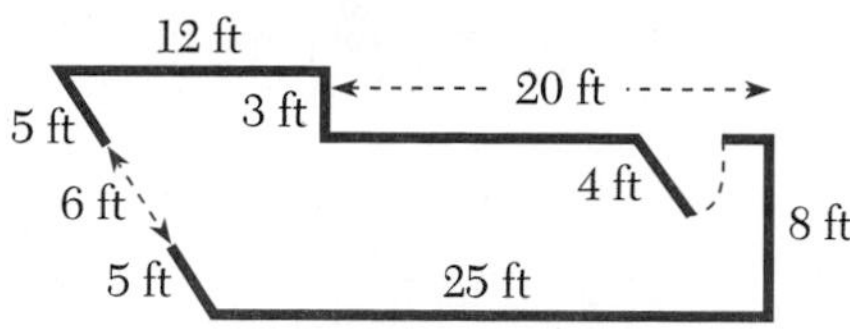

Exercises 35–38 refer to the chapter application (see page 149).

35. What is the shape of the patio?

36. What is the perimeter of the patio?

37. Estimate the perimeter of the flower bed that wraps around the right side of the patio.

38. Draw a line down the center of the back yard. What do you notice about the two halves? Mathematicians call figures like this symmetrical, because a center line divides the figure into "mirror images." Colonial architecture typically makes use of symmetry.

39. Are rectangles symmetrical around a line drawn lengthwise through the center? Are squares symmetrical?

40. Draw a triangle that is symmetrical, and another that is not.

STATE YOUR UNDERSTANDING

41. Explain what perimeter is and how the perimeter of the figure below is determined. What possible units (both English and metric) would the perimeter be likely to be measured in if the figure is a national park? If the figure is a room in a house? If the figure is a scrap of paper?

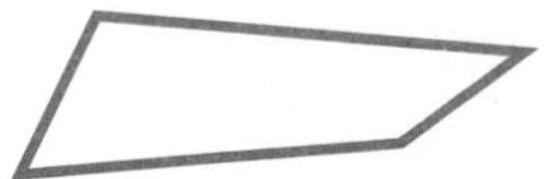

42. Is a rectangle a parallelogram? Why or why not?

43. Explain the difference between a square and a rectangle.

CHALLENGE

44. A farmer wants to build the goat pens pictured below. Each pen will have a gate 2 ft 6 in. wide on one end. What is the total cost of the pens if the fencing is \$3.00 per linear foot and each gate is \$15.00?

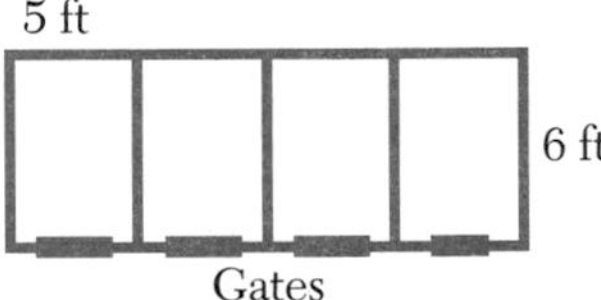

GROUP ACTIVITY

45. Federal Express will accept packages according to the formula:

$L_1 + 2L_2 + 2L_3 \leq 160$ in.

in which L_1 is the longest side, L_2 is the next longest side, and L_3 is the shortest side.

Make a table of dimensions of boxes that can be shipped by Federal Express. Try to find the maximum sizes possible. Use whole numbers.

Longest Side	Next Longest Side	Shortest Side	Total

NAME ______________________ CLASS ____________ DATE ____________

MAINTAIN YOUR SKILLS (SECTIONS 1.1, 1.2, 1.3, 1.4, 1.6, 1.7, 2.1)

46. Write the place value name for nine hundred thousand, fifty.

47. Write the place value name for nine hundred fifty thousand.

48. Round 32,571,600 to the nearest ten thousand.

49. Find the difference between 733 and 348.

50. Find the product of 733 and 348.

51. Find the quotient of 153,204 and 51.

52. Find the sum of 2 lb 3 oz and 3 lb 14 oz.

53. Find the difference of 6 ft and 7 in.

54. Find the average bowling score for Fred if he bowled games of 167, 182, and 146.

2.3

Area

OBJECTIVE

Find the area of common polygons.

VOCABULARY

Area is a measure of surface that is the amount of space inside a two-dimensional figure. It is measured in square units.

The **base** of a geometric figure is a side, parallel to the horizon.

The **altitude** or **height** of a geometric figure is the perpendicular distance from the base to the highest point of the figure.

Table 2.5 shows the base and altitude of some common geometric figures.

TABLE 2.5 BASE AND HEIGHT OF COMMON GEOMETRIC FIGURES

Figure		
Rectangle	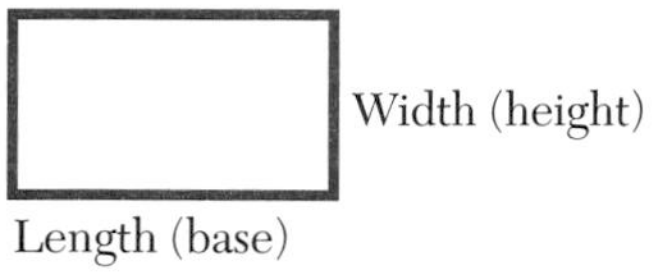 Width (height); Length (base)	
Triangle	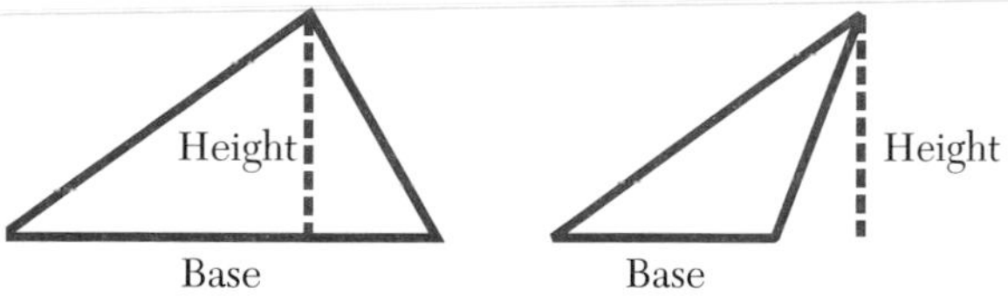 Height; Base	Height; Base
Parallelogram	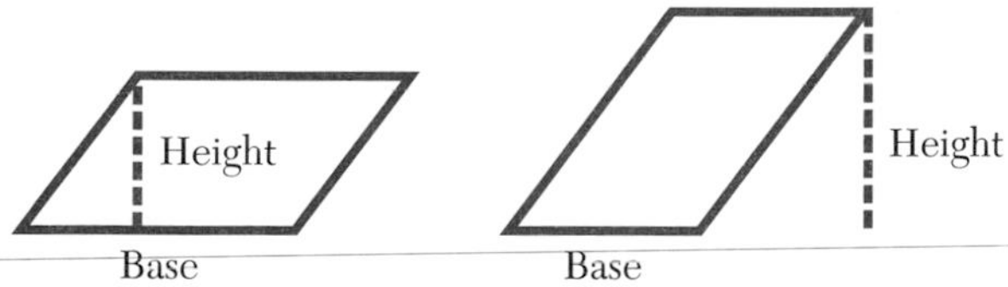 Height; Base	Height; Base
Trapezoid	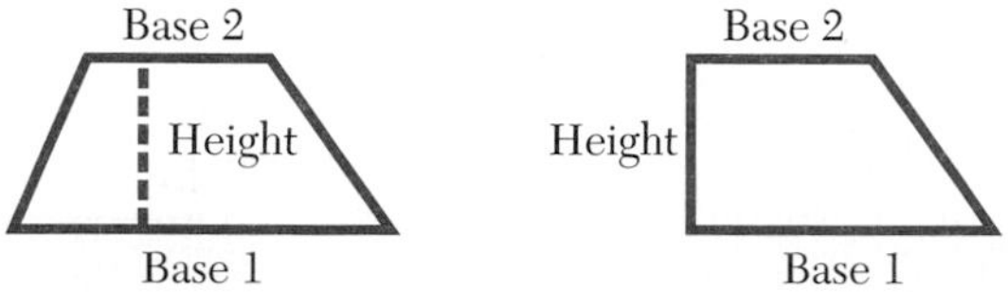 Base 2; Height; Base 1	Base 2; Height; Base 1

HOW AND WHY

Objective

Find the area of common geometric polygons.

Suppose you wish to tile a rectangular bathroom floor that measures 5 ft by 6 ft. The tiles are 1-ft-by-1-ft squares. How many tiles do you need?

Using the picture as a model of the tiled floor, you can count that 30 tiles are necessary.

Area is a measure of the surface, that is the amount of space inside a two-dimensional figure. It is measured in square units. Square units are literally squares that measure one unit of length on each side. For example,

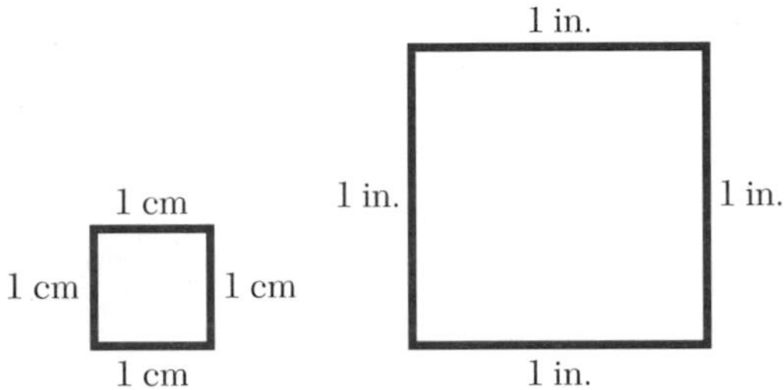

The measure of the square area on the left is 1 square centimeter, abbreviated cm^2. The measure of the square area on the right is 1 square inch, abbreviated in^2. The superscript in the abbreviation is simply part of the unit's name. It does not indicate an exponential operation.

CAUTION

$10\ cm^2 \neq 100\ cm$.

In the bathroom floor example, each tile has an area of 1 ft^2. So, the number of tiles needed is the same as the area of the room, 30 ft^2. We could have arrived at this number by multiplying the length of the room by the width. This is not a coincidence. It works for all rectangles.

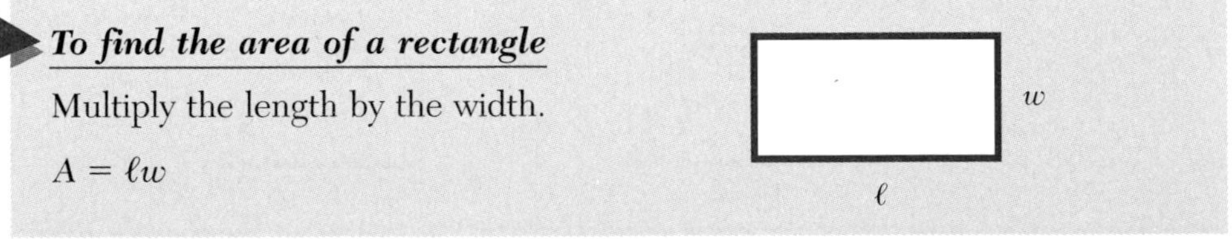

To find the area of a rectangle

Multiply the length by the width.

$A = \ell w$

To find the area of other geometric shapes, we use the area of a rectangle as a reference.

Because a square is a special case of a rectangle with all sides equal, the formula for the area of a square is

$$\begin{aligned} A &= \ell w \\ &= s(s) \\ &= s^2 \end{aligned}$$

To find the area of a square

Square the length of one of the sides.

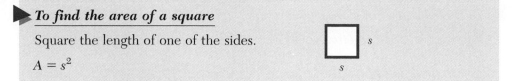

$A = s^2$

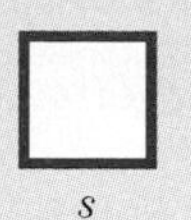

Now let's consider the area of a triangle. We start with a right triangle, that is, a triangle with one 90° angle.

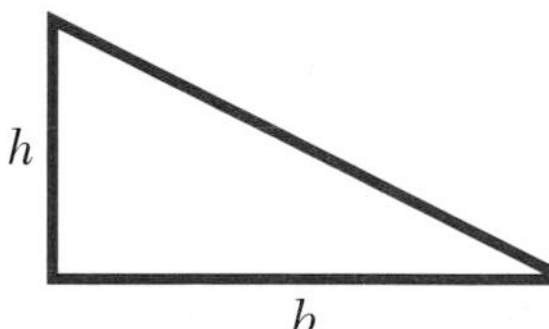

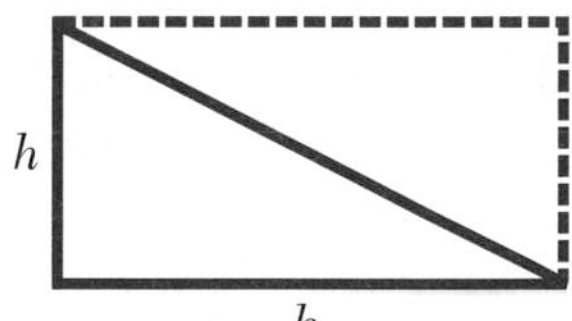

The triangle on the left has base b and height h. The figure on the right is a rectangle with length b and width h. According to the formula for rectangles, the area is $A = \ell w$ or $A = bh$. But the rectangle is made up of two triangles, both of which have a base of b and a height of h. Consequently, it stands to reason that the area of the rectangle (bh) is exactly twice the area of the triangle. So, we conclude that the area of the triangle is $(bh) \div 2$.

Now let's consider a more general triangle.

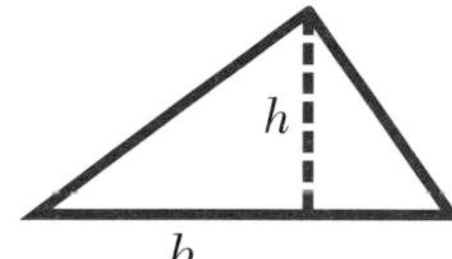

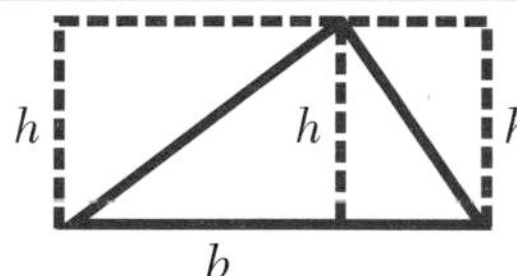

Again, the figure on the left is a triangle with base b and height h. And the figure on the right is a rectangle with length b and width h. Can you see that the rectangle must be exactly twice the area of the original triangle? So again we conclude that the area of the triangle is $A = (bh) \div 2$. It is possible to use this technique with any triangle.

To find the area of a triangle

Multiply the base times the height and divide by 2.

$A = (bh) \div 2$ or $A = \dfrac{bh}{2}$ or $A = \dfrac{1}{2}bh$

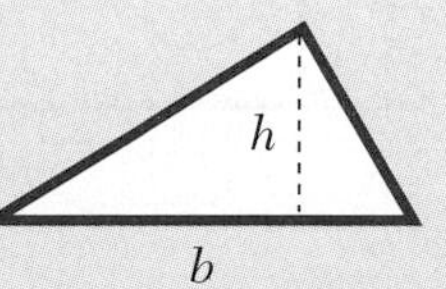

It is possible to use rectangles to find the formulas for the areas of parallelograms and trapezoids. This is left as an exercise.

To find the area of a parallelogram

Multiply the base times the height.

$A = bh$

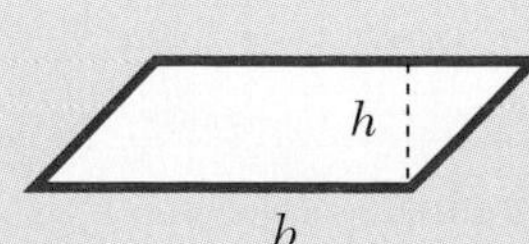

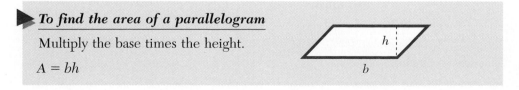

To find the area of a trapezoid

Add the two bases together, multiply by the height, then divide by 2.

$A = (b_1 + b_2)h \div 2$ or $A = \dfrac{(b_1 + b_2)h}{2}$

or

$A = \dfrac{1}{2}(b_1 + b_2)h$

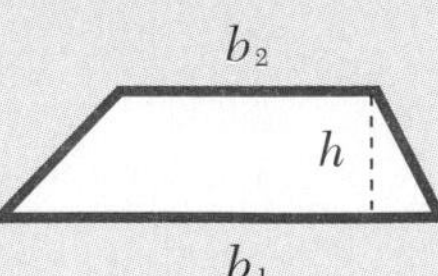

The lengths, widths, bases, sides, and heights must all be measured using the same units before the area formulas may be applied. If the units are different, convert to a common unit before calculating area.

Warm Ups A–F

A. Find the area of a square that is 11 cm on each side.

B. A decorator found a 15 yd^2 remnant of carpet. Will it be enough to carpet a 5-yd by 4-yd playroom?

C. Find the area of this triangle.

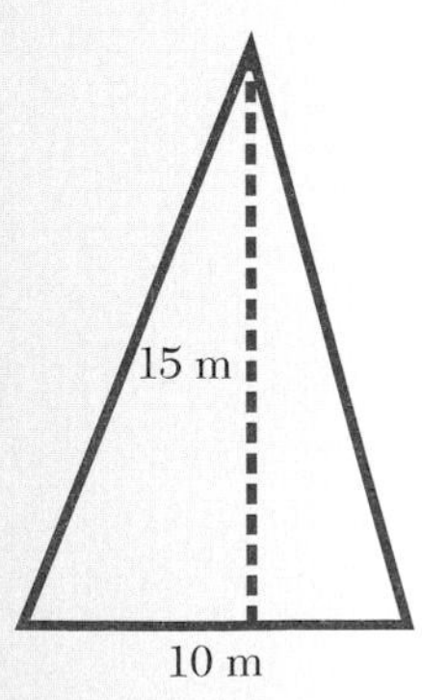

Examples A–F

Directions: Find the area.

Strategy: Use the area formulas.

A. Find the area of a square that is 5 in. on each side.

$A = s^2$ **Formula.**

$= 5^2 = 25$ **Substitute.**

The area is 25 in^2.

B. A gallon of deck paint will cover 400 ft^2. A contractor needs to paint a rectangular deck that is 26 ft long and 15 ft wide. Will one gallon of paint be enough?

$A = \ell w$ **Formula.**

$= 26 \text{ ft } (15 \text{ ft})$ **Substitute.**

$= 390 \text{ ft}^2$

Because $390 \text{ ft}^2 < 400 \text{ ft}^2$ one gallon of paint is enough to paint the deck.

C. Find the area of this triangle.

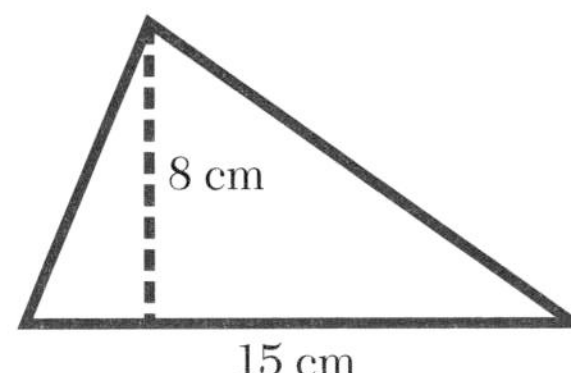

$A = \dfrac{bh}{2}$ **Formula.**

$= \dfrac{(15 \text{ cm})(8 \text{ cm})}{2}$ **Substitute.**

$= \dfrac{120 \text{ cm}^2}{2}$ **Simplify.**

$= 60 \text{ cm}^2$

The area is 60 cm^2.

Answers to Warm Ups A. The area is 121 cm^2. B. No, there will not be enough carpet. C. The area is 75 m^2.

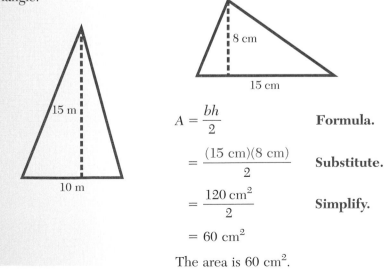

D. Find the area of the parallelogram with a base of 1 ft and a height of 4 in.

$A = bh$	**Formula.**
$= (1 \text{ ft})(4 \text{ in.})$	**Substitute.**
$= (12 \text{ in.})(4 \text{ in.})$	**Convert so that units match.**
$= 48 \text{ in.}^2$	**Simplify.**

The area of the parallelogram is 48 in^2.

D. Find the area of the parallelogram with a base of 6 yd and a height of 5 ft.

E. Find the area of the trapezoid pictured.

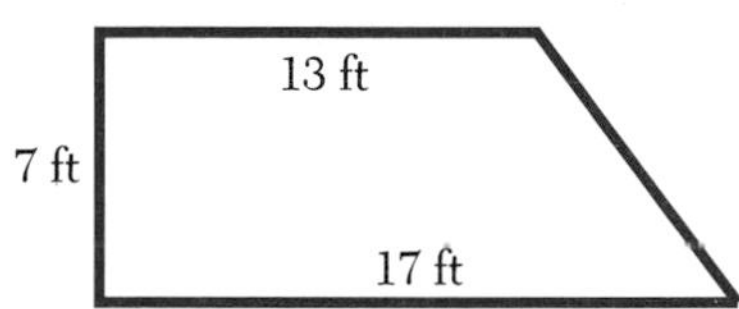

$A = (b_1 + b_2)h \div 2$	**Formula.**
$= (17 \text{ ft} + 13 \text{ ft})(7 \text{ ft}) \div 2$	**Substitute.**
$= (30 \text{ ft})(7 \text{ ft}) \div 2$	**Simplify.**
$= (210 \text{ ft}^2) \div 2$	
$= 105 \text{ ft}^2$	

The area is 105 ft^2.

E. Find the area of the trapezoid with bases of 51 m and 36 m and a height of 12 m.

F. Find the area of the polygon.

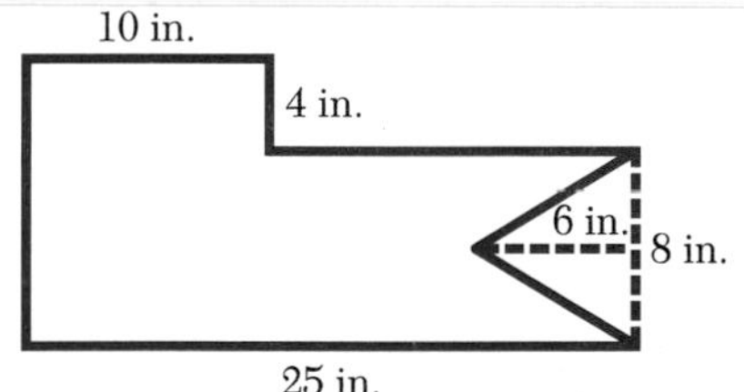

Strategy: To find the area of a polygon that is a combination of two or more common figures, first divide it into the common figure components.

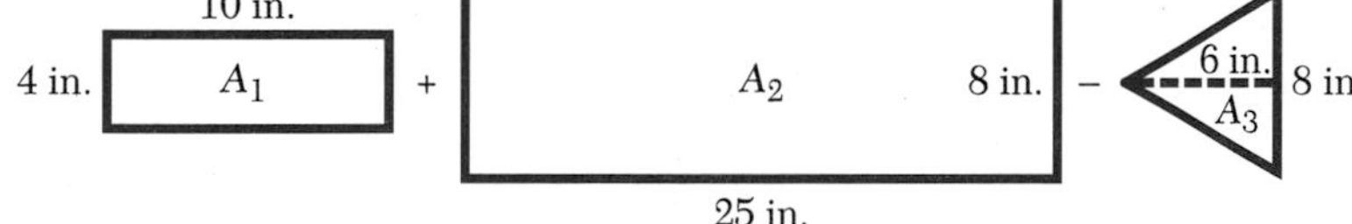

Total area $= A_1 + A_2 - A_3$	**Divide into component figures.**
$A_1 = (10 \text{ in.})(4 \text{ in.}) = 40 \text{ in}^2$	**Compute the areas of each component.**
$A_2 = (8 \text{ in.})(25 \text{ in.}) = 200 \text{ in}^2$	
$A_3 = (6 \text{ in.})(8 \text{ in.}) \div 2$	
$= (48 \text{ in}^2) \div 2 = 24 \text{ in}^2$	
Total area $= 40 \text{ in}^2 + 200 \text{ in}^2 - 24 \text{ in}^2$	**Combine the areas.**
$= 216 \text{ in}^2$	

The area of the figure is 216 in^2.

F. Find the area of the polygon.

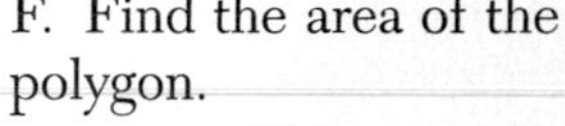

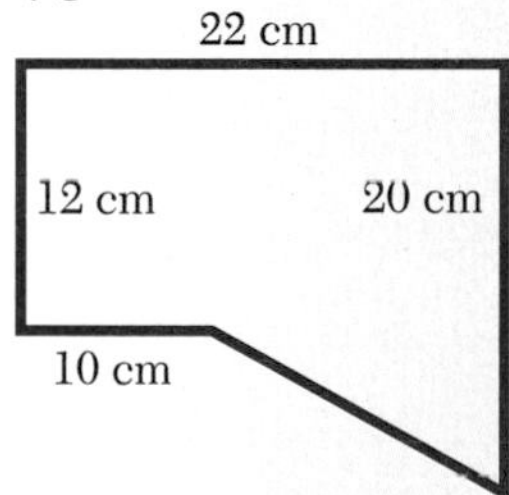

Answers to Warm Ups D. The area is 90 ft^2. E. The area is 522 m^2. F. The area is 312 cm^2.

NAME ______________________ CLASS ____________ DATE ____________

Exercises 2.3

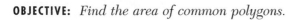

OBJECTIVE: *Find the area of common polygons.*

A.

Find the area of the following figures:

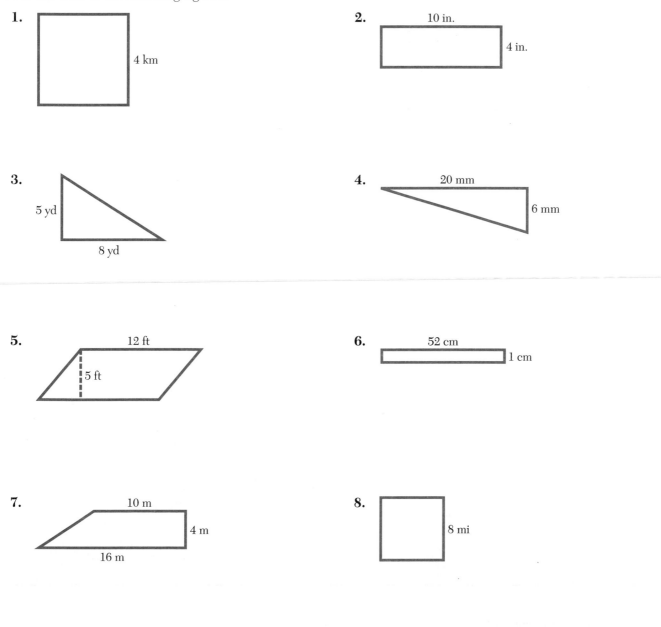

1.

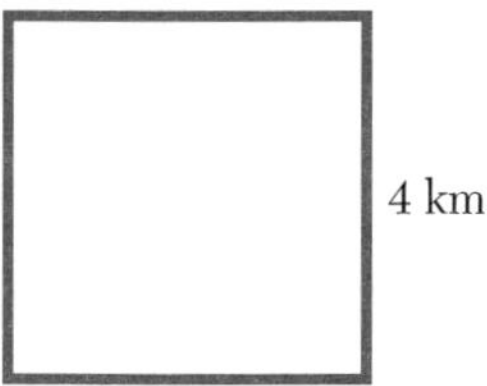

2.

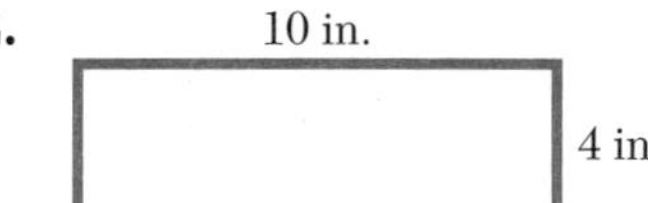

3.

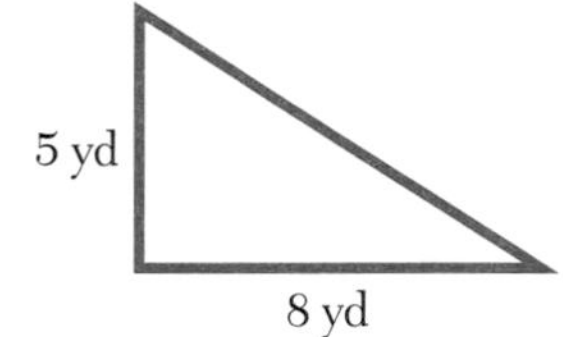

4.

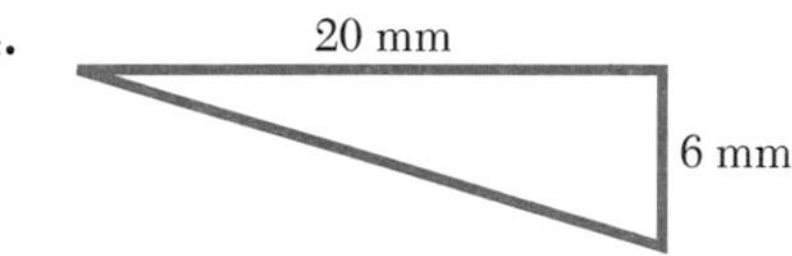

5.

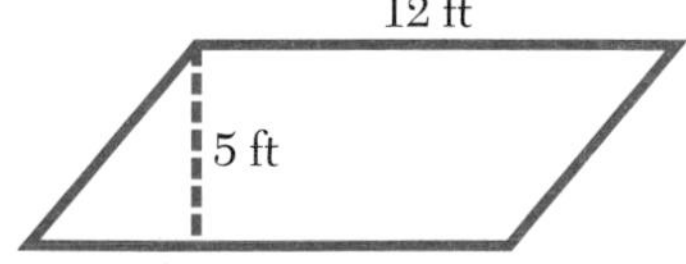

6.

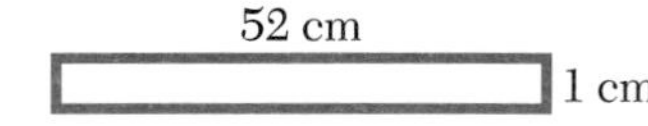

7.

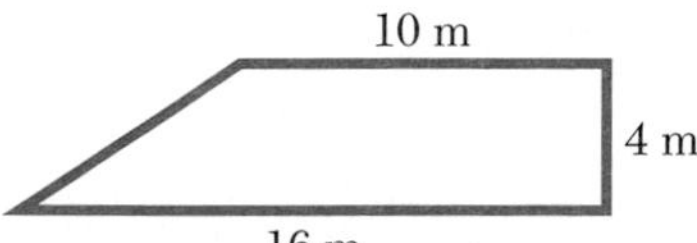

8.

B.

9. Find the area of a rectangle that has a length of 12 km and a width of 11 km.

10. Find the area of a square with sides of 35 cm.

11. Find the area of a triangle with base 28 yd and height 17 yd.

12. Find the area of a parallelogram with base 56 in. and height 32 in.

Find the areas of the polygons:

13.

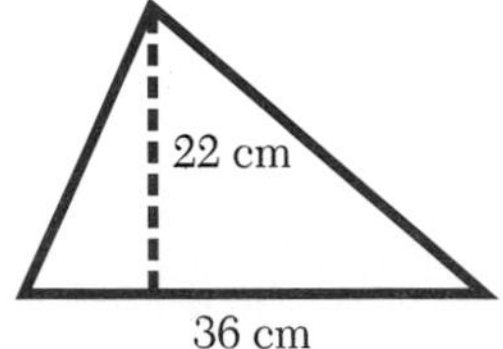

14.

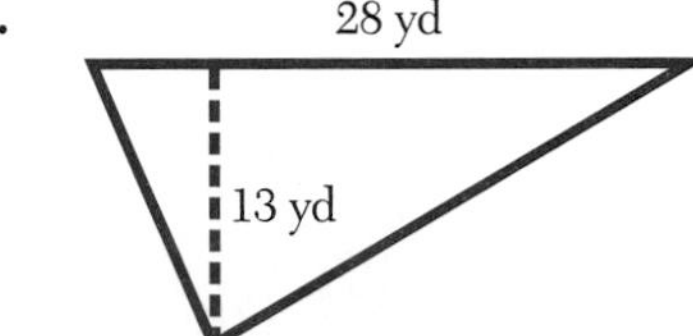

15.

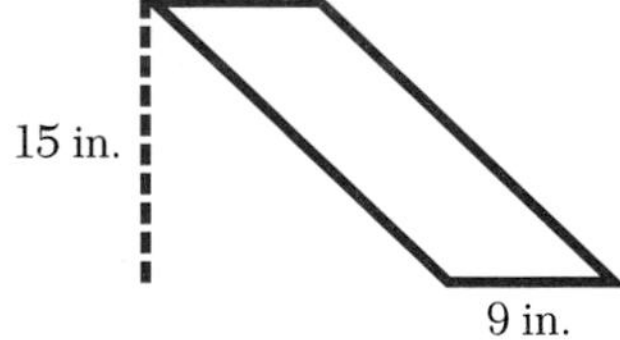

16.

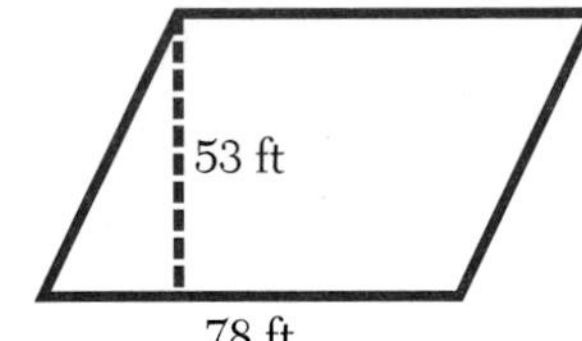

17.

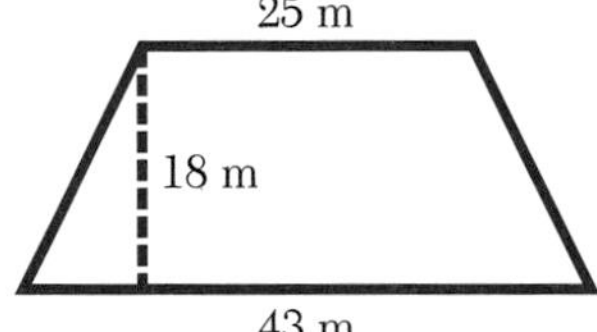

18.

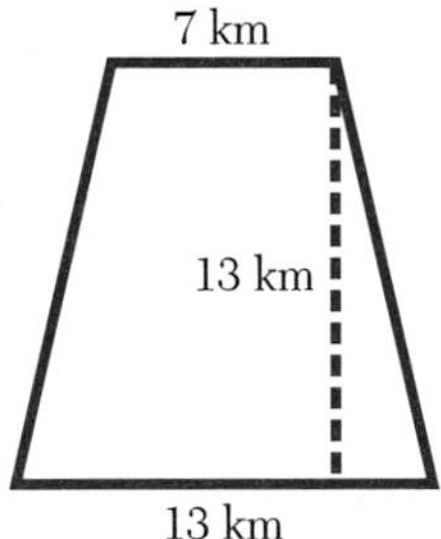

C.

Find the areas of the polygons:

19.

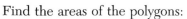

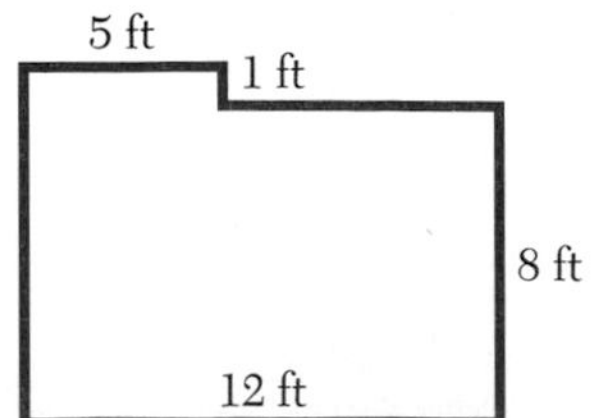

20.

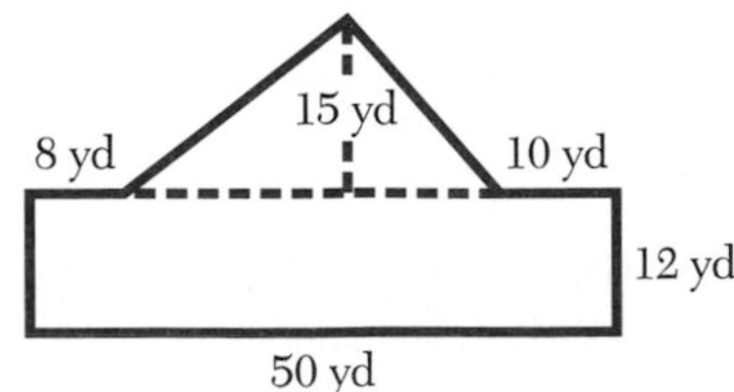

NAME ______________________ CLASS ____________ DATE ____________

21.

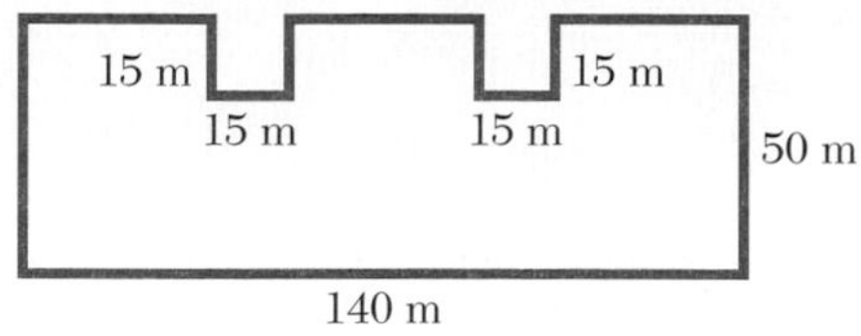

22.

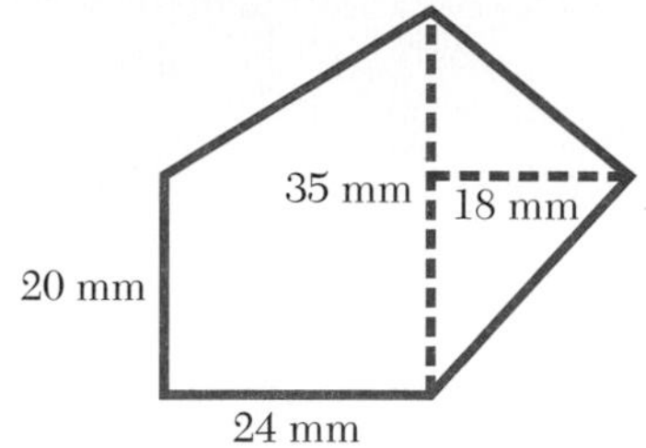

23.

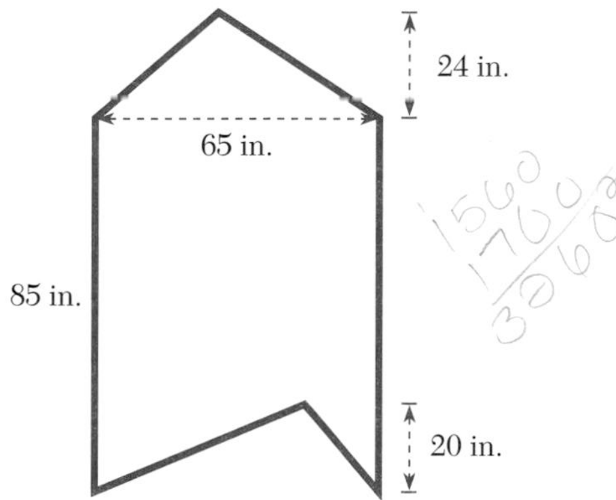

24.

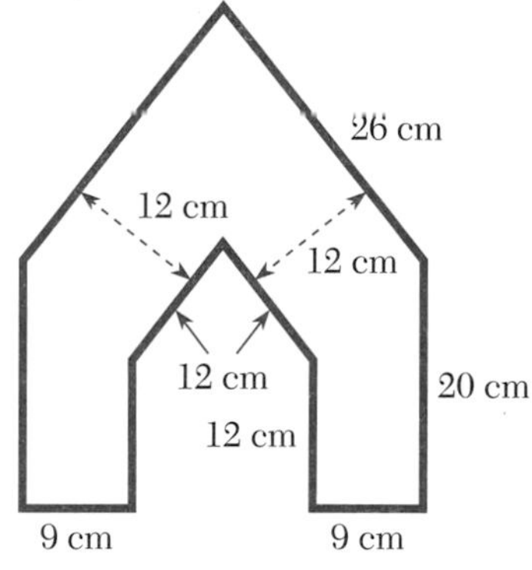

25. Use a picture to decide how many ft^2 there are in one yd^2.

26. Use a picture to decide how many cm^2 there are in one m^2.

27. The side of Jane's house that has no windows measures 35 ft by 22 ft. If 1 gallon of stain will cover 250 ft^2, will 2 gallons of stain be enough to stain this side?

28. The south side of Jane's house measures 85 ft by 22 ft and has two windows, each 4 ft by 6 ft. Will 4 gallons of the stain in Exercise 27 be enough for this side?

29. If 1 ounce of weed killer treats one square meter of lawn, how many ounces of weed killer will Debbie need to treat a rectangular lawn that measures 30 m by 8 m?

30. To the nearest acre, how many acres are contained in a rectangular plot of ground if the length is 1850 ft and the width is 682 ft? (43,560 ft^2 = 1 acre)

31. How much glass is needed to replace a set of 2 sliding glass doors that each measure 3 ft by 6 ft?

32. How many square yards of carpet are needed to cover a rectangular room that is 9 ft by 12 ft?

33. How many square feet of sheathing is needed for the gable end of a house that has a rise of 9 ft and a span of 36 ft? (See the drawing.)

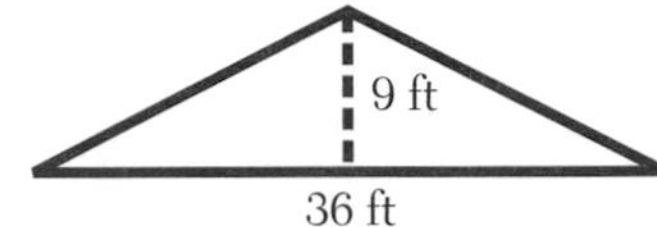

34. A farmer wants to construct a small shed that is 6 ft × 9 ft around the base and 8 ft high. How many gallons of paint will be needed to cover the outside of all four walls of the shed? (Assume 1 gallon covers 250 ft^2.)

35. How much padding is needed to make a pad for the hexagonal table pictured?

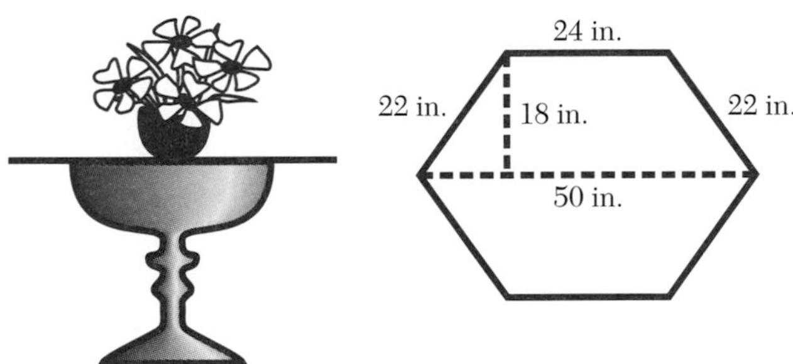

36. A window manufacturer is reviewing the plans of a home to determine the amount of glass needed to fill the order. The number and size of the windows and sliding glass doors are listed in the table. How much glass does he need to fill the order?

	Dimensions	Number Needed
Windows	3 ft × 3 ft	4
	3 ft × 4 ft	7
	3 ft × 5 ft	2
	4 ft × 4 ft	2
	5 ft × 6 ft	1
Sliding Doors	7 ft × 3 ft	2

NAME ______________________ CLASS ____________ DATE ____________

37. One 2-lb bag of wildflower seed will cover 70 ft^2. How many bags of seed are needed to cover the region pictured below? (Do not seed the shaded area.)

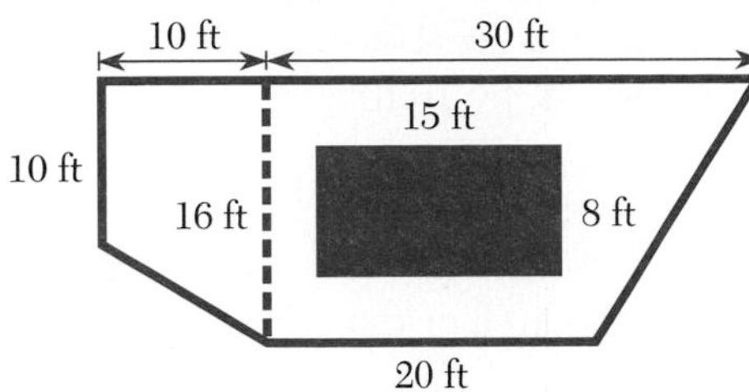

Exercises 38–39 refer to the chapter application (see page 149).

38. The contractor who was hired to build the brick patio begins by pouring a concrete slab. Then he will put the bricks on the slab. The estimate of both the number of bricks and the amount of mortar needed is based on the area of the patio and walkways. Subdivide the patio and the walkways into geometric figures, then calculate the total area to be covered in bricks.

39. The number of bricks and amount of mortar needed also depend on the thickness of the mortar between the bricks. The plans specify a joint thickness of a quarter inch. According to industry standards, this will require 7 bricks per square foot. Find the total number of bricks required for the patio and walkways.

STATE YOUR UNDERSTANDING

40. What kinds of units measure area? Give examples from both systems.

41. Explain how to calculate the area of the figure below. Do not include the shaded portion.

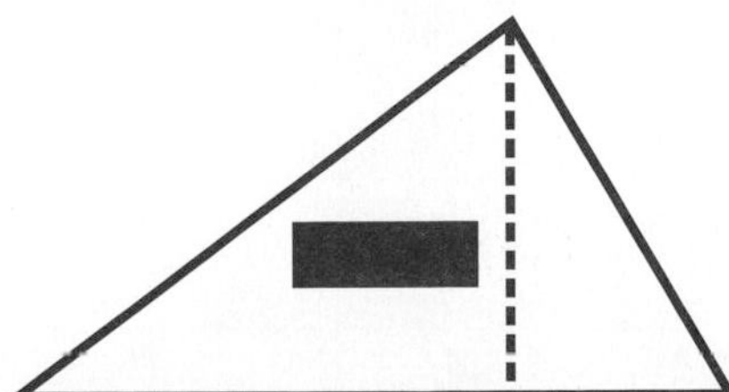

42. Describe how you could approximate the area of a geometric figure using 1-in. squares.

CHALLENGE

43. Joe is going to cover his kitchen floor. Along the outside he will put black squares that are 6 in. on each side. The next (inside) row will be white squares that are 6 in. on each side. The remaining inside rows will alternate between black and white squares that are 1 ft on each side. How many squares of each color will he need for the kitchen floor that measures 9 ft by 10 ft?

44. A rectangular plot of ground measuring 120 ft by 200 ft is to have a cement walk 5 ft wide placed around the inside of the perimeter. How much of the area of the plot will be used by the walk and how much of the area will remain for the lawn?

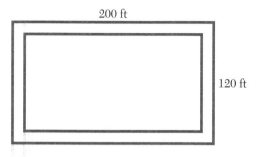

45. Ingrid is going to carpet two rooms in her house. The floor in one room measures 30 ft by 24 ft and the floor in the other room measures 22 ft by 18 ft. If the carpet costs $27.00 per square yard installed, what will it cost Ingrid to have the carpet installed?

GROUP ACTIVITY

46. Use the formula for the area of a rectangle to show how to find the area of a parallelogram and a trapezoid. Draw pictures to illustrate your argument.

47. Determine the coverage of 1 gallon of semi-gloss paint. How much of this paint is needed to paint your classroom, excluding chalkboards, windows, and doors? What would it cost? Compare your results with the other groups in the class. Did all the groups get the same results? Give possible explanations for the differences.

MAINTAIN YOUR SKILLS (SECTIONS 1.1, 1.2, 1.3, 1.4, 1.5, 1.6, 1.7, 1.9, 2.2)

48. Find the product of 47 and 962. Round your answer to the nearest 100.

49. Find the quotient of 295,850 and 97.

50. Simplify: $2(46 - 28) + 50 \div 5$

51. A family of six attends a weaving exhibition. Parking was $4, adult admission to the exhibition $6, senior admission $4, and child admission $3. How much does it cost for 2 parents, 1 grandmother and 3 children to attend the exhibition?

NAME ______________________ CLASS ____________ DATE ____________

52. Simplify: $3^2 + 2^3$

53. Find the sum: 3467 + 12 + 946

54. Find the average: 16, 17, 18, 19, 210

55. Find the perimeter of a rectangle that is 4 in. wide and 2 ft long.

56. The table lists calories burned per hour for various activities. Make a bar graph that summarizes this information.

Activity	Calories Burned/Hr
Sitting	50
Slow Walking	140
Stacking Firewood	360
Hiking	500

57. Find the difference: 10,102 − 7968

2.4 Volume

OBJECTIVE

Find the volume of common geometric shapes.

VOCABULARY

A **cube** is a three-dimensional geometric solid that has six sides (called **faces**), each of which is a square.

Volume is the name given to the amount of space that is contained inside a three-dimensional object.

HOW AND WHY

Objective

Find the volume of common geometric shapes.

Suppose you have a shoe box that measures 12 in. long by 4 in. wide by 5 in. high that you want to use to store toy blocks that are 1 in. by 1 in. by 1 in. How many blocks will fit in the box?

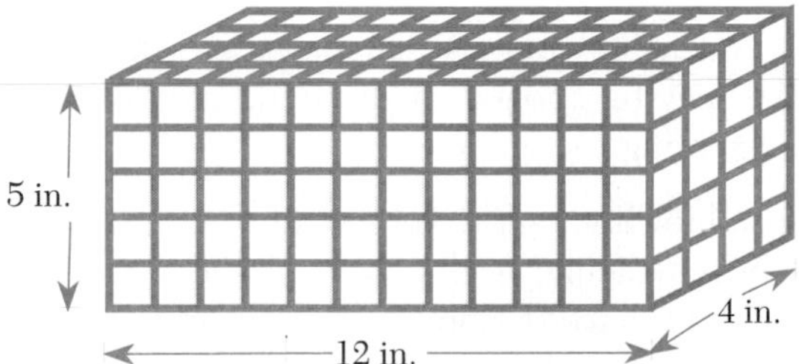

In each layer of blocks there are 12(4) = 48 blocks and there are five layers. Therefore, the box holds 48(5) = 240 blocks.

Volume is a measure of the amount of space that is contained in a three-dimensional object. Often, volume is measured in cubic units. These units are literally *cubes* that measure one unit on each side. For example, pictured below is a cubic inch (1 in^3) and a cubic centimeter (1 cm^3).

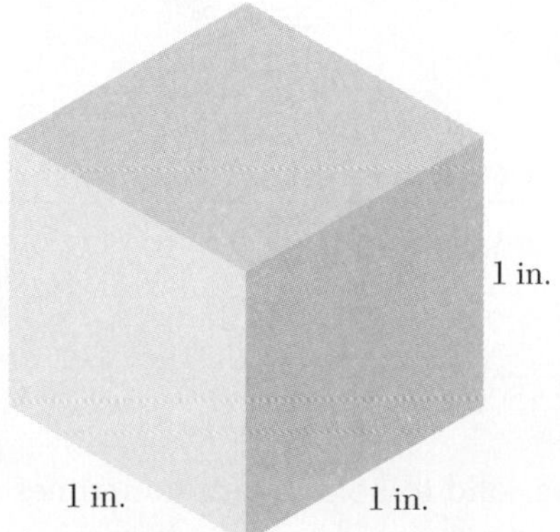

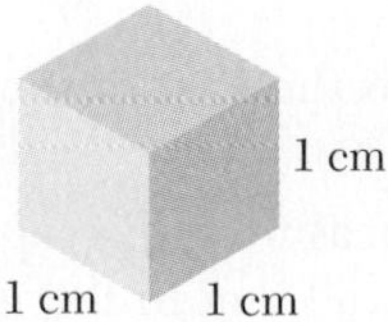

The shoebox discussed previously has a volume of 240 in^3 because exactly 240 blocks, which have volume 1 in^3, can fit inside it and totally fill it up.

In general, volume can be thought of as the number of cubes that fill up a space. If the space is a rectangular solid, like the shoebox, it is a relatively easy matter to determine the volume by making a layer of cubes that covers the bottom, and then deciding how many layers are necessary to fill the box. Note that the number of cubes needed for the bottom layer is the same as the area of the base of the box, ℓw. The number of layers needed is the same as the height of the box, h. So, we come to the following volume formula.

To find the volume of a rectangular solid

Multiply the length by the width by the height.

$V = \ell wh$

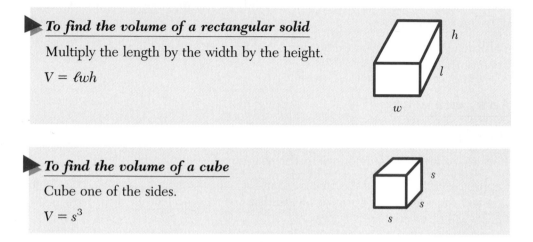

To find the volume of a cube

Cube one of the sides.

$V = s^3$

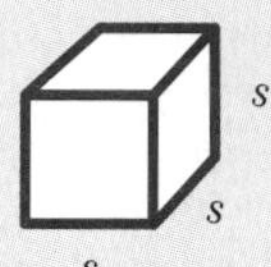

The length, width, and height must all be measured using the same units before the volume formulas may be applied. If the units are different, convert to a common unit before calculating the volume.

The principle used for finding the volume of a box can be extended to any solid with sides that are perpendicular to the base. The area of the base gives the number of cubes necessary to make the bottom layer, and the height gives the number of layers necessary to fill the solid.

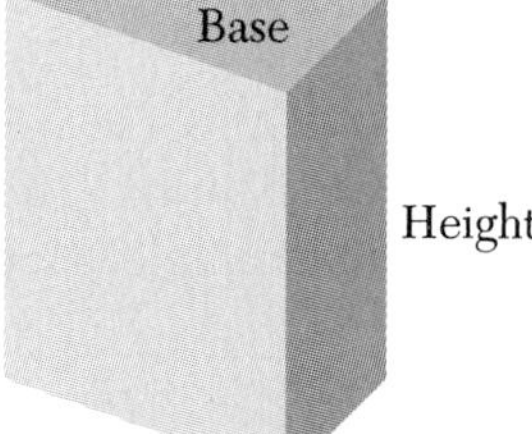

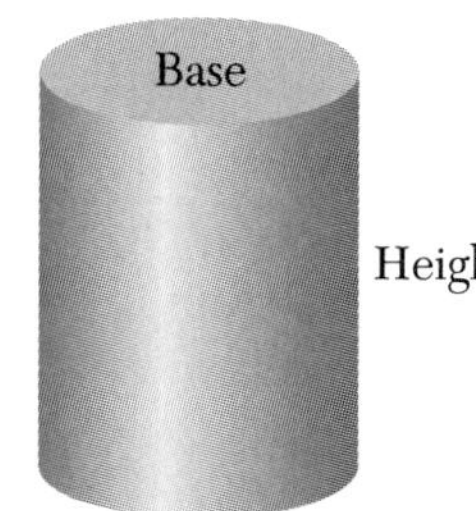

To find the volume of a solid with sides perpendicular to the base

Multiply the area of the base by the height

$V = Bh$

where B is the area of the base.

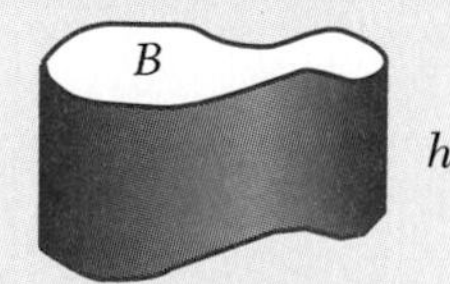

When measuring the capacity of a solid to hold liquid, sometimes special units are used. Recall that in the English system, liquid capacity is measured in ounces, quarts, and gallons. In the metric system, milliliters, liters, and kiloliters are used.

One cubic centimeter measures the same volume as one milliliter. That is, $1 \text{ cm}^3 = 1 \text{ m}\ell$. So, a can whose base has area 10 cm^2 with a height of 5 cm has a volume of

$$\begin{aligned} V &= (10 \text{ cm}^2)(5 \text{ cm}) \\ &= 50 \text{ cm}^3 \\ &= 50 \text{ m}\ell \end{aligned}$$

The can holds 50 mℓ of liquid.

Examples A–D

Directions: Find the volume.

Strategy: Use the volume formulas.

A. Find the volume of a cube that is 7 meters on each edge.

$V = s^3$ **Formula.**

$= (7 \text{ m})^3$ **Substitute.**

$= 343 \text{ m}^3$ **Simplify.**

The volume is 343 m^3.

B. How much concrete is needed to pour a step that is 4 ft long, 3 ft wide, and 6 inches deep?

$V = \ell wh$ **Formula.**

$= (4 \text{ ft})(3 \text{ ft})(6 \text{ in.})$ **Substitute.**

$= (48 \text{ in.})(36 \text{ in.})(6 \text{ in}).$ **Convert to common units.**

$= 10{,}368 \text{ in}^3$ **Simplify.**

The step requires $10{,}368 \text{ in}^3$ of concrete.

C. What is the volume of a can that is 5 in. tall and has a base with area 7 in^2?

$V = Bh$ **Formula.**

$= (7 \text{ in}^2)(5 \text{ in.})$ **Substitute.**

$= 35 \text{ in}^3$ **Simplify.**

The can holds 35 in^3.

Warm Ups A–D

A. Find the volume of a cube that has an edge of length 10 m.

B. How much concrete is needed for a rectangular stepping stone that is 1 ft long, 8 in. wide, and 3 in. deep?

C. What is the volume of a garbage can that is 3 ft tall and has a base with area 4 ft^2?

Answers to Warm Ups A. The volume is 1000 m^3. B. It will take 288 in^3 of concrete. C. The can holds 12 ft^3.

D. How many milliliters of water does this container hold?

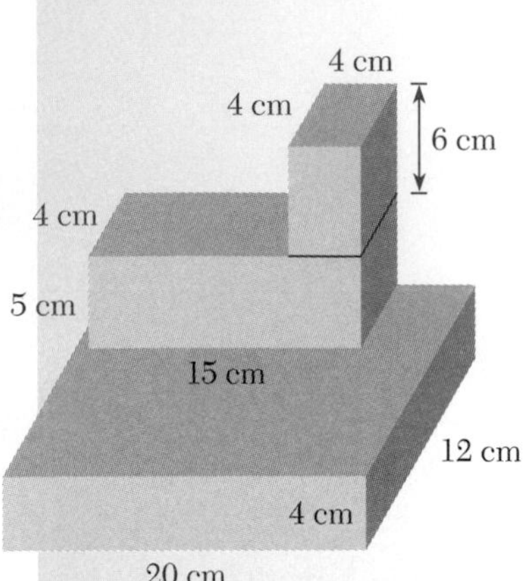

D. How many milliliters of water does this container hold?

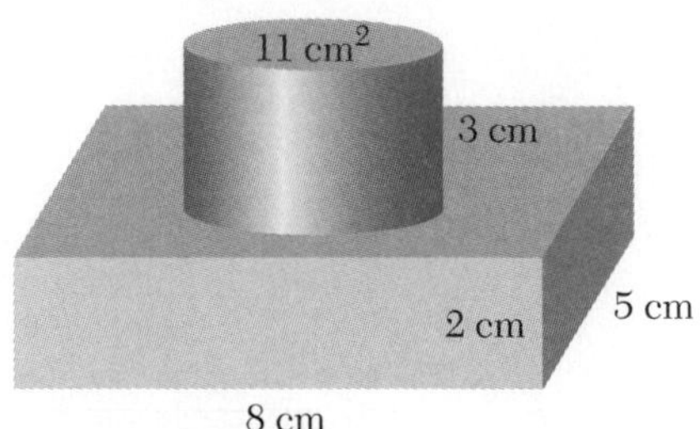

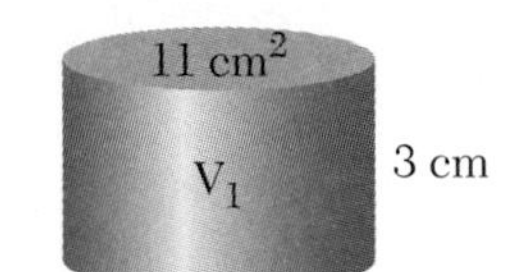

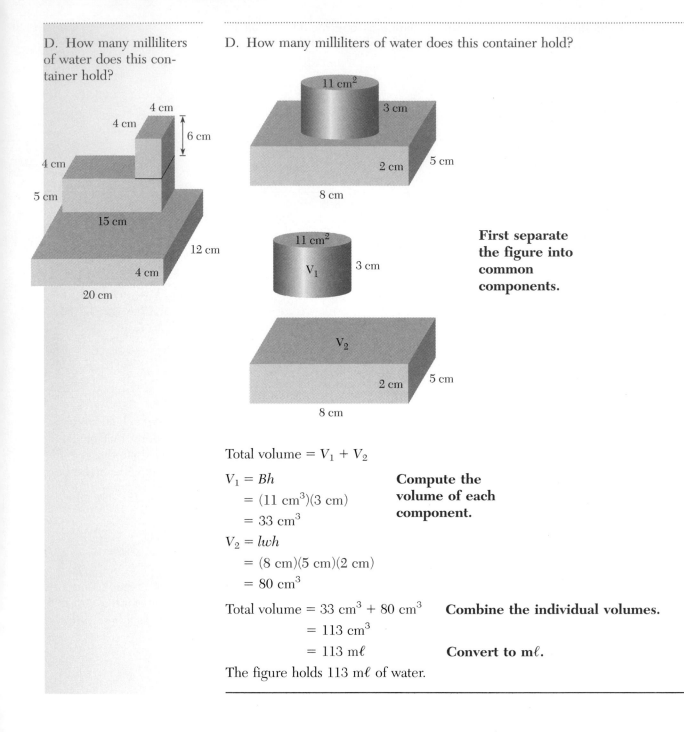

First separate the figure into common components.

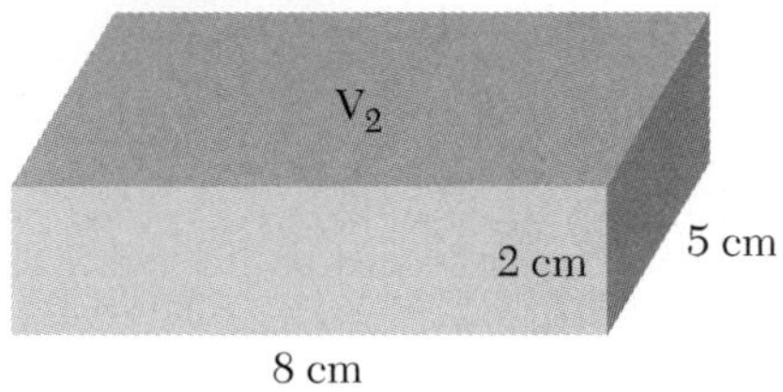

Total volume $= V_1 + V_2$

$V_1 = Bh$

$= (11 \text{ cm}^3)(3 \text{ cm})$

$= 33 \text{ cm}^3$

Compute the volume of each component.

$V_2 = lwh$

$= (8 \text{ cm})(5 \text{ cm})(2 \text{ cm})$

$= 80 \text{ cm}^3$

Total volume $= 33 \text{ cm}^3 + 80 \text{ cm}^3$ **Combine the individual volumes.**

$= 113 \text{ cm}^3$

$= 113 \text{ m}\ell$ **Convert to mℓ.**

The figure holds 113 mℓ of water.

Answer to Warm Up D. It holds 1356 mℓ of water.

NAME ______________________ CLASS __________ DATE __________

Exercises 2.4

OBJECTIVE: *Find the volume of common geometric shapes.*

A.

Find the volume of the figures.

1.

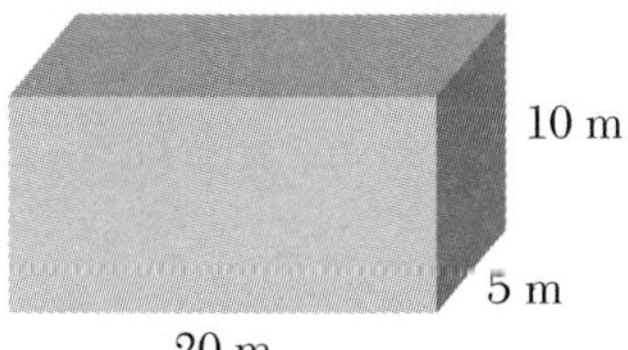

2.

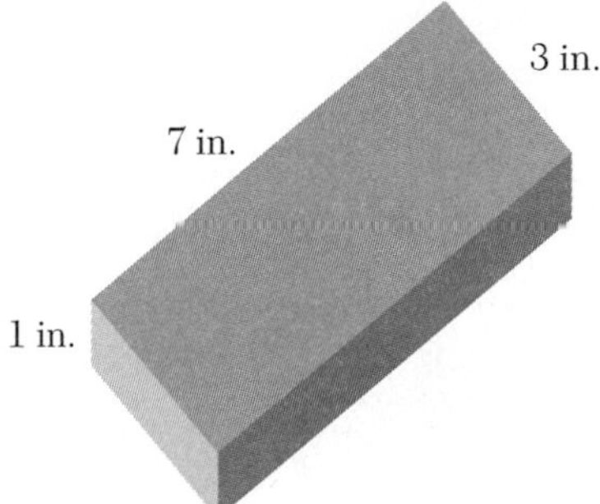

3.

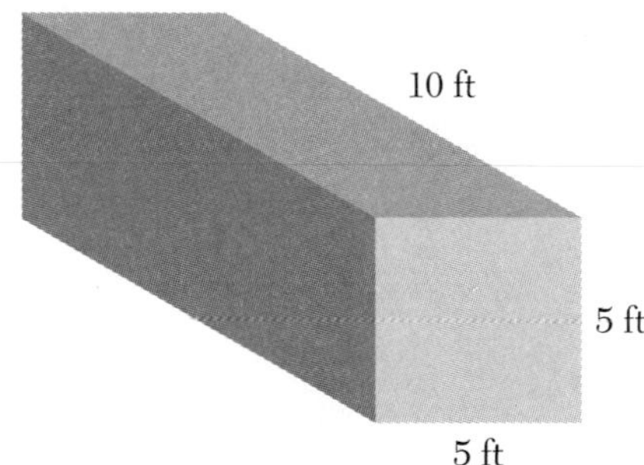

4.

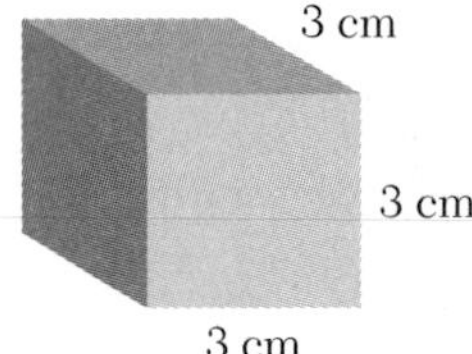

5.

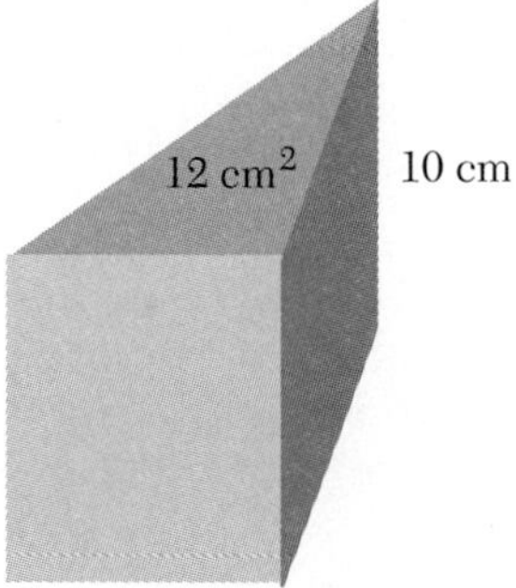

6.

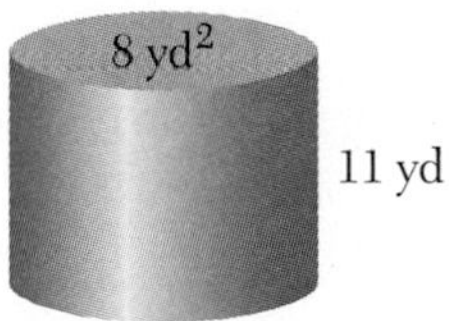

B.

7. How many milliliters of water will fill up a box that measures 52 cm long, 35 cm wide, and 12 cm high?

8. Find the volume of a cube that measures 24 mi on each side.

9. Find the volume of a garbage can that is 4 ft tall and has a base of 12 ft^2.

10. Find the volume of two identical fuzzy dice tied to the mirror of a '57 Chevy if one edge measures 6 in.

Find the volume of the figures.

11.

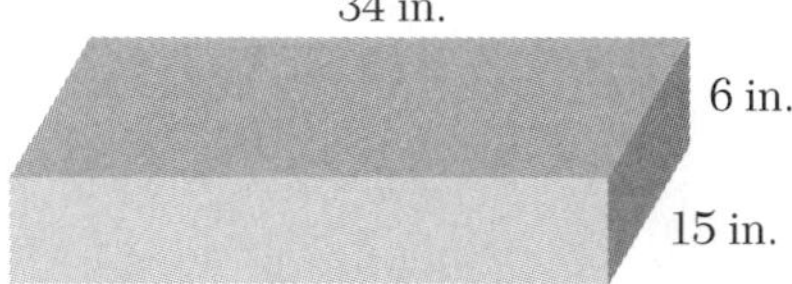

12.

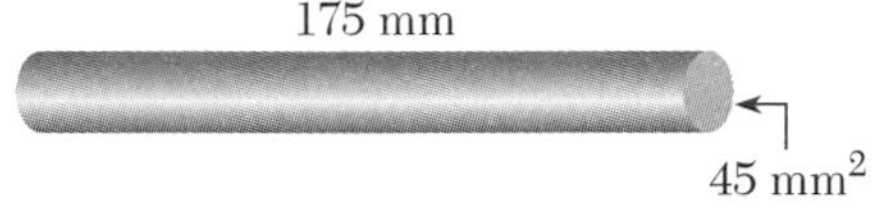

13.

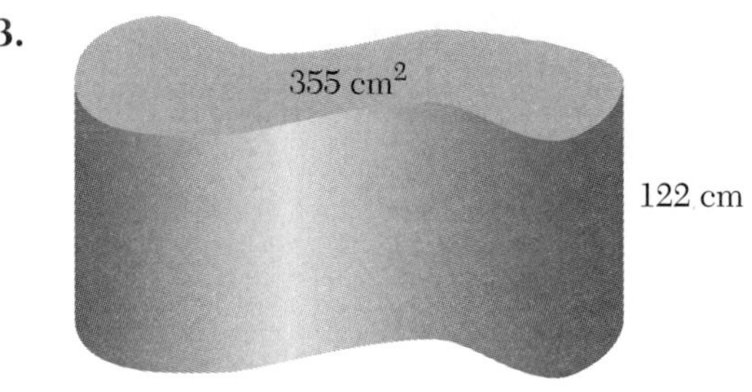

14.

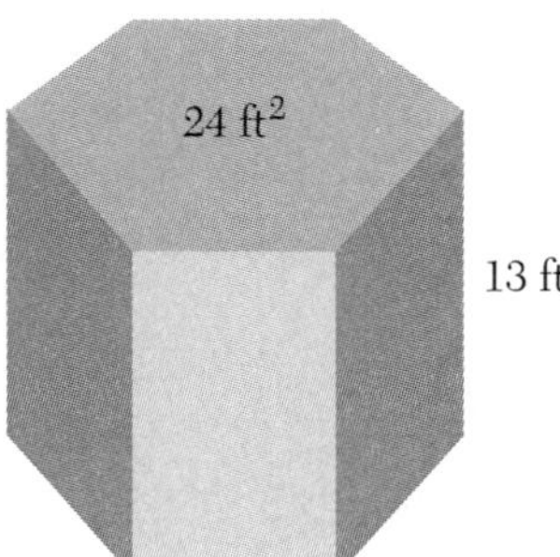

C.

15. Find the volume of a can that has a base with area 245 in^2 and a height of 2 ft.

16. How many cubic inches are there in a cubic foot? How many cubic inches are there in 5 ft^3?

17. How many cubic feet are there in 5184 in^3?

18. How many cubic feet are there in a cubic yard? How many cubic feet are there in 4 yd^3?

19. How many cubic yards are there in 270 ft^3?

20. How many cubic centimeters are there in a cubic meter? How many cubic centimeters are there in 3 m^3?

21. How many cubic inches of concrete are needed to pour a sidewalk that is 3 ft wide, 4 in. deep, and 54 ft long? Concrete is commonly measured in cubic yards, so convert your answer to cubic yards. (*Hint:* convert to cubic feet first, then convert that to cubic yards.)

Find the volume of the figures.

22.

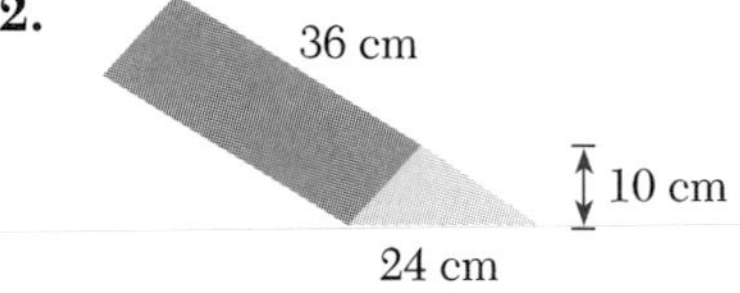

23.

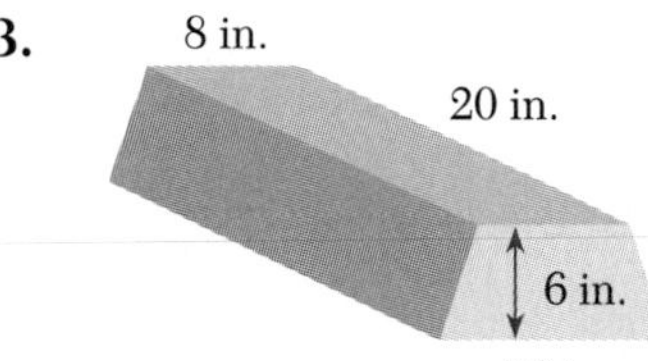

24.

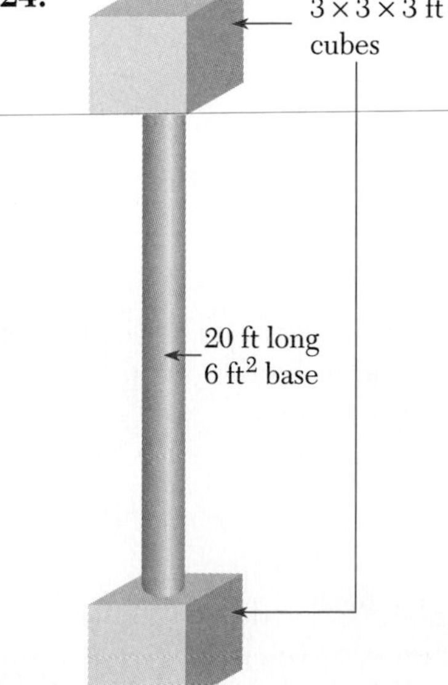

25.

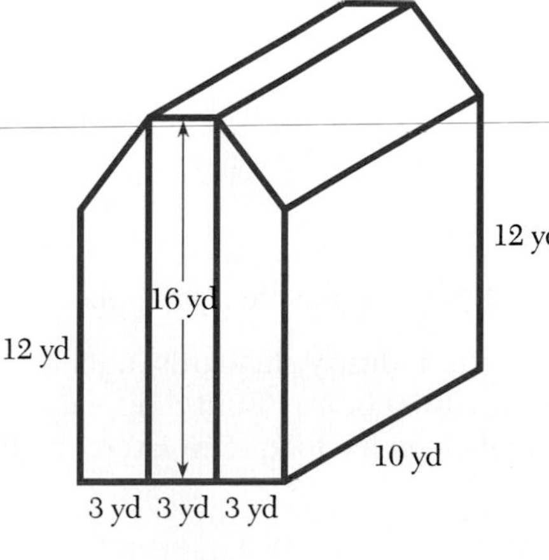

26.

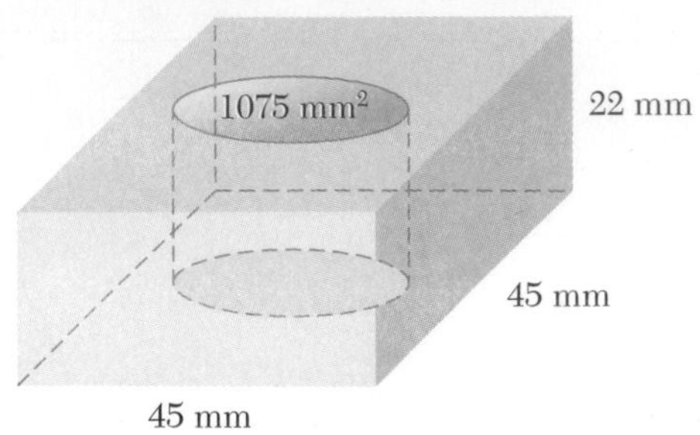

27.

110 cm
15 cm
72 cm
54 cm
72 cm
15 cm

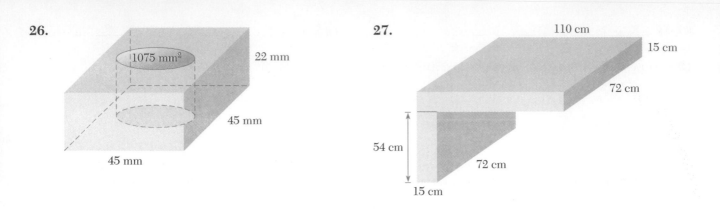

28. A bag of potting soil contains 3500 in^3. How many bags are needed to fill 5 flower boxes, each of which measures 4 ft long, 8 in. high, and 6 in. deep? Remember that you can only buy whole bags of potting soil.

29. An excavation is being made for a basement. The hole is 24 ft wide, 36 ft long, and 7 ft deep. If the bed of a truck holds 378 ft^3, how many truckloads of dirt will need to be hauled away?

30. A farming corporation is building four new grain silos. The inside dimensions of the silos are given in the table. Find the total volume in these silos.

	Area of Base	Height
Silo A	1800 ft^2	60 ft
Silo B	1200 ft^2	75 ft
Silo C	900 ft^2	80 ft
Silo D	600 ft^2	100 ft

Exercises 31–35 refer to the chapter application (see page 149).

31. According to industry standards, a joint thickness of a quarter inch means the bricklayer will need 9 ft^3 of mortar per 1000 bricks. Find the total amount of mortar needed for the patio and walkways. Round to the nearest whole cubic foot. (See Exercise 39 in Section 2.3.)

32. The cement subcontractor orders materials based on the total volume of the slab. The industry standard for patios and walkways is 4 to 5 in. of thickness. Because the slab will be topped with bricks, the contractor decides on a thickness of 4 in. Find the volume of the slab in cubic inches. Convert this answer to cubic feet, rounding up to the next whole cubic foot. Convert this answer to cubic yards. Round up to the next whole cubic yard if necessary.

33. Explain why it is necessary in this circumstance to round up to the next whole unit, rather than using the rounding rule stated in Section 1.1.

34. The cement contractor must first build a wood form that completely outlines the slab. How many linear feet of wood are needed to build the form?

35. The landscaper recommends that Barbara and Paul buy topsoil before planting the garden. They must buy enough to be able to spread the topsoil to a depth of 8 in. How many cubic inches of topsoil do they need? Because soil is usually sold in cubic yards, round this figure up to the next cubic yard.

STATE YOUR UNDERSTANDING

36. Explain what is meant by volume. Name three occasions in the past week when the volume of an object was relevant.

38. Explain why the formula for the volume of a box is a special case of the formula $V = Bh$.

37. Explain how to find the volume of the figure below.

CHALLENGE

39. The Bakers are constructing an inground pool in their back yard. The pool will be 15 ft wide and 30 ft long. It will be 3 ft deep for 10 ft on one end. It will then drop to a depth of 10 ft at the other end. How many cubic feet of water are needed to fill the pool? If the trucks hauling away the dirt dug to make the pool have a capacity of 14 yd^3, how many loads of dirt were hauled away?

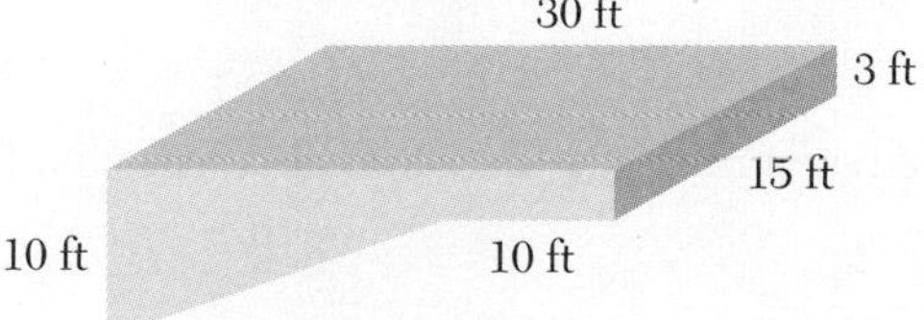

40. Norma is buying mushroom compost to mulch her garden, which is pictured below. How many cubic yards of compost does she need to mulch the entire garden 4 in. deep? (She cannot buy fractional parts of a cubic yard.)

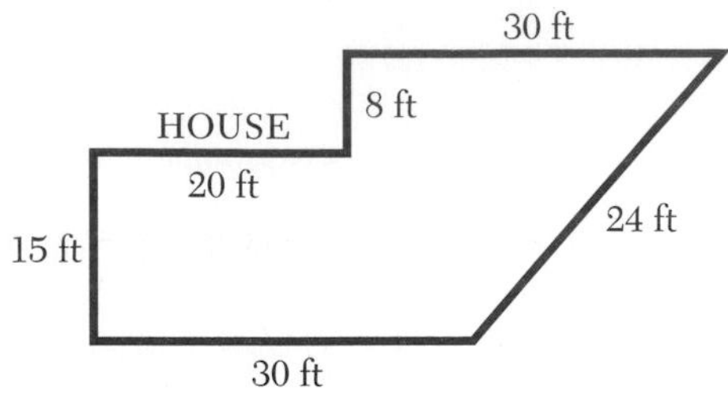

GROUP ACTIVITY

41. The insulating ability of construction materials is measured in R-values. Industry standards for exterior walls are currently R-19. An 8 in. thickness of loose fiberglass is necessary to achieve an R-19 value. Calculate the amount of cubic feet of loose fiberglass needed to insulate the exterior walls of the mountain cabin pictured. All 4 side windows measure 2 ft by 3 ft. Both doors measure 4 ft by 7 ft. The front window measures 5 ft by 3 ft.

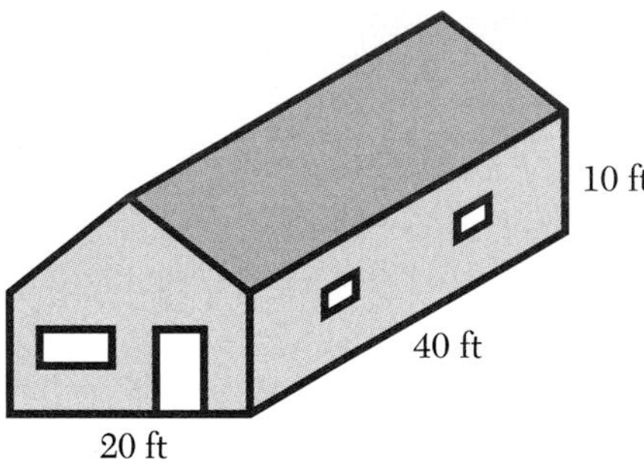

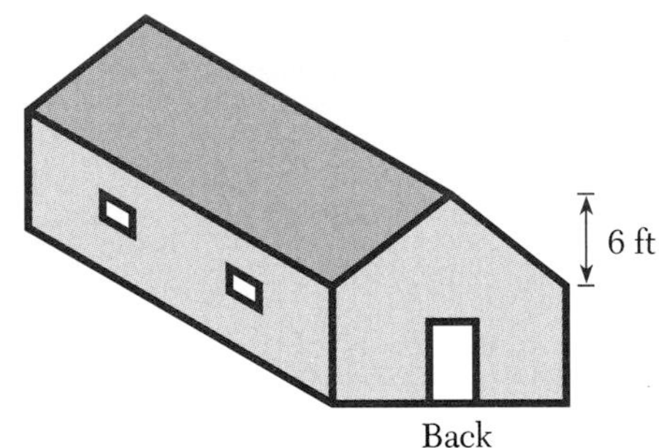

MAINTAIN YOUR SKILLS (SECTIONS 1.1–1.7, 2.1, 2.3)

Evaluate the following:

42. 8^3

43. $4^2 + 5^2$

44. $2^3 + 3^3 + 4^3$

45. $9^2 - 4^2$

46. $(2^2 + 3^2)^2$

47. $3168 \div 24$

48. $(10^2)(2^2)(3^2)$

49. $5(36 \div 12 + 1 - 2)$

50. Find the area of a rectangle that is 3 in. wide and 2 ft long.

51. A warehouse store sells 5-lb bags of Good and Plenty. Bob buys a bag and stores the candy in 20-oz jars. How many jars does he need?

NAME ______________________ CLASS ____________ DATE ____________

CHAPTER 2

Group Project *(2–3 weeks)*

OPTIONAL

You are working for a kitchen design firm that has been hired to design a kitchen for the 10-ft-by-12-ft room pictured below.

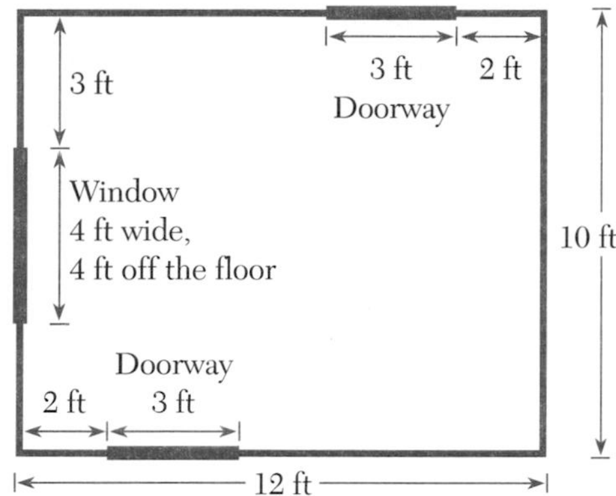

The following table lists appliances and dimensions. Some of the appliances are required and others are optional. All dimensions are in inches.

Appliance	High	Wide	Deep	Required
Refrigerator	68	30 or 33	30	Yes
Range/Oven	30	30	26	Yes
Sink	12	36	22	Yes
Dishwasher	30	24	24	No
Trash compactor	30	15	24	No
Built-in microwave	24	24	24	No

The base cabinets are all 30 in. high and 24 in. deep. The widths can be any multiple of 3 from 12 in. to 36 in. Corner units are 36 in. along the wall in each direction. The base cabinets (and the range, dishwasher, and compactor) will all be installed on 4-in. bases that are 20 in. deep.

The wall (upper) cabinets are all 30 in. high and 12 in. deep. Here too, the widths can be any multiple of 3 from 12 in. to 36 in. Corner units are 24 in. along the wall in each direction.

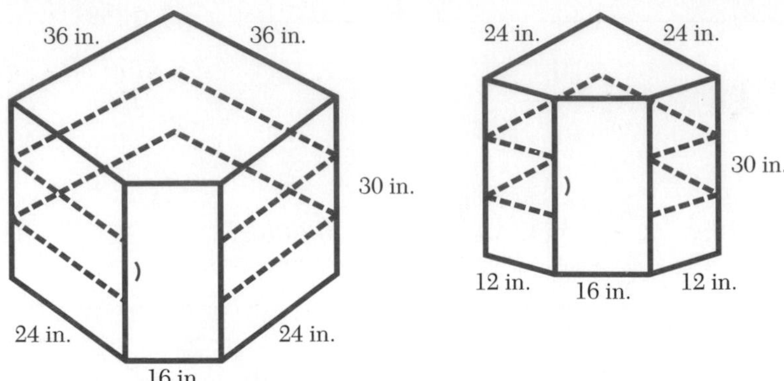

1. The first step is to place the cabinets and appliances. Your client has specified that there must be at least 80 ft^3 of cabinet space. Make a scale drawing of the kitchen and indicate the placement of the cabinets and appliances. Show calculations to justify that your plan satisfies the 80 ft^3 requirement.
2. Countertops measure 25 in. deep with a 4-in. backsplash. The countertops can either be tile or Formica. If the counters are Formica, there will be a 2-in. facing of Formica. If the counters are tile, the facing will be wood that matches the cabinets. (See figure.) Calculate the amount of Formica needed for the counters and the amount of tile and wood needed for the counters.

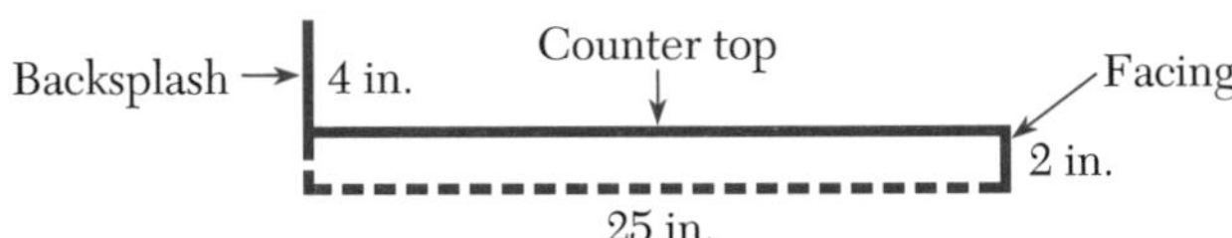

3. The bases under the base cabinets will be covered with a rubber kickplate that is 4 in. high and comes in 8-ft lengths. Calculate the total length of kickplate material needed and the number of lengths of kickplate material necessary to complete the kitchen.
4. Take your plan to a store that sells kitchen cabinets and counters. Your goal is to get the best-quality materials for the least amount of money. Prepare at least two cost estimates for the client. Do not include labor in your estimates, but do include the appliances. Include a rationale with each estimate, explaining the choices you made. Which plan will you recommend to the client and why?

NAME ______________________ CLASS ____________ DATE ____________

CHAPTER 2

True–False Concept Review

Check your understanding of the language of algebra and geometry. Tell whether each of the following statements is true (always true) or false (not always true). For each statement you judge to be false, revise it to make a statement that is true.

ANSWERS

1. English measurements are the most commonly used in the world. **1.** ____________

2. Equivalent measures have different units of measurement. **2.** ____________

3. A liter is a measure of weight. **3.** ____________

4. The perimeter of a square can be found in inches, feet, centimeters, or meters. **4.** ____________

5. Area is the measure of the inside of a solid, such as a box or a can. **5.** ____________

6. The volume of a square is $V = s^3$. **6.** ____________

7. The formula for the area of a trapezoid is $A = \dfrac{(b_1 + b_2)h}{2}$. **7.** ____________

8. It is possible to find equivalent measures without remeasuring the original object. **8.** ____________

9. The metric system utilizes the base-ten place value system. **9.** ____________

10. Volume is the measure of how much a container will hold. **10.** ____________

11. Weight can be measured in pounds, grams, or kilograms. **11.** ____________

12. A parallelogram has three sides. **12.** ____________

13. Volume can be thought of as the number of squares in an object. **13.** ____________

14. One milliliter is equivalent to one cubic centimeter. **14.** ____________

15. Measurements can be added or subtracted only when they are expressed with the same unit of measure. 15. ______

16. The distance around a geometric figure is called the perimeter. 16. ______

17. Volume is always measured in cubic units. 17. ______

18. A trapezoid is a quadrilateral. 18. ______

19. $1 \text{ ft}^2 = 12 \text{ in}^2$ 19. ______

20. The prefix "kilo" means 100. 20. ______

NAME ______________________ CLASS ____________ DATE ____________

CHAPTER 2

Test

ANSWERS

1. 7 m + 454 mm = ? mm — **1.** ________

2. Find the perimeter of a square that is 34 cm on a side. — **2.** ________

3. Find the volume of a drawer that is 4 in. high, 18 in. wide and 24 in. deep. — **3.** ________

4. How much vinyl flooring is needed to cover the room pictured? — **4.** ________

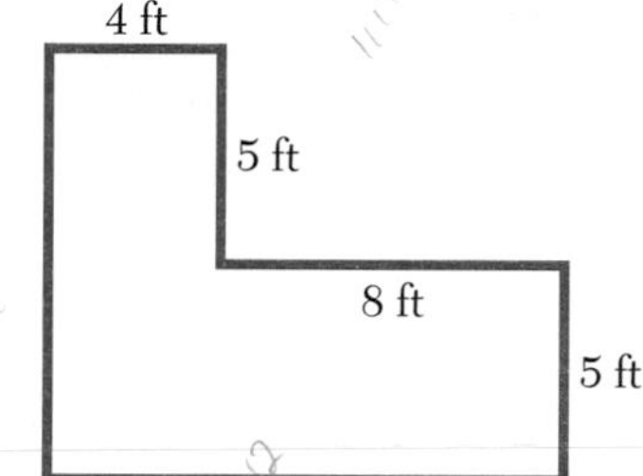

5. Anna has 135 lb of strawberries to divide equally among her 5 children. How many pounds of berries will each one receive? — **5.** ________

6. Find the perimeter of the figure. — **6.** ________

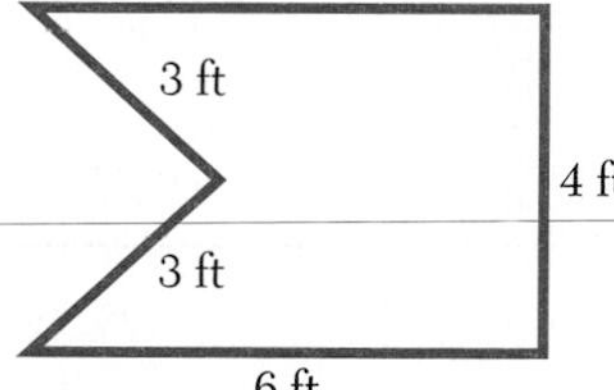

7. Find the area of the triangle. — **7.** ________

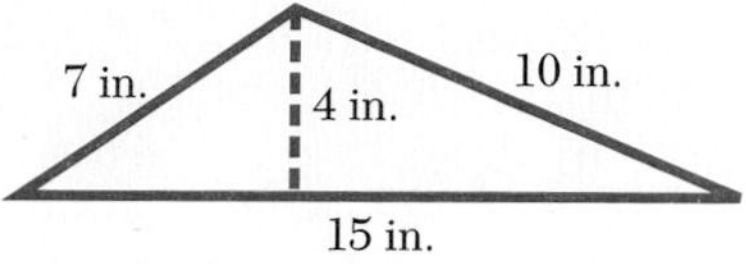

8. Subtract: — **8.** ________

$$\begin{array}{r} 5 \text{ gal } 2 \text{ qt} \\ -\ 3 \text{ gal } 2 \text{ qt } 1 \text{ pt} \\ \hline \end{array}$$

9. Name two units of measure in the English system for weight. Name two units of measure in the metric system for weight.

9. ________________

10. Change 2 ft^2 to in^2.

10. ________________

11. How much molding is needed to trim a picture window 4 ft wide and 5 ft tall and two side windows each measuring 2 ft wide by 5 ft tall?

11. ________________

12. Find the volume of the figure.

12. 

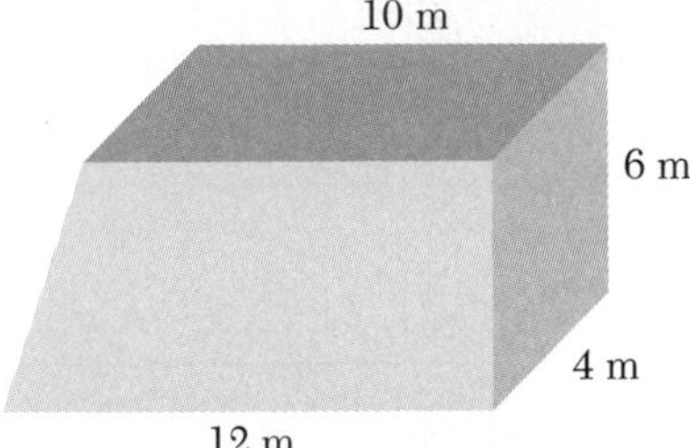

13. Find the area of the parallelogram.

13. 

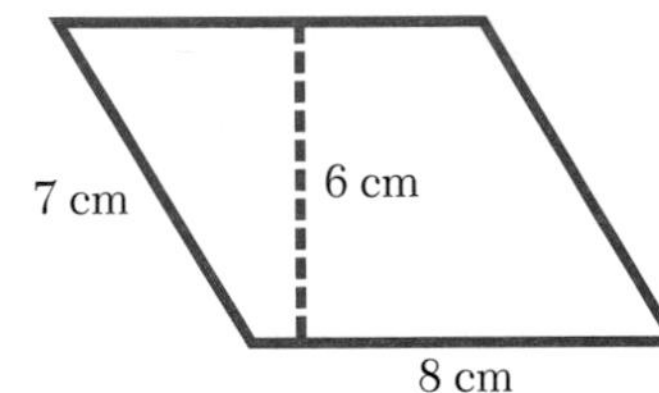

14. Find the volume of a hot water tank that has a circular base with area of 4 ft^2 and a height of 5 ft.

14. ________________

15. The Golden Silver Company has a bar of silver weighing 684 g. If it is melted down to form six bars of equal weight, what will each bar weigh?

15. ________________

16. The cost of heavy duty steel wire fencing is $2 per linear foot. How much will it cost to build the dog runs pictured?

16. ________________

3 ft 3 ft 3 ft

12 ft

17. Name two units of measure in the English system for volume. Name two units of measure in the metric system for volume.

17. ________________

18. Find the area of the figure. **18.** ______________

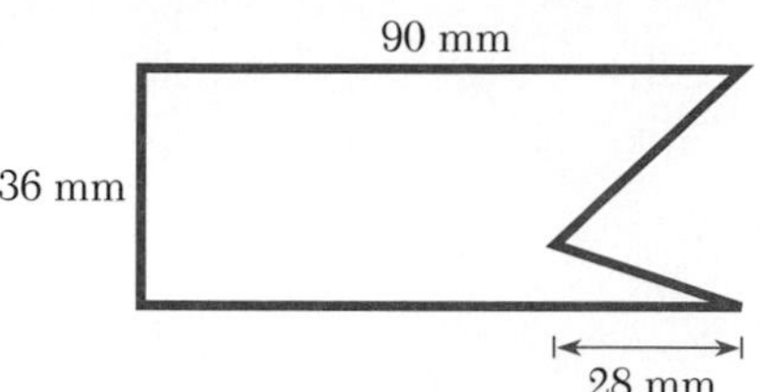

19. A football coach is conditioning his team by having them run around the edge of the field three times each hour. A football field measures 100 yd long by 60 yd wide. How far does each player run in a 2-hr practice? **19.** ______________

20. Both area and volume describe interior space. Explain how they are different. **20.** ______________

21. Jared is lifting weights in preparation for football season. One week he bench presses 180 lb on Monday, 185 lb on Tuesday, 195 lb on Wednesday, 205 lb on Thursday, and 200 lb on Friday. What is his average bench press weight for the week? **21.** ______________

22. Li is buying lace with which she plans to edge a tablecloth. The tablecloth is 60 in. by 108 in. How much lace does she need? If the lace is available in whole yard lengths only, how much must she buy? **22.** ______________

23. Khallil has a 40-lb bag of dog food that is approximately 52 in. long, 16 in. wide, and 5 in. deep. He wants to transfer it to a plastic storage box that is 24 in. long, 18 in. wide, and 12 in. high. Will all the dog food fit into the box? **23.** ______________

24. A developer of an apartment complex is required by the county to preserve and improve the wetlands area diagrammed below. The developer estimates that it will cost $50 per square foot to improve the wetlands. What is the total cost? **24.** ______________

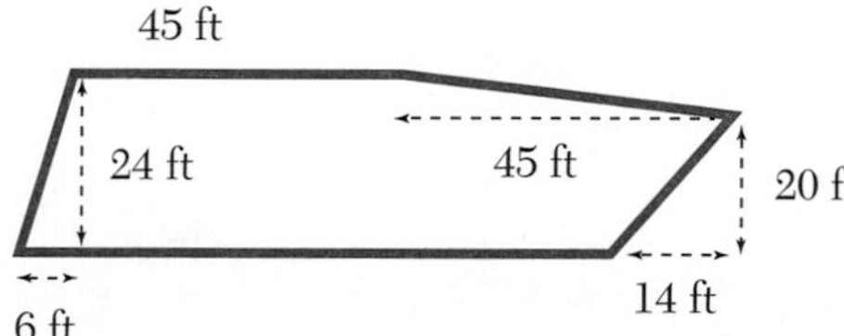

25. Neela has a gift box that is 10 in. by 15 in. by 2 in. What is the minimum amount of wrapping paper needed to wrap the gift? **25.** ______________

Good Advice for Studying

Managing Anxiety

For those of you who become anxious as you begin to study, we recommend that you devote a section in your math notebook to record your thoughts and feelings each time you study. Record your thoughts in the first person, as "I-statements." For example, "Nobody ever uses this stuff in real life," rephrased as an "I" statement, it would be: "I would like to know where I would use this in my life." Research shows that a positive attitude is the single most important key to your success in mathematics. By recognizing negative thoughts and replacing them with positive thoughts, you are beginning to work on changing your attitude. The first step is to become aware of your self-talk.

Negative self-talk falls into three categories of irrational beliefs that you have about yourself and how you view the world. You think that (1) *Worrying helps*. Wrong. Worrying leads to excessive anxiety, which is distracting and hinders performance. (2) *Your worth as a person is determined by your performance*. Wrong. Not being able to solve math problems doesn't mean you won't amount to anything. If you think it does, you are a victim of "catastrophizing." (3) *You are the only one who feels anxious*. Wrong. By thinking that other students have some magical coping skills that allows them to avoid anxiety, you are comparing yourself to some irrational mythical norm.

As you begin your first section, ask yourself such questions as "What am I saying to myself now?" "What is triggering these thoughts?" "How do I feel physically?" and "What emotions am I feeling now?" Your answers will likely reveal a pattern to your thoughts and feelings. You need to analyze your statements and change them into more positive and rational self-talk. You can use a simple technique called "rational emotive therapy."[1] This method, if practiced regularly, can quickly and effectively change the way you think and feel about math. When you find yourself getting upset, watch for words such as *should, must, ought to, never, always,* and *I can't*. They are clues to negative self-talk and signals for you to direct your attention back to math. Use the following ABCD model.

A. *Triggering event:* You start to do your math homework and your mind goes blank.

B. *Negative self-talk* in response to the trigger: "I can't do math. I'll never pass. My life is ruined!"

C. *Anxiety* caused by the negative self-talk: panic, anger, tight neck, etc.

D. *Positive self-talk* to cope with anxiety: "This negative self-talk is distracting. It doesn't help me solve these problems. Focus on the problems."; or say, "I may be uncomfortable, but I'm working on it."

Recognize that negative self-talk (B) is the culprit. It causes the anxiety (C) and must be restructured to positive rational statements (D). Practice with the model using your own self-talk statements (B) and then complete the remaining steps.

[1]R.E.T. was created by Albert Ellis.

3

Primes and Multiples

APPLICATION

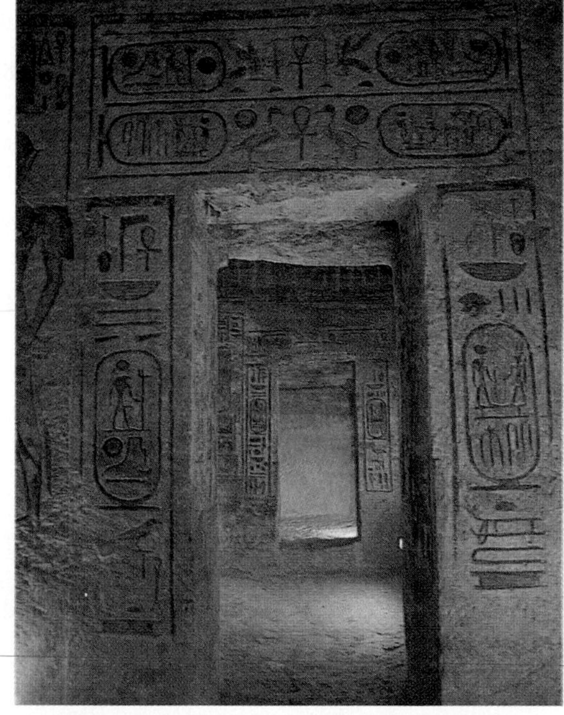

For as long as there has been written communication, people have needed a method to keep some information private. Thus was born the science of cryptology, the study of various methods of writing secret messages, also called codes, ciphers, encryptions, or cryptosystems. Historically, such systems were used almost exclusively for matters of national security by countries at odds with (or at war with) other countries. In more modern times, industrial secrets as well as military secrets are protected by cryptosystems.

Curiously enough, as the use of computers becomes more widespread, the need for secure communications has increased dramatically. Ordinary citizens need to protect certain personal numbers (like Social Security numbers, drivers' license numbers, and credit card account numbers) from interception by unscrupulous persons wanting to steal their money or identity. The safe purchase of items through the Internet is one of the most recent new demands for cryptosystems.

A cryptosystem consists of either a code or a cipher. Both the sender and the receiver must know the key to the system in order to use it. Although both codes and ciphers result in secret messages, they are technically not the same. A code substitutes symbols or words for ordinary text words or phrases. It is like a completely new language, and the only way to break a code is to have access to the code book. The weakness with a code is that once the code book is in the wrong hands, the entire system has to be dropped and a new one developed. For this reason, many writers of secret messages choose ciphers instead. A cipher is a letter-for-letter replacement procedure. The process is known to both the sender and receiver, but the process is easy to store mentally rather than in a code book so there is nothing to be stolen. If an enciphered message is intercepted, the only way to decipher it is to figure out the original cipher process. As we will see, there are some very elementary cipher systems that are relatively easy to break. This is where mathematicians enter the picture. Mathematicians have been useful both in the designing of difficult ciphers and in the breaking of ciphers intercepted from other sources.

For Rail Fence ciphers, see page 216. Other ciphers appear throughout the chapter.

3.1

Divisibility Tests

OBJECTIVES

1. Determine whether a natural number is divisible by 2, 3, or 5.

2. Determine whether a natural number is divisible by 6, 9, or 10.

VOCABULARY

A whole number is **divisible** by another whole number if the quotient of these numbers is a natural number greater than 1 and the remainder is 0. The second number is said to be a **divisor** of the first. Thus, 8 is a divisor of 32, since $32 \div 8 = 4$.

We also say 32 is **divisible** by 8.

The **even digits** are 0, 2, 4, 6, and 8.

The **odd digits** are 1, 3, 5, 7, and 9.

HOW AND WHY

Objective 1

Determine whether a natural number is divisible by 2, 3, or 5.

To ask if a number is divisible by 3 is to ask if the division of the number by 3 comes out even (has no remainder). One can always answer this question by doing the division and checking that there is no remainder. Divisibility tests tell us whether one number will divide another without our actually having to do the division. For some numbers we can test the question of divisibility mentally.

Table 3.1 provides clues for these tests.

TABLE 3.1 PATTERNS OF MULTIPLYING

Some Natural Numbers	Multiply by 2	Multiply by 3	Multiply by 5	Multiply by 9
1	2	3	5	9
2	4	6	10	18
3	6	9	15	27
4	8	12	20	36
5	10	15	25	45
6	12	18	30	54
10	20	30	50	90
16	32	48	80	144
24	48	72	120	216
44	88	132	220	396
66	132	198	330	594

In the Multiply by 2 column, the ones digit of each number is an even digit, that is, either 0, 2, 4, 6, or 8. Since the ones place is even, the number is said to be an even number.

In the Multiply by 3 column, the sum of the digits of each number in the column is divisible by 3. For example, the sum of the digits of 18 is 1 + 8 or 9, and 9 is divisible by 3. Likewise the digits of 198 have a sum that is divisible by 3: 1 + 9 + 8 or 18.

In the Multiply by 5 column the ones digit of each number is either 0 or 5.

To test for divisibility of a natural number by 2, 3, or 5

If the ones-place digit of a natural number is even (0, 2, 4, 6, or 8), the natural number is divisible by 2.

If the sum of the digits of a natural number is divisible by 3, then the natural number is divisible by 3.

If the ones-place digit of a natural number is 0 or 5, the natural number is divisible by 5.

Examples A–D

Directions: Determine whether the natural number is divisible by 2, 3, or 5.

Strategy: First check the ones-place digit. If it is even, the number is divisible by 2. If it is 0 or 5, the number is divisible by 5. Next find the sum of the digits. If the sum is divisible by 3, the number is divisible by 3.

A. Is 48 divisible by 2, 3, or 5?

48 is divisible by 2 **The ones-place digit is 8.**

48 is divisible by 3 **4 + 8, or 12, is divisible by 3.**

48 is not divisible by 5 **The ones-place digit is neither 0 nor 5.**

B. Is 240 divisible by 2, 3, or 5?

240 is divisible by 2 **The ones-place digit is 0.**

240 is divisible by 3 **2 + 4 + 0, or 6, is divisible by 3.**

240 is divisible by 5 **The ones-place digit is 0.**

C. Is 727 divisible by 2, 3, or 5?

727 is not divisible by 2 **The ones-place digit is not even.**

727 is not divisible by 3 **7 + 2 + 7, or 16, is not divisible by 3.**

727 is not divisible by 5 **The ones-place digit is neither 0 nor 5**

D. Georgio and Anna and their three children are on a trip to Okinawa. Anna has a total of ¥ 453 to divide among the children. She wants each child to receive the same amount of yen in whole numbers. Is this possible? Why or why not?

Yes, each child will receive the same amount of yen in whole numbers (¥ 151) because 453 is divisible by 3.

Warm Ups A–D

A. Is 75 divisible by 2, 3, or 5?

B. Is 450 divisible by 2, 3, or 5?

C. Is 488 divisible by 2, 3, or 5?

D. If Anna has ¥ 386 to divide among the children, will each child receive the same amount of yen in whole numbers? Why or why not?

Answers to Warm Ups A. 75 is not divisible by 2; 75 is divisible by 3 and 5 B. 450 is divisible by 2, 3, and 5 C. 488 is divisible by 2; 488 is not divisible by 3 or 5 D. No, each child will not receive the same amount because 386 is not divisible by 3.

HOW AND WHY

Objective 2

Determine whether a natural number is divisible by 6, 9, or 10.

In Table 3.1, some numbers appear in both the Multiply by 2 column and the Multiply by 3 column. These numbers are also divisible by 6 because $6 = 2 \cdot 3$ and every number divisible by 6 must also be divisible by 2 and 3. For example, 132 is divisible by 2 since the ones-place digit is 2 and 132 is also divisible by 3 since $1 + 3 + 2$, or 6, is divisible by 3. Therefore, 132 is divisible by 6.

In the Multiply by 9 column, the sum of the digits of each number in the column is divisible by 9. For example, the sum of the digits in 36 is $3 + 6$, or 9, which is divisible by 9. Also, the sum of the digits in 189 is $1 + 8 + 9$, or 18, which is divisible by 9.

Notice that all natural numbers ending in 0 are divisible by 2 (because 0 is even) and also by 5 (since they end in 0). These numbers are also divisible by 10. These numbers will appear in both the Multiply by 2 column and the Multiply by 5 column. Some examples are 20, 80, and 140.

To test for divisibility of a natural number by 6, 9, or 10

If the natural number is divisible by both 2 and 3, then the natural number is divisible by 6.

If the sum of the digits of a natural number is divisible by 9, then the natural number is divisible by 9.

If the ones-place digit of a natural number is 0, the natural number is divisible by 10.

Warm Ups E–F

E. Is 630 divisible by 6, 9, or 10?

F. Is 1830 divisible by 6, 9, or 10?

Examples E–F

Directions: Determine whether a natural number is divisible by 6, 9, or 10.

Strategy: First check whether the number is divisible by both 2 and 3. If so, the number is divisible by 6. Second, find the sum of the digits. If the sum is divisible by 9, then the number is divisible by 9. Last, check the ones-place digit. If the digit is 0, the number is divisible by 10.

E. Is 810 divisible by 6, 9, or 10?

810 is divisible by 6	**810 is divisible by both 2 and 3.**
810 is divisible by 9	**$8 + 1 + 0$, or 9, is divisible by 9.**
910 is divisible by 10	**The ones-place digit is 0.**

F. Is 1590 divisible by 6, 9, or 10?

1590 is divisible by 6	**1590 is divisible by both 2 and 3.**
1590 is not divisible by 9	**$1 + 5 + 9 + 0$, or 15, is not divisible by 9.**
1590 is divisible by 10	**The ones-place digit is 0.**

Answers to Warm Ups E. 630 is divisible by 6, 9, and 10 F. 1830 is divisible by 6 and 10; 1830 is not divisible by 9

NAME ______________________ CLASS ____________ DATE ____________

Exercises 3.1

OBJECTIVE 1: *Determine whether a natural number is divisible by 2, 3, or 5.*

A.

Is each number divisible by 2?

1. 24 **2.** 22 **3.** 20 **4.** 27

5. 38 **6.** 58

Is each number divisible by 5?

7. 15 **8.** 23 **9.** 45 **10.** 60

11. 551 **12.** 507

Is each number divisible by 3?

13. 15 **14.** 30 **15.** 36 **16.** 27

17. 43 **18.** 53

B.

Determine whether the natural number is divisible by 2, 3, or 5.

19. 2760 **20.** 2670 **21.** 3998 **22.** 3999

23. 4815 **24.** 1845 **25.** 5820 **26.** 6030

27. 11,115 **28.** 11,118

OBJECTIVE 2: *Determine whether a natural number is divisible by 6, 9, or 10.*

A.

Is each number divisible by 6?

29. 108 **30.** 801 **31.** 253 **32.** 333

33. 444 **34.** 450

Is each number divisible by 9?

35. 108 **36.** 801 **37.** 253 **38.** 333

39. 414 **40.** 423

Is each number divisible by 10?

41. 233 **42.** 330 **43.** 670 **44.** 607

45. 1920 **46.** 1290

B.

Determine whether the natural number is divisible by 6, 9, or 10.

47. 4920 **48.** 4896 **49.** 5994 **50.** 3780

51. 5555 **52.** 3333 **53.** 5700 **54.** 7880

55. 6690 **56.** 9066

C.

57. Pedro and two friends plan to buy a used car that is priced at $1221. Is it possible for each of them to contribute the same whole number of dollars? Explain.

58. Janna and four of her friends decide to run a distance of 82 miles in relays. Is it possible for each runner to run the same whole number of miles?

59. Ed and his four partners made a profit of $2060 in their stereo installation business. Can the profits be divided evenly in whole dollars among them? Explain.

60. Four merchants agree to build one new store in each of 320 cities. Is it possible for each merchant to oversee the building of the same number of new stores? Explain.

61. Lucia teaches 120 students in a single class at a community college. She wants to divide the students in small equal-sized groups to work on a group project. Is it possible to have groups of 5, 6, or 9? Explain.

62. A movie theater is to have 250 seats, and the manager wants to arrange them in rows with the same number of seats in each row. Use divisibility tests to determine if it is possible to have rows of 10 seats each. Are 15-chair rows possible?

 Exercises 63–67 relate to the chapter application.

The "Rail Fence" system is a relatively simple cipher. The text is written with every other letter dropped down. The cipher is created by reading all the top letters followed by all the bottom letters. The resulting ciphertext is often grouped in sets of four or five letters to further disguise the text. For example, to encipher the word "mathematics" we write

m t e a i s
 a h m t c

The ciphered form is MTEA ISAH MTCX. Note the extra X at the end of the third group. It is called a null character and was added in order to make each group the same length.

The following messages have been ciphered; decifer them using the Rail Fence system.

63. CNOF ETEO EOIH AYUE LHLV TNGT

64. LVTE NYUE IHOE HOEO RWTL

Gaius Julius Caesar (the one who married Cleopatra) was a great general with a keen interest in cryptography. A family of ciphers known as the Caesar Ciphers are so named because history records that he used one of them to communicate with his army. The Caesar Ciphers are substitution ciphers, in which each letter is substituted for another. In the case of the Caesar Ciphers, the alphabet is simply shifted over for a specified number of letters. The one Caesar is said to have used is referred to as the D-Alphabet because each letter was shifted over three letters, so A was encrypted as D, B was encrypted as E, and so on. Here is the complete D-Alphabet Cipher.

text:	abcde	fghij	klmno	pqrst	uvwxyz
cipher:	defgh	ijklm	nopqr	stuvw	xyzabc

65. Encrypt the following message in the D-Alphabet Cipher.

Love makes the world go round.

66. Decipher the following message written in the G-Alphabet Cipher.

EUABK RUYZ ZNGZ RUBOT LKKROTM

67. Decipher the following message written in one of the Caesar Ciphers.

EHOX BL T FTGR LIEXGWHKXW MABGZ

STATE YOUR UNDERSTANDING

 **68.** Explain what it means when we say "This number is divisible by 5." Give an example of a number that is divisible by 5 and one that is not divisible by 5.

69. Explain why a number that is divisible by 2 and by 3 must also be divisible by 6.

70. Explain the difference in the divisibility tests for 2 and 3.

71. Write a short statement to explain why every number divisible by 9 is also divisible by 3.

CHALLENGE

72. Is 23,898 divisible by 6?

73. Is 12,870 divisible by 15? Write a divisibility test for 15.

74. Is 12,870 divisible by 30? Write a divisibility test for 30.

75. Is 99,333,111,375 divisible by:
   a. 2? __________, because the ______________________________
   b. 3? __________, because the ______________________________
   c. 5? __________, because the ______________________________
   d. 6? __________, because the ______________________________
   e. 9? __________, because the ______________________________
   f. 10? __________, because the ______________________________

GROUP ACTIVITY

76. With your group, find divisibility tests for 4, 8, 20, and 25. Report to and compare your findings with the other groups.

77. Call your local recycling center and determine the amount of newspaper they collect each day for one complete week. Round the average amount of newspaper collected per day to the nearest whole number. Is this amount divisible into 2-, 3-, 5-, 6-, 9-, or 10-pound bins? Which size is the most efficient? Why?

MAINTAIN YOUR SKILLS (SECTIONS 1.1, 2.1, 2.2, 2.3)

78. Round 87,754 to the nearest thousand.

79. Round 87,754 to the nearest ten.

80. Round 8,382,254 to the nearest hundred.

81. Round 345,678,912 to the nearest million.

82. Round 567,891,234 to the nearest ten thousand.

83. True or false: $168 < 159$

84. True or false: $3142 > 3200$

85. Add:

$$\begin{array}{r} 2 \text{ yd } 2 \text{ ft } 10 \text{ in.} \\ + \underline{1 \text{ yd } 2 \text{ ft } \;\; 5 \text{ in.}} \end{array}$$

86. Find the perimeter of a square that is 13 cm on a side.

87. Find the area of a square that is 13 cm on a side.

3.2
Multiples

OBJECTIVES

1. List multiples of a whole number.
2. Determine whether one whole number is a multiple of another whole number.

VOCABULARY

A **multiple** of a whole number is the product of that number and a natural number. For instance,

8×2, or 16, is a multiple of 8
8×11, or 88, is a multiple of 8
8×13, or 104, is a multiple of 8
8×20, or 160, is a multiple of 8

HOW AND WHY

Objective 1

List multiples of a whole number.

To list the multiples of 8, we just multiply 8 by each natural number as shown in Table 3.2.

TABLE 3.2 MULTIPLES OF 8

Natural Number	Multiple of 8
1	8
2	16
3	24
4	32
5	40
6	48
⋮	⋮
15	120
⋮	⋮
40	320

Table 3.2 can be continued forever. We say the first multiple of 8 is 8, the second multiple of 8 is 16, the 15th multiple is 120, the 40th multiple is 320, and so on. To find any particular multiple, say the 23rd, we multiply the number by 23.

Examples A–E

Directions: List the required multiples.

Strategy: Multiply the natural number by the value specified.

Warm Ups A–E

A. List the first five multiples of 7.

A. List the first five multiples of 4.

$1 \cdot 4$ *or* 4 **Multiply 4 by 1, 2, 3, 4, and 5.**

$2 \cdot 4$ *or* 8

$3 \cdot 4$ *or* 12

$4 \cdot 4$ *or* 16

$5 \cdot 4$ *or* 20

The first five multiples are 4, 8, 12, 16, and 20.

B. List the first five multiples of 12.

B. List the first five multiples of 16.

$1 \cdot 16$ *or* 16 **Multiply 16 by 1, 2, 3, 4, and 5.**

$2 \cdot 16$ *or* 32

$3 \cdot 16$ *or* 48

$4 \cdot 16$ *or* 64

$5 \cdot 16$ *or* 80

The first five multiples are 16, 32, 48, 64, and 80.

C. Find the 6th, 24th, 29th, and 468th multiples of 9.

Calculator Example

C. Find the 6th, 24th, 29th, and 468th multiples of 7.

Strategy: Use a calculator to multiply 6, 24, 29, and 468 by 7.

The 6th multiple of 7 is 42, the 24th multiple of 7 is 168, the 29th multiple of 7 is 203, and the 468th multiple of 7 is 3276.

D. List all of the multiples of 8 between 201 and 241.

Calculator Example

D. List all of the multiples of 7 between 200 and 240.

Strategy: Use a calculator to make a quick estimate. Start with, say, the 28th multiple. Multiply 28 by 7.

$28(7) = 196$ **The product, 196, is less than 200.**

$29(7) = 203$ **This is the first multiple needed.**

The 29th multiple is 203 so keep going.

$29(7) = 203$ $30(7) = 210$ $31(7) = 217$

$32(7) = 224$ $33(7) = 231$ $34(7) = 238$

$35(7) = 245$

The multiples of 7 between 200 and 240 are 203, 210, 217, 224, 231, and 238.

E. If Jordan's math instructor assigns homework problems from 1 to 70 that are multiples of 4, which problems should he work?

E. Maria's mathematics teacher assigns homework problems from 1 to 60 that are multiples of 6. Which problems should she work?

Strategy: Find the multiples of 6 from 1 to 60 by multiplying 6 by 1, 2, 3, 4, 5, and so on, until the product is 60 or larger.

$1 \times 6 = 6$ $4 \times 6 = 24$

$2 \times 6 = 12$ $5 \times 6 = 30$

$3 \times 6 = 18$ $6 \times 6 = 36$

Answers to Warm Ups A. 7, 14, 21, 28, 35 B. 12, 24, 36, 48, 60 C. 54, 216, 261, 4212 D. 208, 216, 224, 232, 240 E. Jordan should work problems 4, 8, 12, 16, 20, 24, 28, 32, 36, 40, 44, 48, 52, 56, 60, 64, and 68.

$7 \times 6 = 42$ $9 \times 6 = 54$

$8 \times 6 = 48$ $10 \times 6 = 60$

The multiples of 6 from 1 to 60 are 6, 12, 18, 24, 30, 36, 42, 48, 54, and 60. These are the problems from 1 to 60 that Maria should work.

HOW AND WHY

Objective 2

Determine whether one whole number is a multiple of another whole number.

If one number is a multiple of another number, the first number must be divisible by the second number. To determine whether 852 is a multiple of 6, we check to see if it is divisible by 6. That is, see whether 852 is divisible by both 2 and 3.

852 **Divisible by 2 because the ones-place digit is 2.**

852 **Divisible by 3 because 8 + 5 + 2 = 15, is divisible by 3.**

So 852 is a multiple of 6.

If there is no divisibility test, use long division. For example, is 273 a multiple of 13? To find out divide by 13.

$$\begin{array}{r} 21 \\ 13\overline{)273} \\ \underline{26} \\ 13 \\ \underline{13} \\ 0 \end{array}$$

Because $273 \div 13 = 21$ with no remainder, $273 = 13 \times 21$. Therefore, 273 is a multiple of 13.

Examples F–J

Directions: Determine whether a given whole number is a multiple of another whole number.

Strategy: Use divisibility tests or long division to determine whether the first number is divisible by the second.

F. Is 126 a multiple of 9?

Use the divisibility test for 9.

$1 + 2 + 6 = 9$ **The sum of the digits is divisible by 9.**

So 126 is a multiple of 9.

Warm Ups F–J

F. Is 136 a multiple of 9?

Answer to Warm Up F. No

G. Is 3144 a multiple of 6?

G. Is 1116 a multiple of 6?

Use the divisibility tests for 2 and 3.

1116 is divisible by 2 **The ones-place digit is 6.**

1116 is divisible by 3 **The sum of the digits, 1 + 1 + 1 + 6 = 9, is divisible by 3.**

So 1116 is a multiple of 6.

H. Is 455 a multiple of 13?

H. Is 236 a multiple of 21?

Use long division.

$$\begin{array}{r} 11 \\ 21\overline{)236} \\ \underline{21} \\ 26 \\ \underline{21} \\ 5 \end{array}$$

The remainder is not 0.

No, 236 is not a multiple of 21.

I. Is 540 a multiple of 45?

I. Is 705 a multiple of 15?

Use either the divisibility tests for 3 and 5 or long division.

705 is divisible by 5 **The ones-place digit is 5.**

705 is divisible by 3 **The sum of the digits, 7 + 0 + 5 = 12, is divisible by 3.**

So 705 is a multiple of 15.

Long division gives the same result because $705 \div 15 = 47$.

J. Is 12 a multiple of 24?

J. Is 3 a multiple of 30?

The number 3 is not divisible by 30. The multiples of 30 are 30, 60, 90, 120, . . . and so on.

The smallest multiple of 30 is $30 \cdot 1 = 30$.
No, 3 is not a multiple of 30.

Answers to Warm Ups G. Yes H. Yes I. Yes J. No

NAME ________________ CLASS ________ DATE ________

Exercises 3.2

OBJECTIVE 1: *List multiples of a whole number.*

A.

List the first five multiples of the whole number.

1. 3

2. 7

3. 17

4. 16

5. 12

6. 14

7. 25

8. 35

9. 50

10. 45

B.

11. 48

12. 49

13. 68

14. 78

15. 83

16. 108

17. 123

18. 234

19. 345

20. 367

OBJECTIVE 2: *Determine whether one whole number is a multiple of another whole number.*

A.

Is each number a multiple of 6?

21. 42

22. 38

23. 84

24. 90

25. 89

26. 102

Is each number a multiple of 9?

27. 39

28. 63

29. 89

30. 108

31. 306

32. 109

B.

Is each number a multiple of 7?

33. 56

34. 65

35. 67

36. 105

37. 119

38. 168

Is each number a multiple of 6? of 9? of 15?

39. 648

40. 480

41. 495

42. 675

43. 690

44. 780

Is each number a multiple of 13? of 19?

45. 286

46. 323

47. 247

48. 741

C.

49. Jean is driving home from work one evening when she comes across a police safety inspection team stopping cars. She knows that the team chooses every fifth car to inspect. She counts and determines that she is 16th in line. Will she be selected to have her car safety-checked?

50. Using the information from Exercise 49, will Jean be selected if she is 35th in line?

51. A teacher assigns the problems from 1 to 52 that are multiples of 4. Which problems should the students work?

52. A teacher assigns the problems from 1 to 52 that are multiples of 6. Which problems should the students work?

53. The designer of new production-line equipment at the Bredworthy Jelly plant predicts that 55 small jars of jelly can be produced per minute. The next day a total of 605 jars are produced in 11 minutes. Is the designer's prediction accurate?

54. Referring to Exercise 53, if the plant produces 880 jars in 16 minutes, is the prediction accurate?

55. Joaquim supervises the quality control team at a bottling plant. His team is responsible for checking the quality of a line of soft drinks. The team is assigned to check bottles numbered 800 to 1000 in each batch. He decides that his team will check every bottle that is a multiple of 15. List the bottle numbers the team will check.

56. Minh was recently appointed to a supervisory position at a candy manufacturing plant. She oversees two employees who have been counting the daily total of bonbons. They count one at a time and find that the daily average produced is 1800. To be more efficient, what numbers less than 10 could Minh recommend that the employees use? Explain why you chose these numbers.

57. What is the first year in the 21st century that is a multiple of 35?

58. What was the last year in the 19th century that was a multiple of 36?

59. According to estimates by the Consumer Product Safety Commission, nationwide in 1996 there were approximately 105,000 injuries treated in hospital emergency rooms related to in-line skating. During the same year there were approximately 35,000 injuries due to skateboarding. Is the number of in-line skating injuries a multiple of the number of skateboarding injuries? If you were reporting a comparison of these injuries, how would you write the comparison so that it is easily understood?

Exercises 60–64 relate to the chapter application.

Some mathematicians are interested in numbers relative to how far away they are from the multiples of a certain number. This is called "modular arithmetic." We will use this idea later in the chapter to construct a special cipher.

Suppose that we want to count in the modulus of 5 (usually referred to as mod 5). Then every number is viewed by asking how far away it is from the last multiple of 5. Take 7, for instance. 7 = 2 mod 5 because 7 is 2 more than 5. 16 = 1 mod 5 because 16 is 1 more than 15, a multiple of 5. Here are the first 20 numbers and their mod 5 equivalents.

Number:	1	2	3	4	5	6	7	8	9	10
Mod 5:	1	2	3	4	0	1	2	3	4	0

Number:	11	12	13	14	15	16	17	18	19	20
Mod 5:	1	2	3	4	0	1	2	3	4	0

60. What is the pattern of the numbers in their mod 5 equivalents?

61. What do all the multiples of 5 have in common in their mod 5 equivalents?

62. Find the mod 5 equivalents of 33, 167, 845, and 1234.

63. Find the mod 12 equivalents of 18, 27, 34, 48, and 62.

64. Referring to Exercise 63, what is a method for finding the mod 12 equivalent of any number?

NAME ______________________ CLASS ______________ DATE ______________

STATE YOUR UNDERSTANDING

65. Write a short statement to explain why *every* multiple of 12 is also a multiple of 6.

66. One of the factors of 126 is 9. The number 9 is also a divisor of 126. Every factor of a number is also a divisor of that number. Explain in your own words why you think we have two different words, "factor" and "divisor" for such numbers.

67. Suppose you are asked to list all the multiples of 7 from 77 to 126. Describe a method you could use to be sure all the multiples are listed.

CHALLENGE

68. Is 3645 a multiple of 81?

69. Is 7998 a multiple of 93?

70. Find the largest number less than 5000 that is a multiple of 6, 9, and 17.

71. How many multiples of 3 are there between 1000 and 5000?

GROUP ACTIVITY

72. A high school marching band is practicing. When they march by twos, Elmo, the sousaphone player, has no one to march with. The band director rearranged them into threes, but Elmo was still left over.

a. Will someone be left over if the band marches in fours? Why or why not?

b. When the band marches by fives, Elmo is in a group. Find two possible sizes for the band.

73. Make a large circle on a sheet of paper. Mark nine points that are approximately the same distances apart on the circle and number them. Now list 10 consecutive multiples of your age. The first multiple and the last multiple will be associated with point number 1. Repeatedly add the digits together until you get a single digit. (For example: 75; $7 + 5 = 12$, then $1 + 2 = 3$.) Connect the dots in the order of the sums of the digits. Make one circle for each member of the group. Compare the designs.

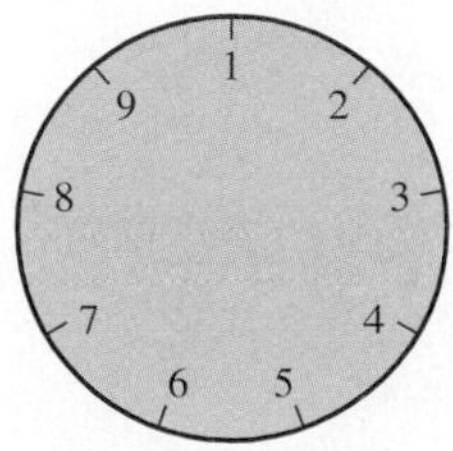

MAINTAIN YOUR SKILLS (SECTIONS 1.2, 1.3, 1.6)

74. Find the sum of 78, 1823, 503, 3, 1007, and 28.

75. Find the sum of 33,485, 8928, 29, and 2092.

76. Find the difference of 10,023 and 5987.

77. Find the difference of 21,003 and 9875.

78. Find the difference of 10,000 and 8923.

79. Find the product of 22, 32, and 9.

80. Find the product of 209 and 12.

81. Find the product of 297 and 98.

82. The Grow-em Good vegetable farm has a total of 12 vehicles that use diesel fuel. The storage tank that contains the supply of diesel fuel holds a total of 2500 gallons. Eighteen gallons are pumped into each of the 12 vehicles per day (average) from the tank. How many full days will the fuel in the tank last? How many gallons remain in the tank?

83. The local Pay and Take grocery has a sale on tomato soup. To ensure an ample supply, they buy 12 cases of tomato soup. Each case contains 48 cans. If the store sells an average of 42 cans per day, how many days will the 12 cases last? How many cans of soup are left?

3.3 Divisors and Factors

OBJECTIVES

1. Write a whole number as the product of two factors in all possible ways.
2. List all of the factors (divisors) of a whole number.

VOCABULARY

If a first number is a multiple of a second number, we say the second is a **factor** of the first. Thus, 7 is a factor of 28 because $4 \cdot 7 = 28$.

Recall that $14^2 = 14 \cdot 14 = 196$, so the **square** of 14 is 196. The number 196 is called a **perfect square.**

When two or more numbers are multiplied, each number is a factor. Therefore, if a first number is a factor of a second number, it is also a **divisor** of the second number.

HOW AND WHY

Objective 1

Write a whole number as the product of two factors in all possible ways.

Finding the factors of a number can be pictured using blocks, pennies, or other small objects. To illustrate, let's examine the factors of 12. Arrange 12 blocks or squares in a rectangle. The rectangle will have an area of 12 square units, and the length and width of the rectangle will each be factors of 12. For instance, the rectangle

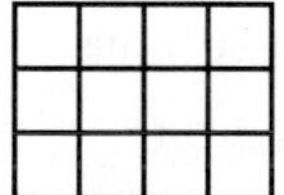

has 3 rows of 4 squares. Because $3 \times 4 = 12$, 3 and 4 are factors of 12. Now rearrange the blocks into a different rectangle.

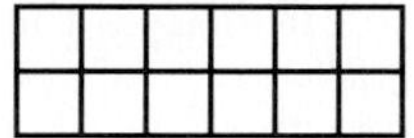

This arrangement shows that $2 \times 6 = 12$. So 2 and 6 are also factors of 12. Another arrangement is

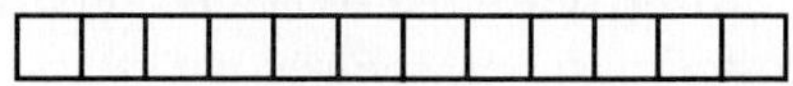

which shows that 1 and 12 are factors of 12. Arrangements with rows of any other length do not result in rectangles. The figures below show the results of trying to use rows of 5, 7, 8, 9, 10, or 11. None of these are rectangles.

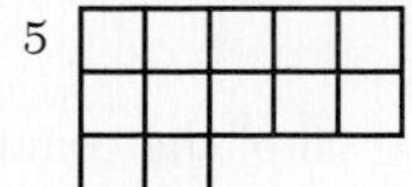

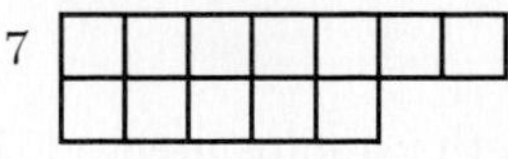

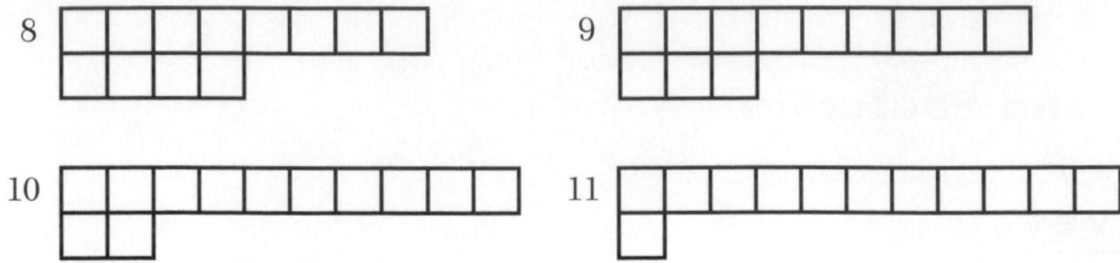

We conclude that the *only* pairs of factors of 12 are $1 \cdot 12$, $2 \cdot 6$, and $3 \cdot 4$.

To write 250 as a product of two factors in all possible ways, we could draw rectangles or divide by every number smaller than 250. So many divisions take too long. The following steps save time.

1. List all the counting numbers from 1 to the first number whose square is larger than 250. Because $15 \times 15 = 225$ and $16 \times 16 = 256$, we stop at 16.

1	6	11	16
2	7	12	
3	8	13	
4	9	14	
5	10	15	

We can stop at 16 since 250 divided by any number larger than 16 gives a quotient that is less than 16. But all the possible factors less than 16 are already in the chart.

2. Divide each of the listed counting numbers into 250. If it divides evenly, write the factors. If not, cross out the number.

$1 \cdot 250$	~~6~~	~~11~~	~~16~~
$2 \cdot 125$	~~7~~	~~12~~	
~~3~~	~~8~~	~~13~~	
~~4~~	~~9~~	~~14~~	
$5 \cdot 50$	$10 \cdot 25$	~~15~~	

These steps give us a list of all the two-factor products. Hence 250 written as a product of two factors in all possible ways is

$1 \cdot 250 \quad 2 \cdot 125 \quad 5 \cdot 50 \quad 10 \cdot 25$

To use the square method to write a whole number as the product of two factors in all possible ways

1. List all the counting numbers from 1 to the first number whose square is larger than the whole number.
2. For each number on the list, test whether the number is a divisor of the whole number.
3. If the number is not a divisor, cross it off the list.
4. If the number is a divisor, write the indicated product of two factors. The first factor is the tested number; the second factor is the quotient of the number and the tested number.

Warm Ups A–C

Examples A–C

Directions: Write the whole number as the product of two factors in all possible ways.

Strategy: Use the square method. Begin by testing all of the counting numbers from one to the first number whole square is larger than the whole number.

A. Write 78 as a product of two factors in all possible ways.

$1 \cdot 78$	~~5~~	~~9~~	**We can stop at 9, since $9^2 = 81$, which is larger than 78.**
$2 \cdot 39$	$6 \cdot 13$		
$3 \cdot 26$	~~7~~		
~~4~~	~~8~~		

The pairs of factors whose product is 78 are $1 \cdot 78$, $2 \cdot 39$, $3 \cdot 26$, and $6 \cdot 13$.

A. Write 48 as a product of two factors in all possible ways.

B. Write 150 as a product of two factors in all possible ways.

$1 \cdot 150$	$6 \cdot 25$	~~11~~	**We can stop at 13, since $13^2 = 169$, which is larger than 150.**
$2 \cdot 75$	~~7~~	~~12~~	
$3 \cdot 50$	~~8~~	~~13~~	
~~4~~	~~9~~		
$5 \cdot 30$	$10 \cdot 15$		

The pairs of factors whose product is 150 are $1 \cdot 150$, $2 \cdot 75$, $3 \cdot 50$, $5 \cdot 30$, $6 \cdot 25$, and $10 \cdot 15$.

B. Write 160 as a product of two factors in all possible ways.

C. A television station has 130 minutes of late-night programming to fill. In what ways can the time be scheduled if each program must last a whole number of minutes and if each schedule must include programs all the same length?

Strategy: List the pairs of factors of 130.

$1 \cdot 130$

1 program that is 130 minutes long or 130 programs that are 1 minute long (probably too short).

$2 \cdot 65$

2 programs that are 65 minutes long or 65 programs that are 2 minutes long.

~~3~~

~~4~~

$5 \cdot 26$

5 programs that are 26 minutes long or 26 programs that are 5 minutes long.

~~6~~

~~7~~

~~8~~

~~9~~

$10 \cdot 13$

10 programs that are 13 minutes long or 13 programs that are 10 minutes long.

~~11~~

~~12~~ **Stop, since $12^2 = 144$, which is larger than 130.**

C. A TV station has 90 minutes of programming to fill. In what ways can the time be scheduled if each program must last a whole number of minutes and if each schedule must include programs all the same length?

Answers to Warm Ups A. $1 \cdot 48$, $2 \cdot 24$, $3 \cdot 16$, $4 \cdot 12$, and $6 \cdot 8$ B. $1 \cdot 160$, $2 \cdot 80$, $4 \cdot 40$, $5 \cdot 32$, $8 \cdot 20$, and $10 \cdot 16$ C. The time can be filled with 1 program of 90 minutes or 90 programs of 1 minute long; 2 programs of 45 minutes or 45 programs 2 minutes long; 3 programs 30 minutes long or 30 programs 3 minutes long; 5 programs 18 minutes long or 18 programs 5 minutes long; 6 programs 15 minutes long or 15 programs 6 minutes long; 9 programs 10 minutes long or 10 programs 9 minutes long.

HOW AND WHY

Objective 2

List all the factors (divisors) of a whole number.

The square method to find the pairs of factors of a whole number also gives us the list of all factors (divisors) of the given whole number. When we found all pairs of factors of 250, we found $1 \cdot 250$, $2 \cdot 125$, $5 \cdot 50$, and $10 \cdot 25$. To make a list of all the factors in order, list the pairs vertically, then read the factors as the arrows indicate.

↓

$1 \cdot 250$ **Read down the left column of factors and then up the right column of factors.**

$2 \cdot 125$

$5 \cdot 50$

$10 \cdot 25$

↓→↑

Reading in the direction of the arrows, we see that the ordered list of all the factors of 250 contains eight numbers. The complete list is

1, 2, 5, 10, 25, 50, 125, and 250.

At the same time we have also listed all of the divisors of 250. Because each is a factor of 250, it is also a divisor of 250.

To list, in order, all the factors (divisors) of a number

1. List all pairs of factors of the number, in vertical form.
2. Read down the left column of factors and up the right column.

Warm Ups D–E

D. List all the factors of 82.

Examples D–E

Directions: List, in order, all the factors of a given whole number.

Strategy: Use the square method to find all the factors. To list the factors in order, read down the left column of factors and up the right column.

D. List all the factors of 84.

$1 \cdot 84$	~~5~~	~~9~~	**Stop at 10, since $10^2 = 100$. This gives us all the pairs of factors whose product is 84.**
$2 \cdot 42$	$6 \cdot 14$	~~10~~	
$3 \cdot 28$	$7 \cdot 12$		
$4 \cdot 21$	~~8~~		

↓

$1 \cdot 84$ **List the pairs and list the factors following the arrows.**

$2 \cdot 42$

$3 \cdot 28$

$4 \cdot 21$

$6 \cdot 14$

$7 \cdot 12$

↓→↑

In order, all the factors of 84 are 1, 2, 3, 4, 6, 7, 12, 14, 21, 28, 42, and 84.

Answer to Warm Up D. 1, 2, 41, and 82

E. List all the factors of 37.

$1 \cdot 37$	~~4~~	~~7~~	**Stop at 7 since $7^2 = 49$.**
~~2~~	~~5~~		
~~3~~	~~6~~		

The list of factors is 1 and 37.

E. List all the factors of 41.

Answer to Warm Up E. 1 and 41

NAME ______________________ CLASS ____________ DATE ____________

Exercises 3.3

OBJECTIVE 1: *Write a whole number as the product of two factors in all possible ways.*

A.

Write the whole number as the product of two factors in all possible ways.

1. 16

2. 18

3. 23

4. 31

5. 20

6. 24

7. 46

8. 50

9. 47

10. 53

B.

11. 98

12. 99

13. 100

14. 104

15. 114

16. 105

17. 115

18. 108

19. 444

20. 445

OBJECTIVE 2: *List all the factors (divisors) of a whole number.*

A.

List all the factors (divisors) of the whole number.

21. 16

22. 18

23. 23

24. 31

25. 30

26. 36

27. 55

28. 56

29. 65

30. 66

B.

31. 72

32. 76

33. 92

34. 90

35. 102

36. 112

37. 122

38. 132

39. 142

40. 152

C.

Write the whole number as the product of two factors in all possible ways.

41. 485

42. 655

43. 650

44. 480

45. 660

46. 720

47. In what ways can a television station schedule 120 minutes of time if each program must last a whole number of minutes, and if each schedule must include programs of all the same length?

Number of Programs	Length of Each
1	120 min
2	60 min
3	40 min
4	30 min
5	24 min
6	20 min
8	15 min
10	12 min
12	10 min
15	8 min
20	6 min
24	5 min
30	4 min
40	3 min
60	2 min
120	1 min

48. In what ways can a television station schedule 75 minutes of time if each program must last a whole number of minutes, and if each schedule must include programs of all the same length?

Number of Programs	Length of Each
1	75 min
3	25 min
5	15 min
15	5 min
25	3 min
75	1 min

49. Child care experts currently recommend that child care facilities have 1 adult for every 3 or 4 infants. The Playtime Daycare has 24 infants. If they staff according to the low end of the recommendation, how many adults do they need? If they staff according to the high end of the recommendation, how many adults do they need?

50. Child care experts currently recommend that child care facilities have 1 caregiver for 7 to 10 preschoolers. KinderCare has 65 preschoolers. If they have 5 caregivers, does this meet the recommendations?

51. In a recent year, the average single-family detached residence consumed $1230 in energy costs. Find the possible costs of energy consumption (whole number amounts) per person. Assume that each household has from 1 to 10 people.

52. A rectangular floor area requires 180 square-foot tiles. List all the possible whole-number dimensions that this floor could measure that would require all the tile.

53. Milwaukee High School has a marching band with 72 members. How many different ways can the members make a rectangular formation?

54. Mathematicians define **perfect numbers** as numbers that are the sum of all their divisors excluding the number itself. The first perfect number is six because $6 = 1 + 2 + 3$. Find another perfect number.

Exercises 55–56 relate to the chapter application.

Another kind of cipher involves substituting two-digit numbers for each letter. The numbers to be substituted are assigned using a grid system. One of the earliest recorded grid substitution systems was recorded by the Greek historian, Polybius, about 220 B.C. Here is the classic Polybius Grid.

	1	2	3	4	5
1	a	b	c	d	e
2	f	g	h	ij	k
3	l	m	n	o	p
4	q	r	s	t	u
5	v	w	x	y	z

The number associated with each letter gets its first digit from the row heading and the second digit from the column heading. So the word "computer" becomes

c	o	m	p	u	t	e	r
13	34	32	35	45	44	15	42

Notice that "i" and "j" share the same space. The number, 24, represents both "i" and "j." This is not a problem. Statistically, the occurrence of j in the English language is relatively rare. Other letters that are statistically rare are k, q, x, and z.

55. Decrypt the following message using the Polybius Grid.

11141 42413 44151 44434 31345 11587

56. A 5-by-5 grid is the standard, but it is not the only size grid that can be used.

a. Create a 4-by-6 grid.

b. How many spaces will have double letters?

c. Use your grid to encrypt the phrase "The power of love."

d. How many other grid sizes are possible using 24 spaces? Name them.

e. Is it possible to have a grid with 23 spaces? Why or why not?

STATE YOUR UNDERSTANDING

57. Explain the difference between a factor and a divisor of a number.

58. Explain how multiples, factors, and divisors are related to each other. Give an example.

CHALLENGE

59. Find the largest factor of 2959 that is less than 2959.

60. Find the largest factor of 3367 that is less than 3367.

GROUP ACTIVITY

61. Have each person in the group select a different whole number. If the number is even, divide by 2. If the number is odd, multiply by 3 and add 1. Record the results. Now do the same thing to the new number and record the results. Do the same thing a third time. Continue until you know you have gone as far as possible. Now compare your results with the other members of the group. You should all have reached the same point, regardless of the number you started with. Mathematicians believe that this will always happen, yet they have been unable to prove it is true for all whole numbers. This is called "The Syracuse Conjecture."

62. In 1990 each American consumed approximately 42 gallons of soft drinks, 25 gallons of milk, and 40 gallons of alcoholic beverages. Determine the number of ounces of soft drinks, milk, and alcoholic beverages each member of your group consumes in one week. Multiply these amounts by 52 to get the annual consumption. Divide by 128 to determine the number of gallons per category per person. Determine a group average for each category. Compare these with the 1990 national averages in a chart or graph.

MAINTAIN YOUR SKILLS (SECTIONS 1.3, 1.4, 1.7)

Multiply.

63. 49(51)

64. 69(403)

65. 88(432)

66. 307(502)

Divide.

67. $78\overline{)2418}$

68. $82\overline{)24,682}$

69. $401\overline{)9664}$

70. $401\overline{)160,805}$

71. How many speakers can be wired from a spool of wire containing 1000 feet if each speaker requires 24 feet of wire? How much wire is left?

72. A consumer magazine tested 15 brands of tires to determine the number of miles they could travel before the tread would be worn away. The results are in the chart below. What was the average mileage of the 15 brands?

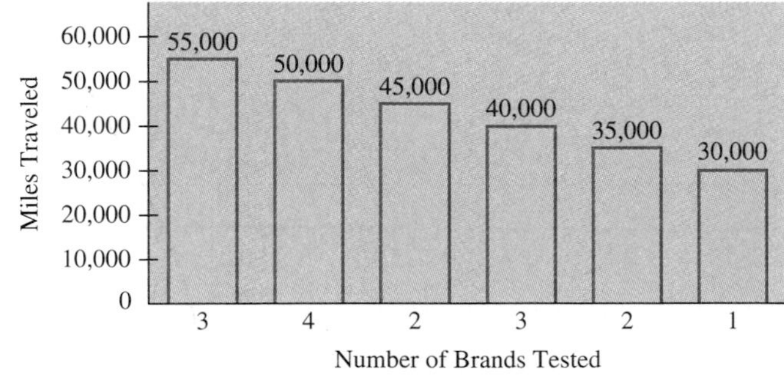

3.4 Primes and Composites

OBJECTIVE

Determine whether a whole number is prime or composite.

VOCABULARY

A **prime number** is a whole number greater than 1 with exactly two different factors (divisors). The factors are only 1 and the number itself.

A **composite number** is a whole number greater than 1 with more than two different factors (divisors).

HOW AND WHY

Objective

Determine whether a whole number is prime or composite.

The whole numbers zero (0) and one (1) are neither prime nor composite. Two (2) is the first prime number ($2 = 1 \cdot 2$), because 2 and 1 are the only factors of 2. Three (3) is a prime number ($3 = 1 \cdot 3$), because 1 and 3 are its only factors. Four (4) is a composite number ($4 = 1 \cdot 4$ and $4 = 2 \cdot 2$), because 4 has more than two factors.

To determine whether a number is prime or composite, list its factors or divisors in a chart like those in Section 3.3. Then count the number of factors. For instance, the chart for 299 is

$1 \cdot 299$	~~6~~	~~11~~	
~~2~~	~~7~~	~~12~~	
~~3~~	~~8~~	$13 \cdot 23$	**We stop here since 299 has at least four factors.**
~~4~~	~~9~~		
~~5~~	~~10~~		

Therefore, 299 is a composite number.

The chart for 307 is

$1 \cdot 307$	~~6~~	~~11~~	~~16~~	
~~2~~	~~7~~	~~12~~	~~17~~	
~~3~~	~~8~~	~~13~~	~~18~~	**Stop here since $18 \cdot 18 = 324$.**
~~4~~	~~9~~	~~14~~		
~~5~~	~~10~~	~~15~~		

We see that 307 has exactly two factors (1 and 307), so 307 is a prime number.

All primes up to a given number may be found by a method called the Sieve of Eratosthenes. Eratosthenes (born ca. 230 B.C.) is remembered for both the Prime Sieve and his method for measuring the circumference of the earth. The accuracy of his measurement, compared with modern methods, is within 50 miles, or six-tenths of one percent.

To use the famous Sieve to find the primes up to 30, list the numbers 2 to 30.

2	3	~~4~~	5	~~6~~	7
~~8~~	9	~~10~~	11	~~12~~	13
~~14~~	15	~~16~~	17	~~18~~	19
~~20~~	21	~~22~~	23	~~24~~	25
~~26~~	27	~~28~~	29	~~30~~	

The number 2 is prime, but all other multiples of 2 are not prime. They are crossed off.

The next number to consider is 3, which is prime.

2	3	~~4~~	5	~~6~~	7
~~8~~	~~9~~	~~10~~	11	~~12~~	13
~~14~~	~~15~~	~~16~~	17	~~18~~	19
~~20~~	~~21~~	~~22~~	23	~~24~~	25
~~26~~	~~27~~	~~28~~	29	~~30~~	

All remaining multiples of 3 are not prime, so they are crossed off.

The next number, 4, has already been eliminated. The next number, 5, is prime.

2	3	~~4~~	5	~~6~~	7
~~8~~	~~9~~	~~10~~	11	~~12~~	13
~~14~~	~~15~~	~~16~~	17	~~18~~	19
~~20~~	~~21~~	~~22~~	23	~~24~~	~~25~~
~~26~~	~~27~~	~~28~~	29	~~30~~	

All remaining multiples of 5 are not prime, so they are crossed off.

The multiples of the other numbers, except themselves, have been crossed off. We need to test divisors only up to the first number whose square ($6 \cdot 6 = 36$) is larger than 30. So, the primes less than 30 are 2, 3, 5, 7, 11, 13, 17, 19, 23, and 29.

From the preceding Sieve we see that we can shorten the factor chart by omitting all the numbers except those that are prime. For instance, is 371 prime or composite?

$1 \cdot 371$
~~2~~
~~3~~
~~5~~
$7 \cdot 53$ **Stop here, since we do not need all factors.**

Since $7 \cdot 53 = 371$, 371 is composite. (It has *at least* four factors: 1, 7, 53, and 371). We know that a number is prime if no smaller prime divides it evenly.

Keep the divisibility tests for 2, 3, and 5 in mind since they are prime numbers.

To tell whether a number is prime or composite (0 and 1 are neither prime nor composite)

Test 1 and all prime numbers whose square is less than the number.

a. If the number has exactly two divisors (factors), it is prime.
b. If the number has more than two divisors (factors), it is composite.

Examples A–F

Directions: Determine whether the number is prime or composite.

Strategy: Test 1 and all the possible prime factors of the number. If there are exactly two factors, the number is prime.

A. Is 103 prime or composite?
Test 1 and the prime numbers from 2 to 11. We stop at 11 since $11^2 > 103$.

$1 \cdot 103$	~~5~~	**The numbers 2, 3, and 5 can be crossed out using the divisibility tests. Eliminate 7 and 11 by division.**
~~2~~	~~7~~	
~~3~~	~~11~~	

The number 103 is a prime number.

B. Is 143 prime or composite?
Test 1 and the prime numbers from 2 to 13.

$1 \cdot 143$	~~7~~	
~~2~~	$11 \cdot 13$	**Stop testing at 11 since we have at least four factors.**
~~3~~		
~~5~~		

The number 143 is a composite number.

C. Is 251 prime or composite?
Test 1 and the prime numbers from 2 to 17.

$1 \cdot 251$	~~7~~	
~~2~~	~~11~~	
~~3~~	~~13~~	
~~5~~	~~17~~	**Stop at 17 since $17^2 > 251$.**

After testing all the prime numbers in the list, we determine that 251 has only two factors. So 251 is a prime number.

D. Is 451 prime or composite?
Test 1 and the prime numbers from 2 to 23.

$1 \cdot 451$	~~7~~
~~2~~	$11 \cdot 41$
~~3~~	
~~5~~	

We stop testing at 11 because we have at least four factors (1, 11, 41, and 451). So 451 is a composite number.

E. Is 124,645 prime or composite?

124,645 is divisible by 5 **The ones-place digit is 5.**

We have at least three factors, 1, 5, and 124,645. So 124,645 is a composite number.

Warm Ups A–F

A. Is 73 prime or composite?

B. Is 117 prime or composite?

C. Is 257 prime or composite?

D. Is 551 prime or composite?

E. Is 314,211 prime or composite?

Answers to Warm Ups A. Prime B. Composite C. Prime D. Composite E. Composite

F. In the same contest John was asked "Is the number 234,425 prime or composite?" What should he have answered?

F. Janna won a math contest at her school. One of the questions in the contest was "Is 234,423 prime or composite?" What should Janna have answered?

Strategy: Test the number for divisibility by prime numbers.

234,423 is not divisible by 2	**The one-place digit is 3.**
234,423 is divisible by 3	**2 + 3 + 4 + 4 + 2 + 3 = 18 which is divisible by 3.**

We have at least three factors, 1, 3, and 234,423. Janna should have answered "composite."

Answer to Warm Up F. Janna should have answered "composite."

NAME ______________________ CLASS ____________ DATE ____________

Exercises 3.4

OBJECTIVE: *Determine whether a whole number is prime or composite.*

A.

Tell whether the number is prime or composite.

1. 8 **2.** 4 **3.** 7 **4.** 11

5. 12 **6.** 14 **7.** 19 **8.** 21

9. 22 **10.** 31 **11.** 29 **12.** 23

13. 24 **14.** 28 **15.** 37 **16.** 41

17. 26 **18.** 36 **19.** 43 **20.** 47

21. 40 **22.** 44

B.

Tell whether the number is prime or composite.

23. 48 **24.** 51 **25.** 61 **26.** 59

27. 88 **28.** 81 **29.** 83 **30.** 87

31. 91 **32.** 93 **33.** 97 **34.** 99

35. 110 **36.** 109 **37.** 133 **38.** 123

39. 147 **40.** 153 **41.** 203 **42.** 201

43. 249 **44.** 287 **45.** 345 **46.** 381

C.

Tell whether the number is prime or composite.

47. 439 **48.** 419 **49.** 449 **50.** 459

51. 825 **52.** 927 **53.** 2345 **54.** 2347

55. The year 1999 is a prime number. What is the next year that is a prime number?

56. What was the last year before 1997 that was a prime number?

57. The Wright brothers made the first sustained airplane flight at Kitty Hawk, North Carolina, in December 1903. Is 1903 a prime number?

58. Tell whether the year of your birth is a prime or composite number.

59. In the motion picture *Contact,* a scientist played by Jodie Foster intercepts a message from outer space. The message begins with a series of impulses grouped according to successive prime numbers. How many impulses came in the group that followed 47?

NAME ______________________ CLASS ____________ DATE ____________

60. Vigorous physical activity can lead to accidental injury. The following table gives estimates by the Consumer Product Safety Commission of nationwide basketball injuries. These are estimates of the number of hospital emergency room injuries.

Year	1993	1994	1995	1996
Injuries	761,171	716,182	692,396	653,675

Three of the four entries in the table are clearly not prime numbers. Identify them and tell how you know they are not prime.

61. How many different rectangular arrangements can be made with 17 blocks?

62. Mathematicians define **perfect numbers** as numbers that are the sum of all their divisors, excluding the number itself. Explain why a prime number cannot be a perfect number.

 Exercises 63–64 relate to the chapter application.

The Rail Fence cipher is an example of a transpose cipher. A transpose cipher contains the actual letters in the word or message, but they are in a different order. A more difficult variation involves writing the message in lines of predetermined length. The cipher is made by reading down the columns. For example, to encipher the message: "But darling, most of all, I love how you love me," write the text in lines of nine letters.

```
b  u  t  d  a  r  l  i  n
g  m  o  s  t  o  f  a  l
l  i  l  o  v  e  h  o  w
y  o  u  l  o  v  e  m  e
```

Now read down each column to make the cipher.

BGLYU MIOTO LUDSO LATVO ROEVL FHEIA OMNLW E

63. Make a different transpose cipher for the same message using seven columns instead of nine.

64. Decrypt the following message, which is a seven-column transpose cipher. A few null characters (letters not in the original message) were added to make the groups come out even.

WVDTO SAIHE OWVEN OAGWH ECDNT OIABO EPSTT TUNMD
LTHST DOMOO ILAHT

STATE YOUR UNDERSTANDING

65. Explain the difference between prime numbers and composite numbers. Give an example of each.

66. What is the minimum number of factors that a composite number can have? Explain why.

67. What is the maximum number of factors that a composite number can have? Explain why.

CHALLENGE

68. Is 26,087 a prime or a composite number?

69. Is 37,789 a prime or a composite number?

70. A sphenic number is a number that is a product of three unequal prime factors. The smallest sphenic number is $2 \cdot 3 \cdot 5 = 30$. Is 4199 a sphenic number?

GROUP ACTIVITY

71. Mathematicians have established that every prime number greater than 3 is either one more than or one less than a multiple of 6. Use the Sieve of Eratosthenes to identify all the prime numbers less than 100. Divide these primes up and verify that each of them is one away from a multiple of 6. Here is a step-by-step confirmation of this fact. As a group, supply the reasons for each step of the confirmation.

Consider a portion of the number line graphed below. The first point, a, is at an even number.

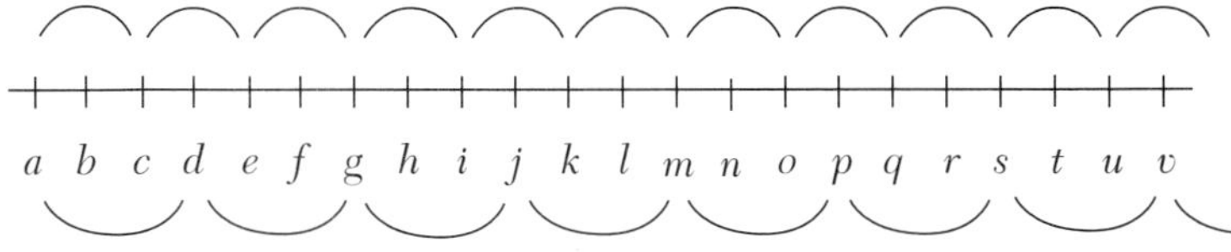

Statement	Reason
a. If a is even, then so are $c, e, g, i, k, m, o, q, s,$ and u.	a.
b. If a is a multiple of 3, then so are $d, g, j, m, p, s,$ and v.	b.
c. $a, g, m,$ and s are multiples of 6.	c.
d. None of the points in steps 1, 2, or 3 can be prime.	d.
e. The only possible places for prime numbers to occur are at $b, f, h, l, n, r,$ and t.	e.
f. All primes are one unit from a multiple of 6.	f.

72. Have each person in your group determine the total of the ages of the people now living in their household. Are these prime numbers? Report back to your class and compute the class total. Is this a prime number?

MAINTAIN YOUR SKILLS (SECTIONS 1.1, 1.5, 1.6)

73. True or false? $82 < 91$

74. True or false? $289 > 300$

75. Find the value of 13^4.

76. Find the value of 5^6.

77. Multiply: 29×10^5

78. Divide: $750{,}000 \div 10^3$

79. Subtract the product of 9 and 12 from 753.

80. Add the quotient of 93 and 3 to 122.

81. Multiply the product of 19 and 18 times 22.

82. Divide the sum of 822 and 90 by 24.

3.5

Prime Factorization

OBJECTIVES

1. Write the prime factorization of a whole number by repeated division.

2. Write the prime factorization of a whole number using the Tree Method.

VOCABULARY

The **prime factorization** of a whole number is the number written as an indicated product of prime numbers. There are two ways of asking the same question.

"What is the prime factorization of this number?"

"Write this number in prime factored form."

$21 = 3 \cdot 7$ and $30 = 2 \cdot 3 \cdot 5$ are prime factorizations.

$30 = 3 \cdot 10$ is not a prime factorization, because 10 is not prime.

Recall that exponents show repeated factors. This can save space in writing.

$2 \cdot 2 \cdot 2 = 2^3$ and $3 \cdot 3 \cdot 3 \cdot 3 \cdot 7 \cdot 7 \cdot 7 = 3^4 \cdot 7^3$

HOW AND WHY

Objective 1

Write the prime factorization of a whole number by repeated division.

In chemistry, we learn that every compound in the world is made up of a particular combination of basic elements. For instance, salt is NaCl and water is H_2O. This means that every molecule of salt contains one atom of sodium (Na) and one atom of chlorine (Cl). Sodium and chlorine are two of the basic elements. Similarly, every molecule of water is made up of two atoms of hydrogen (H) and one of oxygen (O), which are also basic elements. The chemical formula tells how many units of each basic element are needed to make the compound.

In mathematics, the basic elements are the prime numbers. Every whole number (except 0 and 1) is either prime or a unique combination of prime factors. Finding the prime factorization of a composite number is comparable to finding the chemical formula for a compound. The prime factorization simply allows us to see the basic elements of the number.

To find the prime factorization, repeatedly divide by prime numbers until the quotient is 1. Then write the number as the product of these primes. To find the prime factors of 48, divide 48 by 2 and then divide the quotient, 24, by 2, and so on. When the quotient can no longer be divided by 2, we divide by 3, 5, 7, 11, . . . to check for other prime factors. To save time and space we do not rewrite the division problem, but simply divide each quotient, starting at the top and dividing down until the quotient is 1:

2)48
2)24
2)12 $48 = 2 \cdot 2 \cdot 2 \cdot 2 \cdot 3 = 2^4 \cdot 3$
2)6
3)3
1

If there is a large prime factor, you will find it when you have tried each prime whose square is smaller than the number. Consider the number 822.

2)822
3)411
137)137 **137 is not divisible by 2, 3, 5, 7, 11, or 13;**
1 **also $13^2 = 169 > 137$. Therefore, 137 is prime.**

So $822 = 2 \cdot 3 \cdot 137$.

In each case the goal is to keep dividing until the quotient is 1.

To write the prime factorization of a whole number using the repeated division method

1. Divide the whole number and each succeeding quotient by a prime number until the quotient is 1.
2. Write the indicated product of all of the divisors.

Warm Ups A–D

A. Prime factor 28.

B. Prime factor 50.

Examples A–D

Directions: Write the prime factorization of the whole number.

Strategy: Use repeated division by prime numbers.

A. Prime factor 42.

Strategy: The ones-place digit is 2 so start dividing by 2.

2)42
3)21 **21 is divisible by 3.**
7)7 **7 is prime, divided by 7. The quotient is 1.**
1

CHECK: Multiply all of the divisors.

$2 \cdot 3 \cdot 7 = 42$

The prime factorization of 42 is $2 \cdot 3 \cdot 7$.

B. Prime factor 848.

Strategy: 848 is even so start dividing by 2. Continue until 2 is no longer a divisor. Then try another prime number as a divisor.

2)848
2)424
2)212
2)106
53)53 **53 is not divisible by 3, 5, 7, or 11 so 53 is a prime number.**
1 **The quotient is 1.**

CHECK: $2 \cdot 2 \cdot 2 \cdot 2 \cdot 53 = 848$

The prime factorization of 848 is $2^4 \cdot 53$.

Answers to Warm Ups A. $2^2 \cdot 7$ B. $2 \cdot 5^2$

Calculator Example

C. Prime factor 5280.

Strategy: Record the number of times you divide by each prime number.

$2 \cdot 2 \cdot 2 \cdot 2 \cdot 2 \cdot 3 \cdot 5 \cdot 11$

The prime factorization of 5280 is $2^5 \cdot 3 \cdot 5 \cdot 11$.

C. Prime factor 360.

D. When he was adding two fractions, Chris needed to find the Least Common Denominator of $\frac{1}{18}$ and $\frac{5}{24}$; therefore, he needed to write the prime factorization of 18 and 24. What are the prime factors of 18 and 24?

Strategy: Because 18 is divisible by 2, we start by dividing by 2.

$$\begin{array}{r|l} 2 & 18 \\ 3 & 9 \\ 3 & 3 \\ & 1 \end{array}$$

Strategy: Because 24 is also divisible by 2, start by dividing by 2.

$$\begin{array}{r|l} 2 & 24 \\ 2 & 12 \\ 2 & 6 \\ 3 & 3 \\ & 1 \end{array}$$

The prime factors of 18 and 24 are

$18 = 2 \cdot 3 \cdot 3$ or $2 \cdot 3^2$

$24 = 2 \cdot 2 \cdot 2 \cdot 3$ or $2^3 \cdot 3$

D. Christine needed to know the Least Common Denominator of $\frac{7}{36}$ and $\frac{11}{54}$, so she needed the prime factorization of 36 and 54. What are the prime factors of 36 and 54?

HOW AND WHY

Objective 2

Write the prime factorization of a whole number using the Tree Method.

Another method for prime factoring is the Tree Method. We draw "factor branches" using *any* two factors of the number. Additional branches are then formed by using factors of the numbers at the end of each branch. A branch stops splitting when in ends in a prime number. We use the Tree Method to prime factor 48.

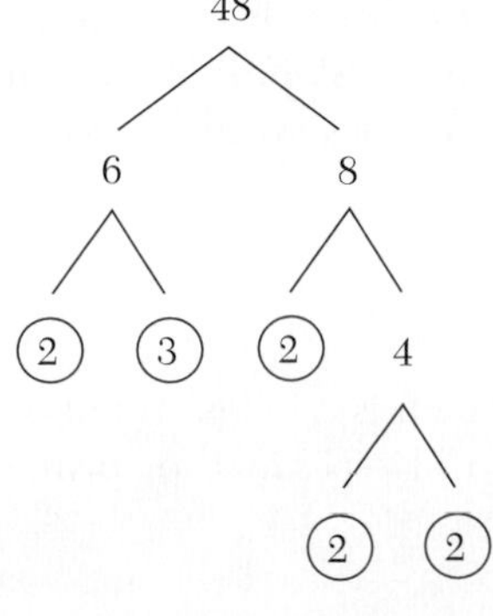

$48 = 2 \cdot 2 \cdot 2 \cdot 2 \cdot 3$ or $2^4 \cdot 3$

Answers to Warm Ups C. $2^3 \cdot 3^2 \cdot 5$ D. The prime factors of 36 are $2^2 \cdot 3^2$ and the prime factors of 54 are $2 \cdot 3^3$.

It is worth noting that there are often several different trees for the same number. Examine these other two trees used to prime factor 48.

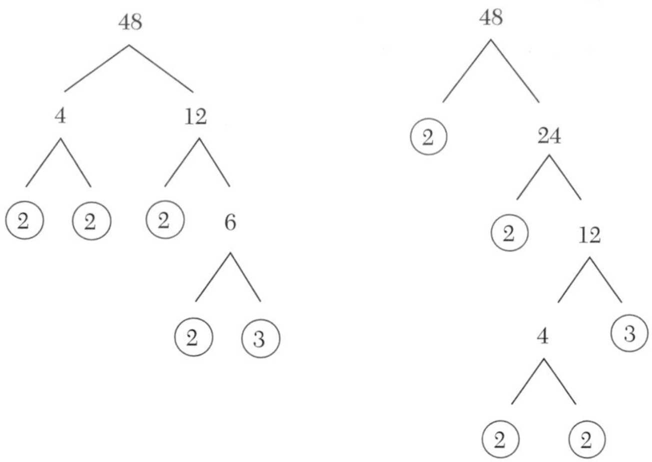

In each case we can see that $48 = 2^4 \cdot 3$. It does not matter which tree we use. As long as you keep branching until you come to a prime number, you will get the same prime factorization. In fact, any tree results in the same prime factorization as the repeated division method.

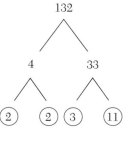

To prime factor a given whole number using the Tree Method

Draw factor branches starting with any two factors of the number. Form additional branches by using factors of the number at the end of each branch. The factoring is complete when the number at the end of each branch is a prime number.

Warm Ups E–F

E. Prime factor 126.

Examples E–F

Directions: Write the prime factorization of the whole number.

Strategy: Use the Tree Method to write the given whole number as a product of primes.

E. Prime factor 132.

Strategy: We start with $4 \cdot 33$ as the pair of factors of 132.

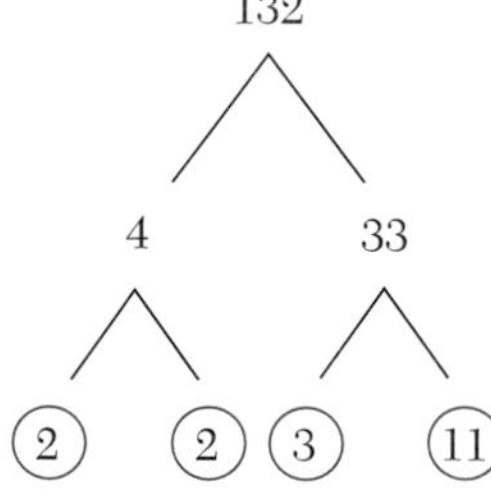

We could have used 2 and 66, 11 and 12, or 3 and 44 as the first pair of factors. The same group of prime factors would have been obtained.

Each number at the end of a branch is a prime number. We compute the product as a check. Because $2 \cdot 2 \cdot 3 \cdot 11 = 132$, the prime factorization of 132 is $2 \cdot 2 \cdot 3 \cdot 11 = 2^2 \cdot 3 \cdot 11$.

Answer to Warm Up E. $2 \cdot 3 \cdot 3 \cdot 7$ or $2 \cdot 3^2 \cdot 7$

F. Prime factor 468.

Strategy: Select any two factors whose product is 468. The tree will be shorter if larger numbers are used.

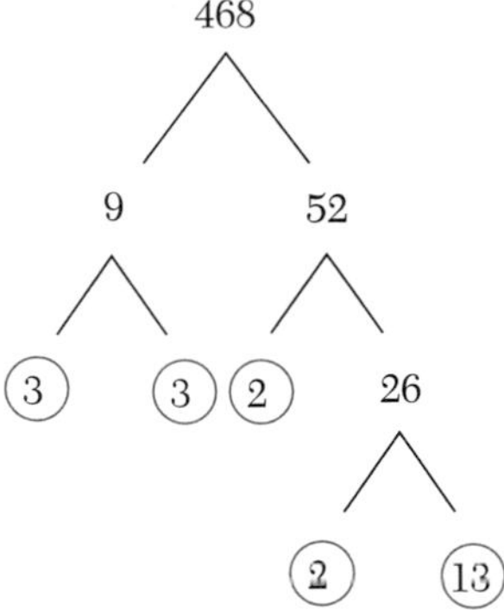

Another pair of factors of 468 are chosen, 4 and 117. Observe that the same set of prime factors is obtained. They are just written in a different order.

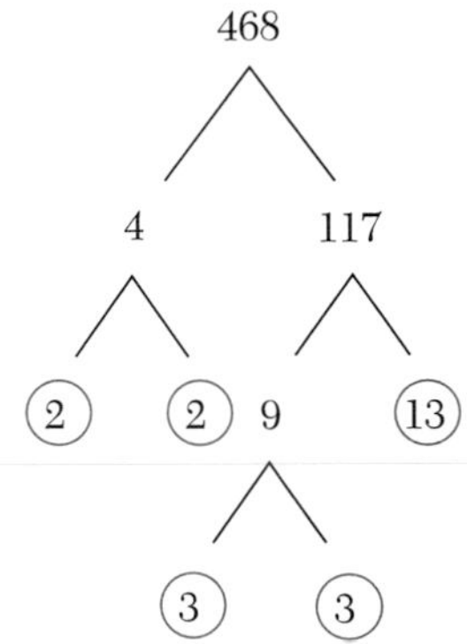

Because $3 \cdot 3 \cdot 2 \cdot 2 \cdot 13 = 2 \cdot 2 \cdot 13 \cdot 3 \cdot 3 = 468$, the prime factorization is $2 \cdot 2 \cdot 3 \cdot 3 \cdot 13$ or $2^2 \cdot 3^2 \cdot 13$.

F. Prime factor 612.

Answer to Warm Up F. $2 \cdot 2 \cdot 3 \cdot 3 \cdot 17$ or $2^2 \cdot 3^2 \cdot 17$

NAME ______________________ CLASS ____________ DATE ____________

Exercises 3.5

OBJECTIVE 1: *Write the prime factorization of a whole number by repeated division.*
OBJECTIVE 2: *Write the prime factorization of a whole number using the Tree Method.*

A.

Write the prime factorization of the whole number using either method.

1. 12 **2.** 14 **3.** 16 **4.** 18

5. 20 **6.** 21 **7.** 24 **8.** 26

9. 28 **10.** 30 **11.** 32 **12.** 36

13. 48 **14.** 50 **15.** 72 **16.** 70

17. 81 **18.** 75 **19.** 90 **20.** 95

B.

Write the prime factorization of the whole number.

21. 91 **22.** 51 **23.** 64 **24.** 96

25. 102 **26.** 104 **27.** 132 **28.** 123

29. 153 **30.** 162 **31.** 180 **32.** 190

33. 182 **34.** 216 **35.** 225 **36.** 230

37. 306 **38.** 324 **39.** 396 **40.** 369

C.

Write the prime factorization of the whole number.

41. 303 **42.** 515 **43.** 323

44. 437 **45.** 209 **46.** 247

47. 401 **48.** 409 **49.** 625

50. 864 **51.** 1218 **52.** 1302

NAME ______________________ CLASS __________ DATE __________

53. Determine the prime factorization of the highest temperature in the United States for today.

54. Determine the prime factorization for the sum of the digits in today's numerical date (month, day, and year).

55. Determine the prime factorization of your birth year.

Exercises 56–57 relate to the chapter application.

Many ciphers used in World War II were column transpose ciphers, made more difficult by scrambling the order of the columns according to some secret keyword that changed daily, for example, the message "Love me tender, love me sweet, never let me go" can be arranged in seven columns.

l	o	v	e	m	e	t
e	n	d	e	r	l	o
v	e	m	e	s	w	e
e	t	n	e	v	e	r
l	e	t	m	e	g	o

But instead of reading the columns in order, we use a seven-letter keyword such as FACTORS. Each letter in the keyword heads one column and the columns are then read in alphabetical order according to the keyword.

3	1	2	7	4	5	6
F	A	C	T	O	R	S
l	o	v	e	m	e	t
e	n	d	e	r	l	o
v	e	m	e	s	w	e
e	t	n	e	v	e	r
l	e	t	m	e	g	o

The encrypted message is: onete vdmnt level mrsve elweg toero eeeem. The keyword has the effect of scrambling the columns unpredictably.

56. Use the keyword "NUMBERS" to encrypt the message "Learning to love yourself—the greatest love of all."

57. Decrypt the following message, which is an eight-column transpose cipher with the keyword MULTIPLY.

EISE ENAO YMYE IHEU RAEM VGUV MICL TBOX

STATE YOUR UNDERSTANDING

58. Explain how to prime factor 660.

59. Explain how to determine whether a number is written in prime factored form.

CHALLENGE

60. Prime factor 1892.

61. Prime factor 2982.

62. Prime factor 2431.

GROUP ACTIVITY

63. The Goldbach Conjecture states that every even number greater than 4 can be written as the sum of two prime numbers. Mathematicians think this is always true, but as yet no one has proved it for *all* even numbers. Fill in the chart below.

Even Number	Sum of Primes	Even Number	Sum of Primes
6	3 + 3	26	
8		28	
10		30	
12		32	
14		34	
16		36	
18		38	
20		40	
22		42	
24		44	

Do you believe the Goldbach Conjecture?

64. Call your local recycling center to determine the total number of pounds of cans, glass, and newspaper that is recycled in your area each month over a 12-month period. Find the prime factorization of each of these numbers.

MAINTAIN YOUR SKILLS (SECTIONS 2.2, 2.3, 2.4, 3.2, 3.3)

65. Is 3003 a multiple of 11?

66. Is 4004 a multiple of 11?

67. Is 1001 a multiple of 11?

68. Is 5005 a multiple of 11?

69. Is 7 a divisor of 5005?

70. Is 13 a divisor of 5005?

Use the following figure for Exercises 71–72.

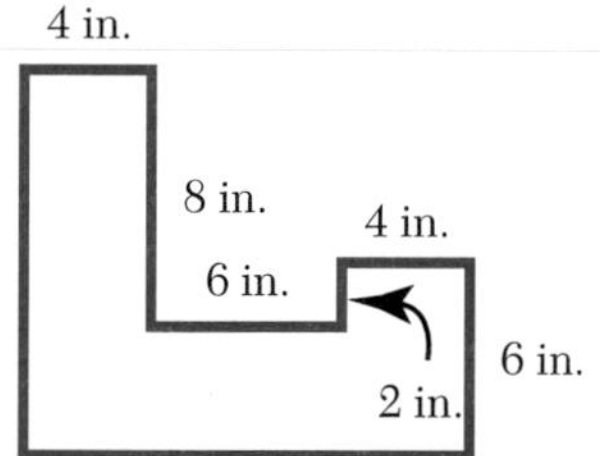

71. Find the perimeter.

72. Find the area.

73. Find the volume of a cube that measures 11 cm on an edge.

74. Find the volume of a rectangular box with length 1 ft 8 in., width 9 in., and height 5 in.

3.6

Least Common Multiple

OBJECTIVES

1. Find the Least Common Multiple of two or more whole numbers using the Individual Prime Factoring Method.
2. Find the Least Common Multiple of two or more whole numbers using the Group Prime Factoring Method.

VOCABULARY

The **Least Common Multiple** of two or more whole numbers is

1. The smallest natural number that is a multiple of each whole number.
2. The smallest natural number that has each whole number as a factor.
3. The smallest natural number that has each whole number as a divisor.
4. The smallest natural number that each whole number will divide evenly.

These statements are four ways of expressing the same idea.

LCM is an abbreviation of least common multiple.

HOW AND WHY

Objective 1

Find the Least Common Multiple of two or more whole numbers using the Individual Prime Factoring Method.

The factorizations of this chapter are most often used to simplify fractions and find LCMs. LCMs are used to compare, add, and subtract fractions. In algebra, LCMs are useful in equation solving.

Find the LCM of 21 and 35. This can be done by listing the multiples of each and finding the smallest one common to both lists:

Multiples of 21: 21, 42, 63, 84, $\boxed{105}$, 126, 147, 168, 189, $\boxed{210}$, 231, . . .
Multiples of 35: 35, 70, $\boxed{105}$, 140, 175, $\boxed{210}$, 245, 280, 315, 350, . . .

The LCM of 21 and 35 is 105, since it is the smallest common multiple in both lists. This fact can be stated in four (equivalent) ways:

1. 105 is the smallest natural number that is a multiple of 21 and 35.
2. 105 is the smallest natural number that has both 21 and 35 as factors.
3. 105 is the smallest natural number that has both 21 and 35 as divisors.
4. 105 is the smallest natural number that both 21 and 35 will divide evenly.

Finding the LCM by this method has one drawback: you might have to list hundreds of multiples. For this reason we look for a shortcut.

To find the LCM of 12 and 18, write the prime factorization of each. Write these prime factors in columns, so that the prime factors of 18 are under the prime factors of 12. Leave blank spaces for prime factors that do not match.

		Primes with Largest Exponents
$12 = 2 \cdot 2 \cdot 3$	$= 2^2 \cdot 3^1$	
$18 = 2 \cdot 3 \cdot 3$	$= 2^1 \cdot 3^2$	2^2 and 3^2

Since 12 must divide the LCM, the LCM must have $2 \cdot 2 \cdot 3$ as part of its factors; and since 18 must divide the LCM, the LCM must also have $2 \cdot 3 \cdot 3$ as part of its factors. Thus, the LCM is $2 \cdot 2 \cdot 3 \cdot 3$, or 36. Note that the LCM is the product of the highest power of each prime factor.

What is the LCM of 16, 10, and 24?

		Primes with Largest Exponents
$16 = 2 \cdot 2 \cdot 2 \cdot 2$	$= 2^4$	
$10 = 2 \cdot 5$	$= 2^1 \cdot 5^1$	$2^4, 3^1, 5^1$
$24 = 2 \cdot 2 \cdot 2 \cdot 3$	$= 2^3 \cdot 3^1$	

The LCM $= 2 \cdot 2 \cdot 2 \cdot 2 \cdot 3 \cdot 5 = 2^4 \cdot 3^1 \cdot 5^1$.
$= 16 \cdot 15 = 240$

To find the LCM of two or more whole numbers using the Individual Prime Factoring Method

1. Write each number in prime factored form using exponents.
2. Find the product of the highest power of each prime factor.

Warm Ups A–C

A. Find the LCM of 12 and 20.

B. Find the LCM of 10, 12, and 30.

Examples A–C

Directions: Find the LCM of two or more whole numbers.

Strategy: Use the Individual Prime Factoring Method.

A. Find the LCM of 16 and 20.

Strategy: Prime factor 16 and 20; write each in exponent form.

$16 = 2 \cdot 2 \cdot 2 \cdot 2 = 2^4$
$20 = 2 \cdot 2 \cdot 5 = 2^2 \cdot 5^1$

The different prime factors are 2 and 5. The largest exponent of 2 is 4 and the largest exponent of 5 is 1. Multiply the powers.

The LCM of 16 and 20 is $2^4 \cdot 5 = 80$.

B. Find the LCM of 18, 24, and 30.

Strategy: Prime factor 18, 24, and 30; write each in exponent form.

$18 = 2 \cdot 3 \cdot 3 = 2^1 \cdot 3^2$
$24 = 2 \cdot 2 \cdot 2 \cdot 3 = 2^3 \cdot 3^1$
$30 = 2 \cdot 3 \cdot 5 = 2^1 \cdot 3^1 \cdot 5^1$

The different prime factors needed are 2, 3, and 5. The largest exponent of 2 is 3, of 3 is 2 and of 5 is 1. Multiply the powers.

The LCM of 18, 24, and 30 is $2^3 \cdot 3^2 \cdot 5 = 360$.

Answers to Warm Ups A. 60 B. 60

C. Find the LCM of 12, 16, 24, and 36.

Strategy: Prime factor and write each in exponent form.

$12 = 2 \cdot 2 \cdot 3 = 2^2 \cdot 3^1$

$16 = 2 \cdot 2 \cdot 2 \cdot 2 = 2^4$

$24 = 2 \cdot 2 \cdot 2 \cdot 3 = 2^3 \cdot 3^1$

$36 = 2 \cdot 2 \cdot 3 \cdot 3 = 2^2 \cdot 3^2$

The different prime factors needed are 2 and 3. The largest exponent of 2 is 4 and of 3 is 2. Multiply the powers.

The LCM of 12, 16, 24, and 36 is $2^4 \cdot 3^2 = 144$.

C. Find the LCM of 8, 12, 18, and 24.

HOW AND WHY

Objective 2

Find the Least Common Multiple of two or more whole numbers using the Group Prime Factoring Method.

A second method for finding the LCM is sometimes referred to as the Group Prime Factoring Method. To use this method with two numbers, find a prime number that will divide both. If there is no prime number that will divide both, then multiply the two numbers and this is the LCM. Keep doing this process until no prime number will divide the two numbers. The product of all of the divisors and the remaining quotients is the LCM.

We use this method to find the LCM of 18 and 24.

2)	18	24	**Divide both numbers by 2.**
3)	9	12	**Divide the quotients by 3.**
	3	4	

There is no common prime factor (divisor) in 3 and 4. Therefore, the LCM is the product of the divisors and the remaining quotients. The LCM of 18 and 24 is $2 \cdot 3 \cdot 3 \cdot 4$, or 72.

To find the LCM of three or more numbers, find a prime number that will divide at least two of the numbers. Divide all numbers, if possible. Otherwise, bring down the numbers that cannot be divided evenly. Continue until no common factors remain. We find the LCM of 12, 18, and 24.

2)	12	18	24	**Divide each by 2.**
2)	6	9	12	**Divide 6 and 12 by 2.**
3)	3	9	6	**Bring down the 9.**
	1	3	2	**Divide each by 3.**

The remaining quotients have no common factors, so the LCM is the product of the remaining quotients and the divisors. The LCM of 12, 18, and 24 is $2 \cdot 2 \cdot 3 \cdot 1 \cdot 3 \cdot 2$, or 72.

To find the LCM of two or more numbers using the Group Prime Factoring Method

1. Divide at least two of the numbers by any common prime number factor (divisor). Continue dividing the remaining quotients in the same manner until no two quotients have a common divisor. When a number cannot be divided, bring it down as a remaining quotient.
2. Write the product of the remaining quotients and the divisors.

Answer to Warm Up C. 72

Warm Ups D–G

D. Find the LCM of 30 and 35.

E. Find the LCM of 12, 20, 25, and 48.

F. Find the LCM of the denominators of $\frac{1}{3}, \frac{3}{5}, \frac{7}{12}$, and $\frac{9}{10}$.

G. If Jane had saved nickels and Robin had saved quarters, what is the least each could pay for the same item?

Examples D–G

Directions: Find the LCM of two or more whole numbers.

Strategy: Use the Group Prime Factoring Method.

D. Find the LCM of 36 and 40.

2) 36 40		**Divide by 2.**
2) 18 20		**Divide by 2.**
9 10		

There is no common prime factor in the remaining quotients 9 and 10. Therefore, multiply them by the divisors. The LCM of 36 and 40 is $2 \cdot 2 \cdot 9 \cdot 10$, or 360.

E. Find the LCM of 12, 20, 24, and 50.

2) 12 20 24 50	**Divide each by 2.**
2) 6 10 12 25	**Divide 6, 10, and 12 by 2.**
3) 3 5 6 25	**Divide 3 and 6 by 3.**
5) 1 5 2 25	**Divide 5 and 25 by 5.**
1 1 2 5	

There is no common prime factor in the remaining quotients, so the LCM is the product of the remaining quotients and the divisors. The LCM of 12, 20, 24, and 50 is $2 \cdot 2 \cdot 3 \cdot 5 \cdot 1 \cdot 1 \cdot 2 \cdot 5$, or 600.

F. Find the LCM of the denominators of $\frac{1}{6}, \frac{4}{9}, \frac{5}{12}$, and $\frac{7}{18}$.

The denominators are 6, 9, 12, and 18.

3) 6 9 12 18	**Divide each by 3.**
2) 2 3 4 6	**Divide 2, 4, and 6 by 2.**
3) 1 3 2 3	**Divide the 3's by 3.**
1 1 2 1	

There is no common prime factor in the remaining quotients. The LCM is the product of the remaining quotients and the divisors. The LCM of 6, 9, 12, and 18 is $3 \cdot 2 \cdot 3 \cdot 1 \cdot 1 \cdot 2 \cdot 1$, or 36.

G. Jane and Robin have each been saving coins. Jane saved dimes and Robin saved quarters. The girls went shopping together and they bought the same item, each spending all of her coins. What is the least amount the item could cost?

Strategy: The least amount the item could cost is the smallest number that is divisible by 10 and 25. That number is the LCM of 10 and 25. We use the Group Prime Factoring Method.

5) 10 25	**Divide by 5.**
2 5	

Since 2 and 5 are each prime, multiply the remaining quotients and the divisor. The result is the LCM of 10 and 25, which is $5 \cdot 2 \cdot 5$ or 50. The least the item could cost is 50¢—five dimes or two quarters.

Answers to Warm Ups D. 210 E. 1200 F. 60 G. The least each could have paid is 25¢.

NAME ______________________ CLASS ____________ DATE ____________

Exercises 3.6

OBJECTIVE 1: *Find the Least Common Multiple of two or more whole numbers using the Individual Prime Factoring Method.*

OBJECTIVE 2: *Find the Least Common Multiple of two or more whole numbers using the Group Prime Factoring Method.*

A.

Find the LCM of each group of whole numbers using either method.

1. 4, 8 **2.** 3, 6 **3.** 5, 10 **4.** 6, 12

5. 2, 10 **6.** 3, 9 **7.** 7, 14 **8.** 8, 16

9. 4, 6 **10.** 6, 9 **11.** 8, 12 **12.** 9, 12

13. 2, 4, 6 **14.** 2, 6, 12 **15.** 3, 6, 9 **16.** 2, 5, 10

17. 4, 6, 8 **18.** 2, 8, 12 **19.** 2, 6, 10 **20.** 4, 6, 12

B.

Find the LCM of each group of whole numbers using either method.

21. 10, 15 **22.** 12, 18 **23.** 16, 24 **24.** 14, 21

25. 8, 20

26. 12, 20

27. 12, 16

28. 16, 20

29. 18, 24

30. 30, 45

31. 6, 8, 10

32. 8, 12, 16

33. 12, 16, 24

34. 10, 12, 15

35. 2, 6, 12, 24

36. 4, 12, 10, 15

37. 20, 25, 40

38. 16, 24, 48

39. 8, 9, 12

40. 6, 10, 24

C.

Find the LCM of each group of whole numbers using either method.

41. 12, 18, 24, 36

42. 8, 14, 28, 32

43. 12, 17, 51, 68

44. 14, 35, 49, 56

45. 35, 50, 56, 70, 175

46. 15, 20, 25, 30, 50, 75

Find the LCM of the denominators for each set of fractions.

47. $\frac{2}{3}, \frac{1}{4}, \frac{5}{8}$

48. $\frac{1}{6}, \frac{3}{5}, \frac{7}{9}$

49. $\frac{4}{15}, \frac{15}{16}, \frac{1}{12}$

50. $\frac{21}{30}, \frac{14}{15}, \frac{5}{12}$

51. Open your math book two different times. Write down the left page number the first time and the right page number the second time. Find the least common multiple of the two numbers.

52. Find the Least Common Multiple for the highest and lowest temperatures in the United States yesterday.

53. Melissa goes to the bank and withdraws some 20-dollar bills. Sean goes to the bank and withdraws some 50-dollar bills. Luigi goes to the bank and withdraws some 5-dollar bills. If they all intend to buy the same item, what is the least price they could pay?

Exercises 54–55 relate to the chapter application.

Column transpose ciphers can be broken without knowing the keyword, but it takes much trial and error. One way to tackle such a cipher is to concentrate on common letter combinations such as qu, th, er, re, en. Common factors can be used to determine the column length of a cipher. For example:

letqs aieen evsee mrgho utoih utrat cqs

Observe that there are exactly two q's and two u's in the cipher. It is reasonable to assume that the letter combination "qu" appears twice in the message. We will use this to try to break the cipher. We identify the first and second q and u, then count the number of letters from each q to each u. To count from q2, go to the end of the cipher and continue to count by wrapping around to the beginning. Here is a table of our results.

letqs aieen evsee mrgho utoih utrat cqs

1 1 2 2

	u1	u2
q1	17	22
q2	22	27

Once we have the table, we look for common factors. Because 22 appears twice, and 17 and 27 have no common factors, we conclude that the number of rows in the original message is a factor of 22, either 2 or 11. Since there are 33 letters in the message, we will guess 11 rows of length 3. Rewrite the cipher in groups of 11, each of which will be one column.

letqsaieene vseemrghout oihutratcqs

1 1 2 2

Now we must decide in what order to write the columns because we assume they have been scrambled by a keyword. Note that the table gives us the additional information that q1 goes with u2 and q2 goes with u1. This is because the "22s" appear in these locations. So we arrange the columns to make the qu's line up. Putting q1 with u2 means that the first column is followed by the third. Putting q2 with u1 means that the second column follows the third, giving us this arrangement.

l	o	v
e	i	s
t	h	e
q	u	e
s	t	m
a	r	r
i	a	g
e	t	h
e	c	o
n	q	u
e	s	t

The message is: Love is the quest, marriage the conquest. (*Historical note:* This method was successfully employed by the British M18 during WW I to break intercepted ciphers from the Germans. The messages were, of course, in German. The British decipherers used the fact that in German, c is always followed by h.)

54. Use the "qu's" and the factor method to break the following cipher.

ieict vqrhe oeatu eures lhmeq ttgnl ssaom

55. Use the factoring method to break the following cipher. (*Hint:* Since there are no q's, try looking at the "th" combinations.)

hhine iarlo wsoot sifoh nvhwt edels nzgtr imeee

STATE YOUR UNDERSTANDING

 **56.** Explain how to find the LCM of 20, 24, and 45.

57. Identify the error in the following problem and correct it. Determine the LCM of 100, 75, and 50.

$100 = 2 \cdot 2 \cdot 5 \cdot 5$

$75 = 3 \cdot 5 \cdot 5$

$50 = 2 \cdot 5 \cdot 5$

The LCM is $2 \cdot 3 \cdot 5 \cdot 5 = 150$.

NAME ________________________________ CLASS ____________ DATE ____________

CHALLENGE

58. Find the LCM of 144, 180, and 240.

59. Find the LCM of 128, 256, and 192.

60. Find the LCM of 1728, 960, and 864.

61. Find the LCM of 1800, 1500, and 1200.

GROUP ACTIVITY

62. A classic Hindu puzzle from the seventh century: A woman carrying a basket of eggs is frightened by a horse galloping by. She drops the basket and breaks all the eggs. Concerned passersby ask how may eggs she lost, but she can't remember. She does remember that there was one egg left over when she counted by twos, 2 eggs left over when she counted by threes, 3 eggs leftover when she counted by fours, and 4 eggs left over when she counted by fives. How many eggs did she have? Show calculations that prove your answer fits the conditions of the puzzle.

MAINTAIN YOUR SKILLS (SECTIONS 1.6, 1.7, 3.3)

63. List all the factors of 375.

64. List all the factors of 275.

65. List all the divisors of 488.

66. List all the divisors of 480.

67. Is 8008 divisible by 56?

68. Is 8008 divisible by 143?

69. Is 8008 divisible by 16?

70. Is 8008 divisible by 44?

71. The Sav-Mor Department Store has made a profit on appliances of $112 so far this week. If their profit on each appliance is $7, how many more appliances must they sell so that the profit for the entire week will be more than $170?

72. A marketing researcher checked the weekly attendance at 14 theaters. The results are shown in the table.

Number of Theaters	Attendance
1	900
2	1000
2	1200
3	1300
2	1400
2	1600
1	1800
1	1900

What was the average attendance at the 14 theaters?

CHAPTER 3

Group Project *(1–2 weeks)*

OPTIONAL

The problem with conventional ciphers is that once a person possesses an encrypted message, he or she can discover the encryption process and then reverse the process to decrypt the message. Throughout history, people have had to abandon ciphers because an unwanted third party gained access to an encrypted message. Further, as mentioned earlier, the widespread use of computers and in particular the Internet has created a need for a vast number of secret communications. So cryptologists began looking for a way to encrypt messages in which access to the encrypted message would not compromise the secrecy.

This concept is called "public key cryptology." Simply put, each person is given a public and a private "key." The public "key" is published so that anyone wishing to send a message to a particular person uses the public key to encrypt it. The receiver then uses her private "key" to decrypt the message. The only way to decrypt the message is through the private "key." The process is such that an encrypted message and the public "key" do not provide enough information to break the cipher. One of the most well-known public key systems is based on the RSA Algorithm, which was developed in 1977. (R, S, and A are the initials of the three developers of the system, Ronald Rivest, Adi Shamir, and Leonard Adleman.) The algorithm is based on prime numbers and the difficulty in factoring very large numbers, even with the use of high-speed computers.

The RSA Algorithm works as follows. Select two prime numbers, and call them $p1$ and $p2$. Let $a = (p1)(p2)$ and $b = (p1 - 1)(p2 - 1)$. Now select another prime number that is not a factor of b and call it c. This, together with the value of a, is the public "key." The private "key" is the number d that satisfies the equation:

$$1 = (c)(d) - (e)(b)$$

The RSA developers used number theory to prove that this is possible and why it works. For now, let's go through an example.

Pick two prime numbers, 5 and 11. So $a = 55$ and $b = (5 - 1)(11 - 1) = (4)(10) = 40$. Now we pick a prime that is not a factor of 40, say 23. So 55 and 23 are the public "key." To find our private "key," we have to find a whole number d that satisfies the equation:

$$1 = 23(d) - 40(e)$$

This will involve some trial and error. We want $23(d)$ to be one more than a multiple of 40. Make a table of multiples of 23.

1	2	3	4	5	6	7	8
23	46	69	92	115	138	161	

Since the multiples of 40 are 40, 80, 120, 160, . . . we see that $d = 7$ is the value we are looking for, $1 = 7(23) - 4(40)$. So 7 is our private "key."

Now let's turn our attention to encryption. The RSA Algorithm requires that we take the plain text and convert it to numbers. Then each number is raised to the power that is the public "key," in our case 23. This power is then expressed modulo a, in our case modulo 55. (For a review of modulo arithmetic, see Exercises 60–64 in Section 3.2.)

For example, to encrypt the letter L, we assign it a numerical value. The easiest way of doing this is to let $a = 1$, $b = 2$, and so on. So $L = 12$. Using a calculator to find 12^{23} mod 55, we note that 12^{23} is so large a number (it has 25 digits) that the calculator must use scientific notation to write it. While this is not a problem for computers, it is a problem for us. We can solve it by encrypting the 1 and the 2 separately.

To encrypt 1 we evaluate 1^{23} mod 55.

1^{23} mod 55 = 1 mod 55   **$1^{23} = 1$**

= 1   **1 in any mod is still 1.**

To encrypt 2 we evaluate 2^{23} mod 55.

2^{23} mod = 8388608 mod 55   **Use a calculator for 2^{23}.**

= 8   **$55\overline{)8388608}$ = 152520 R8**

So L becomes 12, which is encrypted to 1-8. To decrypt the message, take each number in the encryption and raise it to the power of the private "key," which in our case is 7. Again, decrypting 1 is relatively simple.

1^7 mod 55 = 1 mod 55   **$1^7 = 1$**

= 1   **1 any mod = 1**

8^7 mod 55 = 2097152 mod 55   **Use a calculator for 8^7.**

= 2   **$55\overline{)2097152}$ = 38130 R2**

So the encrypted 1-8 becomes 12, or L.

Now suppose you have intercepted the message 1-8 and you know that the public key is 23 and the mod is 55. It is fairly obvious that the prime factorization of 55 = 5(11). Once we know the two initial primes, we can calculate a value for b and then use it and 23 to find the private key d (as we did to find d in the first place). Since both 23 and 55 are published, it would seem that this is not a very secure system. What makes the public key systems based on the RSA Algorithm secure is the fact that the primes that are actually used are VERY large—more than 100 digits each. So when their product is published, the number a is so large that no computer can factor it. This is one reason that number theorists are still actively looking for ever larger prime numbers. The larger the prime numbers, the larger a is, and the more secure the system is.

Now it's your turn.

1. You are given a public key of 7, with modulus value of 33. Encrypt the message "Love Child," showing all work.
2. Pick another song title and encrypt it using the public key of 7 with modulus 33. Exchange your encrypted message with that of another group.
3. Break the code and find the private key for your system.
4. Use the private key to decrypt both "Love Child" and the song from the other group.
5. Following is an encrypted message from another system. 18-3, 1-3, 18-1, 9, 1, 1-16, 18-0, 2, 18-1, 1-2, 1-2, 18-3, 1-18, 1-3, 18-18, 3. Break the cipher using the public key of 7 and a modulus of 22.
6. Write a paragraph discussing why it is so difficult to prime factor very large numbers while it is relatively easy to prime factor small numbers.

NAME ______________________ CLASS ____________ DATE ____________

CHAPTER 3

True–False Concept Review

Check your understanding of the language of basic mathematics. Tell whether each of the following statements is True (always true) or False (not always true). For those statements that you judge to be false, revise them to make them true.

ANSWERS

1. Every multiple of 6 ends with the digit 6. **1.** ____________

2. Every multiple of 10 ends with the digit 0. **2.** ____________

3. Every multiple of 13 is divisible by 13. **3.** ____________

4. Every multiple of 7 is the product of 7 and some natural number. **4.** ____________

5. Every whole number, except the number 1, has at least two different factors. **5.** ____________

6. Every factor of 200 is also a divisor of 200. **6.** ____________

7. Every multiple of 200 is also a factor of 200. **7.** ____________

8. The square of 200 is 100. **8.** ____________

9. Every natural number ending in 4 is divisible by 4. **9.** ____________

10. Every natural number ending in 6 is divisible by 2. **10.** ____________

11. Every natural number ending in 9 is divisible by 3. **11.** ____________

12. The number 123,321,231 is divisible by 3. **12.** ____________

13. The number 123,321,234 is divisible by 4. **13.** ____________

14. The number 123,321,235 is divisible by 5. **14.** ____________

15. All prime numbers are odd. 15. ____________

16. Every composite number ends in 1, 3, 7, or 9. 16. ____________

17. Every composite number has four or more factors. 17. ____________

18. Every prime number has exactly two multiples. 18. ____________

19. It is possible for a composite number to have exactly three divisors. 19. ____________

20. All of the prime factors of a natural number are smaller than the number. 20. ____________

21. The least common multiple (LCM) of three different prime numbers is the product of the three numbers. 21. ____________

22. Some natural numbers have exactly five different prime factors. 22. ____________

23. The largest divisor of the Least Common Multiple (LCM) of three numbers is the largest of the three numbers. 23. ____________

24. It is possible for some groups of numbers to have two LCMs. 24. ____________

25. If any number in a group of numbers is an even number, then the least common multiple (LCM) of the group is an even number. 25. ____________

NAME ______________________ CLASS ____________ DATE ____________

CHAPTER 3

Test

		ANSWERS
1.	Is 234,114 divisible by 6?	**1.** ________
2.	List all of the factors (divisors) of 110.	**2.** ________
3.	Is 8736 divisible by 7?	**3.** ________
4.	Is 189,561 divisible by 3?	**4.** ________
5.	What is the LCM of 10 and 32?	**5.** ________
6.	Write 75 as the product of two factors in as many ways as possible.	**6.** ________
7.	Write the prime factorization of 260.	**7.** ________
8.	Find the LCM of 9, 42, and 84.	**8.** ________
9.	Write the prime factorization of 846.	**9.** ________
10.	Write all of the multiples of 13 between 100 and 150.	**10.** ________
11.	Is 200 a multiple of 400?	**11.** ________
12.	Is 107 a prime or a composite number?	**12.** ________
13.	Is 221 a prime or a composite number?	**13.** ________
14.	Write the prime factorization of 847.	**14.** ________
15.	What is the smallest natural number that 18, 21, and 56 will divide evenly?	**15.** ________

16. What is the smallest prime number? **16.** ____________

17. What is the largest composite number that is less than 200? **17.** ____________

18. What is the smallest natural number that 6, 18, 24, and 30 will divide evenly? **18.** ____________

19. Can two different numbers have the same prime factorization? Explain. **19.** ____________

20. List two sets of three different numbers whose LCM is 30. **20.** ____________

Good Advice for Studying

Planning Makes Perfect

Now is the time to formalize a study plan. Set aside a time of day, every day, to focus all of your attention on math. For some students, finding a quiet place in the library to study regularly for one hour is far more efficient than studying two hours at home where there are constant distractions. For others, forming a study group where you can talk about what you have learned is helpful. Decide which works best for you.

Try to schedule time as close to the class session as possible while the concepts are fresh in your mind. If you wait several hours to practice what seemed clear during class, you may find that what was clear earlier may no longer be meaningful. This may mean planning a schedule of classes that includes an hour after class to study.

If there are some days that you cannot devote one or two hours to math, find at least a few minutes and review one thing—perhaps read your notes, reread the section objectives, or if you want to do a few problems, do the section warmups. This helps to keep the concepts fresh in your memory.

If you choose to form or join a study group with other math-anxious students, don't use your time together to gripe. Instead use it to discuss and recognize the content of your negative self-talk and to write positive coping statements.

Plan, too, for your physical health. Notice how your anxious thought patterns trigger physical tension. When you wrinkle your forehead, squint your eyes, make a trip to the coffee machine, or light up a cigarette, you are looking for a way to release these tensions. Learning relaxation techniques, specifically progressive relaxation, is a healthier alternative to controlling body tension. Briefly, relaxation training involves alternately tensing and relaxing all of the major muscles in the body with the goal of locating your specific muscle tension and being able to relax it away. Use a professionally prepared progressive relaxation tape, or take a stress management class to properly learn this technique. Allow at least twenty to thirty minutes for this exercise daily. The time it takes for you to deeply relax will become briefer as you become more skilled. Soon relaxation will be as automatic as breathing and when you find yourself feeling math anxious, you can stop, take control, and relax.

Fractions and Mixed Numbers

APPLICATION

Sandy and Hannah have decided to start a gourmet cookie business. They each have several family recipes that they consider excellent. Since they are just starting their business, they bake only when they have orders to fill. The women need to decide how to price their cookies. This will depend on their costs and on the local market conditions.

They begin by estimating the cost of making their cookies. They determine that they will use only the finest and freshest ingredients, including real butter, fresh eggs, high-quality chocolate chips and pecans. Hannah's pecan chocolate chip recipe makes 2 dozen 4-in. cookies, which cost about 20¢ each. Sandy's dipped butterfingers recipe makes 3 dozen 3-in. cookies, which cost 18¢ each. The women buy small boxes for 38¢ each. The boxes hold 20 pecan chocolate chip cookies or 30 dipped butterfingers each. Each box is lined with a sheet of tissue paper that costs 2¢.

Calculate the total costs for a box of each type of cookie. Go to or call a bakery and find the price per cookie of comparable cookies in your area. Determine the cost of boxes of bakery cookies the same number and type of cookie as Hannah's and Sandy's boxes of cookies. Make a recommendation for the selling price of each type of box of cookies. How much profit will the women make on each type of box of cookies? Fill in the table below with your information.

	Pecan-Chocolate Chip	Dipped Butterfinger
Cost per cookie		
Cost of box and tissue		
Total cost per box		
Bakery cost per cookie		
Total cost per bakery cookie box		
Selling price per box		
Profit per box		

4.1

Proper and Improper Fractions; Mixed Numbers

OBJECTIVES

1. Write a fraction to describe the parts of a unit.
2. Select proper or improper fractions from a list of fractions.
3. Change improper fractions to mixed numbers.
4. Change mixed numbers to improper fractions.

VOCABULARY

A **fraction** $\left(\text{such as } \frac{5}{9}\right)$ is a name for a number. The upper numeral (5) is the **numerator.** The lower numeral (9) is the **denominator.**

A **proper fraction** is one in which the numerator is less than the denominator $\left(\text{such as } \frac{5}{9}\right)$.

An **improper fraction** is one in which the numerator is not less than the denominator $\left(\text{such as } \frac{9}{5} \text{ or } \frac{7}{7}\right)$.

A **mixed number** is the sum of a whole number and a fraction $\left(2 + \frac{4}{5}\right)$ with the plus sign left out $\left(2\frac{4}{5}\right)$. The fraction part is usually a proper fraction.

HOW AND WHY

Objective 1

Write a fraction to describe the parts of a unit.

A unit (here we use a rectangle) may be divided into smaller parts of equal size in order to picture a fraction. The rectangle in Figure 4.1 is divided into seven parts, and six of the parts are shaded. The fraction $\frac{6}{7}$ represents the shaded part. The denominator (7) tells the number of parts in the unit. The numerator (6) tells the number of shaded parts.

Figure 4.1

Since fractions are another way of writing division and since division by zero is not defined, the denominator can never be zero. There will be at least one part in a unit.

The unit may also be shown on a ruler. The fraction $\frac{6}{10}$ represents the distance from 0 to the arrowhead in Figure 4.2.

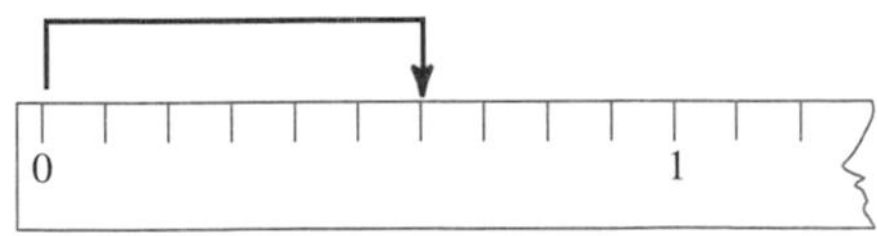

Figure 4.2

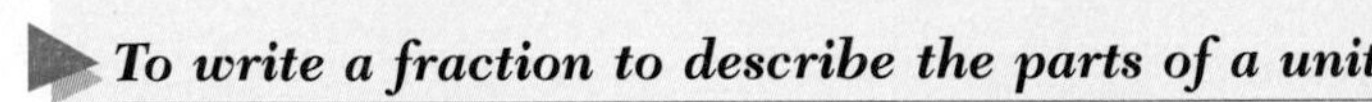

To write a fraction to describe the parts of a unit

Write the fraction:

$$\frac{\text{numerator}}{\text{denominator}} = \frac{\text{number of shaded parts}}{\text{total number of parts in one unit}}$$

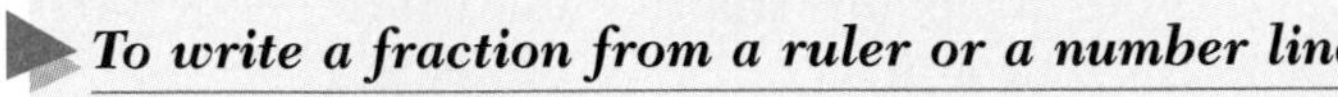

To write a fraction from a ruler or a number line

Write the fraction:

$$\frac{\text{numerator}}{\text{denominator}} = \frac{\text{number of spaces between zero and end of arrow}}{\text{number of spaces between zero and one}}$$

Examples A–E

Directions: Write the fraction represented by the figure.

Strategy: First count the number of parts that are shaded or marked. This number is the numerator. Now count the total number of parts in the unit. This number is the denominator.

A. Write the fraction represented by

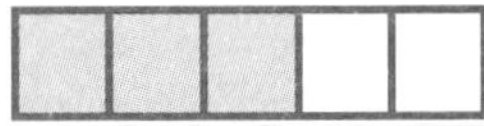

The figure represents $\frac{3}{5}$. **The count of shaded parts is 3. The total count is 5.**

B. Write the fraction represented by

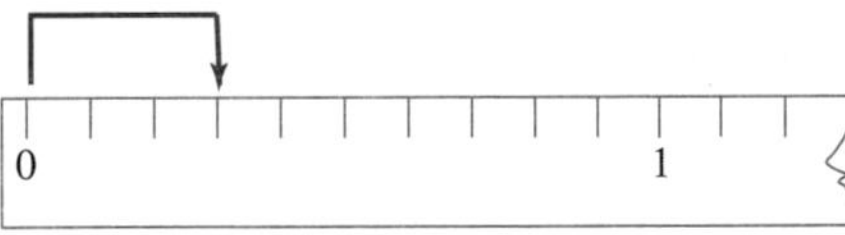

The figure represents $\frac{3}{10}$. **There are 3 spaces between 0 and the arrowhead. There are 10 spaces between 0 and 1.**

C. Write the fraction represented by

The figure represents $\frac{6}{6}$, or 1. **The number of shaded parts is the same as the total number of parts. The whole unit, or 1, is shaded.**

Warm Ups A–E

A. Write the fraction represented by

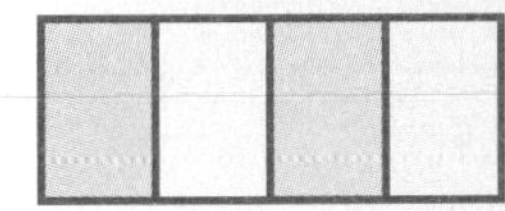

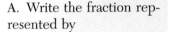

B. Write the fraction represented by

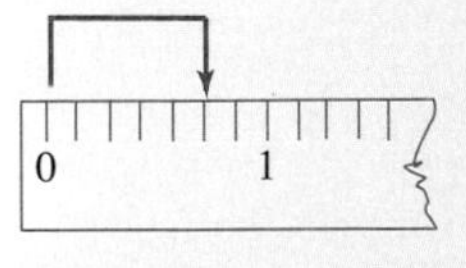

C. Write the fraction represented by

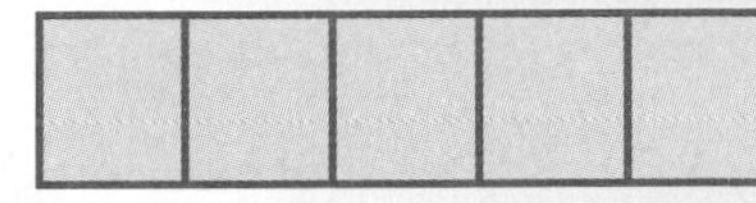

Answers to Warm Ups A. $\frac{2}{4}$ B. $\frac{5}{7}$ C. $\frac{5}{5}$ or 1

D. Write the fraction represented by

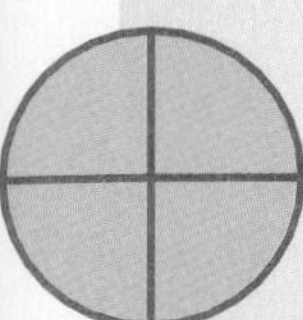
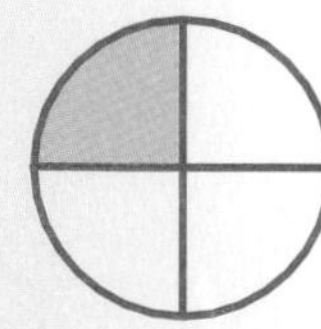

D. Write the fraction represented by

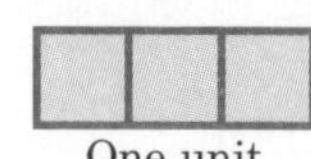

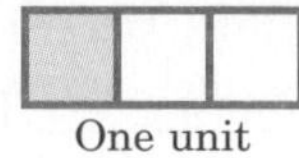

We have two units. The denominator is the number of parts in one unit.

The figure represents $\frac{4}{3}$. **Four parts are shaded and there are 3 parts in each unit.**

E. In a second low-cholesterol margarine spread, one tablespoon contains 10 g of fat. Each fat gram has 9 calories. The total number of calories in a tablespoon is 97. Write a fraction that represents the portion of calories from fat.

E. One tablespoon of a low-cholesterol margarine spread contains 11 g of fat. Each fat gram has 9 calories, so there are 99 calories from fat in a tablespoon of margarine. The total number of calories in a tablespoon of margarine is 100. Write a fraction that represents the portion of calories from fat.

The fraction is the number of calories from fat over the total number of calories.

The fraction is $\frac{99}{100}$. **99 calories are from fat, and total number is 100.**

HOW AND WHY

Objective 2

Select proper or improper fractions from a list of fractions.

Fractions are called proper if the numerator is smaller than the denominator. If the numerator is equal to or greater than the denominator, the fractions are called improper. So in the list

$$\frac{5}{6}, \frac{11}{9}, \frac{19}{19}, \frac{10}{11}, \frac{14}{16}, \frac{18}{11}, \frac{24}{29}$$

the proper fractions are $\frac{5}{6}$, $\frac{10}{11}$, $\frac{14}{16}$, and $\frac{24}{29}$. The improper fractions are $\frac{11}{9}$, $\frac{19}{19}$, and $\frac{18}{11}$.

If the numerator and the denominator are equal, as in $\frac{19}{19}$, the value of the fraction is 1. This is easy to see from a picture because the entire unit is shaded. Improper fractions have a value that is greater than or equal to 1. Proper fractions have a value that is less than 1 because some part of the unit is not shaded (see Table 4.1).

To determine if a fraction is proper or improper

1. Compare the size of the numerator and the denominator.
2. If the numerator is smaller, the fraction is proper. Otherwise the fraction is improper.

Answers to Warm Ups D. $\frac{5}{4}$ E. The fraction of calories from fat is $\frac{90}{97}$.

TABLE 4.1 REGIONS WHICH SHOW PROPER AND IMPROPER FRACTIONS

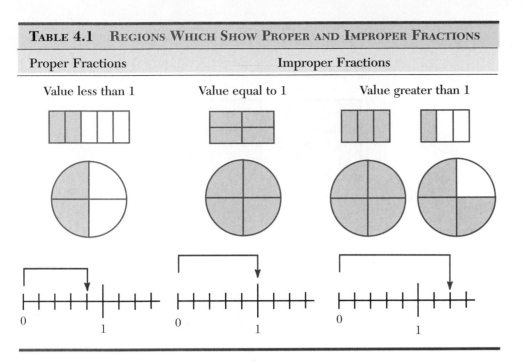

Proper Fractions	Improper Fractions	
Value less than 1	Value equal to 1	Value greater than 1

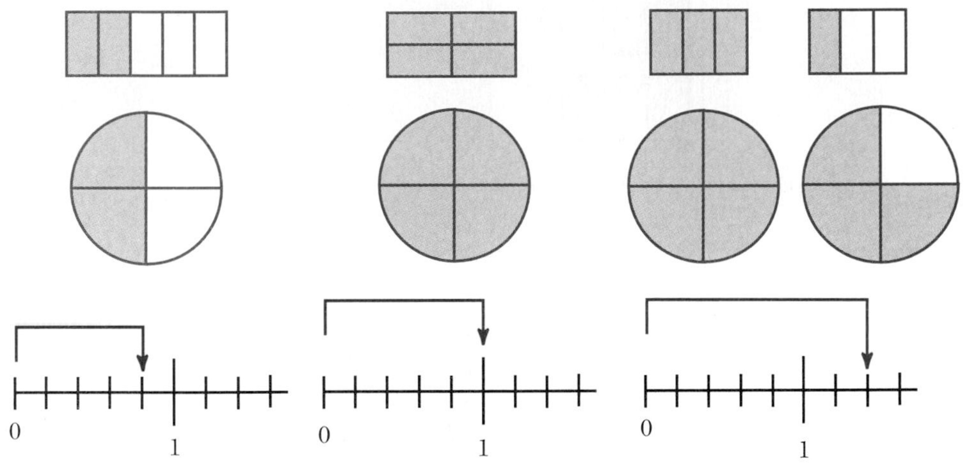

0 1 0 1 0 1

Example F

Directions: Identify the proper and improper fractions in the list.

Strategy: Compare the numerator and the denominator. If the numerator is smaller, the fraction is proper. If not, the fraction is improper.

F. $\frac{4}{5}, \frac{3}{5}, \frac{6}{6}, \frac{8}{7}, \frac{22}{23}, \frac{24}{24}, \frac{25}{24}, \frac{30}{40}$

The proper fractions are

$\frac{4}{5}, \frac{3}{5}, \frac{22}{23}, \frac{30}{40}$ **The numerators are smaller than the denominators.**

The improper fractions are

$\frac{6}{6}, \frac{8}{7}, \frac{24}{24}, \frac{25}{24}$ **The numerators are not smaller than the denominators.**

Warm Up F

F. $\frac{9}{9}, \frac{14}{15}, \frac{7}{8}, \frac{11}{9}, \frac{6}{5}, \frac{15}{19}, \frac{8}{5}, \frac{19}{19}$

HOW AND WHY

Objective 3

Change improper fractions to mixed numbers.

An improper fraction is equal either to a whole number or to a mixed number (the sum of a whole number and a fraction). The figures show the conversions.

Answer to Warm Up F. The proper fractions are $\frac{14}{15}, \frac{7}{8}, \frac{15}{19}$. The improper fractions are $\frac{9}{9}, \frac{11}{9}, \frac{6}{5}, \frac{8}{5}, \frac{19}{19}$.

An improper fraction changed to a whole number:

$$\frac{6}{3} = 2$$

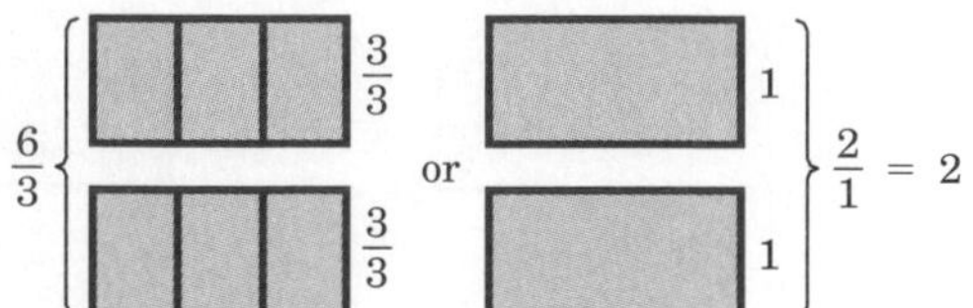

An improper fraction changed to a mixed number:

$$\frac{9}{7} = 1\frac{2}{7}$$

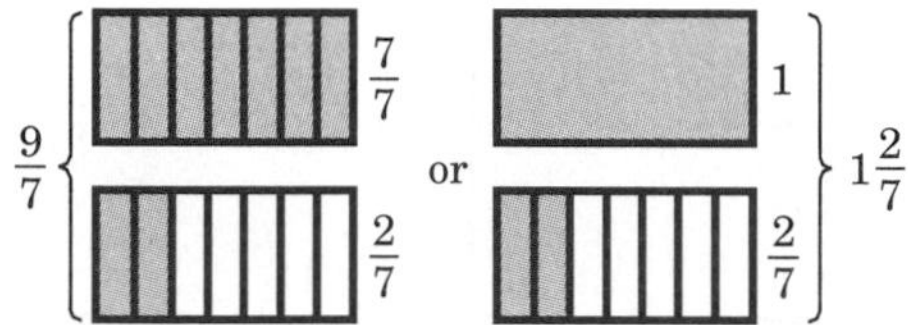

The shortcut for changing an improper fraction to a mixed number is to divide:

$$\frac{14}{7} = 14 \div 7 = 2$$

$$\frac{10}{7} = 10 \div 7 = 7\overline{)10} = 1\frac{3}{7}$$

$$\begin{array}{r} 1 \\ 7\overline{)10} \\ \underline{7} \\ 3 \end{array}$$

To change an improper fraction to a mixed number

1. Divide the numerator by the denominator.
2. If there is a remainder, write the whole number and then write the fraction: $\dfrac{\text{remainder}}{\text{divisor}}$.

CAUTION

Do not confuse the process of changing an improper fraction to a mixed number with "simplifying." Simplifying fractions is a totally different procedure. See Section 4.2.

Warm Ups G–I

Examples G–I

Directions: Change the improper fraction to a mixed number.

Strategy: Divide the numerator by the denominator to find the whole number. If there is a remainder, write it over the denominator to form the fraction part.

G. Change $\frac{11}{3}$ to a mixed number.

$$\frac{11}{3} = 3\overline{)11} = 3\frac{2}{3}$$
$$\begin{array}{r} 3 \\ 3\overline{)11} \\ \underline{9} \\ 2 \end{array}$$

Divide 11 by 3.

G. Change $\frac{19}{6}$ to a mixed number.

H. Change $\frac{114}{6}$ to a mixed number.

$$\frac{114}{6} = 6\overline{)114} = 19$$
$$\begin{array}{r} 19 \\ 6\overline{)114} \\ \underline{6} \\ 54 \\ \underline{54} \end{array}$$

Divide 114 by 6. There is no remainder, so the fraction is equal to a whole number.

H. Change $\frac{115}{5}$ to a mixed number.

Calculator Example

I. Change $\frac{348}{7}$ to a mixed number.

If your calculator has a key for fractions, refer to the manual to see how you can use it to change fractions to mixed numbers. If your calculator does not have a fraction key divide 348 by 7. The quotient $\frac{348}{7} \approx 49.714$, so the whole number part is 49. Now subtract to find the remainder: $348 - 7(49) = 5$.

$348 \div 7 = 49$ R 5, so $\frac{348}{7} = 49\frac{5}{7}$.

I. Change $\frac{319}{14}$ to a mixed number.

HOW AND WHY

Objective 4

Change mixed numbers to improper fractions.

Despite the value judgment attached to the name "improper," in many cases improper fractions are a more convenient and useful form than mixed numbers. Thus, it is important to be able to convert from mixed numbers to improper fractions.

Every mixed number can be changed to an improper fraction:

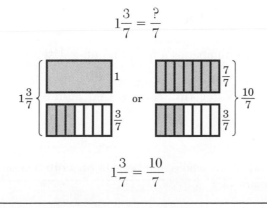

$$1\frac{3}{7} = \frac{?}{7}$$

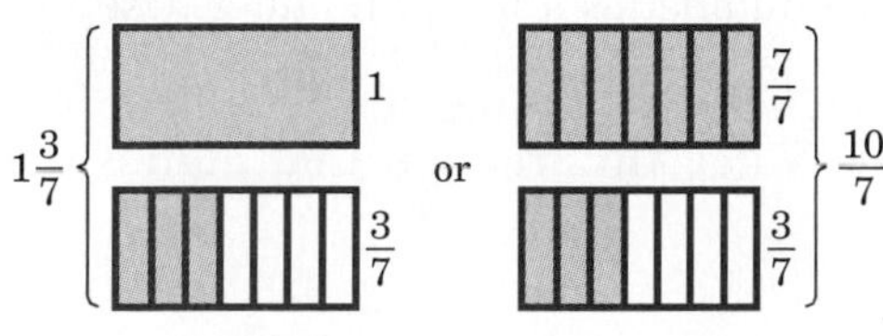

$$1\frac{3}{7} = \frac{10}{7}$$

Answers to Warm Ups G. $3\frac{1}{6}$ H. 23 I. $22\frac{11}{14}$

The shortcut uses multiplication and addition:

$$1\frac{3}{7} = \frac{7 \cdot 1 + 3}{7} = \frac{7 + 3}{7} = \frac{10}{7}$$

To change a mixed number to an improper fraction

1. Multiply the denominator times the whole number.
2. Add the numerator to the product in step 1.
3. Place the sum from step 2 over the denominator.

Examples J–L

Directions: Change each mixed number to an improper fraction.

Strategy: Multiply the whole number by the denominator. Add the numerator. Write the sum over the denominator.

J. Change $2\frac{4}{5}$ to an improper fraction.

$$2\frac{4}{5} = \frac{2(5) + 4}{5}$$

Multiply the whole number by the denominator, add the product to the numerator, and place the sum over the denominator.

$$= \frac{14}{5}$$

K. Change $4\frac{5}{9}$ to an improper fraction.

$$4\frac{5}{9} = \frac{4(9) + 5}{9}$$

Multiply the whole number by the denominator, add the product to the numerator, and place the sum over the denominator.

$$= \frac{41}{9}$$

L. Change 7 to an improper fraction.

First, rewrite the whole number as a mixed number. Use the fraction $\frac{0}{1}$.

$$7 = 7\frac{0}{1} = \frac{7(1) + 0}{1}$$

Note: **Any fraction that equals 0 could be used.**

$$= \frac{7}{1}$$

As example L illustrates, any whole number can be written as an improper fraction by using a denominator of 1.

Warm Ups J–L

J. Change $3\frac{5}{6}$ to an improper fraction.

K. Change $5\frac{5}{8}$ to an improper fraction.

L. Change 8 to an improper fraction.

Answers to Warm Ups J. $\frac{23}{6}$ K. $\frac{45}{8}$ L. $\frac{8}{1}$

NAME ______________________ CLASS ____________ DATE ____________

Exercises 4.1

OBJECTIVE 1: *Write a fraction to describe the parts of a unit.*

A.

Write the fraction represented by the figure.

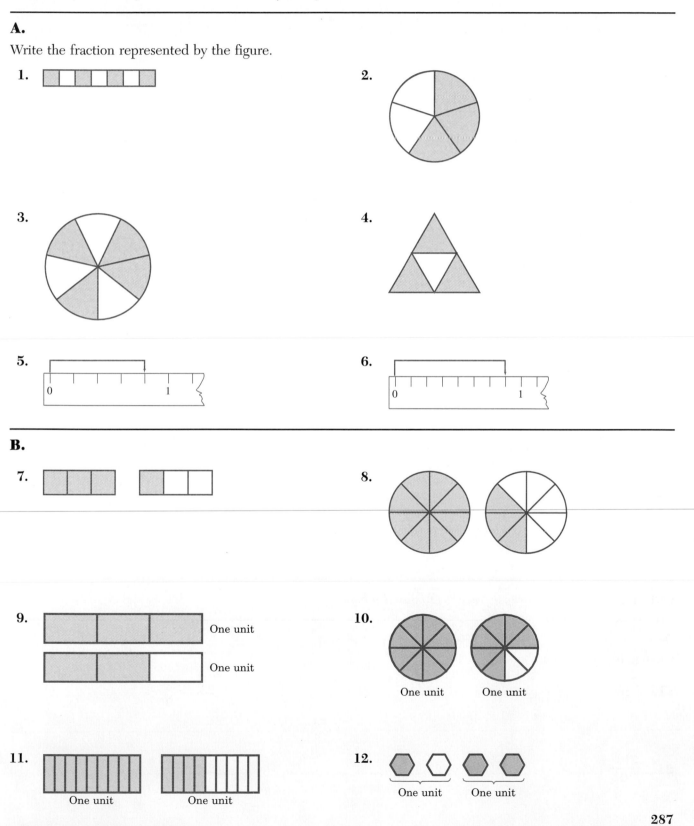

1.

2. 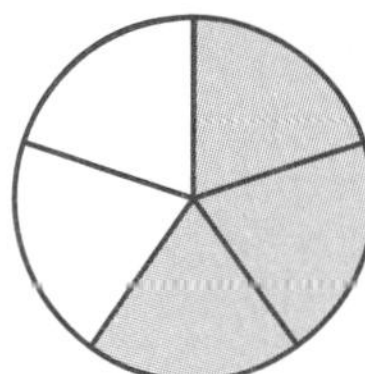

3. 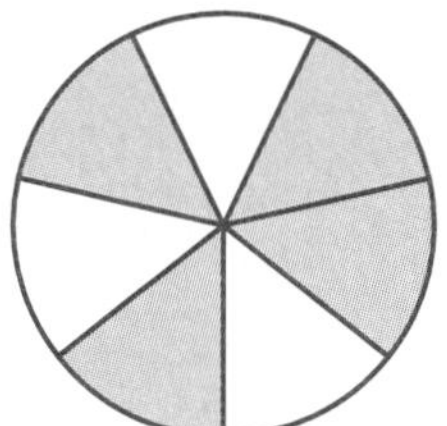

4. 

5. 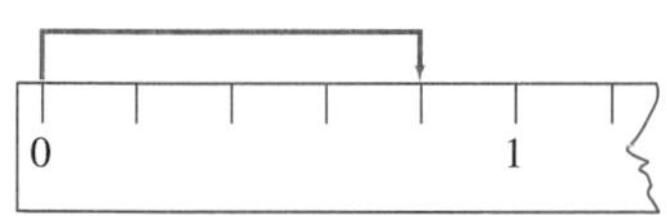

6. 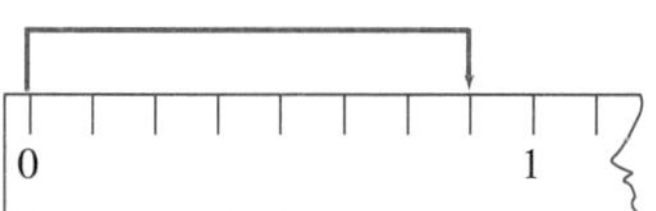

B.

7. 

8. 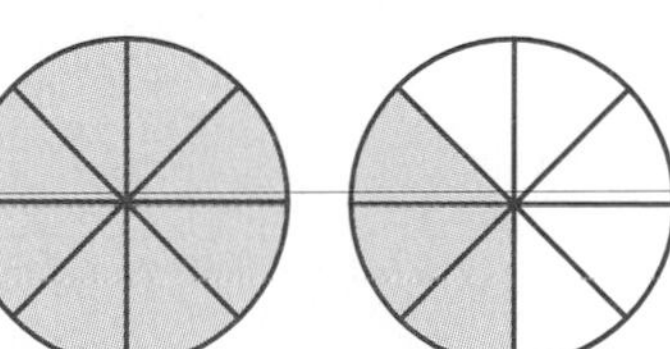

9. 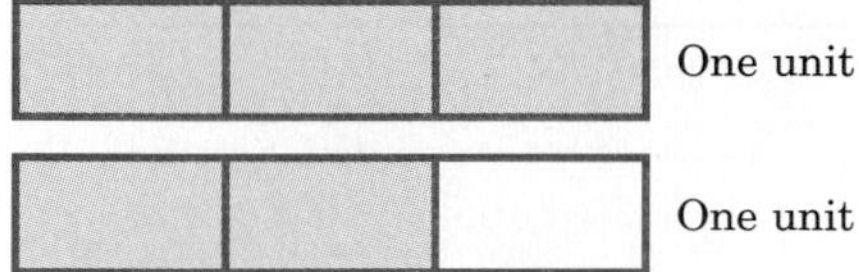

10. 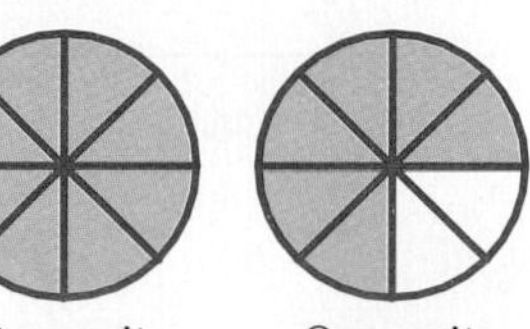

11. 

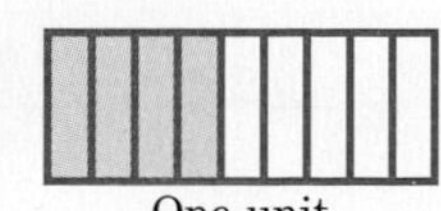

12. 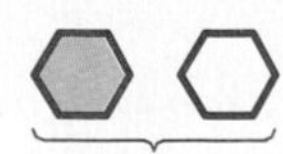
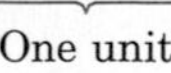

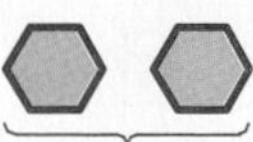

OBJECTIVE 2: *Select proper or improper fractions from a list of fractions.*

A.

Identify the proper and improper fractions from the list.

13. $\frac{3}{7}, \frac{4}{7}, \frac{5}{7}, \frac{6}{7}, \frac{7}{7}, \frac{8}{7}, \frac{9}{7}$

14. $\frac{5}{6}, \frac{8}{7}, \frac{14}{15}, \frac{16}{18}, \frac{17}{17}, \frac{23}{25}$

15. $\frac{7}{13}, \frac{8}{15}, \frac{10}{13}, \frac{11}{15}, \frac{12}{23}$

16. $\frac{9}{11}, \frac{9}{10}, \frac{9}{9}, \frac{9}{8}, \frac{9}{7}$

17. $\frac{7}{4}, \frac{10}{11}, \frac{13}{13}, \frac{20}{19}, \frac{3}{5}$

18. $\frac{6}{11}, \frac{10}{8}, \frac{11}{6}, \frac{10}{12}, \frac{9}{9}$

B.

19. $\frac{5}{5}, \frac{9}{10}, \frac{18}{18}, \frac{102}{103}, \frac{147}{147}$

20. $\frac{5}{6}, \frac{6}{7}, \frac{7}{8}, \frac{8}{9}, \frac{10}{10}, \frac{11}{12}, \frac{12}{13}, \frac{13}{13}, \frac{15}{15}$

OBJECTIVE 3: *Change improper fractions to mixed numbers.*

A.

Change to a mixed number.

21. $\frac{213}{5}$

22. $\frac{213}{4}$

23. $\frac{213}{8}$

24. $\frac{213}{10}$

25. $\frac{112}{9}$

26. $\frac{103}{6}$

B.

27. $\frac{98}{13}$ **28.** $\frac{83}{21}$ **29.** $\frac{329}{22}$

30. $\frac{400}{13}$ **31.** $\frac{387}{17}$ **32.** $\frac{453}{23}$

OBJECTIVE 4: *Change mixed numbers to improper fractions.*

A.

Change to an improper fraction.

33. $5\frac{4}{7}$ **34.** $4\frac{3}{8}$ **35.** 12

36. 13 **37.** $7\frac{3}{4}$ **38.** $6\frac{5}{6}$

B.

39. $29\frac{2}{3}$ **40.** $37\frac{1}{2}$ **41.** $12\frac{5}{6}$

42. $17\frac{4}{7}$ **43.** $19\frac{3}{8}$ **44.** $21\frac{7}{9}$

C.

Fill in the boxes so the statement is true. Explain your answer.

45. The fraction $\frac{0}{\square}$ is a proper fraction.

46. $55\frac{\square}{7} = \frac{387}{7}$

47. Find the error in the statement: $15\frac{2}{7} = \frac{17}{7}$

48. Find the error in the statement: $\frac{37}{8} = 3\frac{7}{8}$

Write the fraction represented by the figure.

49.

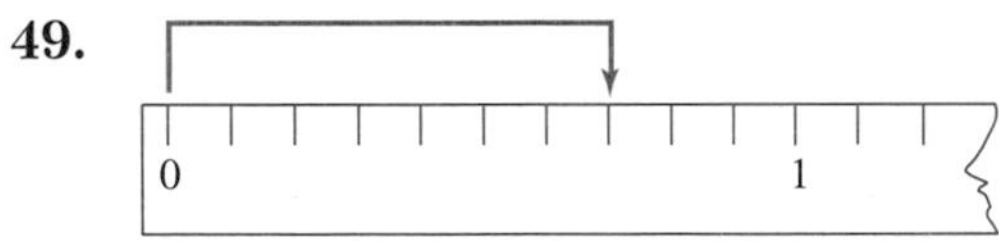

50.

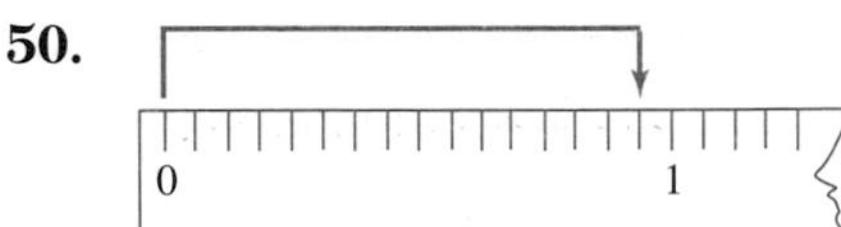

51.

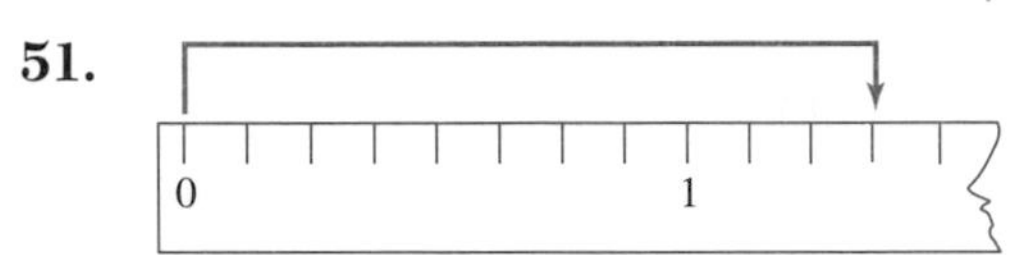

52.

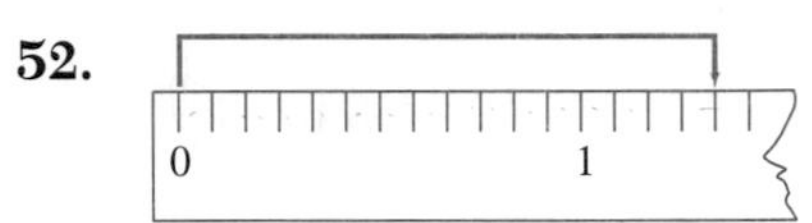

53. Draw a unit divided into equal parts that shows $\frac{4}{5}$.

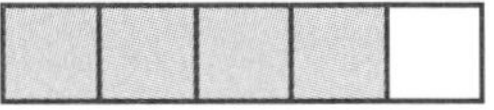

54. Draw units divided into equal parts that show $\frac{8}{5}$.

55. In a class of 21 students there are 13 women. What fraction represents the part of the class that is female?

56. In a history class there are 45 students; 23 of them are male. What fraction represents the part of the class that is male?

57. If a six-cylinder motor has one cylinder that is not firing, what fractional part of the cylinders is firing?

58. The Adams family budgets \$1050 for food and housing. They spend \$700 per month for housing. What fractional part of this budget is spent for food?

59. What fraction of a full tank of gas is indicated by the gas gauge?

60. What fraction of a full tank of gas is indicated by the gas gauge?

61. A scale is marked with a whole number at each pound. What whole number mark is closest to the weight of $\frac{50}{16}$ lb?

62. A ruler is marked with a whole number at each centimeter. What whole number mark is closest to a length of $\frac{87}{10}$ cm?

63. Sue's construction company must place section barriers, each $\frac{1}{352}$ mile long, between two sides of a freeway. How many such sections are needed for $7\frac{31}{352}$ miles of freeway?

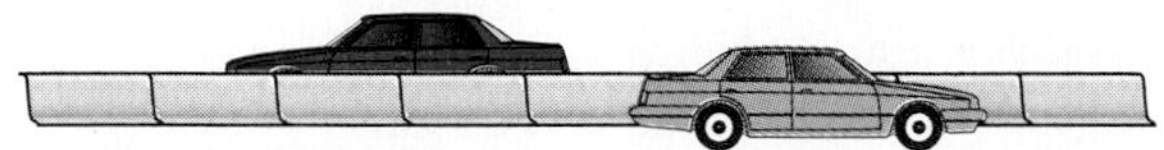

64. How many sections from Exercise 63 are needed for $37\frac{197}{352}$ miles of freeway?

65. A food processor packs 24 cans of beans in a case for shipping. Write as a mixed number the number of cases that can be made from 65,345 cans of beans.

66. A food processor packs 16 cans of juice in a case for shipping. Write as a mixed number the number of cases that can be made from 23,789 cans of juice.

Exercises 67–69 relate to the chapter application.

67. Sandy doesn't like to dirty a lot of dishes when she bakes, so she often uses a $\frac{1}{2}$-c measure for all of her flour and sugar. The dipped butterfinger recipe calls for $2\frac{1}{2}$ c of flour. How many $\frac{1}{2}$-c measures can she use?

68. The recipe in Exercise 67 also calls for 2 c of sugar. How many $\frac{1}{2}$-c measures can she use?

69. The recipe in Exercise 67 also calls for $1\frac{3}{4}$ tsp of baking soda. If Sandy only has a $\frac{1}{4}$-tsp measure, how many $\frac{1}{4}$-tsp measures of baking soda can she use in the cookie dough?

STATE YOUR UNDERSTANDING

70. Explain how to change $\frac{34}{5}$ to a mixed number. Explain how to change $7\frac{3}{8}$ to an improper fraction.

71. Tell why mixed numbers and improper fractions are both useful. Give examples of the use of each.

72. Explain why a proper fraction cannot be changed into a mixed number.

CHALLENGE

73. Write the whole number 13 as an improper fraction with (a) the numerator 117, and (b) the denominator 117.

74. Write the whole number 14 as an improper fraction with (a) the numerator 154, and (b) the denominator 154.

75. The Swete Tuth candy company packs 30 pieces of candy in its special Mother's Day box. Write, as a mixed number, the number of special boxes that can be made from 67,892 pieces of candy. The company then packs 25 of the special boxes that are filled in a carton for shipping. Write, as a mixed number, the number of cartons that can be filled. If it costs Swete Tuth $45 per carton to ship the candy, what is the shipping cost for the number of full cartons that can be shipped?

76. Jose has $21\frac{3}{4}$ yd of rope to use in a day-care class. If the rope is cut into $\frac{1}{4}$-yd pieces, how many pieces will there be? If there are 15 children in the class, how many pieces of rope will each child get? How many pieces will be left over?

GROUP ACTIVITY

77. Many times data about a population are presented as a pie chart. Pie charts are based on fractions. If the fractions are easy to draw, a pie chart can be drawn quickly. Sometimes the drawing, which is a reasonable estimate, is adequate. A survey revealed that half of the class preferred pepperoni pizza, one quarter of them preferred cheese pizza, one eighth preferred Canadian bacon pizza, and one eighth preferred sausage pizza. Make a pie chart that illustrates this survey. Explain your strategy.

78. According to a radio advertising survey, 77 of 100 people in the United States say they listen to the radio daily. Make a pie chart that illustrates this survey. Explain your strategy.

79. Using the following figure, write a fraction to name each of the division marks. Now divide each section in half and write new names for each mark. Divide the sections in half again and also find new names. You now have three names for each original mark; what can you say about these names?

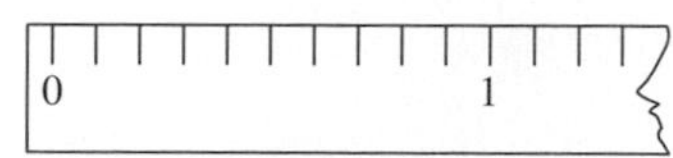

80. Express $\frac{45,679}{37}$ as a mixed number using a calculator. Find a way to use the calculator to find the numerator of the fraction part of the mixed number. Be prepared to show the class how it can be done.

MAINTAIN YOUR SKILLS (SECTIONS 1.1, 1.3, 1.5, 1.7, 2.1, 3.1, 3.4, 3.5)

81. Divide: $90{,}700{,}000 \div 10^4$

82. Prime factor 796.

83. Find the value of 4^8.

84. Is 269 prime or composite?

85. Round 4573 to the nearest hundred.

86. Is 45,067,233 divisible by 3?

87. Bonnie's motorcycle averages 55 miles per gallon of gasoline. On a recent trip she used 17 gallons of gas. How far did she travel?

88. After a tune-up, Bonnie went 377 miles, 375 miles, 383 miles, and 385 miles, on four tanks of gas. If her tank holds 10 gallons of gas, what was her average miles per tank? What was her average miles per gallon?

89. An electric motor turns 65,000 revolutions in 8 minutes. How many revolutions does it make in one minute?

90. An electric motor turns 7225 revolutions in 1 minute. How many revolutions does it make in 10 minutes?

4.2

Simplifying Fractions

OBJECTIVE

Simplify a fraction.

VOCABULARY

Equivalent fractions are fractions that are the different names for the same number.

Simplifying a fraction is the process of renaming it by using a smaller numerator and denominator. A fraction is **completely simplified** when its numerator and denominator have no common factors other than one. For instance, $\frac{12}{18} = \frac{2}{3}$.

HOW AND WHY

Objective

Simplify a fraction.

Fractions are *equivalent* if they represent the same quantity. When we compare the two units in Figure 4.3, we see that each is divided into four parts. The shaded part on the left is named $\frac{2}{4}$, whereas the shaded part on the right is labeled $\frac{1}{2}$. It is clear that the two are the same size, and therefore we say $\frac{2}{4} = \frac{1}{2}$.

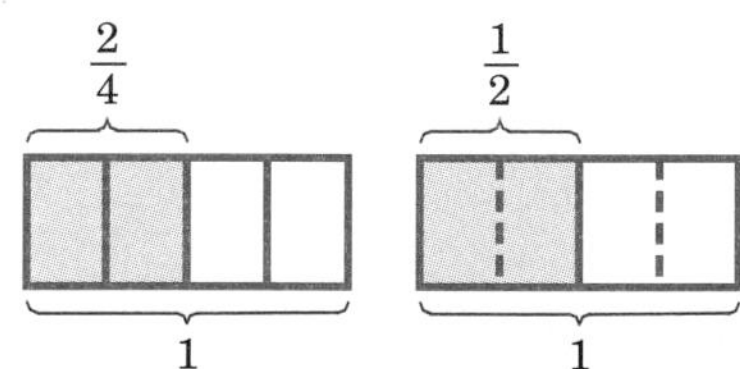

Figure 4.3

The arithmetical way of showing that $\frac{2}{4} = \frac{1}{2}$ is to eliminate the common factors by dividing:

$$\frac{2}{4} = \frac{2 \div 2}{4 \div 2} = \frac{1}{2}$$

The division can also be shown by eliminating the common factors.

$$\frac{2}{4} = \frac{1 \cdot \overset{1}{\cancel{2}}}{2 \cdot \underset{1}{\cancel{2}}} = \frac{1}{2}$$

This method works for all fractions. To simplify $\frac{15}{21}$:

$$\frac{15}{21} = \frac{5 \cdot \overset{1}{\cancel{3}}}{7 \cdot \underset{1}{\cancel{3}}} = \frac{5}{7}$$ **Eliminate the common factors.**

or

$$\frac{15}{21} = \frac{15 \div 3}{21 \div 3} = \frac{5}{7}$$ **Divide out the common factors.**

When all common factors have been eliminated (divided out), the fraction is completely simplified.

$$\frac{24}{40} = \frac{12}{20} = \frac{6}{10} = \frac{3}{5}$$ **Completely simplified.**

If the common factors are not discovered easily, they can be found by writing the numerator and denominator in prime-factored form. See Example E.

To simplify a fraction completely

Eliminate all common factors, other than 1, in the numerator and the denominator.

Warm Ups A–H

A. Simplify: $\frac{24}{36}$

B. Simplify: $\frac{18}{45}$

C. Simplify: $\frac{64}{48}$

Examples A–H

Directions: Simplify completely.

Strategy: Eliminate the common factors in the numerator and the denominator.

A. Simplify: $\frac{21}{28}$

$$\frac{21}{28} = \frac{3 \cdot \overset{1}{\cancel{7}}}{4 \cdot \underset{1}{\cancel{7}}} = \frac{3}{4}$$ **The common factor is 7. Eliminate the common factor by dividing.**

B. Simplify: $\frac{16}{24}$

$$\frac{16}{24} = \frac{\overset{1}{\cancel{2}} \cdot 8}{\underset{1}{\cancel{2}} \cdot 12} = \frac{8}{12}$$ **A common factor of 2 is eliminated.**

$$= \frac{\overset{1}{\cancel{2}} \cdot 4}{\underset{1}{\cancel{2}} \cdot 6} = \frac{4}{6}$$ **There is still a factor of 2 in the numerator and the denominator.**

$$= \frac{\overset{1}{\cancel{2}} \cdot 2}{\underset{1}{\cancel{2}} \cdot 3} = \frac{2}{3}$$ **Again, a common factor of 2 is eliminated by dividing.**

or

$$\frac{16}{24} = \frac{2 \cdot \overset{1}{\cancel{8}}}{3 \cdot \underset{1}{\cancel{8}}} = \frac{2}{3}$$ **Rather than divide by 2 three times, divide by 8 once, and the fraction is simplified completely.**

C. Simplify: $\frac{54}{36}$

$$\frac{54}{36} = \frac{54 \div 18}{36 \div 18} = \frac{3}{2}$$ **Divide both the numerator and denominator by 18.**

or

$$\frac{54}{36} = \frac{\overset{3}{\cancel{54}}}{\underset{2}{\cancel{36}}} = \frac{3}{2}$$

Answers to Warm Ups A. $\frac{2}{3}$ B. $\frac{2}{5}$ C. $\frac{4}{3}$

D. Simplify: $\frac{100}{600}$

D. Simplify: $\frac{4000}{14{,}000}$

$\frac{100}{600} = \frac{100 \div 100}{600 \div 100} = \frac{1}{6}$ **Divide both numerator and denominator by 100.**

or

$\frac{100}{600} = \frac{1}{6}$ **Divide by 100 mentally using the shortcut for dividing by powers of 10; $100 = 10^2$.**

E. Simplify: $\frac{126}{144}$

E. Simplify: $\frac{160}{256}$

Strategy: Since the numbers are large, write them in prime-factored form.

$\frac{126}{144} = \frac{\overset{1}{\cancel{2}} \cdot \overset{1}{\cancel{3}} \cdot \overset{1}{\cancel{3}} \cdot 7}{\underset{1}{\cancel{2}} \cdot 2 \cdot 2 \cdot 2 \cdot \underset{1}{\cancel{3}} \cdot \underset{1}{\cancel{3}}}$ **Eliminate the common factors.**

$= \frac{7}{2 \cdot 2 \cdot 2}$

$= \frac{7}{8}$ **Multiply.**

F. Simplify: $\frac{8}{9}$

F. Simplify: $\frac{16}{25}$

$\frac{8}{9} = \frac{2 \cdot 2 \cdot 2}{3 \cdot 3} = \frac{8}{9}$ **There are no common factors. The fraction is already completely simplified.**

 **Calculator Example**

G. Simplify: $\frac{493}{551}$

G. Simplify: $\frac{703}{851}$

$\frac{493}{551} = \frac{17}{19}$ **Use the fraction key on your calculator.**

H. Mel washes cars on Saturday to earn extra money. On a particular Saturday he has 12 cars to wash. After he has washed 8 of them, what fraction of the total has he washed? Simplify the fraction completely.

H. In Example H, if Mel washes only 4 cars, what fraction of the total are washed?

Strategy: Form the fraction: $\frac{\text{number of cars washed}}{\text{total number of cars}}$

$\frac{8}{12}$ **Eight cars are washed out of 12.**

$\frac{8}{12} = \frac{2 \cdot \overset{1}{\cancel{4}}}{3 \cdot \underset{1}{\cancel{4}}} = \frac{2}{3}$ **Simplify.**

So $\frac{2}{3}$ of the cars are washed.

Answers to Warm Ups D. $\frac{2}{7}$ E. $\frac{5}{8}$ F. $\frac{16}{25}$ G. $\frac{19}{23}$ H. He has washed $\frac{1}{3}$ of the cars.

NAME ______________________ CLASS ____________ DATE ____________

Exercises 4.2

OBJECTIVE: *Simplify a fraction.*

A.

Simplify completely.

1. $\frac{6}{12}$ **2.** $\frac{6}{15}$ **3.** $\frac{6}{9}$ **4.** $\frac{8}{12}$ **5.** $\frac{10}{25}$ **6.** $\frac{16}{18}$

7. $\frac{30}{50}$ **8.** $\frac{40}{70}$ **9.** $\frac{12}{16}$ **10.** $\frac{18}{24}$ **11.** $\frac{12}{20}$ **12.** $\frac{20}{22}$

13. $\frac{32}{40}$ **14.** $\frac{30}{40}$ **15.** $\frac{60}{36}$ **16.** $\frac{55}{22}$ **17.** $\frac{14}{18}$ **18.** $\frac{28}{36}$

19. $\frac{21}{35}$ **20.** $\frac{25}{45}$ **21.** $\frac{20}{5}$ **22.** $\frac{40}{4}$

B.

23. $\frac{63}{27}$ **24.** $\frac{60}{35}$ **25.** $\frac{14}{42}$ **26.** $\frac{30}{45}$ **27.** $\frac{12}{36}$ **28.** $\frac{20}{36}$

29. $\frac{27}{36}$ **30.** $\frac{32}{36}$ **31.** $\frac{29}{36}$ **32.** $\frac{23}{36}$ **33.** $\frac{50}{75}$ **34.** $\frac{30}{75}$

35. $\frac{55}{75}$ **36.** $\frac{15}{75}$ **37.** $\frac{600}{800}$ **38.** $\frac{500}{900}$ **39.** $\frac{45}{80}$ **40.** $\frac{65}{80}$

41. $\frac{72}{96}$ **42.** $\frac{88}{92}$ **43.** $\frac{72}{12}$ **44.** $\frac{96}{16}$ **45.** $\frac{75}{125}$ **46.** $\frac{64}{120}$

47. $\frac{96}{126}$ **48.** $\frac{72}{100}$ **49.** $\frac{99}{132}$ **50.** $\frac{84}{120}$

C.

51. Which of these fractions is completely simplified?

$\frac{14}{21}$ $\frac{26}{39}$ $\frac{34}{49}$ $\frac{38}{57}$

52. Which of these fractions is completely simplified?

$\frac{28}{49}$ $\frac{44}{75}$ $\frac{52}{91}$ $\frac{76}{133}$

Simplify completely.

53. $\frac{120}{144}$

54. $\frac{84}{144}$

55. $\frac{268}{402}$

56. $\frac{546}{910}$

57. $\frac{630}{1050}$

58. $\frac{294}{1617}$

59. Maria does a tune-up on her automobile. She finds that 4 of the 6 spark plugs are fouled. What fraction represents the number of fouled plugs? Simplify.

60. The float on a tank registers 12 feet. If the tank is full when it registers 28 feet, what fraction of the tank is full? Simplify.

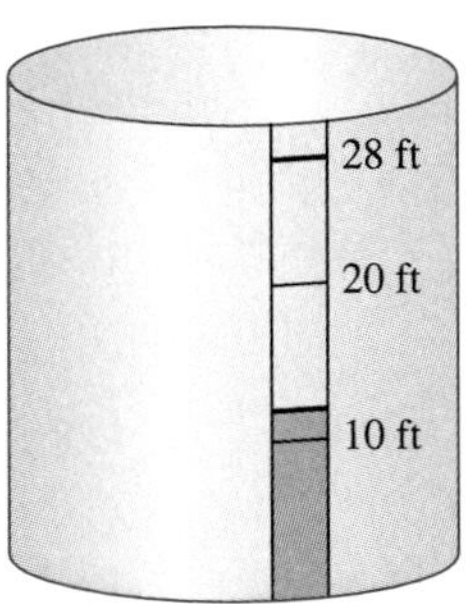

61. Gyrid completes 28 hr out of her weekly shift of 40 hr. What fraction of her weekly shift remains? Simplify.

62. Cardis has worked 12 hr out of his weekly part-time job of 30 hr. What fraction of his weekly shift remains? Simplify.

63. The Bonneville Dam fish counter shows 4320 salmon passing over the dam. Of this number, 220 are cohos. What fraction of the salmon are cohos? Simplify.

64. One hundred thirty elk are counted at the Florence Refuge. Forty-five of the elk are bulls. What fraction of the elk are bulls? Simplify.

65. On a math test a student answers 42 items correctly and 18 incorrectly. What fraction of the items are answered correctly? Simplify.

66. A district attorney successfully prosecutes 45 cases and is unsuccessful on another 20. What fraction of the cases are successfully prosecuted? Simplify.

67. The energy used to produce 1 lb of virgin rubber is 15,800 BTUs. Producing 1 lb of recycled rubber requires only 4600 BTUs. What fraction of the BTUs needed to produce 1 lb of virgin rubber is used to produce 1 lb of recycled rubber? Simplify.

Exercises 68–69 refer to the chapter application (see page 279).

68. Hannah and Sandy intend to sell a box of pecan chocolate chip cookies for $6.00. Write a simplified fraction that represents the portion of the selling price that is profit.

69. The women also agree to price a box of dipped butterfingers at $7.20. Write a simplified fraction that represents the portion of the selling price that is profit.

STATE YOUR UNDERSTANDING

70. Draw a picture that illustrates that $\frac{9}{12} = \frac{6}{8} = \frac{3}{4}$.

 **71.** Explain how to simplify $\frac{525}{1125}$.

72. Explain why $\frac{12}{16}$ and $\frac{15}{20}$ are equivalent fractions.

CHALLENGE

73. Are these four fractions equivalent? Justify.

$\frac{495}{1188}$ $\frac{660}{1584}$ $\frac{1090}{2616}$ $\frac{885}{2124}$

74. Are these four fractions equivalent? Justify.

$\frac{1170}{2925}$ $\frac{864}{2160}$ $\frac{672}{1440}$ $\frac{1134}{2430}$

GROUP ACTIVITY

75. In the figure, the circle is divided into halves. Divide each of the halves in half. Now use the figure to answer the question "What is $\frac{1}{2}$ of $\frac{1}{2}$?" Devise a rule for finding the product without using the circle.

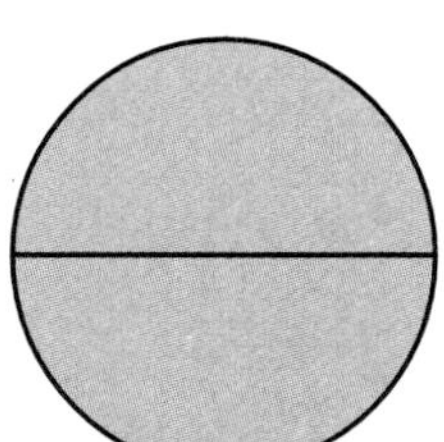

76. In the figure, the circle is divided into thirds. Divide each of the thirds in half. Now use the figure to answer the question. "What is $\frac{1}{2}$ of $\frac{1}{3}$?" Devise a rule for finding the product without using the circle.

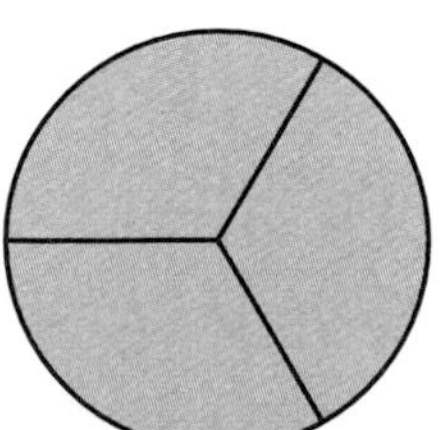

NAME ______________________ CLASS ______________ DATE ______________

MAINTAIN YOUR SKILLS (SECTIONS 1.2, 1.6, 3.5)

77. How much less than 6002 is 4096?

78. How much more is 9367 than 654?

79. Find the difference of 5130 and 2918.

80. Subtract 7798 from 10,000.

81. Find the sum of 1099, 789, 38, and 89.

82. Find the total of 61,234, 4190, 37, and 788.

83. Prime factor 320.

84. Prime factor 374.

85. Trudy has to read all of her psychology textbook before the final week of the semester. If the book is 980 pages long and Trudy does not want to read during weekends, how many pages must she read each day, on the average, to read the whole text before the start of the final week? Assume that the semester is 15 weeks long.

86. Franco has to read all of his economics textbook before the final week of the term. If the book is 715 pages long and Franco does not want to read during weekends, how many pages must he read each day, on the average, to read the whole text before the start of the final week? Assume that the term is 12 weeks long.

4.3

Multiplying and Dividing Fractions

OBJECTIVES

1. Multiply fractions.
2. Find the reciprocal of a number.
3. Divide fractions.

VOCABULARY

A **product** is the answer to a multiplication problem. If two fractions have a product of 1, each fraction is called the **reciprocal** of the other. For example, $\frac{2}{3}$ is the reciprocal of $\frac{3}{2}$.

HOW AND WHY

Objective 1

Multiply fractions.

The word "of" often indicates multiplication. For example, what is $\frac{1}{2}$ of $\frac{1}{3}$ or $\frac{1}{2} \cdot \frac{1}{3} = ?$ See Figure 4.4. The rectangle is divided into three parts. One part, $\frac{1}{3}$, is shaded blue. To find $\frac{1}{2}$ of the shaded third, divide each of the thirds into two parts (halves). Figure 4.5 shows the rectangle divided into six parts. So, $\frac{1}{2}$ of the shaded third is $\frac{1}{6}$ of the rectangle, which is shaded yellow:

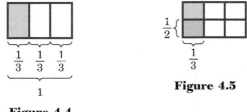

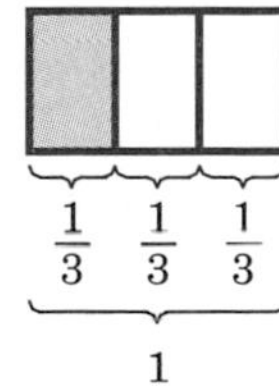

Figure 4.4

1/2 {
1/3

Figure 4.5

$$\frac{1}{2} \text{ of } \frac{1}{3} = \frac{1}{2} \cdot \frac{1}{3} = \frac{1}{6} = \frac{\text{number of parts shaded yellow}}{\text{total number of parts}}$$

What is $\frac{1}{4}$ of $\frac{3}{4}$? $\left(\frac{1}{4} \cdot \frac{3}{4} = ?\right)$ In Figure 4.6 the rectangle has been divided into four parts, and $\frac{3}{4}$ is represented by the parts that are shaded blue. To find $\frac{1}{4}$ of the $\frac{3}{4}$, divide each of the fourths into four parts. The rectangle is now divided into 16 parts, so that $\frac{1}{4}$ of each of the three original fourths is shaded yellow and represents $\frac{3}{16}$. (See Fig. 4.7.)

$$\frac{1}{4} \text{ of } \frac{3}{4} = \frac{1}{4} \cdot \frac{3}{4} = \frac{3}{16}$$

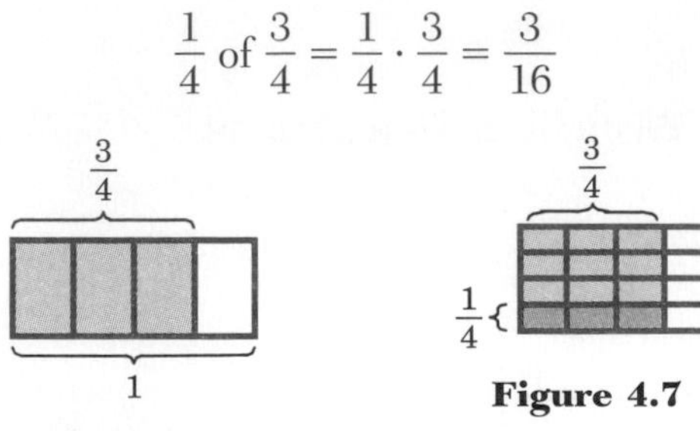

Figure 4.6

Figure 4.7

We have seen that $\frac{1}{2} \cdot \frac{1}{3} = \frac{1}{6}$ and $\frac{1}{4} \cdot \frac{3}{4} = \frac{3}{16}$. The shortcut is to multiply the numerators and multiply the denominators.

To multiply two or more fractions, write the product of the numerators over the product of the denominators. So

$$\frac{12}{35} \cdot \frac{25}{18} = \frac{300}{630}$$

$= \dfrac{300 \div 10}{630 \div 10}$ **Simplify by dividing both the numerator and denominator by 10.**

$= \dfrac{30}{63}$

$= \dfrac{30 \div 3}{63 \div 3}$ **Simplify by dividing both the numerator and denominator by 3.**

$= \dfrac{10}{21}$ **The fraction is completely simplified.**

Multiplying two or more fractions, such as the preceding ones, can be done more quickly by simplifying before multiplying.

$\dfrac{12}{35} \cdot \dfrac{25}{18} = \dfrac{\overset{6}{\cancel{12}}}{35} \cdot \dfrac{25}{\underset{9}{\cancel{18}}}$ **Simplify by dividing 12 and 18 by 2.**

$= \dfrac{\overset{2}{\cancel{6}}}{\underset{7}{\cancel{35}}} \cdot \dfrac{\overset{5}{\cancel{25}}}{\underset{3}{\cancel{9}}}$ **Divide 25 and 35 by 5. Divide 6 and 9 by 3.**

$= \dfrac{10}{21}$ **Multiply.**

The next example shows all of the reducing done in one step.

$\dfrac{\overset{\overset{1}{\cancel{3}}}{\cancel{24}}}{\underset{\underset{1}{\cancel{4}}}{\cancel{32}}} \cdot \dfrac{\overset{2}{\cancel{8}}}{\underset{3}{\cancel{9}}} = \dfrac{2}{3}$ **Divide 24 and 32 by 8, then divide 3 and 9 by 3, and finally divide 8 and 4 by 4. Then multiply.**

If the numbers are large, prime factor each numerator and denominator. (See Example E.)

To multiply fractions

1. Simplify.
2. Write the product of the numerators over the product of the denominators.

CAUTION

Simplifying before doing the operation works only for multiplication because it is based on multiplying by one. It *does not* work for addition, subtraction, or division.

Examples A–G

Directions: Multiply. Simplify completely.

Strategy: Simplify and then multiply.

A. Multiply and simplify: $\frac{1}{2} \cdot \frac{3}{4} \cdot \frac{5}{8}$

$\frac{1}{2} \cdot \frac{3}{4} \cdot \frac{5}{8} = \frac{1 \cdot 3 \cdot 5}{2 \cdot 4 \cdot 8}$ **No common factors. Write the product of the numerators over the product of the denominators.**

$= \frac{15}{64}$

B. Multiply and simplify: $\frac{5}{2} \cdot 3$

Strategy: First write 3 as an improper fraction.

$\frac{5}{2} \cdot 3 = \frac{5}{2} \cdot \frac{3}{1}$ $3 = \frac{3}{1}$

$= \frac{5 \cdot 3}{2 \cdot 1}$ **Write the product of the numerators over the product of the denominators.**

$= \frac{15}{2}$ or $7\frac{1}{2}$

C. Multiply and simplify: $\frac{3}{4} \cdot \frac{6}{7}$

$\frac{3}{4} \cdot \frac{6}{7} = \frac{3}{\cancel{4}_{2}} \cdot \frac{\cancel{6}^{3}}{7} = \frac{9}{14}$ **Eliminate the common factor of 2 in 4 and 6. Multiply.**

D. Multiply and simplify: $\frac{7}{9} \cdot \frac{18}{5} \cdot \frac{10}{21}$

$\frac{7}{9} \cdot \frac{18}{5} \cdot \frac{10}{21} = \frac{\cancel{7}^{1}}{\cancel{9}_{1}} \cdot \frac{\cancel{18}^{2}}{\cancel{5}_{1}} \cdot \frac{\cancel{10}^{2}}{\cancel{21}_{3}}$ **The common factors of 9, 5, and 7 are eliminated.**

$= \frac{4}{3}$ **Multiply.**

Warm Ups A–G

A. Multiply and simplify: $\frac{1}{3} \cdot \frac{4}{7} \cdot \frac{8}{5}$

B. Multiply and simplify: $4 \cdot \frac{8}{3}$

C. Multiply and simplify: $\frac{7}{8} \cdot \frac{4}{5}$

D. Multiply and simplify: $\frac{8}{9} \cdot \frac{12}{18} \cdot \frac{15}{16}$

Answers to Warm Ups A. $\frac{32}{105}$ B. $\frac{32}{3}$ or $10\frac{2}{3}$ C. $\frac{7}{10}$ D. $\frac{5}{9}$

E. Multiply and simplify: $\frac{32}{45} \cdot \frac{35}{24}$

E. Multiply and simplify: $\frac{20}{30} \cdot \frac{15}{88}$

Strategy: Prime factor the numbers because there are so many common factors it is not easy to see all of them.

$$\frac{20}{30} \cdot \frac{15}{88} = \frac{2 \cdot 2 \cdot 5}{2 \cdot 3 \cdot 5} \cdot \frac{3 \cdot 5}{2 \cdot 2 \cdot 2 \cdot 11}$$
$$= \frac{\cancel{2} \cdot \cancel{2} \cdot \cancel{5}}{\cancel{2} \cdot \cancel{3} \cdot \cancel{5}} \cdot \frac{\cancel{3} \cdot 5}{\cancel{2} \cdot 2 \cdot 2 \cdot 11}$$
$$= \frac{5}{44}$$

F. Multiply and simplify: $\frac{18}{35} \cdot \frac{28}{45}$

Calculator Example

F. Multiply and simplify: $\frac{16}{75} \cdot \frac{45}{56}$

$\frac{16}{75} \cdot \frac{45}{56} = \frac{6}{35}$ **Use the fraction key. The calculator will automatically simplify the product.**

G. During one year $\frac{2}{3}$ of all the tires sold by the Tire Factory were highway tread tires. If $\frac{1}{40}$ of the highway treads had to be repaired, what fraction of the tires sold were highway treads that had to be repaired?

G. During one year, $\frac{7}{8}$ of all the cars sold by Trust-em Used Cars had automatic transmissions. Of the cars sold with automatic transmissions, $\frac{1}{70}$ had to be repaired before they were sold. What fraction of the cars sold had automatic transmissions and had to be repaired?

Strategy: Multiply the fraction of the cars with automatic transmissions by the fraction with transmissions that had to be repaired.

$$\frac{7}{8} \cdot \frac{1}{70} = \frac{\overset{1}{\cancel{7}}}{8} \cdot \frac{1}{\underset{10}{\cancel{70}}} = \frac{1}{80}$$

So $\frac{1}{80}$ of the total cars sold had automatic transmissions that had to be repaired.

HOW AND WHY

Objective 2

Find the reciprocal of a number.

Finding a reciprocal is often called "inverting" a fraction. For instance, the reciprocal of $\frac{3}{7}$ is $\frac{7}{3}$. We check by showing that the product is 1.

$$\frac{3}{7} \cdot \frac{7}{3} = \frac{21}{21} = 1$$

Answers to Warm Ups E. $\frac{28}{27}$ or $1\frac{1}{27}$ F. $\frac{8}{25}$ G. Of the tires sold, $\frac{1}{60}$ were highway treads that needed repair.

To find the reciprocal of a fraction

Interchange the numerator and the denominator.

The reciprocal of whole numbers or mixed numbers can be found by first writing them as improper fractions. The reciprocal of 32 $\left(32 = \frac{32}{1}\right)$ is $\frac{1}{32}$ and the reciprocal of $4\frac{1}{5}$ $\left(4\frac{1}{5} = \frac{21}{5}\right)$ is $\frac{5}{21}$.

CAUTION

The number zero (0) does not have a reciprocal.

Examples H–I

Directions: Find the reciprocal.

Strategy: Interchange the numerator and the denominator, or "invert" the fraction.

H. Find the reciprocal of $\frac{7}{10}$.

$\frac{10}{7}$ **Exchange the numerator and the denominator.**

CHECK: $\frac{7}{10} \cdot \frac{10}{7} = \frac{70}{70} = 1.$

The reciprocal of $\frac{7}{10}$ is $\frac{10}{7}$ or $1\frac{3}{7}$.

I. Find the reciprocal of $1\frac{4}{9}$.

Strategy: First write $1\frac{4}{9}$ as an improper fraction.

$1\frac{4}{9} = \frac{13}{9}$

$\frac{9}{13}$ **Invert the fraction.**

CHECK: $\frac{13}{9} \cdot \frac{9}{13} = \frac{117}{117} = 1$

The reciprocal of $1\frac{4}{9}$ is $\frac{9}{13}$.

Warm Ups H–I

H. Find the reciprocal of $\frac{8}{11}$.

I. Find the reciprocal of $2\frac{5}{6}$.

HOW AND WHY

Objective 3

Divide fractions.

It is pointed out in Chapter 1 that division is the inverse of multiplication. That is, the answer to a division problem is the number that is multiplied times the divisor (second number), which will give the first number as an answer. Another way of

Answers to Warm Ups H. $\frac{11}{8}$ or $1\frac{3}{8}$ I. $\frac{6}{17}$

thinking of division is to ask, "How many groups of a certain size are contained in a number?"

	Think	*Answer*
$6 \div 2$	How many twos in six?	3
$\frac{4}{5} \div \frac{1}{10}$	How many one-tenths in four-fifths?	See Figure 4.8.

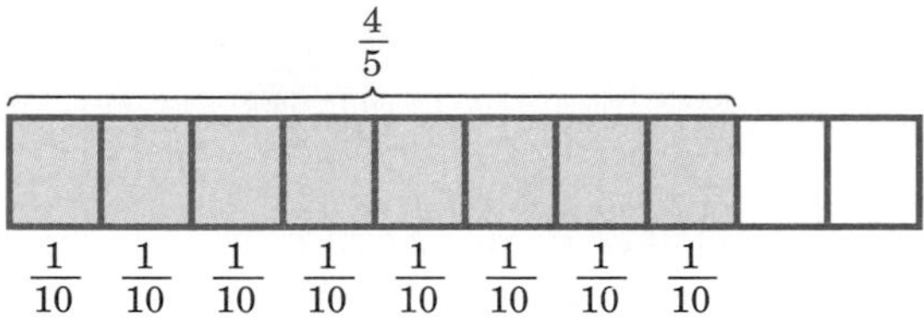

Figure 4.8

In Figure 4.8 we see that there are eight one-tenths in four fifths. Therefore, we can say

$$\frac{4}{5} \div \frac{1}{10} = 8$$

Since $8 \cdot \frac{1}{10} = \frac{8}{10} = \frac{4}{5}$, we know that the answer is correct.

The answer can also be obtained from the fractions by multiplying $\frac{4}{5}$ by the reciprocal of $\frac{1}{10}$.

$$\frac{4}{5} \div \frac{1}{10} = \frac{4}{5} \cdot \frac{10}{1} = \frac{40}{5} = 8$$

To divide fractions

Multiply the first fraction by the reciprocal of the divisor; that is, invert and multiply.

CAUTION

Do not simplify the fractions before changing the division to multiplication; that is, invert before simplifying.

Warm Ups J–N

Examples J–N

Directions: Divide. Simplify completely.

Strategy: Multiply by the reciprocal of the divisor.

J. Divide: $\frac{7}{23} \div \frac{16}{23}$

$\frac{7}{23} \div \frac{16}{23} = \frac{7}{\cancel{23}} \cdot \frac{\cancel{23}}{16}$ **Multiply by the reciprocal of the divisor.**

$= \frac{7}{16}$

J. Divide: $\frac{11}{19} \div \frac{15}{19}$

K. Divide: $\frac{8}{3} \div \frac{4}{5}$

$\frac{8}{3} \div \frac{4}{5} = \frac{\overset{2}{\cancel{8}}}{3} \cdot \frac{5}{\underset{1}{\cancel{4}}}$ **Invert the divisor and multiply.**

$= \frac{10}{3}$ or $3\frac{1}{3}$

K. Divide: $\frac{9}{2} \div \frac{6}{5}$

L. Divide: $\frac{1}{12} \div \frac{3}{5}$

$\frac{1}{12} \div \frac{3}{5} = \frac{1}{12} \cdot \frac{5}{3}$ **Invert the divisor and multiply.**

$= \frac{5}{36}$

L. Divide: $\frac{5}{11} \div \frac{2}{3}$

Calculator Example

M. Divide and simplify: $\frac{8}{21} \div \frac{24}{77}$

$\frac{8}{21} \div \frac{24}{77} = \frac{11}{9}$ **Use the fraction key. The calculator will automatically invert the divisor and simplify the quotient.**

M. Divide and simplify: $\frac{9}{64} \div \frac{27}{80}$

N. If the distance a nut moves on a bolt with one turn is $\frac{3}{16}$ inch, how many turns will it take to move the nut $\frac{3}{4}$ inch?

Strategy: To find the number of turns needed to move the nut the required distance, divide the required distance by the distance the nut moves in one turn.

$\frac{3}{4} \div \frac{3}{16} = \frac{\overset{1}{\cancel{3}}}{\underset{1}{\cancel{4}}} \cdot \frac{\overset{4}{\cancel{16}}}{\underset{1}{\cancel{3}}} = 4$ **Invert the divisor and multiply.**

It takes 4 turns to move the nut $\frac{3}{4}$ inch.

N. If in Example N the distance the nut moves on the bolt with one turn is $\frac{3}{32}$ inch, how many turns will it take to move the nut $\frac{3}{4}$ inch?

Answers to Warm Ups J. $\frac{11}{15}$ K. $\frac{15}{4}$ or $3\frac{3}{4}$ L. $\frac{15}{22}$ M. $\frac{5}{12}$ N. Eight turns are needed to move the nut $\frac{3}{4}$ in.

NAME ______________________ CLASS ____________ DATE ____________

Exercises 4.3

OBJECTIVE 1: *Multiply fractions.*

A.

Multiply. Simplify completely.

1. $\frac{1}{5} \cdot \frac{2}{5}$

2. $\frac{7}{8} \cdot \frac{1}{6}$

3. $\frac{3}{2} \cdot \frac{5}{14}$

4. $\frac{3}{5} \cdot \frac{6}{11}$

5. $\frac{5}{4} \cdot \frac{7}{10}$

6. $\frac{5}{3} \cdot \frac{1}{15}$

7. $\frac{2}{3} \cdot \frac{3}{8}$

8. $\frac{5}{2} \cdot \frac{2}{15}$

9. $\frac{5}{24} \cdot \frac{8}{10}$

10. $\frac{4}{9} \cdot \frac{3}{8}$

B.

11. $\frac{4}{6} \cdot \frac{9}{30} \cdot \frac{10}{6}$

12. $\frac{2}{3} \cdot \frac{4}{15} \cdot \frac{18}{7}$

13. $7 \cdot \frac{1}{4} \cdot \frac{8}{21}$

14. $\frac{21}{2} \cdot 8 \cdot \frac{1}{7}$

15. $\frac{21}{5} \cdot \frac{5}{4} \cdot \frac{4}{21}$

16. $\frac{24}{30} \cdot \frac{3}{8} \cdot \frac{4}{9}$

17. $\frac{5}{6} \cdot \frac{6}{7} \cdot \frac{7}{8} \cdot \frac{8}{9}$

18. $\frac{11}{12} \cdot \frac{12}{13} \cdot \frac{13}{14} \cdot \frac{14}{15}$

19. $\frac{5}{3} \cdot \frac{4}{25} \cdot 5 \cdot \frac{0}{3}$

20. $\frac{7}{9} \cdot \frac{5}{0} \cdot \frac{27}{16} \cdot 4$

OBJECTIVE 2: *Find the reciprocal of a number.*

A.

Find the reciprocal.

21. $\frac{3}{8}$

22. $\frac{5}{11}$

23. 5

24. 0

25. $4\frac{1}{2}$

B.

26. $3\frac{7}{11}$

27. $\frac{17}{14}$

28. 23

29. $\frac{1}{8}$

30. 10

OBJECTIVE 3: *Divide fractions.*

A.

Divide. Simplify completely.

31. $\frac{3}{7} \div \frac{4}{9}$

32. $\frac{9}{8} \div \frac{5}{7}$

33. $\frac{7}{20} \div \frac{14}{15}$

34. $\frac{8}{13} \div \frac{2}{13}$

35. $\frac{8}{9} \div \frac{8}{3}$

36. $\frac{5}{6} \div \frac{5}{3}$

37. $\frac{6}{33} \div \frac{3}{11}$

38. $\frac{8}{12} \div \frac{5}{12}$

39. $\frac{12}{16} \div \frac{3}{4}$

40. $\frac{24}{36} \div \frac{8}{9}$

B.

41. $\frac{28}{25} \div \frac{21}{80}$

42. $\frac{30}{49} \div \frac{20}{42}$

43. $\frac{15}{18} \div \frac{30}{27}$

44. $\frac{12}{15} \div \frac{15}{8}$

45. $\frac{9}{100} \div \frac{3}{14}$

46. $\frac{35}{32} \div \frac{10}{3}$

47. $\frac{90}{55} \div \frac{9}{5}$

48. $\frac{32}{45} \div \frac{8}{9}$

49. $\frac{25}{40} \div \frac{5}{8}$

50. $\frac{12}{15} \div \frac{20}{25}$

NAME ______________________ CLASS ____________ DATE ____________

C.

Multiply. Simplify completely.

51. $\frac{36}{55} \cdot \frac{33}{54} \cdot \frac{35}{24}$

52. $\frac{21}{15} \cdot \frac{18}{14} \cdot \frac{15}{42}$

53. $\left(\frac{24}{35}\right)\left(\frac{40}{44}\right)\left(\frac{77}{96}\right)$

54. $\left(\frac{243}{1000}\right)\left(\frac{25}{81}\right)\left(\frac{40}{45}\right)$

Divide. Simplify completely.

55. $\frac{25}{55} \div \frac{75}{10}$

56. $\frac{81}{100} \div \frac{54}{150}$

57. $\frac{16}{81} \div \frac{8}{108}$

58. $\frac{75}{90} \div \frac{50}{72}$

Fill in the boxes with a single number so the statement is true. Explain your answer.

59. $\frac{2}{3} \cdot \frac{\square}{16} = \frac{5}{24}$

60. $\frac{3}{\square} \cdot \frac{7}{12} = \frac{7}{64}$

61. Find the error(s) in the statement: $\frac{3}{8} \cdot \frac{4}{5} = \frac{15}{32}$. Correct the statement. Explain how you would avoid this error.

62. Find the error(s) in the statement: $\frac{3}{8} \div \frac{4}{5} = \frac{32}{15}$. Correct the statement. Explain how you would avoid this error.

63. A container holds $\frac{7}{5}$ gallons. How much does it hold when it is $\frac{3}{4}$ full?

64. Lois spends half of the family income on rent, utilities, and food. She pays $\frac{2}{7}$ of this amount for rent. What fraction of the family income goes for rent?

65. An article is priced to sell for $96 at the Aquarium Gift Store. It is sale-priced at $\frac{1}{3}$ off. What is its sale price?

66. The National Zoo has T-shirts that regularly sell for $30. The shirts are marked $\frac{1}{5}$ off during the Labor Day sale. What is the sale price of a T-shirt?

67. As part of her job at a pet store, Becky feeds each gerbil $\frac{1}{8}$ cup of seeds each day. If the seeds come in packages of $\frac{5}{4}$ cups, how many gerbils can be fed from one package?

68. The Green Thumb Nursery advises that when planting spinach you should use $\frac{1}{16}$ cup of seed for a 50-foot row. How many rows can be planted using $\frac{7}{8}$ cup of seed?

69. Underinflation of car tires can waste up to $\frac{1}{20}$ of a car's fuel by increasing the "rolling resistance." If Melvin uses 820 gallons of gas in a year, how many gallons could potentially be saved by proper tire inflation?

70. Underinflation of truck tires can waste up to $\frac{1}{15}$ of a truck's fuel by increasing the "rolling resistance." If a trucker uses 500 gallons of gas while not aware that the tires are low, how many gallons could potentially be wasted.

Exercises 71–73 refer to the chapter application (see page 279).

71. The pecan chocolate chip recipe calls for $\frac{1}{2}$ tsp of salt and $\frac{3}{4}$ cup of brown sugar. Sandy has an order for seven boxes.

a. How many cookies does she need?

b. How many times must she make the recipe to fill the order?

c. How much salt does she need for the entire order?

d. How much brown sugar does she need for the entire order?

72. Hannah has an order for six boxes of dipped butterfingers. Her recipe calls for $\frac{3}{8}$ cup of bittersweet chocolate.

a. How many cookies does she need?

b. How many times must she make the recipe to fill the order?

c. How much bittersweet chocolate does she need for the entire order?

73. Sally has been hired to cater a dinner party for 15 people. She is serving lasagna, but her recipe yields 40 servings.

a. Assuming that each guest will get one serving, write a simplified fraction that represents the part of the recipe Sally needs to make for the party.

b. The recipe calls for 3 pounds of ricotta cheese. How many pounds does she need for the party?

STATE YOUR UNDERSTANDING

74. Explain how to find the product of $\frac{35}{24}$ and $\frac{40}{14}$.

75. Explain how to find the quotient of $\frac{35}{24}$ and $\frac{40}{14}$.

76. Evalynne's supervisor tells her that her salary is to be divided by one-half. Should she quit her job? Explain.

CHALLENGE

77. Simplify: $\left(\frac{81}{75} \cdot \frac{96}{99} \cdot \frac{55}{125}\right) \div \frac{128}{250}$

78. The In-n-Out Grocery has a standard work week of 40 hours. Jane works $\frac{3}{4}$ of a standard week, Jose works $\frac{5}{8}$ of a standard week, Aria works $\frac{9}{8}$ of a standard week, and Bill works $\frac{6}{5}$ of a standard week. How many hours did each employee work? In-n-Out pays an average salary of $7 an hour. What is the week's payroll for the above four employees?

GROUP ACTIVITY

79. Divide your group into two subgroups. Have the two subgroups take opposite sides to discuss the following statement. "Since computers and calculators use decimals more often than fractions, the use of fractions will eventually disappear." Have each subgroup write down their arguments for or against the statement.

To make accurate pie charts, it is necessary to use a measuring instrument for angles. One such instrument is called a protractor. It is also necessary to convert the fractions of the components into equivalent fractions with denominators of 360 because there are 360° in a complete circle. Starting with a circle of any radius, use the protractor to measure the correct angle for each component.

80. An aggressive investment strategy allocates $\frac{3}{4}$ of a portfolio to stocks, $\frac{1}{5}$ to bonds, and $\frac{1}{20}$ to money market funds. Make an accurate pie chart for this strategy.

81. A moderate investment strategy allocates $\frac{3}{5}$ of a portfolio to stocks, $\frac{3}{10}$ to bonds, and $\frac{1}{10}$ to money market funds. Make an accurate pie chart for this strategy.

82. A conservative investment strategy allocates $\frac{2}{5}$ of a portfolio to stocks, $\frac{9}{20}$ to bonds, and $\frac{3}{20}$ to money market funds. Make an accurate pie chart for this strategy.

MAINTAIN YOUR SKILLS (SECTIONS 1.3, 1.4, 1.6, 3.6, 4.1)

83. Find the LCM of 24, 15, and 10.

84. Find the LCM of 24, 36, and 48.

85. Multiply and round to the nearest thousand: (678)(351)

86. Divide 172,676 by 49 and round to the nearest ten.

87. Change $\frac{79}{13}$ to a mixed number.

88. Change $26\frac{2}{3}$ to an improper fraction.

89. Geri needs 75 building blocks, each weighing 19 lb. Her pickup can safely carry 1200 lb. Will she be able to haul all the blocks in one load?

90. Kevin needs a load of gravel for a drainage field. His truck can safely carry a load of $1\frac{3}{4}$ tons (3500 lb). If the gravel sells in 222-lb scoops, how many scoops can Kevin safely haul?

91. In Mr. Smart's math class 15 students bring their books and 12 students do not bring their books. What fraction of the class bring their books?

92. Roberta is taking classes at a community college. She budgets $17 per week for transportation for the school year. What is her total transportation budget for the year? Assume 34 weeks, three terms of ten weeks each and two weeks between each term. After five weeks she has $488 left in her transportation budget. If her travel cost remains the same, will she be over or under her budget at the end of the year? By how much?

4.4

Multiplying and Dividing Mixed Numbers

OBJECTIVES

1. Multiply mixed numbers.
2. Divide mixed numbers.

HOW AND WHY

Objective 1

Multiply mixed numbers.

In Section 4.1 we changed mixed numbers to improper fractions. For instance,

$$6\frac{5}{8} = \frac{6(8) + 5}{8} = \frac{53}{8}$$

To multiply mixed numbers we change them to improper fractions and then multiply.

$$\left(4\frac{1}{2}\right)\left(3\frac{2}{3}\right) = \left(\frac{9}{2}\right)\left(\frac{11}{3}\right)$$ **Change to improper fractions.**

$$= \frac{99}{6}$$ **Multiply.**

$$= \frac{33}{2} = 16\frac{1}{2}$$ **Simplify and write as a mixed number.**

Products can be left as improper fractions or mixed numbers; either is acceptable. In this textbook we write mixed numbers. In algebra, improper fractions are preferred.

To multiply whole numbers and/or mixed numbers

1. Change them to improper fractions.
2. Multiply.

Examples A–D

Directions: Multiply. Write as a mixed number.

Strategy: Change the mixed numbers and whole numbers to improper fractions. Multiply and simplify. Write the answer as a mixed number.

A. Multiply: $\frac{4}{5}\left(1\frac{1}{3}\right)$

$$\frac{4}{5}\left(1\frac{1}{3}\right) = \frac{4}{5}\cdot\frac{4}{3}$$ $1\frac{1}{3} = \frac{4}{3}$.

$$= \frac{16}{15}$$ **Multiply.**

$$= 1\frac{1}{15}$$ **Write as a mixed number.**

Warm Ups A–D

A. Multiply: $\frac{3}{7}\left(2\frac{1}{2}\right)$

Answer to Warm Up A. $1\frac{1}{14}$

B. Multiply: $\left(2\frac{2}{3}\right)\left(2\frac{1}{4}\right)$

B. Multiply: $\left(3\frac{3}{4}\right)\left(2\frac{2}{5}\right)$

$$\left(3\frac{3}{4}\right)\left(2\frac{2}{5}\right) = \frac{15}{4} \cdot \frac{12}{5}$$

$$= \frac{\overset{3}{\cancel{15}}}{\underset{1}{\cancel{4}}} \cdot \frac{\overset{3}{\cancel{12}}}{\underset{1}{\cancel{5}}}$$ **Simplify.**

$$= \frac{9}{1} = 9$$ **Multiply and write as a whole number.**

C. Multiply: $4\left(3\frac{3}{5}\right)\left(\frac{5}{9}\right)$

C. Multiply: $6\left(\frac{3}{4}\right)\left(2\frac{1}{3}\right)$

CAUTION

$6\left(\frac{3}{4}\right)$ means $6 \cdot \frac{3}{4}$, but $6\frac{3}{4}$ means $6 + \frac{3}{4}$.

$$6\left(\frac{3}{4}\right)\left(2\frac{1}{3}\right) = \frac{\overset{3}{\cancel{6}}}{1} \cdot \frac{\overset{1}{\cancel{3}}}{\underset{1}{\cancel{4}}} \cdot \frac{7}{\underset{1}{\cancel{3}}}$$

$$= \frac{21}{2} = 10\frac{1}{2}$$ **Multiply and write as a mixed number.**

Calculator Example

D. Multiply: $5\frac{3}{8}\left(6\frac{1}{2}\right)$

D. Multiply: $4\frac{2}{3}\left(8\frac{5}{6}\right)$

$$4\frac{2}{3}\left(8\frac{5}{6}\right) = 41\frac{2}{9}$$ **On a calculator with fraction keys it is not necessary to change the mixed numbers to improper fractions first. The calculator is programmed to operate with simple fractions or with mixed numbers.**

HOW AND WHY

Objective 2

Divide mixed numbers.

Division of mixed numbers is also done by changing to improper fractions first.

$$\left(6\frac{1}{6}\right) \div \left(2\frac{1}{2}\right) = \left(\frac{37}{6}\right) \div \left(\frac{5}{2}\right)$$ **Change to improper fractions.**

$$= \left(\frac{37}{6}\right)\left(\frac{2}{5}\right)$$ **Multiply by the reciprocal of the divisor.**

$$= \frac{37}{15}$$ **Simplify and multiply.**

$$= 2\frac{7}{15}$$ **Write as a mixed number.**

Answers to Warm Ups B. 6 C. 8 D. $34\frac{15}{16}$

To divide whole numbers and/or mixed numbers

1. Change them to improper fractions.
2. Divide.

Examples E–I

Directions: Divide. Write as a mixed number.

Strategy: Change the mixed numbers and whole numbers to improper fractions. Divide and simplify completely. Write the answer as a mixed number.

E. Divide: $3\frac{1}{3} \div 6\frac{7}{8}$

$3\frac{1}{3} \div 6\frac{7}{8} = \frac{10}{3} \div \frac{55}{8}$ **$3\frac{1}{3} = \frac{10}{3}$ and $6\frac{7}{8} = \frac{55}{8}$.**

$= \frac{\overset{2}{\cancel{10}}}{3} \cdot \frac{8}{\underset{11}{\cancel{55}}}$ **Invert the divisor and multiply.**

$= \frac{16}{33}$

F. Divide: $12\frac{3}{16} \div 21\frac{3}{5}$

$12\frac{3}{16} \div 21\frac{3}{5} = \frac{195}{16} \div \frac{108}{5}$

$= \frac{\overset{65}{\cancel{195}}}{16} \cdot \frac{5}{\underset{36}{\cancel{108}}}$ **Invert the divisor and multiply.**

$= \frac{325}{576}$

G. Divide: $4 \div 3\frac{1}{3}$

$4 \div 3\frac{1}{3} = \frac{4}{1} \div \frac{10}{3}$

$= \frac{\overset{2}{\cancel{4}}}{1} \cdot \frac{3}{\underset{5}{\cancel{10}}}$ **Invert the divisor and multiply.**

$= \frac{6}{5} = 1\frac{1}{5}$ **Write as a mixed number.**

Warm Ups E–I

E. Divide: $4\frac{5}{6} \div 2\frac{11}{12}$

F. Divide: $15\frac{5}{8} \div 12\frac{1}{2}$

G. Divide: $8\frac{5}{6} \div 5$

Answers to Warm Ups E $1\frac{23}{35}$ F. $1\frac{1}{4}$ G. $1\frac{23}{30}$

H. Divide: $13\frac{4}{5} \div 12\frac{3}{10}$

Calculator Example

H. Divide: $25\frac{1}{4} \div 30\frac{3}{4}$

$25\frac{1}{4} \div 30\frac{3}{4} = \frac{101}{123}$

On a calculator with fraction keys it is not necessary to change the mixed numbers to improper fractions first. The calculator is programmed to operate with simple fractions or with mixed numbers.

I. The Elko Aluminum Company also produces ingots that are $7\frac{1}{2}$ inches thick. What is the height in feet of a stack of 28 ingots?

I. The Elko Aluminum Company produces ingots that are $6\frac{3}{4}$ inches thick. What is the height in feet of a stack of 15 ingots?

Strategy: To find the height of the stack, multiply the thickness of one ingot by the number of ingots.

$15\left(6\frac{3}{4}\right) = \frac{15}{1} \cdot \frac{27}{4}$

$= \frac{405}{4} = 101\frac{1}{4}$ **Multiply and write as a mixed number.**

The height of the stack is $101\frac{1}{4}$ inches. To find the height in feet, divide by 12. (There are 12 inches in one foot.)

$101\frac{1}{4} \div 12 = \frac{405}{4} \div \frac{12}{1}$

$= \frac{\overset{135}{\cancel{405}}}{4} \cdot \frac{1}{\underset{4}{\cancel{12}}}$ **Invert the divisor and multiply.**

$= \frac{135}{16} = 8\frac{7}{16}$ **Write as a mixed number.**

The stack of ingots is $8\frac{7}{16}$ feet high.

Answers to Warm Ups H. $1\frac{5}{41}$ I. The height is $17\frac{1}{2}$ feet.

NAME ______________________ CLASS ____________ DATE ____________

Exercises 4.4

OBJECTIVE 1: *Multiply mixed numbers.*

A.

Multiply. Simplify completely and write as a mixed number if possible.

1. $\left(\frac{3}{4}\right)\left(1\frac{3}{4}\right)$

2. $\left(\frac{3}{7}\right)\left(2\frac{5}{7}\right)$

3. $\left(4\frac{1}{2}\right)\left(2\frac{2}{3}\right)$

4. $\left(3\frac{1}{3}\right)\left(1\frac{4}{5}\right)$

5. $3\left(2\frac{2}{3}\right)$

6. $4\left(1\frac{1}{4}\right)$

7. $\left(3\frac{3}{5}\right)\left(1\frac{1}{4}\right)(5)$

8. $\left(\frac{3}{5}\right)\left(2\frac{1}{3}\right)(3)$

9. $\left(7\frac{1}{2}\right)\cdot\left(3\frac{1}{5}\right)$

10. $\left(4\frac{1}{6}\right)\left(4\frac{4}{5}\right)$

11. $\left(3\frac{4}{7}\right)\left(\frac{14}{15}\right)$

12. $\left(4\frac{4}{9}\right)\left(\frac{12}{25}\right)$

B.

13. $\left(2\frac{5}{8}\right)\left(1\frac{1}{6}\right)$

14. $\left(4\frac{2}{5}\right)\left(1\frac{3}{8}\right)$

15. $\left(7\frac{3}{4}\right)\left(\frac{2}{3}\right)(0)$

16. $\left(3\frac{5}{9}\right)\left(6\frac{3}{8}\right)(0)$

17. $\left(3\frac{1}{3}\right)(6)\left(3\frac{3}{4}\right)$

18. $\left(4\frac{1}{5}\right)(5)\left(2\frac{7}{9}\right)$

19. $\left(3\frac{2}{3}\right)\left(\frac{15}{22}\right)\left(7\frac{1}{2}\right)$

20. $\left(4\frac{3}{4}\right)\left(3\frac{1}{5}\right)\left(5\frac{5}{8}\right)$

21. $\left(4\frac{1}{5}\right)\left(1\frac{1}{3}\right)\left(6\frac{2}{7}\right)$

22. $\left(12\frac{1}{4}\right)\left(1\frac{1}{7}\right)\left(2\frac{1}{3}\right)$

23. $(14)\left(6\frac{1}{2}\right)\left(1\frac{2}{13}\right)$

24. $(12)\left(1\frac{4}{15}\right)\left(6\frac{1}{4}\right)$

OBJECTIVE 2: *Divide mixed numbers.*

A.

Divide. Simplify completely and write as a mixed number if possible.

25. $3 \div 1\frac{1}{2}$

26. $4 \div 1\frac{1}{4}$

27. $3\frac{7}{8} \div 5\frac{1}{6}$

28. $8\frac{2}{5} \div 2\frac{1}{3}$

29. $2\frac{1}{2} \div 1\frac{1}{5}$

30. $2\frac{1}{2} \div 1\frac{1}{3}$

31. $1\frac{5}{8} \div 2\frac{1}{4}$

32. $1\frac{5}{8} \div 1\frac{3}{4}$

33. $3\frac{1}{3} \div 2\frac{1}{2}$

34. $5\frac{1}{3} \div 1\frac{1}{7}$

35. $4\frac{4}{15} \div 6\frac{2}{5}$

36. $6\frac{1}{4} \div 7\frac{1}{2}$

B.

37. $\frac{7}{8} \div 3\frac{3}{4}$

38. $\frac{11}{15} \div 2\frac{4}{5}$

39. $6\frac{2}{3} \div 10$

40. $5\frac{3}{7} \div 19$

41. $3\frac{2}{3} \div \frac{1}{5}$

42. $8\frac{3}{4} \div 2\frac{1}{3}$

43. $3\frac{3}{4} \div \frac{7}{15}$

44. $5\frac{5}{6} \div \frac{20}{9}$

45. $31\frac{1}{3} \div 1\frac{1}{9}$

46. $21\frac{3}{7} \div 8\frac{1}{3}$

47. $10\frac{2}{3} \div 2\frac{2}{7}$

48. $22\frac{2}{3} \div 6\frac{6}{7}$

C.

Multiply. Simplify completely and write as a mixed number if possible.

49. $\left(5\frac{1}{3}\right)\left(2\frac{3}{14}\right)(7)\left(2\frac{1}{4}\right)$

50. $\left(3\frac{3}{5}\right)\left(4\frac{1}{3}\right)(3)\left(2\frac{7}{9}\right)$

Divide. Simplify completely and write as a mixed number if possible.

51. $20\frac{5}{6} \div 3\frac{4}{7}$

52. $18\frac{2}{5} \div 17\frac{1}{4}$

53. Find the error(s) in the statement: $1\frac{2}{3} \cdot 1\frac{1}{2} = 1\frac{1}{3}$. Correct the statement. Explain how you would avoid this error.

54. Find the error(s) in the statement: $6\frac{2}{9} \div 2\frac{2}{3} = 3\frac{1}{3}$. Correct the statement. Explain how you would avoid this error.

55. The iron content in a water sample at Lake Hieda is eight parts per million. The iron content in Swan Lake is $2\frac{3}{4}$ times greater than the content in Lake Hieda. What is the iron content in Swan Lake in parts per million?

56. A recent wildlife survey finds that there are $3\frac{1}{3}$ times as many Canada geese as there are brant geese. If the survey counts 2322 brants, how many Canada geese are counted?

57. A glass of reconstituted juice contains $\frac{3}{16}$ oz of concentrate. How many glasses of juice can Karla make from 36 oz of the concentrate?

58. The Top Notch Candy Company packages jelly beans in $1\frac{3}{8}$ lb bags. How many bags can be made from 660 pounds of jelly beans?

59. The water pressure during a bad fire is reduced to $\frac{5}{9}$ its original pressure at the hydrant. What is the reduced pressure if the original pressure was $70\frac{1}{5}$ pounds per square inch?

60. A wheat farmer in Iowa averages $55\frac{3}{4}$ bushels of wheat per acre on 150 acres of wheat. How many bushels of wheat does she harvest?

61. The amount of CO_2 a car emits is directly related to the amount of gas it uses. Cars give off 20 lb of CO_2 for every gallon of gas used. A car averaging 27 mpg will emit 2000 lb in 2700 miles. A car averaging 18 mpg will emit $1\frac{1}{2}$ times more CO_2 in the same distance. How many pounds of CO_2 does the less-efficient car emit in the 2700 miles?

62. Floors over unheated crawl spaces are generally recommended to be insulated at the R-19 level. It takes a thickness of $8\frac{3}{4}$ in. of fiberglass to produce an R-19 insulation value. How many cubic inches of fiberglass are needed to properly insulate a one-story home of 1100 square feet?

63. What is the area of a triangle that has a height of $3\frac{3}{4}$ in. and a base of $5\frac{1}{2}$ in.?

64. What is the area of a triangle whose base is $12\frac{3}{4}$ cm and whose height is $15\frac{1}{2}$ cm?

Exercises 65–67 refer to the chapter application (see page 279).

65. Sandy is making 10 recipes of her pecan chocolate chip cookies.

a. The recipe calls for $2\frac{1}{4}$ c of flour. How much flour does she need for 10 recipes?

b. The recipe also calls for $1\frac{1}{2}$ c of sugar. How much sugar does she need for 10 recipes?

66. Hannah has an order for 3 boxes of dipped butterfingers.

a. How many cookies does she need?

b. How many times does she need to increase her recipe in order to get exactly the correct number of cookies?

c. The recipe calls for $1\frac{1}{2}$ c of flour. How much flour does she need for the order?

d. The recipe calls for $1\frac{1}{2}$ tsp of vanilla. How much vanilla does she need for the order?

e. The recipe calls for $\frac{1}{4}$ tsp baking powder. How much baking powder does she need for the order?

67. Quan really wants to make brownies, but his recipe calls for three eggs and he only has two.

a. Write a fraction that represents the part of the recipe Quan can make with only two eggs.

b. His recipe calls for 6 tbs of butter. How much butter should Quan use?

c. His recipe also calls for $1\frac{1}{2}$ tsp of vanilla. How much vanilla should he use?

NAME ______________________ CLASS ____________ DATE ____________

STATE YOUR UNDERSTANDING

68. Explain how to simplify $5\frac{1}{4} \div 1\frac{7}{8}$.

69. When a number is multiplied by $1\frac{1}{2}$ the result is larger than the original number. But when you divide by $1\frac{1}{2}$ the result is smaller. Explain why.

70. Why is it helpful to change mixed numbers to improper fractions before multiplying or dividing?

CHALLENGE

71. Multiply and simplify completely $\left(\frac{2}{15}\right)\left(1\frac{6}{7}\right)\left(2\frac{2}{49}\right)\left(16\frac{2}{5}\right)\left(8\frac{3}{4}\right)\left(5\frac{1}{4}\right)\left(3\frac{3}{13}\right)$.

72. The Celtic Candy Company has two packs of mints that they sell in discount stores. One pack contains $1\frac{1}{4}$ lb of mints and the other contains $3\frac{1}{3}$ lb of mints. If the smaller pack sells for \$2 and the larger pack for \$5, which size should they use to get the most income from 3000 lb of mints? How much more is the income?

GROUP ACTIVITY

73. Some people are saying that mixed numbers are no longer being used because of the ease of using approximations with a calculator. See how many examples of the use of mixed numbers your group can find. Share the results with the class.

74. See if your group can find a way to multiply two mixed numbers without changing them to improper fractions. Report your method in class.

MAINTAIN YOUR SKILLS (SECTIONS 1.2, 1.3, 1.4, 1.7, 3.6, 4.1)

75. Add: 2 + 157 + 9854 + 765 + 19 + 4356

76. Subtract: 73,021 − 56,489

77. Find the product of 180 and 231.

78. Find the quotient of 187,701 and 267.

79. Find the LCM of 18, 20, and 24.

80. Find the LCM of 22, 24, and 25.

81. Change $\frac{553}{15}$ to a mixed number.

82. Change $\frac{628}{17}$ to a mixed number.

83. For you to be eligible for a drawing at the Flick movie house, your ticket stub number must be a multiple of three. If Jean's ticket number is 234572, is she eligible for the drawing?

84. The sales of the Goodstone Company totaled $954,000 last year. During the first six months of last year the monthly sales were $72,400, $68,200, $85,000, $89,500, $92,700, and $87,200. What were the average monthly sales for the rest of the year?

Getting Ready for Algebra

OBJECTIVE

Solve an equation of the form $\frac{ax}{b} = \frac{c}{d}$, where a, b, c, and d are whole numbers.

HOW AND WHY

We have previously solved equations in which variables (letters) were either multiplied or divided by whole numbers. We performed the inverse operations to solve for the variable. To eliminate multiplication, we divided by the number being multiplied. To eliminate division, we multiplied by the number that is the divisor. Now we solve some equations in which variables are multiplied by fractions. Recall from Chapter 1 that if a number is multiplied times a variable, there is usually no multiplication sign between them. That is, $2x$ is understood to mean 2 times x, and $\frac{2}{3}x$ means $\frac{2}{3}$ times x. However, we usually do not write $\frac{2}{3}x$. Instead, we write this as $\frac{2x}{3}$. We can do this because

$$\frac{2}{3}x = \frac{2}{3} \cdot x = \frac{2}{3} \cdot \frac{x}{1} = \frac{2x}{3}$$

Therefore, we will write $\frac{2}{3}x$ as $\frac{2x}{3}$. Remember, however, for convenience we may use either of these forms. Recall that $\frac{2x}{3}$ means 2 times x with that product divided by 3.

To solve an equation of the form $\frac{ax}{b} = \frac{c}{d}$

1. Multiply both sides of the equation by b to eliminate the division on the left side.
2. Divide both sides by a to isolate the variable.

Examples A–C

Directions: Solve.

Strategy: Multiply both sides by the denominator of the fraction containing the variable. Solve as before.

A. Solve: $\frac{3x}{4} = 2$

$4\left(\frac{3x}{4}\right) = 4(2)$ **To eliminate the division, multiply both sides by 4.**

$3x = 8$ **Simplify.**

$\frac{3x}{3} = \frac{8}{3}$ **To eliminate the multiplication, divide both sides by 3.**

$x = \frac{8}{3}$

Warm Ups A–C

A. Solve: $\frac{7x}{8} = 14$

Answer to Warm Up A. $x = 16$

CHECK:

$\frac{3}{4} \cdot \frac{8}{3} = 2$ **Substitute $\frac{8}{3}$ for x in the original equation; recall that**

$2 = 2$ **$\frac{3x}{4} = \frac{3}{4} \cdot x$.**

The solution is $x = \frac{8}{3}$ or $x = 2\frac{2}{3}$.

B. Solve: $\frac{9x}{8} = \frac{7}{2}$

B. Solve: $\frac{3}{4} = \frac{5x}{6}$

$6\left(\frac{3}{4}\right) = 6\left(\frac{5x}{6}\right)$ **To eliminate the division by 6, multiply both sides by 6.**

$\frac{9}{2} = 5x$

$\frac{9}{2(5)} = \frac{5x}{5}$ **To eliminate the multiplication by 5, divide both sides by 5.**

$\frac{9}{10} = x$

The solution is $x = \frac{9}{10}$. **The check is left for the student.**

C. It takes approximately two fifths the amount of energy to make "new" paper from recycled paper as from trees. If the amount of energy needed to make an amount of paper from recycled paper is equivalent to 1500 BTUs, how much energy would be needed to make the same amount from trees?

C. One third of all the mileage put on a private car is from commuting to work. If Nancy averages 510 miles per month in commuting to work, how many miles does she put on her car in a month?

Strategy: First write the English statement of the equation:

one third(total miles) = commute miles

Let x represent the total miles.

$\frac{1}{3}x = 510$ **The commute miles are 510.**

$3\left(\frac{1}{3}x\right) = 3(510)$ **Multiply both sides by 3 to eliminate the division.**

$x = 1530$

CHECK: Is one third of 1530 equal to 510?

$\frac{1}{3}(1530) = 510$ **Yes.**

Nancy puts 1530 miles on her car per month.

Answers to Warm Ups B. $x = \frac{28}{9}$ or $x = 3\frac{1}{9}$ C. It would take 3750 BTUs.

NAME ______________________ CLASS ____________ DATE ____________

Exercises

Solve.

1. $\frac{2x}{3} = \frac{1}{2}$

2. $\frac{2x}{5} = \frac{2}{3}$

3. $\frac{3y}{4} = \frac{4}{5}$

4. $\frac{7y}{8} = \frac{5}{6}$

5. $\frac{4z}{5} = -\frac{3}{4}$

6. $\frac{5z}{4} = \frac{8}{9}$

7. $\frac{17}{9} = \frac{8x}{9}$

8. $\frac{29}{10} = \frac{9x}{5}$

9. $\frac{7a}{4} = \frac{5}{2}$

10. $\frac{15a}{4} = \frac{24}{5}$

11. $\frac{47}{4} = \frac{47b}{6}$

12. $\frac{13}{3} = \frac{52b}{9}$

13. $\frac{41z}{6} = \frac{41}{3}$

14. $\frac{9b}{23} = \frac{23}{3}$

15. $\frac{2a}{15} = \frac{11}{4}$

16. $\frac{119x}{12} - \frac{119}{8}$

17. Garth walks $\frac{1}{3}$ of the distance from his home to school. If he walks $\frac{1}{2}$ mile, what is the distance from his home to school?

18. Mark cuts a board into nine pieces of equal length. If each piece is $1\frac{4}{9}$ ft long, what was the length of the board?

19. If $3\frac{1}{2}$ times more pounds of glass are recycled than tin, how many pounds of tin are recycled when 630 lb of glass are recycled?

20. Washing machines use about $\frac{7}{50}$ of all the water consumed in the home. If Myra uses 280 gallons of water per month to operate her washing machine, how many gallons of water does she use in a month to run her household?

4.5

Conversion of Units in the Same System

OBJECTIVE

Convert units within the English system or within the metric system.

HOW AND WHY

Objective

Convert units within the English system or within the metric system.

Because 12 in. = 1 ft, they are equivalent measures. When we write 12 in. in place of 1 ft, or vice versa, we say that we have "converted the units." The division: (12 in.) ÷ (1 ft) asks "how many units of measure 1 ft does it take to make 12 in.?" Since they measure the same length, the answer is 1.

$$\frac{12 \text{ in.}}{1 \text{ ft}} = \frac{1 \text{ ft}}{12 \text{ in.}} = 1$$

See inside back cover for a table of equivalent measures.

This idea, along with the multiplication property of one, is used to convert from one measure to another. The units of measure are treated like factors and are simplified before multiplying. For example, to convert 48 in. to feet we multiply.

$48 \text{ in.} = (48 \text{ in.}) \cdot 1$ **Multiply by 1.**

$48 \text{ in.} = \frac{48 \cancel{\text{in.}}}{1} \cdot \frac{1 \text{ ft}}{12 \cancel{\text{in.}}}$ **Substitute $\frac{1 \text{ ft}}{12 \text{ in.}}$ for 1.**

$= \frac{48}{12} \text{ ft}$ **Multiply.**

$= 4 \text{ ft}$ **Simplify.**

In the second step we chose to multiply by $\frac{1 \text{ ft}}{12 \text{ in.}}$ deliberately. We want the unit "inches" to simplify, or divide out so we are left with the unit "feet." Therefore, 48 in. can be converted to 4 ft.

In some cases, it is necessary to multiply by various names for the number one as when we convert 7200 sec to hours.

$7200 \text{ sec} = (7200) \cdot 1 \cdot 1$ **Multiply by 1 twice.**

$7200 \text{ sec} = \frac{7200 \cancel{\text{sec}}}{1} \cdot \frac{1 \cancel{\text{min}}}{60 \cancel{\text{sec}}} \cdot \frac{1 \text{ hr}}{60 \cancel{\text{min}}}$ **Substitute.**

$= \frac{7200(1)(1) \text{ hr}}{1(60)(60)}$

$= \frac{7200}{3600} \text{ hr}$ **Multiply.**

$= 2 \text{ hr}$ **Simplify.**

Hence, 7200 sec is equivalent to 2 hr.

To convert the units of a measurement

1. Multiply by fractions formed by equivalent measurements, that is, names for 1. Choose these fractions so that after simplifying the desired units remain.
2. Simplify.

The examples include converting measures within the metric system. See Chapter 6 for an alternative method of converting metric units.

Remember when calculating areas we get square units of measure such as 5 square meters. We write this as 5 m^2. The symbol "5 m^2" literally means "5 · 1 m · 1 m." See Example E.

Warm Ups A–G

A. Convert 5 hr to seconds.

B. Convert 4 m to centimeters.

C. How many feet are in 2 miles?

Examples A–G

Directions: Convert units of measure.

Strategy: Multiply the given unit of measure by fractions formed by equivalent measures. Use a conversion chart if necessary.

A. Convert 5 gallons to pints.

Strategy: Multiply by quarts/gallon to get quarts and then by pints/quart to get pints.

$$\begin{aligned} 5 \text{ gallons} &= (5 \text{ gallons}) \cdot 1 \cdot 1 \\ &= \frac{5 \cancel{\text{gallons}}}{1} \cdot \frac{4 \cancel{\text{quarts}}}{1 \cancel{\text{gallon}}} \cdot \frac{2 \text{ pints}}{1 \cancel{\text{quart}}} \\ &= 5 \cdot 4 \cdot 2 \text{ pints} \\ &= 40 \text{ pints} \end{aligned}$$

Multiply by 1 twice.

5 gallons = 40 pints

B. Convert 91 kg to grams.

$$\frac{91 \cancel{\text{kg}}}{1} \cdot \frac{1000 \text{ g}}{1 \cancel{\text{kg}}} = 91 \cdot 1000 \text{ g}$$

91 kg = 91,000 g

C. Five hundred meters is how many kilometers?

$$\begin{aligned} \frac{500 \cancel{\text{m}}}{1} \cdot \frac{1 \text{ km}}{1000 \cancel{\text{m}}} &= \frac{500}{1000} \text{ km} \\ &= \frac{1}{2} \text{ km} \end{aligned}$$

$$500 \text{ m} = \frac{1}{2} \text{ km}$$

Answers to Warm Ups A. 18,000 sec B. 400 cm C. 10,560 ft

D. Convert 60 mph to feet per second (fps).

$$60 \text{ mph} = \frac{\overset{1}{\cancel{60}}\,\cancel{\text{mi}}}{1\,\cancel{\text{hr}}} \cdot \frac{1\,\cancel{\text{hr}}}{\underset{1}{\cancel{60}}\,\cancel{\text{min}}} \cdot \frac{1\,\cancel{\text{min}}}{60 \text{ sec}} \cdot \frac{5280 \text{ ft}}{1\,\cancel{\text{mi}}}$$

$$= \frac{1 \cdot 1 \cdot 1 \cdot 5280 \text{ ft}}{1 \cdot 1 \cdot 60 \text{ sec} \cdot 1}$$

$$= \frac{88 \text{ ft}}{\text{sec}}$$

60 mph = 88 fps

D. Convert 66 fps (feet per second) to miles per hour.

E. How many square inches are in 3 ft^2?

Strategy: When we write "in^2" the exponent means to use the factor twice, so 1 in^2 = 1 in. · 1 in. and 1 ft^2 = 1 ft · 1 ft. So 3 ft^2 = 3 · 1 ft · 1 ft.

$$3 \text{ ft}^2 = \frac{3 \cdot 1\,\cancel{\text{ft}} \cdot 1\,\cancel{\text{ft}}}{1} \cdot \frac{12 \text{ in.}}{1\,\cancel{\text{ft}}} \cdot \frac{12 \text{ in.}}{1\,\cancel{\text{ft}}}$$ **Multiply by $\frac{12 \text{ in.}}{1 \text{ ft}}$ twice.**

$$= 3(12)(12)(\text{in.})(\text{in.})$$

$$= 432 \text{ in}^2$$

3 ft^2 is equivalent to 432 in^2.

E. How many square feet are in 6 yd^2?

F. Convert 480 g per liter to milligrams per kiloliter.

$$\frac{480 \text{ g}}{\ell} = \frac{480\,\cancel{\text{g}}}{1\,\cancel{\ell}} \cdot \frac{1000 \text{ mg}}{1\,\cancel{\text{g}}} \cdot \frac{1000\,\cancel{\ell}}{1 \text{ k}\ell}$$

$$= \frac{480,(1000),(1000) \text{ mg}}{1(1)(1) \text{ k}\ell}$$

$$= \frac{480\ 000\ 000 \text{ mg}}{\text{k}\ell}$$

480 g per liter is equivalent to 480,000,000 mg per kiloliter.

F. Convert 50 meters per minute to kilometers per hour.

G. Belinda's employer wants to change her hourly wage of \$9.00 per hour to an equivalent piecework wage. If she makes an average of 4 circuit boards per hour, what will be the equivalent wage per board?

Strategy: Multiply by $\frac{1 \text{ hr}}{4 \text{ boards}}$ because, in one hour, Belinda makes 4 boards.

$$\frac{\$9.00}{1 \text{ hr}} = \frac{\$9.00}{1\,\cancel{\text{hr}}} \cdot \frac{1\,\cancel{\text{hr}}}{4 \text{ boards}}$$

$$= \frac{\$9}{4 \text{ boards}}$$

$$= 2\frac{1}{4} \text{ dollars per board}$$

The equivalent piecework wage is \$2.25 per circuit board.

G. Lucas can type an average of 70 words per minute. If a page averages 600 words, how many pages can he type in 1 hr?

Answers to Warm Ups D. 45 mph E. 54 ft^2 F. 3 kph G. Lucas can type 7 pages per hour.

NAME ______________________ CLASS __________ DATE __________

Exercises 4.5

OBJECTIVE: *Convert units within the English system or within the metric system.*

A.

Convert the units of measurement.

1. 2 wk = ? days

2. 120 sec = ? min

3. 2 yr = ? mo

4. 21 days = ? wk

5. 2 ft = ? in.

6. 36 in. = ? yd

7. 1 mile = ? ft

8. 1760 yd = ? miles

9. 2 meters = ? cm

10. 2 km = ? m

11. 1 yd^2 = ? ft^2

12. 1 m^2 = ? cm^2

13. 80 oz = ? lb

14. 8 quarts = ? gallons

15. 15 cm = ? mm

16. 4 g = ? cg

17. 40 pints = ? gallons

18. 12 ft = ? yd

19. 6000 g = ? kg

20. 9 ft = ? in.

21. 10 kg = ? g

22. 3 lb = ? oz

23. 108 in. = ? yd

24. 9 yd = ? ft

B.

25. 5 m = ? cm

26. 1080 in. = ? yd

27. $\frac{144 \text{ lb}}{1 \text{ ft}} = \frac{? \text{ lb}}{1 \text{ in.}}$

28. $\frac{\$660}{\text{hr}} = \frac{\$?}{\text{min}}$

29. 1000 mm = ? m

30. 1000 cm = ? m

31. 32 in. = ? ft

32. 16 ft = ? yd

33. 3 yds 1 ft = ? in.

34. 4 ft 6 in. = ? yd

35. 3000 lb = ? tons

36. 2640 ft = ? mile

37. 10,080 min = ? days

38. 371 days = ? wk

39. $6\frac{1}{2}$ tons = ? lb

40. $5\frac{1}{2}$ days = ? hr

41. $\frac{45 \text{ miles}}{1 \text{ hr}} = \frac{? \text{ ft}}{1 \text{ sec}}$

42. $\frac{\$180}{1 \text{ ton}} = \frac{? \text{ cents}}{1 \text{ lb}}$

43. $\frac{9 \text{ tons}}{1 \text{ ft}} = \frac{? \text{ lb}}{1 \text{ in.}}$

44. $\frac{10 \text{ oz}}{1 \text{ c}} = \frac{? \text{ lb}}{1 \text{ gallon}}$

45. $\frac{180 \text{ km}}{1 \text{ hr}} = \frac{? \text{ m}}{1 \text{ sec}}$

46. $\frac{24000 \text{ miles}}{1 \text{ hr}} = \frac{? \text{ ft}}{1 \text{ sec}}$

47. $\dfrac{24 \text{ g}}{1 \text{ m}^2} = \dfrac{? \text{ kg}}{1 \text{ km}^2}$

48. $\dfrac{\$36}{1 \text{ gallon}} = \dfrac{? \text{ cents}}{1 \text{ pint}}$

C.

49. A secretary can type 90 words per minute $\left(\dfrac{90 \text{ words}}{1 \text{ min}}\right)$. How many words can he type in 1 sec?

50. Mark is on a diet that causes him to lose 8 oz every day. At this rate, how many pounds will he lose in 6 wk?

51. A physician orders 0.3 g of Elixir Chlor-Trimeton. The available dose has a label that reads 2 mg per cubic centimeter (cc). How many cubic centimeters are needed to fill the doctor's order?

52. Larry's family eats approximately 78 kg of Wheat Bran flakes in a year. If Wheat Bran is sold in boxes containing 500 g of wheat bran, what is the average number of boxes of Wheat Bran that Larry's family consumes in 1 wk?

53. During a fund-raiser to preserve land to save the African elephant, the Jungle Society agrees to donate 12¢ a milligram for the largest bass caught during the club's annual Bass Contest. If the largest bass caught weighs 2 kg, how much does the Jungle Society donate to save the African elephant?

54. During a fund-raiser to save the black rhino, Greg's Sports Shop agrees to donate 5¢ for every yard Rita jogs during 1 wk. Rita jogs 2 miles a day for 5 weekdays. How much does Greg's Sports Shop donate?

55. A snail can crawl $\frac{5}{8}$ in. in 1 min. How long will it take the snail to move 3 in.?

3 in.

56. If Dan averages 50 mph during an 8-hr day of driving, how many days will it take him to drive 2000 miles?

57. A metal alloy weighs 108 lb per cubic foot. How many ounces does 1 in^3 weigh?

58. During a bicycle trip the Trong family averages a speed of 22 fps. At this rate, how many miles will they average in a 7-hr day of cycling? How many days will it take the family to cover 350 miles?

59. An insect repellent weighs 100 kg per cubic meter. The Non-Sting Company sells the repellent in bottles containing 300 cc. How many grams of repellent are sold in one bottle? If the repellent is priced at 20¢ per gram, what is the price of one bottle?

STATE YOUR UNDERSTANDING

60. Explain how to convert 12,000 sec to hours.

61. Explain how to change $\frac{x\text{ ft}^2}{\text{sec}}$ to $\frac{\text{mi}^2}{\text{hr}}$.

62. Explain the role that the Multiplication Property of One plays in converting units.

CHALLENGE

63. A pharmacy pays \$5 per gram for a new drug. They resell the drug at $7\frac{1}{2}$¢ per centigram. What is the profit on the sale of 1 kg of the drug?

64. A chain made of precious metals is priced at \$7.50 per inch. What is the cost of 4 yd 2 ft 4 in. of the chain?

GROUP ACTIVITY

65. A cubic foot of water weighs about 1000 oz. Iron is $7\frac{4}{5}$ times heavier than water. An open rectangular tank of $\frac{1}{4}$-in. thick iron has outside dimensions of 4 ft long, 2 ft 6 in. wide, and 2 ft high.

a. Find the volume of the iron used to construct the tank.
b. Find the weight of the tank, to the nearest pound.
c. Suppose the tank is filled with water. Find the weight of the filled tank, to the nearest pound.

MAINTAIN YOUR SKILLS (SECTIONS 2.2, 2.4, 2.5, 3.5, 4.3)

66. Find the product of 2 and $\frac{3}{8}$.

67. Find the product of 3 and $\frac{2}{15}$.

68. Find the quotient of 3 and $\frac{15}{4}$.

69. Find the quotient of $\frac{15}{4}$ and 3.

70. Prime factor 2143.

71. Prime factor 2142.

Exercises 72–73 refer to the following figure.

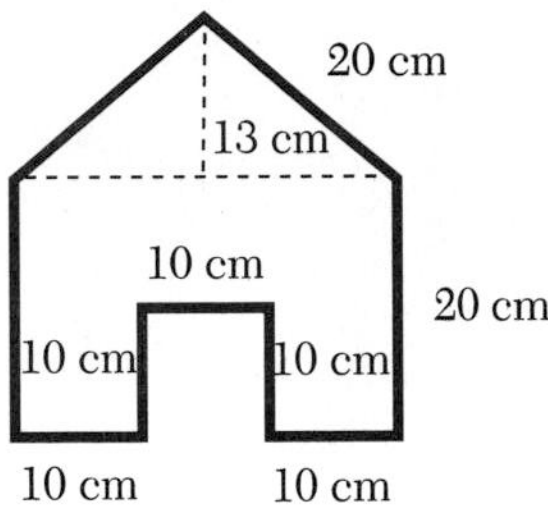

72. Find the perimeter.

73. Find the area.

74. Find the volume of wood in a rectangular block measuring 10 in. × 5 in. × 7 in. if the block has a square plug 3 in. on a side cut out of it.

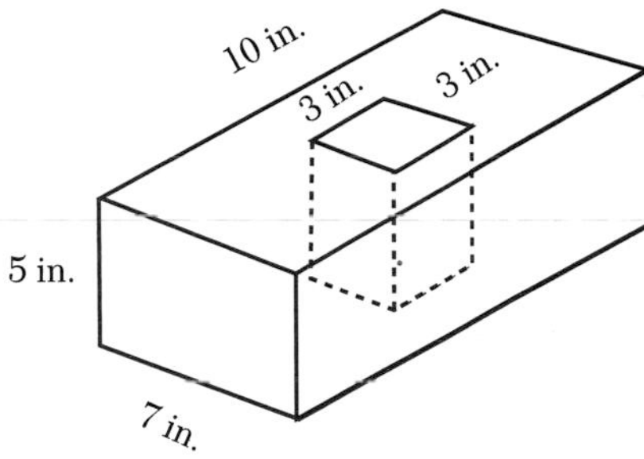

75. Find the volume of the solid with a base area of 7 ft^2 and a height of 18 in.

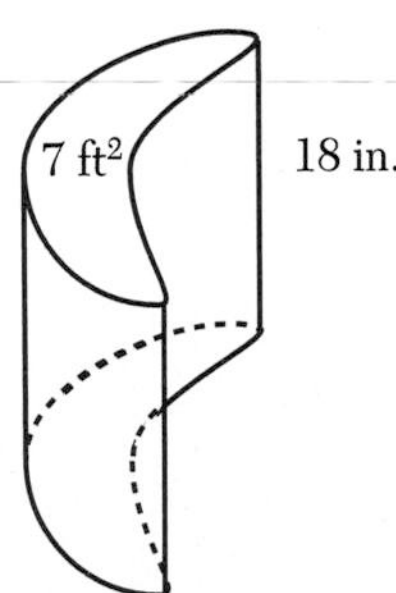

4.6

Building Fractions; Listing in Order; Inequalities

OBJECTIVES

1. Rename fractions by multiplying by 1 in the form $\frac{a}{a}$.
2. Build a fraction by finding the missing numerator.
3. List a group of fractions from smallest to largest.

VOCABULARY

Recall that **equivalent fractions** are fractions that are different names for the same number. For instance, $\frac{3}{6}$ and $\frac{15}{30}$ are equivalent because both represent one half $\left(\frac{1}{2}\right)$ of a unit.

Two or more fractions have a **common denominator** when they have the same denominator.

HOW AND WHY

Objective 1

Rename fractions by multiplying by 1 in the form $\frac{a}{a}$.

The process of renaming fractions is often referred to as "building fractions." Building a fraction means renaming the fraction by multiplying both numerator and denominator by a common factor. This process is often necessary when we add and subtract fractions. We "build fractions" to a common denominator so they can be compared, added, or subtracted. Building fractions is the opposite of "simplifying" fractions.

Simplifying a Fraction	Building a Fraction
$\frac{6}{10} = \frac{3 \cdot \cancel{2}}{5 \cdot \cancel{2}} = \frac{3}{5}$	$\frac{3}{5} = \frac{3 \cdot 2}{5 \cdot 2} = \frac{6}{10}$

Visually, we have

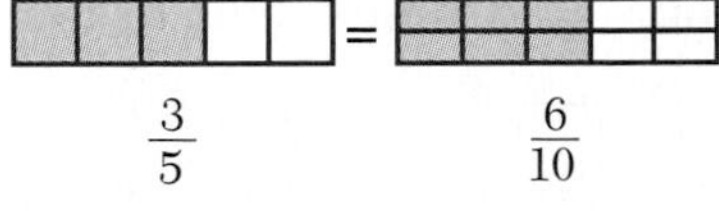

$\frac{3}{5}$ $\frac{6}{10}$

Table 4.2 shows five fractions built to equivalent fractions.

TABLE 4.2 EQUIVALENT FRACTIONS

	Multiply Numerator and Denominator by 2	3	4	6	10	15
$\frac{3}{5}$	$\frac{6}{10}$ =	$\frac{9}{15}$ =	$\frac{12}{20}$ =	$\frac{18}{30}$ =	$\frac{30}{50}$ =	$\frac{45}{75}$
$\frac{1}{2}$	$\frac{2}{4}$ =	$\frac{3}{6}$ =	$\frac{4}{8}$ =	$\frac{6}{12}$ =	$\frac{10}{20}$ =	$\frac{15}{30}$
$\frac{5}{8}$	$\frac{10}{16}$ =	$\frac{15}{24}$ =	$\frac{20}{32}$ =	$\frac{30}{48}$ =	$\frac{50}{80}$ =	$\frac{75}{120}$
$\frac{4}{9}$	$\frac{8}{18}$ =	$\frac{12}{27}$ =	$\frac{16}{36}$ =	$\frac{24}{54}$ =	$\frac{40}{90}$ =	$\frac{60}{135}$
$\frac{7}{3}$	$\frac{14}{6}$ =	$\frac{21}{9}$ =	$\frac{28}{12}$ =	$\frac{42}{18}$ =	$\frac{70}{30}$ =	$\frac{105}{45}$

To rename a fraction

Multiply both the numerator and the denominator of the fraction by the same number, that is multiply the fraction by 1 in the form $\frac{a}{a}$, $a \neq 1$, and $a \neq 0$.

Warm Ups A–B

A. Rename $\frac{8}{9}$ using $\frac{6}{6}$ for 1.

B. Write three fractions equivalent to $\frac{5}{11}$ using $\frac{2}{2}, \frac{3}{3}$, and $\frac{4}{4}$.

Examples A–B

Directions: Rename the fraction.

Strategy: Multiply the fraction by 1 in the form $\frac{a}{a}$.

A. Rename $\frac{3}{10}$ using $\frac{7}{7}$ for 1.

$\frac{3}{10} \cdot \frac{7}{7} = \frac{21}{70}$ **The new fraction, $\frac{21}{70}$, is equivalent to the original.**

B. Write three fractions equivalent to $\frac{6}{7}$ using $\frac{2}{2}, \frac{3}{3}$, and $\frac{4}{4}$.

$\frac{6}{7} \cdot \frac{2}{2} = \frac{12}{14}$

$\frac{6}{7} \cdot \frac{3}{3} = \frac{18}{21}$

$\frac{6}{7} \cdot \frac{4}{4} = \frac{24}{28}$

Answers to Warm Ups A. $\frac{48}{54}$ B. $\frac{10}{22}, \frac{15}{33}, \frac{20}{44}$

HOW AND WHY

Objective 2

Build a fraction by finding the missing numerator.

To find the missing numerator in

$$\frac{4}{5} = \frac{?}{30}$$

divide 30 by 5 to find out what form of 1 to multiply by.

$30 \div 5 = 6$

The correct multiplier is $\frac{6}{6}$. So

$$\frac{4}{5} = \frac{4}{5} \cdot \frac{6}{6} = \frac{24}{30}$$

The shortcut is to write 30, the target denominator. Then multiply the original numerator 4 by 6, the quotient of the target denominator and the original denominator.

$$\frac{4}{5} = \frac{6 \cdot 4}{30} = \frac{24}{30}$$

The fractions $\frac{4}{5}$ and $\frac{24}{30}$ are equivalent. Either fraction can be used in place of the other.

To find the missing numerator when building fractions

1. Divide the target denominator by the original denominator.
2. Multiply this quotient by the original numerator.

Examples C–E

Directions: Find the missing numerator.

Strategy: Multiply the given numerator by the same factor used in the denominator.

C. Find the missing numerator: $\frac{3}{5} = \frac{?}{75}$

$75 \div 5 = 15$ **Divide the denominators.**

$\frac{3}{5} = \frac{3(15)}{75} = \frac{45}{75}$ **Multiply the quotient, 15, by the given numerator.**

D. Find the missing numerator: $\frac{3}{7} = \frac{?}{42}$

$42 \div 7 = 6$ **Divide the denominators.**

$\frac{3}{7} = \frac{3(6)}{42} = \frac{18}{42}$ **Multiply the quotient, 6, by the given numerator.**

Warm Ups C–E

C. Find the missing numerator:

$\frac{3}{7} = \frac{?}{70}$

D. Find the missing numerator:

$\frac{7}{9} = \frac{?}{54}$

Answers to Warm Ups C. 30 D. 42

E. Find the missing numerator:

$\frac{22}{15} = \frac{?}{135}$

E. Find the missing numerator: $\frac{13}{10} = \frac{?}{140}$

$140 \div 10 = 14$

$\frac{13}{10} = \frac{13(14)}{140} = \frac{182}{140}$ **Building is the opposite of simplifying. A built-up fraction can be simplified to the original fraction.**

HOW AND WHY

Objective 3

List a group of fractions from smallest to largest.

If two fractions have the same denominator, the one with the smaller numerator has the smaller value. Figure 4.9 shows that $\frac{2}{5}$ is smaller than $\frac{4}{5}$, that is, $\frac{2}{5} < \frac{4}{5}$.

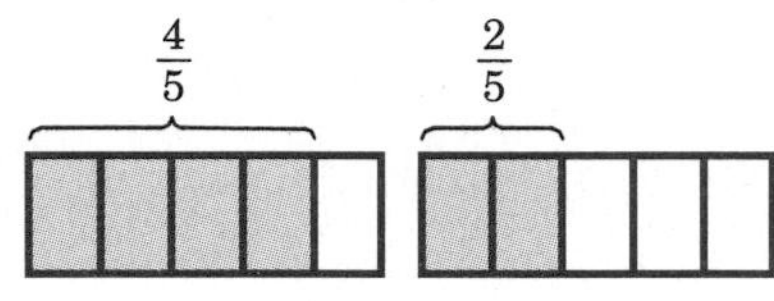

Figure 4.9

$\frac{2}{5} < \frac{4}{5}$ means "$\frac{2}{5}$ is less than $\frac{4}{5}$"

$\frac{4}{5} > \frac{1}{5}$ means "$\frac{4}{5}$ is greater than $\frac{1}{5}$"

If fractions to be compared do not have a common denominator, then one or more must be renamed so that all have a common denominator. The preferred common denominator is the least common multiple (LCM) of all the denominators.

To list $\frac{5}{8}$, $\frac{7}{16}$, $\frac{1}{2}$, and $\frac{9}{16}$ from the smallest to largest, we write each with a common denominator and then compare the numerators. We note that the LCM of all the denominators is 16. Therefore, we build each fraction so that it has a denominator of 16:

$\frac{5}{8} = \frac{10}{16}$ $\frac{7}{16} = \frac{7}{16}$ $\frac{1}{2} = \frac{8}{16}$ $\frac{9}{16} = \frac{9}{16}$ **Each fraction now has a denominator of 16.**

We arrange those fractions whose denominators are 16 in order from the smallest to largest:

$\frac{7}{16} < \frac{8}{16} < \frac{9}{16} < \frac{10}{16}$ **They are now listed in order from the smallest to largest with a common denominator of 16.**

We replace each fraction by the original, so

$\frac{7}{16} < \frac{1}{2} < \frac{9}{16} < \frac{5}{8}$ **They are now listed in order from the smallest to largest.**

Answer to Warm Up E. 198

To list fractions from smallest to largest

1. Build the fractions so that they have a common denominator. Use the LCM of the denominators.
2. List the fractions (with common denominators) with numerators from smallest to largest.
3. Simplify.

Examples F–K

Directions: Tell which fraction is larger.

Strategy: Write the fractions with a common denominator. The fraction with the larger numerator is the larger.

F. Which is larger $\frac{6}{11}$ or $\frac{1}{2}$?

Strategy: The LCM of 11 and 2 is 22. Build each fraction so it has 22 for a denominator.

$\frac{6}{11} = \frac{12}{22}$ and $\frac{1}{2} = \frac{11}{22}$

$\frac{6}{11}$ is larger. **12 > 11. That is, $\frac{6}{11} > \frac{1}{2}$.**

Directions: List the group of fractions from smallest to largest.

Strategy: Build each of the fractions to a common denominator. List the fractions from smallest to largest by the value of the numerator. Simplify.

G. List from smallest to largest: $\frac{2}{3}$, $\frac{3}{8}$, and $\frac{3}{4}$

$\frac{2}{3} = \frac{16}{24}$ $\frac{3}{8} = \frac{9}{24}$ **The LCM of 3, 8, and 4 is 24. Build the fractions to the denominator 24.**

$\frac{3}{4} = \frac{18}{24}$

$\frac{9}{24}, \frac{16}{24}, \frac{18}{24}$ **List the fractions in the order of the numerators: 9 < 16 < 18**

The list is $\frac{3}{8}$, $\frac{2}{3}$, and $\frac{3}{4}$. **Simplify.**

Warm Ups F–K

F. Which is larger $\frac{8}{13}$ or $\frac{3}{5}$?

G. List from smallest to largest: $\frac{5}{6}$, $\frac{7}{8}$, and $\frac{4}{5}$

Answers to Warm Ups F. $\frac{8}{13}$ G. $\frac{4}{5}$, $\frac{5}{6}$, and $\frac{7}{8}$

H. List from smallest to largest: $4\frac{3}{5}$, $4\frac{4}{9}$, and $4\frac{2}{3}$

H. List from smallest to largest: $3\frac{1}{2}$, $3\frac{5}{6}$, and $3\frac{5}{8}$

Strategy: Since the whole number part is the same in each mixed number, they can be listed in the order of the fractions.

$3\frac{1}{2} = 3\frac{12}{24}$ $\quad 3\frac{5}{6} = 3\frac{20}{24}$ **The LCM of 2, 6, and 8 is 24. Write with common denominators.**

$3\frac{5}{8} = 3\frac{15}{24}$

$3\frac{12}{24}, 3\frac{15}{24}, 3\frac{20}{24}$ **Write the numbers in the order of the numerators, smallest to largest.**

The list is $3\frac{1}{2}$, $3\frac{5}{8}$, and $3\frac{5}{6}$. **Simplify.**

I. The Acme Hardware Store also sells "rebar" with diameters of $\frac{3}{4}$, $\frac{7}{8}$, $\frac{7}{16}$, $\frac{15}{32}$, $\frac{15}{64}$, and $\frac{7}{12}$ in. List the diameters from smallest to largest.

I. The Acme Hardware Store sells bolts with diameters of $\frac{5}{16}$, $\frac{3}{8}$, $\frac{1}{2}$, $\frac{5}{8}$, $\frac{1}{4}$, and $\frac{7}{16}$ in. List the diameters from smallest to largest.

$\frac{5}{16} = \frac{5}{16}$ $\quad \frac{3}{8} = \frac{6}{16}$ $\quad \frac{1}{2} = \frac{8}{16}$ **Write each diameter using the common denominator, 16.**

$\frac{5}{8} = \frac{10}{16}$ $\quad \frac{1}{4} = \frac{4}{16}$ $\quad \frac{7}{16} = \frac{7}{16}$

$\frac{4}{16}, \frac{5}{16}, \frac{6}{16}, \frac{7}{16}, \frac{8}{16}, \frac{10}{16}$ **List the diameters in order of the numerators, smallest to largest.**

From smallest to largest, the diameters are

$\frac{1}{4}$, $\frac{5}{16}$, $\frac{3}{8}$, $\frac{7}{16}$, $\frac{1}{2}$, and $\frac{5}{8}$ in. **Simplify.**

Directions: Tell whether the statement is true or false.

Strategy: Build each fraction to a common denominator and compare the numerators.

J. True or false? $\frac{1}{16} > \frac{13}{20}$

J. True or false? $\frac{3}{4} > \frac{7}{9}$

$\frac{3}{4} = \frac{27}{36}$ $\quad \frac{7}{9} = \frac{28}{36}$ **The common denominator is 36.**

The statement is false. **Compare the numerators, $27 < 28$.**

K. True or false? $\frac{13}{16} < \frac{22}{27}$

K. True or false? $\frac{12}{25} < \frac{17}{35}$

$\frac{84}{175} < \frac{85}{175}$ **The common denominator is 175.**

The statement is true. **Compare the numerators.**

Answers to Warm Ups H. $4\frac{4}{9}$, $4\frac{3}{5}$, and $4\frac{2}{3}$ I. From smallest to largest, the diameters are $\frac{15}{64}$, $\frac{7}{16}$, $\frac{15}{32}$, $\frac{7}{12}$, $\frac{3}{4}$, and $\frac{7}{8}$ in. J. False K. True

NAME ______________________ CLASS ____________ DATE ____________

Exercises 4.6

OBJECTIVE 1: *Rename fractions by multiplying by 1 in the form* $\frac{a}{a}$.

A.

Write four fractions equivalent to each of the given fractions by multiplying by $\frac{2}{2}$, $\frac{3}{3}$, $\frac{4}{4}$, and $\frac{5}{5}$.

1. $\frac{2}{3}$

2. $\frac{3}{5}$

3. $\frac{7}{8}$

4. $\frac{5}{6}$

5. $\frac{4}{9}$

6. $\frac{7}{10}$

B.

7. $\frac{4}{11}$

8. $\frac{3}{14}$

9. $\frac{7}{3}$

10. $\frac{6}{5}$

OBJECTIVE 2: *Build a fraction by finding the missing numerator.*

A.

Find the missing numerator.

11. $\frac{1}{2} = \frac{?}{10}$

12. $\frac{3}{4} = \frac{?}{16}$

13. $\frac{2}{3} = \frac{?}{15}$

14. $\frac{4}{7} - \frac{?}{14}$

15. $\frac{4}{5} = \frac{?}{20}$

16. $\frac{7}{8} = \frac{?}{32}$

17. $\frac{3}{4} = \frac{?}{36}$

18. $\frac{2}{3} = \frac{?}{24}$

19. $\frac{1}{5} = \frac{?}{75}$

20. $\frac{5}{9} = \frac{?}{45}$

B.

21. $\frac{?}{12} = \frac{2}{3}$

22. $\frac{?}{66} = \frac{3}{11}$

23. $\frac{23}{6} = \frac{?}{12}$

24. $\frac{9}{5} = \frac{?}{100}$

25. $\frac{?}{300} = \frac{7}{15}$

26. $\frac{6}{9} = \frac{?}{108}$

27. $\frac{?}{126} = \frac{19}{42}$

28. $\frac{?}{147} = \frac{16}{7}$

29. $\frac{15}{18} = \frac{?}{144}$

30. $\frac{11}{16} = \frac{?}{144}$

OBJECTIVE 3: *List a group of fractions from smallest to largest.*

A.

List the fractions from smallest to largest.

31. $\frac{4}{7}, \frac{5}{7}, \frac{3}{7}$

32. $\frac{5}{11}, \frac{4}{11}, \frac{2}{11}$

33. $\frac{1}{2}, \frac{1}{4}, \frac{3}{8}$

34. $\frac{1}{2}, \frac{3}{5}, \frac{7}{10}$

35. $\frac{1}{2}, \frac{3}{8}, \frac{1}{3}$

36. $\frac{2}{3}, \frac{8}{15}, \frac{3}{5}$

Are the following statements true or false?

37. $\frac{1}{4} < \frac{3}{4}$

38. $\frac{5}{9} > \frac{7}{9}$

39. $\frac{11}{16} > \frac{7}{8}$

40. $\frac{9}{16} > \frac{5}{8}$

41. $\frac{3}{10} < \frac{7}{15}$

42. $\frac{7}{8} > \frac{6}{7}$

NAME ______________________ CLASS ____________ DATE ____________

B.

List the fractions from smallest to largest.

43. $\frac{4}{5}, \frac{2}{3}, \frac{3}{4}$

44. $\frac{5}{8}, \frac{7}{10}, \frac{3}{4}$

45. $\frac{13}{15}, \frac{4}{5}, \frac{5}{6}, \frac{9}{10}$

46. $\frac{7}{9}, \frac{2}{3}, \frac{3}{4}, \frac{5}{6}$

47. $\frac{11}{24}, \frac{17}{36}, \frac{35}{72}$

48. $\frac{3}{5}, \frac{8}{25}, \frac{31}{50}, \frac{59}{100}$

49. $\frac{13}{28}, \frac{17}{35}, \frac{6}{14}$

50. $\frac{11}{15}, \frac{17}{20}, \frac{9}{12}$

51. $2\frac{3}{4}, 2\frac{7}{8}, 2\frac{5}{6}$

52. $1\frac{3}{8}, 1\frac{5}{16}, 1\frac{1}{4}$

Are the following statements true or false?

53. $\frac{5}{8} < \frac{47}{80}$

54. $\frac{8}{25} < \frac{59}{100}$

55. $\frac{15}{20} > \frac{55}{75}$

56. $\frac{19}{40} > \frac{31}{60}$

57. $\frac{11}{30} < \frac{7}{18}$

58. $\frac{11}{27} > \frac{29}{36}$

C.

59. Find the LCM of the denominators of $\frac{1}{2}, \frac{2}{3}, \frac{1}{6}$, and $\frac{5}{8}$. Build the four fractions so that each has the LCM as the denominator.

60. Find the LCM of the denominators of $\frac{1}{4}$, $\frac{4}{13}$, and $\frac{5}{26}$. Build the four fractions so that each has the LCM as the denominator.

61. Janie answers $\frac{4}{5}$ of the problems correctly on her Chapter I test. If there are 40 problems on the Chapter III test, how many must she get correct to answer the same fractional amount?

62. The night nurse at Malcolm X Community Hospital finds bottles containing codeine tablets out of the usual order. The bottles contain tablets having the following strengths of codeine: $\frac{1}{8}$, $\frac{3}{32}$, $\frac{5}{16}$, $\frac{3}{8}$, $\frac{9}{16}$, $\frac{1}{2}$, and $\frac{1}{4}$ grain, respectively. Arrange the bottles in order of the strength of codeine from the smallest to largest.

63. Joe, an apprentice, is given the task of sorting a bin of bolts according to their diameters. The bolts have the following diameters: $\frac{11}{16}$, $\frac{7}{8}$, $1\frac{1}{16}$, $\frac{3}{4}$, $1\frac{1}{8}$, and $1\frac{3}{32}$ in. How should he list the diameters from the smallest to largest?

64. Four pickup trucks are advertised in the local car ads. The load capacities listed are $\frac{3}{4}$ ton, $\frac{5}{8}$ ton, $\frac{7}{16}$ ton, and $\frac{1}{2}$ ton. Which capacity is the smallest and which is the largest?

65. A container of a chemical is weighed by three people. Mary records the weight as $3\frac{1}{8}$ lb. George reads the weight as $3\frac{3}{16}$ lb. Chang reads the weight as $3\frac{1}{4}$ lb. Whose measurement is heaviest?

66. Three rulers are marked in inches. On the first ruler the spaces are divided into tenths, on the second they are divided into sixteenths, and on the third they are divided into eighths. All are used to measure a line on a scale drawing. The nearest mark on the first ruler is $5\frac{7}{10}$, the nearest mark on the second is $5\frac{11}{16}$, and the nearest mark on the third is $5\frac{6}{8}$. Which is the largest (longest) measurement?

67. Three fourths of a serving of Tostie-Os is fiber. How many ounces of fiber are there in 120 ounces of Tostie-Os?

68. During one week on her diet, Samantha ate five servings of chicken, each containing $\frac{3}{16}$ oz of fat. During the same period her brother ate four servings of beef, each containing $\frac{6}{25}$ oz of fat. Who ate the greatest amount of fat from these entrees?

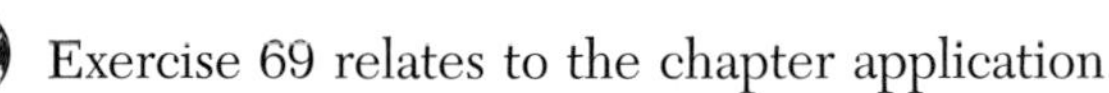

Exercise 69 relates to the chapter application.

69. **a.** If Sandy and Hannah sell a box of pecan chocolate chip cookies for \$6.00, then $\frac{4}{15}$ of the selling price is profit. If they sell a box of dipped butterfingers for \$7.20, then $\frac{1}{6}$ of the selling price is profit. Which kind of cookies gives them a bigger profit?

b. Hannah is doubling a recipe that calls for 5 tbs of oil. Rather than taking the time to measure out 10 single tablespoons, she pours $\frac{3}{4}$ c of oil. Is this more or less than she needs? (*Hint:* 1 c = 16 tbs)

STATE YOUR UNDERSTANDING

70. Explain why it is easier to compare the size of two fractions if they have common denominators.

71. What is the difference between simplifying fractions and building fractions?

CHALLENGE

72. List $\frac{12}{25}$, $\frac{14}{29}$, $\frac{29}{60}$, $\frac{35}{71}$, $\frac{39}{81}$, and $\frac{43}{98}$ from smallest to largest.

73. Build $\frac{5}{7}$ so that it has denominators 70, 91, 161, 784, and 4067.

74. Fernando and Filipe are hired to sell tickets for the holiday raffle. Fernando sells $\frac{14}{17}$ of his quota of 765 tickets. Filipe sells $\frac{19}{23}$ of his quota of 759 tickets. Who sells the most of his quota? Who sells the most tickets?

GROUP ACTIVITY

75. We saw that $\frac{2}{5}$ was less than $\frac{4}{5}$ by looking at a rectangle representing each fraction. Show the sum, $\frac{2}{5} + \frac{4}{5}$, visually using a rectangle divided into five parts. Similarly, show $\frac{2}{7} + \frac{3}{7}$.

MAINTAIN YOUR SKILLS (SECTIONS 1.1, 1.2, 1.6, 3.3, 3.4, 3.6, 4.1, 4.2)

76. Find the sum of 3796, 43, 296, 4099, and 5310.

77. Is 611 a prime or a composite number?

78. List all the factors of 996.

79. Round 65,458,999 to the nearest ten thousand.

80. Find the LCM of 25, 35, 45, and 63.

81. Simplify: $\frac{812}{928}$

82. Change $27\frac{5}{8}$ to an improper fraction.

83. Change $\frac{217}{12}$ to a mixed number.

84. The Forest Service rents a two-engine plane at \$625 per hour and a single-engine plane at \$365 per hour to drop fire retardant. During a forest fire, the two-engine plane was used for 4 hr and the single-engine plane was used for 2 hr. What was the cost of using the two planes?

85. Ms. Wallington is taking one capsule containing 250 mg of a drug every 8 hr. Beginning next Wednesday her doctor's instructions are to increase the dosage to 500 mg every 6 hr. How many 250-mg capsules should the pharmacist give her for the following week (7 days)?

4.7

Adding Fractions

OBJECTIVES

1. Add like fractions.
2. Add unlike fractions.

VOCABULARY

Like fractions are fractions with common denominators. **Unlike fractions** are fractions with different denominators.

HOW AND WHY

Objective 1

Add like fractions.

What is the sum of $\frac{1}{5} + \frac{2}{5}$? The denominators tell the number of parts in the unit. The numerator tells us how many of these parts are shaded. By adding the numerators we find the total number of shaded parts. The common denominator keeps track of the size of the parts. (See Fig. 4.10.)

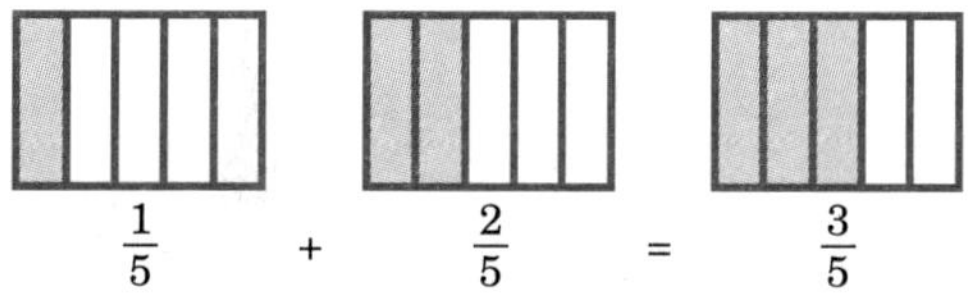

$\frac{1}{5} + \frac{2}{5} = \frac{3}{5}$

Figure 4.10

To add like fractions

1. Add the numerators.
2. Write the sum over the common denominator.

CAUTION

Do not add the denominators.

$$\frac{5}{9} + \frac{2}{9} \neq \frac{7}{18}$$

Warm Ups A–D

A. Add: $\frac{5}{9} + \frac{2}{9}$

B. Add: $\frac{5}{6} + \frac{1}{6} + \frac{5}{6}$

C. Add: $\frac{3}{10} + \frac{3}{10} + \frac{2}{10}$

D. The previous week the Eastin Corp. stock rose $\frac{1}{8}$ point on Monday, $\frac{1}{8}$ point on Tuesday, $\frac{1}{8}$ point on Wednesday, $\frac{7}{8}$ point on Thursday, and $\frac{3}{8}$ point on Friday. What was the total rise of the stock for the week?

Examples A–D

Directions: Add and simplify.

Strategy: Add the numerators and write the sum over the common denominator. Simplify.

A. Add: $\frac{4}{7} + \frac{2}{7}$

$\frac{4}{7} + \frac{2}{7} = \frac{6}{7}$ **Add the numerators. Keep the common denominator.**

B. Add: $\frac{1}{3} + \frac{2}{3} + \frac{2}{3}$

$\frac{1}{3} + \frac{2}{3} + \frac{2}{3} = \frac{5}{3}$ **Add.**

$= 1\frac{2}{3}$ **Write as a mixed number.**

C. Add: $\frac{1}{6} + \frac{2}{6} + \frac{1}{6}$

$\frac{1}{6} + \frac{2}{6} + \frac{1}{6} = \frac{4}{6}$ **Add.**

$= \frac{2}{3}$ **Simplify.**

D. The stock of the Eastin Corp. rose $\frac{3}{8}$ point on Monday, $\frac{1}{8}$ point on Tuesday, $\frac{5}{8}$ point on Wednesday, $\frac{1}{8}$ point on Thursday, and $\frac{5}{8}$ point on Friday. What was the total rise of the stock for the week?

$\frac{3}{8} + \frac{1}{8} + \frac{5}{8} + \frac{1}{8} + \frac{5}{8} = \frac{15}{8}$ **Add the gains.**

$= 1\frac{7}{8}$ **Write as a mixed number.**

The stock rose $1\frac{7}{8}$ points during the week.

Answers to Warm Ups A. $\frac{7}{9}$ B. $\frac{11}{6}$ or $1\frac{5}{6}$ C. $\frac{4}{5}$ D. The stock rose a total of $\frac{13}{8}$, or $1\frac{5}{8}$, points.

HOW AND WHY

Objective 2

Add unlike fractions.

The sum $\frac{1}{2} + \frac{1}{5}$ cannot be worked in this form. A look at Figure 4.11 shows that the parts are not the same size.

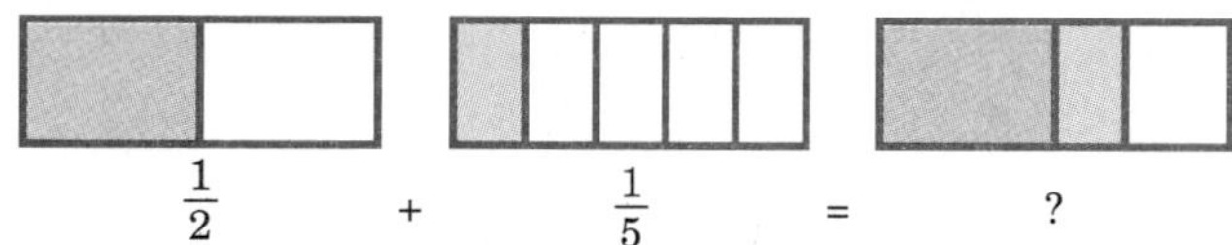

Figure 4.11

To add, rename $\frac{1}{2}$ and $\frac{1}{5}$ as like fractions. The LCM (Least Common Multiple) of the two denominators serves as the least common denominator. The LCM of 2 and 5 is 10. We can now write

$$\frac{1}{2} = \left(\frac{1}{2}\right)\left(\frac{5}{5}\right) = \frac{5}{10} \quad \text{and} \quad \frac{1}{5} = \left(\frac{1}{5}\right)\left(\frac{2}{2}\right) = \frac{2}{10}$$

and the problem can now be seen in Figure 4.12.

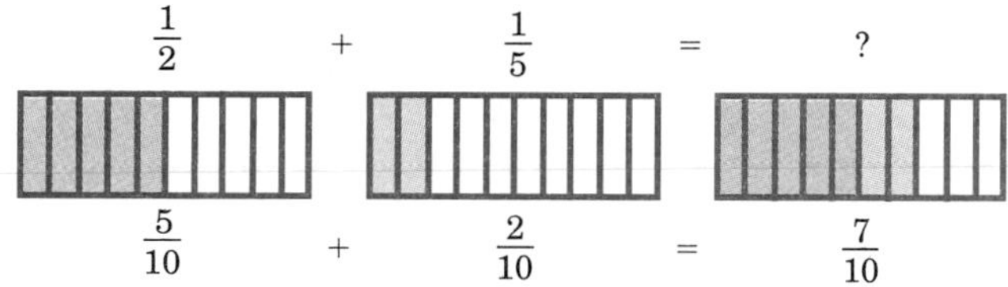

Figure 4.12

To add unlike fractions

1. Build the fractions so that they have a common denominator.
2. Add and simplify.

Examples E–J

Directions: Add and simplify.

Strategy: Build each of the fractions to a common denominator.

E. Add: $\frac{3}{8} + \frac{1}{4}$

$\frac{3}{8} + \frac{1}{4} = \frac{3}{8} + \frac{2}{8}$ **The LCM of 8 and 4 is 8.**

$= \frac{5}{8}$ **Add.**

Warm Ups E–J

E. Add: $\frac{2}{3} + \frac{1}{6}$

Answer to Warm Up E. $\frac{5}{6}$

F. Add: $\frac{5}{8} + \frac{1}{6}$

F. Add: $\frac{5}{12} + \frac{2}{9}$

$\frac{5}{12} + \frac{2}{9} = \frac{15}{36} + \frac{8}{36}$ **The LCM of 12 and 9 is 36.** $\frac{5}{12} \cdot \frac{3}{3} = \frac{15}{36}$

$\frac{2}{9} \cdot \frac{4}{4} = \frac{8}{36}$

$= \frac{23}{36}$ **Add.**

G. Add: $\frac{2}{3} + \frac{1}{12}$

G. Add: $\frac{1}{6} + \frac{7}{10}$

$\frac{1}{6} + \frac{7}{10} = \frac{5}{30} + \frac{21}{30}$ **The LCM of 6 and 10 is 30.**

$= \frac{26}{30}$ **Add.**

$= \frac{13}{15}$ **Simplify.**

H. Add: $\frac{13}{45} + \frac{28}{75}$

H. Add: $\frac{11}{96} + \frac{35}{72}$

Strategy: Prime factor the denominators to help find their LCM.

$\frac{11}{96} + \frac{35}{72} = \frac{11(3)}{288} + \frac{35(4)}{288}$ **$96 = 2^5 \cdot 3$, $72 = 2^3 \cdot 3^2$**
$\text{LCM} = 2^5 \cdot 3^2 = 288$

$= \frac{33}{288} + \frac{140}{288}$

$= \frac{173}{288}$ **Add.**

I. Add: $\frac{14}{35} + \frac{13}{28}$

Calculator Example

I. Add: $\frac{37}{60} + \frac{15}{84}$

$\frac{37}{60} + \frac{15}{84} = \frac{167}{210}$

On a calculator with fraction keys it is not necessary to find the common denominator. The calculator is programmed to add and simplify.

Answers to Warm Ups F. $\frac{19}{24}$ G. $\frac{3}{4}$ H. $\frac{149}{225}$ I. $\frac{121}{140}$

J. Sheila Fankowski is assembling a composting bin for her lawn and garden debris. She needs a bolt that will reach through a $\frac{1}{32}$-in.-thick washer, a $\frac{3}{16}$-in.-thick plastic bushing, a $\frac{3}{4}$-in.-piece of steel tubing, a second $\frac{1}{32}$-in.-thick washer, and a $\frac{1}{4}$-in.-thick nut. How long a bolt does she need?

Strategy: Add the thicknesses of each part together to find the length needed.

$\frac{1}{32}+\frac{3}{16}+\frac{3}{4}+\frac{1}{32}+\frac{1}{4}$ **The LCM of 32, 16, and 4 is 32.**

$\frac{1}{32}+\frac{6}{32}+\frac{24}{32}+\frac{1}{32}+\frac{8}{32}$ **Build each fraction to the denominator, 32.**

$=\frac{40}{32}$ **Add.**

$=\frac{5}{4}=1\frac{1}{4}$ **Simplify and write as a mixed number.**

The bolt must be $1\frac{1}{4}$ in. long.

J. A nail must reach through three thicknesses of wood and penetrate the fourth thickness $\frac{1}{4}$ in. If the first piece of wood is $\frac{5}{16}$ in., the second is $\frac{3}{8}$ in., and the third is $\frac{9}{16}$ in., how long must the nail be?

Answer to Warm Up J. The nail must be $1\frac{1}{2}$ inches long.

NAME ______________________ CLASS ____________ DATE ____________

Exercises 4.7

OBJECTIVE 1: *Add like fractions.*

A.

Add. Simplify completely.

1. $\frac{4}{11} + \frac{5}{11}$

2. $\frac{7}{12} + \frac{4}{12}$

3. $\frac{2}{9} + \frac{2}{9} + \frac{2}{9}$

4. $\frac{4}{8} + \frac{1}{8} + \frac{1}{8}$

5. $\frac{3}{4} + \frac{5}{4}$

6. $\frac{6}{7} + \frac{8}{7}$

7. $\frac{2}{10} + \frac{5}{10} + \frac{1}{10}$

8. $\frac{4}{12} + \frac{3}{12} + \frac{3}{12}$

9. $\frac{4}{13} + \frac{5}{13} + \frac{1}{13}$

10. $\frac{5}{11} + \frac{2}{11} + \frac{1}{11}$

11. $\frac{5}{12} + \frac{5}{12} + \frac{5}{12}$

12. $\frac{9}{16} + \frac{7}{16} + \frac{4}{16}$

B.

13. $\frac{3}{16} + \frac{3}{16} + \frac{2}{16}$

14. $\frac{7}{32} + \frac{8}{32} + \frac{5}{32}$

15. $\frac{5}{48} + \frac{7}{48} + \frac{3}{48}$

16. $\frac{3}{16} + \frac{2}{16} + \frac{5}{16}$

17. $\frac{8}{30} + \frac{9}{30} + \frac{1}{30}$

18. $\frac{13}{50} + \frac{7}{50} + \frac{2}{50}$

19. $\frac{5}{24} + \frac{7}{24} + \frac{9}{24}$

20. $\frac{3}{20} + \frac{9}{20} + \frac{3}{20}$

OBJECTIVE 2: *Add unlike fractions.*

A.

Add. Simplify completely.

21. $\frac{1}{6} + \frac{5}{8}$

22. $\frac{2}{3} + \frac{1}{6}$

23. $\frac{3}{8} + \frac{7}{24}$

24. $\frac{4}{15} + \frac{1}{3}$

25. $\frac{7}{16} + \frac{3}{8}$

26. $\frac{4}{9} + \frac{5}{18}$

27. $\frac{1}{10} + \frac{1}{5} + \frac{1}{3}$

28. $\frac{1}{20} + \frac{1}{5} + \frac{1}{4}$

29. $\frac{3}{5} + \frac{3}{10}$

30. $\frac{2}{15} + \frac{2}{5}$

B.

31. $\frac{3}{35} + \frac{8}{21}$

32. $\frac{9}{14} + \frac{5}{21}$

33. $\frac{3}{10} + \frac{9}{20} + \frac{11}{30}$

34. $\frac{7}{8} + \frac{7}{12} + \frac{1}{6}$

35. $\frac{1}{10} + \frac{2}{5} + \frac{5}{6} + \frac{1}{15}$

36. $\frac{1}{2} + \frac{3}{10} + \frac{3}{5} + \frac{1}{4}$

37. $\frac{3}{4} + \frac{3}{8} + \frac{3}{16}$

38. $\frac{15}{36} + \frac{1}{6} + \frac{5}{12}$

39. $\frac{5}{6} + \frac{5}{8} + \frac{3}{4} + \frac{5}{12}$

40. $\frac{7}{10} + \frac{4}{5} + \frac{11}{15} + \frac{8}{35}$

41. $\frac{13}{24} + \frac{17}{36} + \frac{11}{48} + \frac{21}{72}$

42. $\frac{12}{27} + \frac{5}{9} + \frac{32}{81} + \frac{2}{3}$

43. $\frac{7}{9} + \frac{4}{5} + \frac{4}{15} + \frac{11}{30}$

44. $\frac{11}{15} + \frac{7}{12} + \frac{9}{10} + \frac{17}{20}$

C.

Add. Simplify completely.

45. $\frac{17}{100} + \frac{31}{100} + \frac{9}{100} + \frac{3}{100}$

46. $\frac{29}{50} + \frac{3}{50} + \frac{3}{50} + \frac{11}{50}$

47. $\frac{7}{120} + \frac{9}{120} + \frac{25}{120} + \frac{15}{120}$

48. $\frac{29}{144} + \frac{17}{144} + \frac{3}{144} + \frac{25}{144}$

49. $\frac{25}{36} + \frac{19}{48}$

50. $\frac{72}{85} + \frac{69}{102}$

51. $\frac{11}{30} + \frac{7}{15} + \frac{21}{45}$

52. $\frac{25}{72} + \frac{9}{108} + \frac{19}{144}$

53. $\frac{7}{48} + \frac{17}{30}$

54. Find the error(s) in the statement: $\frac{1}{5} + \frac{2}{5} = \frac{3}{10}$. Correct the statement. Explain how you would avoid this error.

55. Find the error(s) in the statement: $\frac{1}{2} + \frac{4}{7} = \frac{5}{14}$. Correct the statement. Explain how you would avoid this error.

56. Chef Ramon prepares a punch for the stockholders' meeting of the Northern Corporation. The punch calls for $\frac{1}{4}$ gallon lemon juice, $\frac{3}{4}$ gallon raspberry juice, $\frac{2}{4}$ gallon cranberry juice, $\frac{1}{4}$ gallon lime juice, $\frac{5}{4}$ gallons 7-Up, and $\frac{3}{4}$ gallon vodka. How many gallons of punch does the recipe make?

57. A physical therapist advises Belinda to swim $\frac{5}{16}$ mile on Monday and increase this distance by $\frac{1}{16}$ mile each day from Tuesday through Friday. What is the total number of miles she advises?

58. In order to complete a project, Perry needs $\frac{1}{10}$ in. of foam, $\frac{3}{10}$ in. of metal, $\frac{4}{10}$ in. of wood, and $\frac{7}{10}$ in. of plexiglass. What will be the total thickness of this project when these materials are piled up?

59. Jonnie is assembling a rocking horse for his granddaughter. He needs a bolt to reach through a $\frac{7}{8}$-in. piece of steel tubing, a $\frac{1}{16}$-in. bushing, a $\frac{1}{2}$-in. piece of tubing, a $\frac{1}{8}$-in.-thick washer, and a $\frac{1}{4}$-in.-thick nut. How long a bolt does he need?

60. On the American Stock Exchange, Joan's stock rose $\frac{1}{8}$ point the first hour and an additional $\frac{3}{16}$ point during the remainder of the day. What was the total rise for the day?

61. What is the total distance (perimeter) around this triangle?

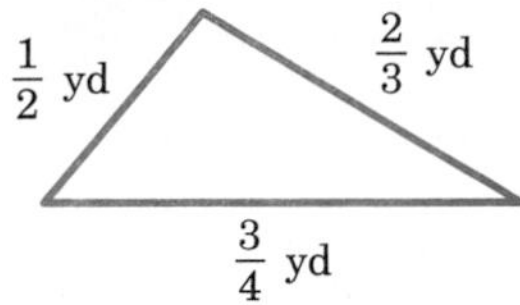

62. Find the length of this pin:

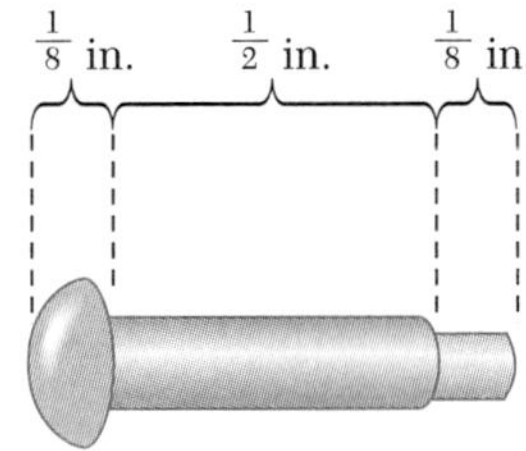

63. An elephant-ear bamboo grew $\frac{1}{2}$ in. on Tuesday, $\frac{3}{8}$ in. on Wednesday, and $\frac{1}{4}$ in. on Thursday. How much did it grow in the three days?

64. The Sandoz family spends $\frac{2}{15}$ of their income on rent, $\frac{1}{4}$ on food, $\frac{1}{20}$ on clothes, $\frac{1}{10}$ on transportation, and $\frac{5}{24}$ on taxes. What fraction of their income is spent on these costs?

65. Find the length of the rod in the figure. Assume that the grooves and teeth are uniform in length.

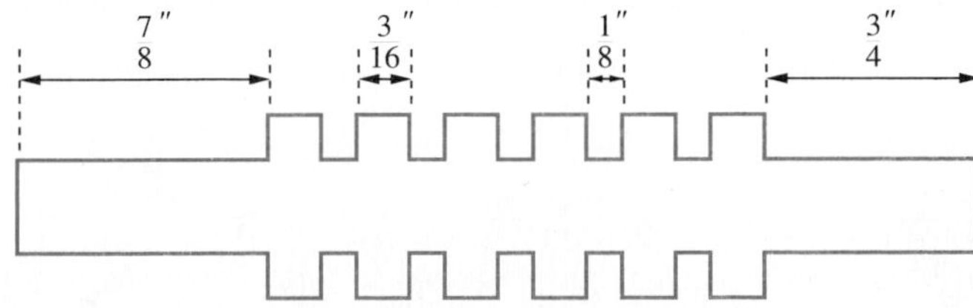

66. An electronics mogul leaves $\frac{2}{9}$, $\frac{1}{5}$, and $\frac{4}{15}$ shares of her estate to her three children. What share of the estate will the children receive?

Exercise 67 relates to the chapter application.

67. Sandy and Hannah propose to make a Christmas sampler box with nut balls, thumbprints, and lemon squares. One recipe of each type of cookie makes enough to fill four boxes.

a. Nutballs require $\frac{1}{4}$ tsp of baking soda, thumbprints require $\frac{1}{2}$ tsp, and lemon squares require $\frac{1}{2}$ tsp. How much baking soda is needed for all three cookie recipes?

b. How much baking soda is required to fill an order for 12 sampler boxes?

c. Nutballs require $\frac{3}{4}$ tsp of salt, thumbprints require $\frac{1}{2}$ tsp, and lemon squares need $\frac{3}{4}$ tsp. How much salt is needed for all three recipes?

d. How much salt is needed to fill an order for 6 sampler boxes?

STATE YOUR UNDERSTANDING

68. Explain how to find the sum of $\frac{5}{12}$ and $\frac{3}{20}$.

69. Why is it important to write fractions with a common denominator before adding?

CHALLENGE

70. Find the sum of $\frac{107}{372}$ and $\frac{41}{558}$.

71. Find the sum of $\frac{67}{124} + \frac{27}{868}$.

72. Janet gave $\frac{1}{7}$, $\frac{3}{14}$, and $\frac{1}{6}$ of her estate to Bob, Greta, and Joe Guerra. She also left $\frac{1}{8}$, $\frac{5}{16}$, and $\frac{1}{9}$ of the estate to Pele, Rhonda, and Shauna Contreras. Which family received the greater share of the estate?

73. Jim is advised by his doctor to limit fat intake in his diet. For breakfast he has the following foods with the indicated fat content: a bagel, $\frac{3}{4}$ g; a banana, $\frac{13}{16}$ g; cereal, $\frac{17}{10}$ g; milk, $\frac{9}{8}$ g; jelly, 0 g; and coffee, 0 g. Rounded to the nearest whole number, how many grams of fat does Jim consume at breakfast? If each gram of fat represents 9 calories and the total calories for the meal is 330, what fraction represents the calories from fat? (Use the rounded whole number grams of fat.)

GROUP ACTIVITY

74. In the next section we add mixed numbers. Devise a procedure for adding $3\frac{3}{4} + 5\frac{4}{5}$. Also, find some applications in which addition of mixed numbers is used. Be prepared to share the results with the class.

MAINTAIN YOUR SKILLS (SECTIONS 1.1, 1.6, 1.7, 3.5, 3.6, 4.4)

75. Prime factor 248.

76. Find the LCM of 12, 14, 16, and 20.

77. Round 678,223,500,000 to the nearest million.

78. Find the average of 1803, 72, 630, 1433, 1009, 2000, and 753.

Perform the indicated operations.

79. $45 - 3 \cdot 8 + 3 \cdot 9 - 6$

80. $(21 - 4)6 - 3(19 - 11)$

81. $(13 - 2^3)^3 - 7(11)$

82. Mrs. Teech has five classes to teach this term. The enrollment in the classes is 42, 36, 56, 32, and 34. What is the average class size?

83. A retail furniture store buys nine sofas for $326 each. If the store wants to make a profit of $1215 on the sofas, how much should each sell for?

84. In a metal benchwork class that has 36 students, each student is allowed $11\frac{5}{8}$ in. of wire solder. How many inches of wire must the instructor provide for the class?

4.8

Adding Mixed Numbers

OBJECTIVE

Add mixed numbers.

HOW AND WHY

Objective

Add mixed numbers.

What is the sum of $3\frac{1}{6}$ and $5\frac{1}{4}$? Pictorially we can show the sum by drawing rectangles.

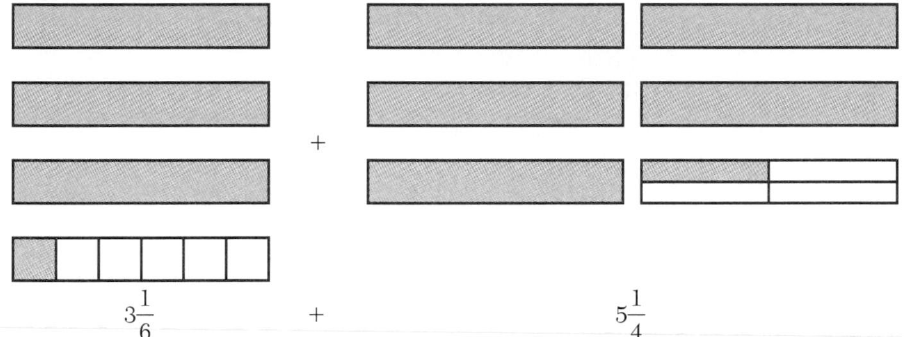

$3\frac{1}{6}$ + $5\frac{1}{4}$

It is easy to see that the sum contains eight whole units. The sum of the fraction parts requires finding a common denominator. The LCM of 6 and 4 is 12.

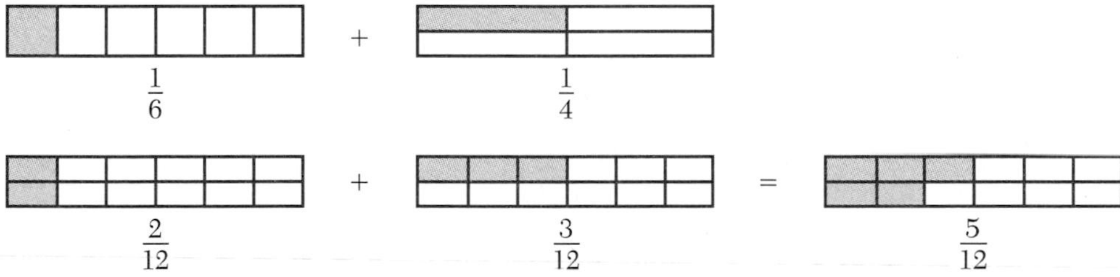

So the sum is $8\frac{5}{12}$.

Mixed numbers can be added horizontally or in columns. We show both methods.

$$\begin{aligned}\left(3 + \frac{1}{6}\right) + \left(5 + \frac{1}{4}\right) &= (3 + 5) + \left(\frac{1}{6} + \frac{1}{4}\right)\\ &= 8 + \left(\frac{2}{12} + \frac{3}{12}\right)\\ &= 8\frac{5}{12}\end{aligned}$$

When we write the sum vertically, the grouping of the whole numbers and the fractions takes place naturally.

$$\begin{array}{rcl} 3\frac{1}{6} & = & 3\frac{2}{12} \\ +\ 5\frac{1}{4} & = & 5\frac{3}{12} \\ \hline & & 8\frac{5}{12} \end{array}$$

Sometimes the sum of the fractions is greater than 1. In this case, change the fraction sum to a mixed number and add it to the whole number part.

$$\begin{array}{rcl} 11\frac{7}{10} & = & 11\frac{21}{30} \\ +\ 23\frac{8}{15} & = & 23\frac{16}{30} \\ \hline & & 34\frac{37}{30} \end{array}$$

Write the fractional parts with a common denominator.

Add the whole-number parts and add the fraction parts.

$= 34 + 1\frac{7}{30}$ **The fraction is improper, so rewrite it as a mixed number.**

$= 35\frac{7}{30}$ **Add the mixed number to the whole number.**

To add mixed numbers

1. Add the whole numbers.
2. Add the fractions. If the sum of the fractions is more than 1, change the fraction to a mixed number and add again.
3. Simplify.

Warm Ups A–E

A. Add: $8\frac{5}{16} + 3\frac{1}{2}$

Examples A–E

Directions: Add. Write as a mixed number.

Strategy: Add the whole numbers and add the fractions. If the sum of the fractions is greater than 1, rewrite as a mixed number and add the whole numbers. Simplify.

A. Add: $5\frac{5}{12} + 1\frac{3}{4}$

Strategy: Write the mixed numbers in a column to group the whole numbers and to group the fractions.

$$\begin{array}{rcl} 5\frac{5}{12} & = & 5\frac{5}{12} \\ +\ 1\frac{3}{4} & = & 1\frac{9}{12} \\ \hline & & 6\frac{14}{12} \end{array}$$

Build the fractions to the common denominator, 12. The LCM of 12 and 4 is 12.

Add.

Answer to Warm Up A. $11\frac{13}{16}$

$6\frac{14}{12} = 6 + 1\frac{2}{12}$ Write the improper fraction as a mixed number.

$= 7\frac{2}{12}$ Add the whole numbers.

$= 7\frac{1}{6}$ Simplify.

B. Add: $25\frac{7}{8} + 13\frac{5}{9} + 7\frac{1}{6}$

B. Add: $12\frac{8}{9} + 14\frac{3}{4} + 6\frac{2}{3}$

$$
\begin{array}{rcr}
25\frac{7}{8} & = & 25\frac{63}{72} \\
13\frac{5}{9} & = & 13\frac{40}{72} \\
+ \quad 7\frac{1}{6} & = & 7\frac{12}{72} \\
\hline
 & & 45\frac{115}{72}
\end{array}
$$

The LCM of 8, 9, and 6 is 72.

Add.

$45\frac{115}{72} = 45 + 1\frac{43}{72}$ Write the improper fraction as a mixed number.

$= 46\frac{43}{72}$

C. Add: $8 + 5\frac{13}{15}$

C. Add: $7\frac{11}{12} + 16$

$8 + 5\frac{13}{15} = 13\frac{13}{15}$ Add the whole numbers.

Calculator Example

D. Add: $31\frac{3}{8} + 62\frac{13}{15}$

D. Add: $51\frac{7}{8} + 36\frac{5}{12}$

$31\frac{3}{8} + 62\frac{13}{15} = \frac{11309}{120} = 94\frac{29}{120}$

On a calculator with fraction keys it is not necessary to find the common denominator. The calculator is programmed to add and simplify. Some calculators may not change the improper sum to a mixed number.

Answers to Warm Ups B. $34\frac{11}{36}$ C. $23\frac{11}{12}$ D. $88\frac{7}{24}$

E. The report also indicates that the city recycles $10\frac{1}{3}$ tons of paper, $3\frac{3}{8}$ tons of aluminum, and $4\frac{5}{12}$ tons of glass each month. How many tons of these materials are recycled each month?

E. In a report by Environmental Hazards Management, a city of 100,000 discharges the following amounts of hazardous material into city drains each month: $3\frac{3}{4}$ tons of toilet bowl cleaner, $13\frac{3}{4}$ tons of liquid household cleaners, and $3\frac{2}{5}$ tons of motor oil. How many tons of these materials are discharged each month?

Strategy: Add the amount of each hazardous material.

$$\begin{aligned} 3\frac{3}{4} &= 3\frac{15}{20} \\ 13\frac{3}{4} &= 13\frac{15}{20} \\ +\ 3\frac{2}{5} &= 3\frac{8}{20} \\ \hline & \ 19\frac{38}{20} \end{aligned}$$

The LCM of 4 and 5 is 20.

Add.

$$19\frac{38}{20} = 20\frac{9}{10}$$

Change the improper fraction to a mixed number and simplify.

The residents discharge $20\frac{9}{10}$ tons of this material each month.

Answer to Warm Up E. There are $18\frac{1}{8}$ tons recycled.

NAME ________________ CLASS ________ DATE ________

Exercises 4.8

OBJECTIVE: *Add mixed numbers.*

A.

Add. Write the results as mixed numbers where possible. Simplify.

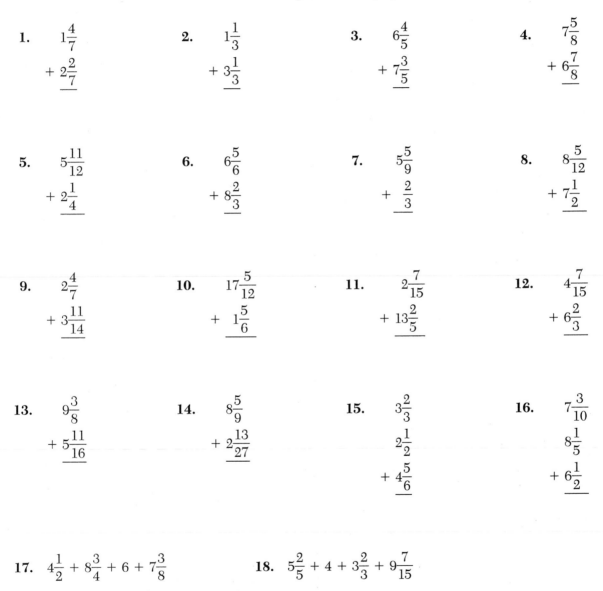

1. $1\frac{4}{7} + 2\frac{2}{7}$

2. $1\frac{1}{3} + 3\frac{1}{3}$

3. $6\frac{4}{5} + 7\frac{3}{5}$

4. $7\frac{5}{8} + 6\frac{7}{8}$

5. $5\frac{11}{12} + 2\frac{1}{4}$

6. $6\frac{5}{6} + 8\frac{2}{3}$

7. $5\frac{5}{9} + \frac{2}{3}$

8. $8\frac{5}{12} + 7\frac{1}{2}$

9. $2\frac{4}{7} + 3\frac{11}{14}$

10. $17\frac{5}{12} + 1\frac{5}{6}$

11. $2\frac{7}{15} + 13\frac{2}{5}$

12. $4\frac{7}{15} + 6\frac{2}{3}$

13. $9\frac{3}{8} + 5\frac{11}{16}$

14. $8\frac{5}{9} + 2\frac{13}{27}$

15. $3\frac{2}{3}$, $2\frac{1}{2}$, $+ 4\frac{5}{6}$

16. $7\frac{3}{10}$, $8\frac{1}{5}$, $+ 6\frac{1}{2}$

17. $4\frac{1}{2} + 8\frac{3}{4} + 6 + 7\frac{3}{8}$

18. $5\frac{2}{5} + 4 + 3\frac{2}{3} + 9\frac{7}{15}$

B.

19. $7\frac{3}{8} + 5\frac{5}{6}$

20. $15\frac{4}{15} + 5\frac{5}{6}$

21. $3\frac{1}{10} + 2\frac{3}{5} + 4\frac{7}{15}$

22. $7\frac{1}{8} + 3\frac{5}{12} + 8\frac{5}{6}$

23. $14\frac{7}{20} + 11\frac{3}{16}$

24. $11\frac{7}{24} + 32\frac{7}{18}$

25. $21\frac{5}{7} + 15\frac{9}{14} + 12\frac{10}{21}$

26. $18\frac{3}{4} + 17\frac{7}{8} + 23\frac{1}{6}$

27. $213\frac{5}{18} + 506\frac{7}{12}$

28. $213\frac{5}{6} + 347\frac{3}{10}$

29. $47\frac{1}{5} + 23\frac{2}{3} + 15\frac{1}{2}$

30. $47\frac{3}{8} + 23 + 42\frac{5}{12}$

31. $25\frac{2}{3} + 16\frac{1}{6} + 18\frac{3}{4}$

32. $29\frac{7}{8} + 19\frac{5}{12} + 32\frac{3}{4}$

33. $62 + 18\frac{5}{9} + 37\frac{7}{15}$

34. $41 + 29\frac{9}{14} + 3\frac{1}{6}$

35. $37\frac{18}{35} + 29\frac{9}{14} + 36$

36. $12\frac{11}{12} + 22\frac{5}{8} + 8$

37.
$$\begin{array}{r} 15\frac{5}{6} \\ 12\frac{9}{10} \\ 16 \\ +\ 17\frac{5}{12} \\ \hline \end{array}$$

38.
$$\begin{array}{r} 22\frac{11}{18} \\ 19 \\ 16\frac{5}{9} \\ +\ 10\frac{1}{3} \\ \hline \end{array}$$

C.

Add. Simplify.

39. $2\frac{2}{5} + 7\frac{1}{6} + \frac{4}{15} + 3\frac{1}{10}$

40. $1\frac{1}{5} + 3\frac{7}{10} + \frac{1}{3} + 7\frac{16}{25}$

41.
$$\begin{array}{r} 28\frac{7}{15} \\ +\ 19\frac{13}{20} \\ \hline \end{array}$$

42.
$$\begin{array}{r} 82\frac{7}{9} \\ +\ 67\frac{4}{5} \\ \hline \end{array}$$

43.
$$\begin{array}{r} 82\frac{7}{24} \\ 18\frac{7}{12} \\ +\ 5\frac{7}{36} \\ \hline \end{array}$$

44.
$$\begin{array}{r} 100\frac{37}{100} \\ 31\frac{7}{50} \\ +\ 15\frac{17}{20} \\ \hline \end{array}$$

45. $12\frac{3}{5} + 7\frac{1}{8} + 29\frac{3}{4} + 14\frac{9}{10}$

46. $14\frac{17}{25} + 4\frac{3}{5} + 25\frac{2}{15} + 10$

47. Juanita worked the following hours at her part-time job during the month of October. How many hours did she work during October?

Week	Oct. 1–7	Oct. 8–14	Oct. 15–21	Oct. 22–28	Oct. 29–31
Hours	$25\frac{1}{2}$	$19\frac{2}{3}$	10	$16\frac{5}{6}$	$4\frac{3}{4}$

48. Michelle worked the following hours at her part-time job during the week. How many hours did she work during the week?

Day	Mon.	Tues.	Wed.	Thurs.	Fri.	Sat.
Hours	$4\frac{1}{2}$	$4\frac{1}{4}$	$4\frac{1}{2}$	$2\frac{1}{4}$	0	$8\frac{1}{4}$

49. Find the perimeter of a rectangle that has length $24\frac{1}{2}$ ft and width $17\frac{3}{4}$ ft.

50. Find the perimeter of a triangle with sides $12\frac{5}{6}$ in., $16\frac{3}{4}$ in., and $14\frac{5}{12}$ in.

51. What is the overall length of this bolt?

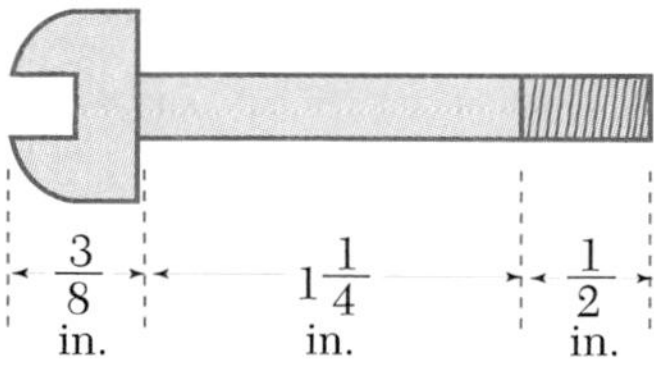

52. The graph displays the average yearly rainfall for five cities. What is the total amount of rain that falls in a given year in Salem, Forest Hills, and Westview?

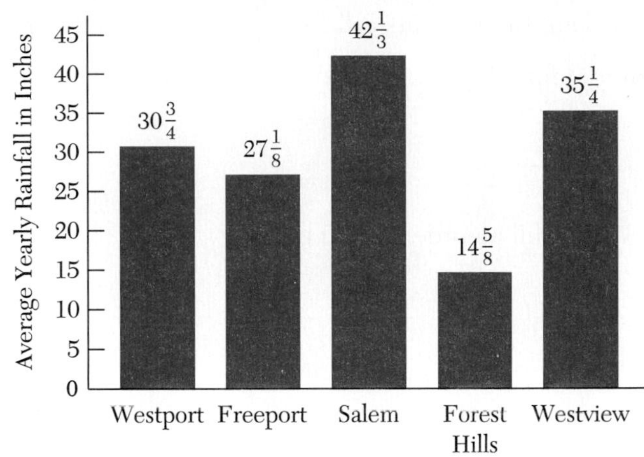

53. The rain gauge at the water reservoir recorded the following rainfall during a 6-month period: $1\frac{2}{3}$ in.; $4\frac{5}{6}$ in.; $3\frac{3}{4}$ in.; $\frac{2}{3}$ in.; $7\frac{7}{8}$ in., and $2\frac{1}{4}$ in. What is the total rainfall recorded during the 6 months? If an inch of rain means a gain of 2,400,000 gallons of water in the reservoir, what is the water gain during the 6 months?

54. The State Department of Transportation must resurface part of seven roads this summer. The distances to be paved are $6\frac{11}{16}$ miles, $8\frac{3}{5}$ miles, $9\frac{3}{4}$ miles, $17\frac{1}{2}$ miles, $5\frac{1}{8}$ miles, $12\frac{4}{5}$ miles, and $\frac{7}{8}$ mile. How many miles of highway are to be resurfaced? If it costs $15,000 to resurface one mile, what is the cost of the resurfacing project? Round to the nearest thousand dollars.

Exercises 55–56 relate to the chapter application.

55. Sandy and Hannah can fill four Christmas sampler boxes with one recipe each of nutballs, thumbprints, and lemon squares.

a. Nutballs call for $1\frac{1}{2}$ c of flour, thumbprints call for $2\frac{3}{4}$ c of flour, and lemon squares call for 2 c of flour. How much flour is required for all three recipes?

b. How much flour is needed to fill an order for 20 boxes?

c. Nutballs call for $\frac{3}{4}$ c of sugar, thumbprints call for $1\frac{1}{2}$ c, and lemon squares call for $1\frac{1}{4}$ c. How much sugar is needed for all three recipes?

d. How much sugar is needed to fill an order for 20 boxes?

56. **a.** Sandy's recipe for pecan chocolate chip cookies makes 2 dozen, but a box holds only 20 cookies. Write a simplified fraction that represents the number of boxes one recipe makes.

b. Hannah's dipped butterfinger recipe makes 3 dozen but a box only holds 30 cookies. Write a simplified fraction that represents the total number of boxes one recipe makes.

STATE YOUR UNDERSTANDING

57. Explain why it is sometimes necessary to rename the sum of two mixed numbers after adding the whole numbers and the fractional parts. Give an example in which this happens.

58. Add $8\frac{4}{5} + 7\frac{3}{8}$ by the procedures of this section. Then add by changing each mixed number to an improper fraction before adding. Be sure you get the same result for both. Which method do you prefer? Why?

NAME ______________________ CLASS ____________ DATE ____________

CHALLENGE

59. Is $3\left(4\frac{1}{4}\right) + 2\frac{2}{3} + 5\left(2\frac{5}{6}\right) = 3\left(3\frac{1}{8}\right) + 7\frac{1}{12}$ a true statement?

60. Is $6\left(6\frac{4}{9}\right) + 5\left(2\frac{5}{6}\right) + 7\left(4\frac{1}{3}\right) = 6\left(5\frac{1}{3}\right) + 8\left(3\frac{2}{3}\right) + 7\left(3\frac{5}{42}\right)$ a true statement?

61. During the month of January the rangers at Yellowstone National Park record the following snowfall: week 1, $6\frac{9}{10}$ in.; week 2, $8\frac{3}{4}$ in.; week 3, $13\frac{5}{6}$ in.; and week 4, $9\frac{2}{3}$ in. How many inches of snow have fallen during the month? If the average snowfall for the month is $38\frac{17}{32}$ in., does this January exceed the average?

GROUP ACTIVITY

62. We know that a mixed number can be changed to an improper fraction, so $5\frac{3}{7} = \frac{38}{7}$. How many ways can you find to express $5\frac{3}{7}$ as a mixed number using improper fractions?

MAINTAIN YOUR SKILLS (SECTIONS 1.1, 1.2, 3.5, 4.2, 4.3)

Multiply.

63. $\frac{7}{8} \cdot \frac{8}{9} \cdot \frac{3}{7}$

64. $\frac{18}{5} \cdot \frac{15}{4} \cdot \frac{7}{9}$

65. $\frac{12}{15} \cdot \frac{14}{33} \cdot \frac{22}{42}$

66. $\frac{35}{36} \cdot \frac{16}{21} \cdot \frac{38}{57}$

67. Prime factor 111.

68. Prime factor 1323.

Simplify.

69. $\frac{1950}{4095}$

70. $\frac{385}{847}$

71. Par on the first nine holes of the Richet Country Club golf course is 36. If Millie records scores of 5, 4, 6, 2, 3, 3, 3, 5, and 4 on those nine holes, what is her total score? Is she under or over par for the first nine holes?

72. Dried prunes weigh one third the weight of fresh prunes. How many pounds of fresh prunes must be dried to make up 124 half-pound packages of dried prunes?

4.9

Subtracting Fractions

OBJECTIVE

Subtract fractions.

HOW AND WHY

Objective

Subtract fractions.

What is the difference of $\frac{2}{3}$ and $\frac{1}{3}$? In Figure 4.13 we can see that we subtract the numerators and keep the common denominator (subtract the yellow region from the blue region).

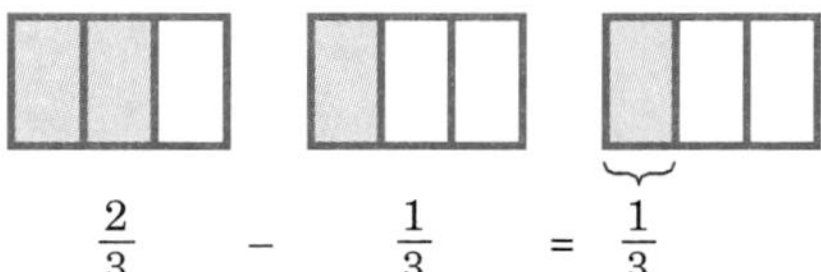

Figure 4.13

What is the answer to $\frac{3}{4} - \frac{1}{3} = ?$ (See Figure 4.14.)

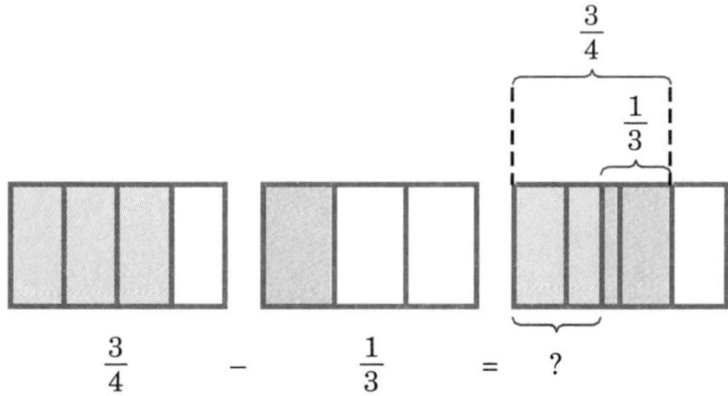

Figure 4.14

The blue-shaded area with the question mark cannot be named immediately, because the original parts are not the same size. If the fractions had common denominators, we could subtract as before. Using the common denominator 12, we see in Figure 4.15 that the answer is $\frac{5}{12}$:

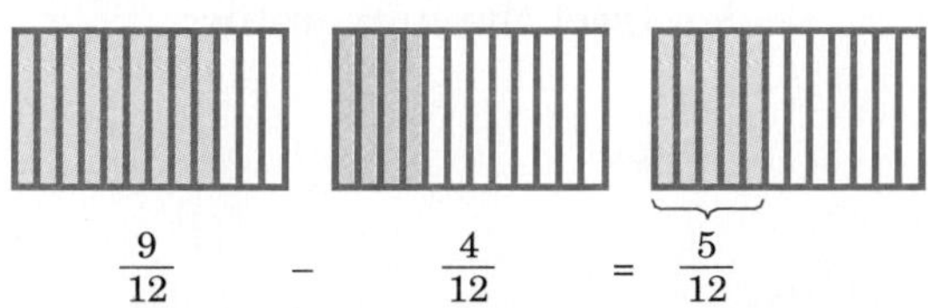

Figure 4.15

$$\frac{3}{4} - \frac{1}{3} = \frac{9}{12} - \frac{4}{12} = \frac{5}{12}$$

The process for subtracting fractions is similar to that for adding fractions.

To subtract fractions

1. Build each fraction to a common denominator.
2. Subtract the numerators and write the difference over the common denominator.
3. Simplify, if possible.

Warm Ups A–F

A. Subtract: $\frac{15}{16} - \frac{11}{16}$

B. Subtract: $\frac{11}{12} - \frac{4}{5}$

C. Subtract: $\frac{15}{32} - \frac{1}{3}$

D. Subtract: $\frac{71}{72} - \frac{31}{90}$

Examples A–F

Directions: Subtract and simplify.

Strategy: Build each fraction to a common denominator.

A. Subtract: $\frac{11}{20} - \frac{5}{20}$

$\frac{11}{20} - \frac{5}{20} = \frac{6}{20}$ **Subtract the numerators.**

$= \frac{3}{10}$ **Simplify.**

B. Subtract: $\frac{7}{8} - \frac{2}{3}$

$\frac{7}{8} - \frac{2}{3} = \frac{21}{24} - \frac{16}{24}$ **The LCM of 8 and 3 is 24.**

$= \frac{5}{24}$ **Subtract the numerators.**

C. Subtract: $\frac{7}{15} - \frac{1}{4}$

$\frac{7}{15} - \frac{1}{4} = \frac{28}{60} - \frac{15}{60}$ **The LCM of 15 and 4 is 60.**

$= \frac{13}{60}$ **Subtract the numerators.**

D. Subtract: $\frac{25}{48} - \frac{13}{80}$

$\frac{25}{48} - \frac{13}{80} = \frac{125}{240} - \frac{39}{240}$ **The LCM of 48 and 80 is 240.**

$= \frac{86}{240}$ **Subtract the numerators.**

$= \frac{43}{120}$ **Simplify.**

Answers to Warm Ups A. $\frac{1}{4}$ B. $\frac{7}{60}$ C. $\frac{13}{96}$ D. $\frac{77}{120}$

Calculator Example

E. Subtract: $\frac{39}{50} - \frac{8}{15}$

$\frac{39}{50} - \frac{8}{15} = \frac{37}{150}$ **On a calculator with fraction keys it is not necessary to find the common denominator. The calculator is programmed to subtract and simplify.**

E. Subtract: $\frac{43}{48} - \frac{23}{32}$

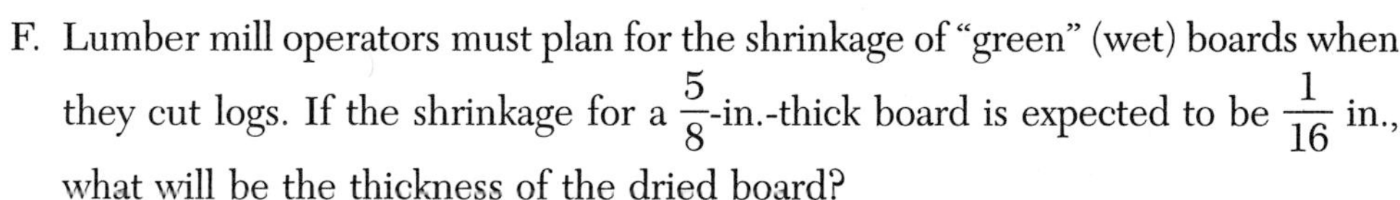

F. Lumber mill operators must plan for the shrinkage of "green" (wet) boards when they cut logs. If the shrinkage for a $\frac{5}{8}$-in.-thick board is expected to be $\frac{1}{16}$ in., what will be the thickness of the dried board?

Strategy: To find the thickness of the dried board, subtract the shrinkage from the thickness of the green board.

$\frac{5}{8} - \frac{1}{16} = \frac{10}{16} - \frac{1}{16}$ Build $\frac{5}{8}$ to have a denominator of 16.

$= \frac{9}{16}$

The dried board will be $\frac{9}{16}$ in. thick.

F. Mike must plane $\frac{3}{32}$ in. from the thickness of a board. If the board is now $\frac{3}{8}$ in. thick, how thick will it be after he has planed it?

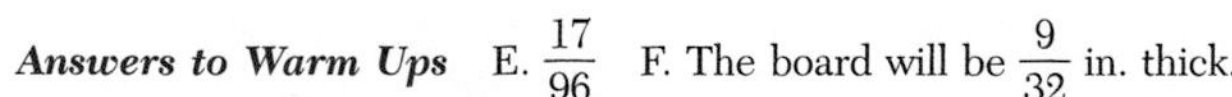

Answers to Warm Ups E. $\frac{17}{96}$ F. The board will be $\frac{9}{32}$ in. thick.

NAME ______________________ CLASS __________ DATE __________

Exercises 4.9

OBJECTIVE: *Subtract fractions.*

A.

Subtract. Simplify completely.

1. $\frac{5}{8} - \frac{3}{8}$ **2.** $\frac{7}{9} - \frac{2}{9}$ **3.** $\frac{3}{4} - \frac{2}{4}$ **4.** $\frac{5}{7} - \frac{3}{7}$ **5.** $\frac{17}{30} - \frac{7}{30}$

6. $\frac{14}{15} - \frac{11}{15}$ **7.** $\frac{5}{7} - \frac{3}{14}$ **8.** $\frac{11}{15} - \frac{2}{5}$ **9.** $\frac{3}{4} - \frac{5}{16}$ **10.** $\frac{8}{9} - \frac{5}{18}$

11. $\frac{3}{15} - \frac{2}{45}$ **12.** $\frac{5}{18} - \frac{2}{9}$ **13.** $\frac{5}{6} - \frac{1}{3}$ **14.** $\frac{5}{6} - \frac{1}{2}$ **15.** $\frac{17}{18} - \frac{2}{3}$

16. $\frac{19}{24} - \frac{3}{8}$ **17.** $\frac{17}{20} - \frac{1}{5}$ **18.** $\frac{19}{30} - \frac{1}{5}$ **19.** $\frac{23}{40} - \frac{1}{8}$ **20.** $\frac{17}{36} - \frac{1}{4}$

B.

21. $\frac{7}{8} - \frac{5}{6}$ **22.** $\frac{2}{3} - \frac{3}{8}$ **23.** $\frac{7}{16} - \frac{1}{12}$ **24.** $\frac{7}{15} - \frac{3}{20}$ **25.** $\frac{6}{14} - \frac{5}{21}$

26. $\frac{5}{6} - \frac{7}{15}$ **27.** $\frac{9}{16} - \frac{1}{6}$ **28.** $\frac{5}{6} - \frac{4}{5}$ **29.** $\frac{8}{9} - \frac{5}{6}$ **30.** $\frac{12}{21} - \frac{5}{14}$

31. $\frac{5}{8} - \frac{1}{12}$ 32. $\frac{7}{10} - \frac{7}{15}$ 33. $\frac{7}{10} - \frac{1}{4}$ 34. $\frac{8}{15} - \frac{5}{12}$ 35. $\frac{8}{9} - \frac{3}{4}$

36. $\frac{11}{12} - \frac{11}{15}$ 37. $\frac{18}{25} - \frac{7}{15}$ 38. $\frac{21}{32} - \frac{15}{48}$ 39. $\frac{13}{16} - \frac{11}{24}$ 40. $\frac{13}{18} - \frac{7}{12}$

C.

Subtract. Simplify completely.

41. $\frac{17}{18} - \frac{11}{12}$ 42. $\frac{9}{10} - \frac{7}{8}$ 43. $\frac{3}{10} - \frac{1}{15}$

44. $\frac{23}{48} - \frac{21}{80}$ 45. $\frac{14}{15} - \frac{11}{20}$ 46. $\frac{47}{60} - \frac{17}{24}$

47. If the shrinkage of a $\frac{5}{4}$-in.-thick "green" board is $\frac{1}{8}$ in., what will be the thickness of the dried board?

48. The reservoir at Bull Run was at $\frac{7}{8}$ capacity. During the last month it lost $\frac{3}{64}$ of its capacity due to evaporation. What fraction of its capacity does it now hold?

49. Wanda finds $\frac{3}{4}$ oz of gold during a day of panning along the Snake River. She gives a $\frac{1}{3}$-ounce nugget to Jose, her guide. What fraction of an ounce of gold does she have left?

50. The fat content of 1 oz of hotdog relish is $\frac{29}{35}$ g. The fat content of one slice of smoked chicken breast is $\frac{38}{45}$ g. Which food has less fat? How much less?

51. A water sample from Lake Tuscumba contains 61 parts per million of phosphate. A sample from Lost Lake contains six parts per hundred thousand. Which lake has the greatest phosphate content? By how much?

52. A population survey finds that $\frac{34}{75}$ of the residents of Ukiah are white and that $\frac{17}{40}$ are black. What fraction represents the difference in population between whites and blacks?

53. At the end of one year John has grown $\frac{5}{12}$ in. During the same time his sister has grown $\frac{3}{8}$ in. Who has grown the most? By how much?

54. The diameter at the large end of a tapered pin is $\frac{7}{8}$ in. and at the smaller end it is $\frac{3}{16}$ in. What is the difference between the diameters?

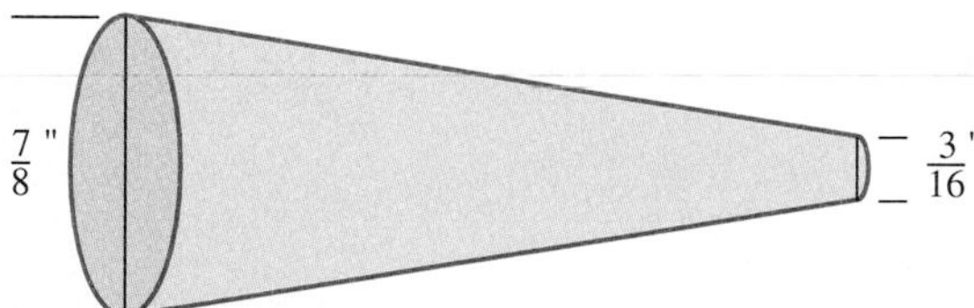

55. A carpenter planes the thickness of a board from $\frac{13}{16}$ in. to $\frac{5}{8}$ in. How much is removed?

56. A machinist needs a bar that is $\frac{5}{8}$ in. thick. If he cuts it from a bar that is $\frac{27}{32}$ in. thick, how much must he cut off?

57. A landscaper is building a brick border, one brick wide, around a formal rose garden. The garden is a 10-ft-by-6-ft rectangle. Standard bricks are 8 in. by $3\frac{3}{4}$ in. by $2\frac{1}{4}$ in., and the landscaper is planning to use a $\frac{3}{8}$-in.-wide mortar in the joints. How many whole bricks are needed for the project? Explain your reasoning.

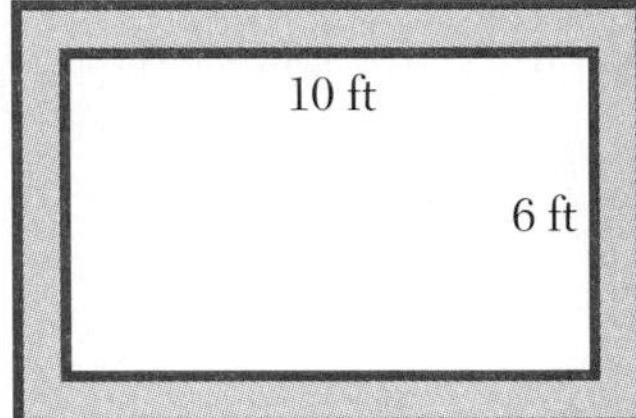

Exercise 58 relates to the chapter application.

58. How can Hannah measure $\frac{3}{8}$ c of nuts with only a $\frac{1}{2}$-c measure and a tablespoon measure? (*Hint:* 1 c = 16 tbs)

STATE YOUR UNDERSTANDING

59. Explain in writing how you would teach a child to subtract fractions.

60. Explain why $\frac{3}{4} - \frac{1}{2}$ is not equal to $\frac{2}{2}$.

NAME ______________________ CLASS __________ DATE __________

CHALLENGE

Subtract.

61. $\frac{213}{560} - \frac{89}{430}$

62. $\frac{5487}{7375} - \frac{247}{625}$

63. A donor agrees to donate \$1000 for each foot Skola outwalks Sheila in 13 min. Skola walks $\frac{19}{24}$ mile. Sheila walks $\frac{47}{60}$ mile. Does Skola out-distance Sheila? By what fraction of a mile? How much does the donor contribute? (A mile equals 5280 ft.)

GROUP ACTIVITY

64. With your group, find a way to subtract $4\frac{7}{8}$ from $6\frac{3}{4}$. Write the procedure down and share it with the class.

MAINTAIN YOUR SKILLS (SECTIONS 1.2, 1.3, 1.4, 1.6, 4.2, 4.3, 4.4)

65. Add: 483 + 3111 + 3450 + 9612 + 5998

66. Subtract: 41,695 − 3969

67. Divide: 251,782 ÷ 121

68. Multiply: (2510)(115)

69. Simplify: $\frac{130}{182}$

70. Simplify: $\frac{255}{357}$

71. Divide: $\frac{2}{7} \div \frac{8}{9}$

72. Divide: $\frac{2}{7} \div 5$

73. Three bricklayers can each lay 795 bricks per day, on the average. How many bricks can they lay in 5 days?

74. If a retaining wall requires 19,080 bricks, how many days will it take the three bricklayers in Exercise 73 to build the wall?

4.10

Subtracting Mixed Numbers

OBJECTIVE

Subtract mixed numbers.

HOW AND WHY

Objective

Subtract mixed numbers.

A subtraction problem may be written in horizontal or vertical form. Horizontally:

$$8\frac{7}{9} - 3\frac{2}{9} = (8 - 3) + \left(\frac{7}{9} - \frac{2}{9}\right)$$
$$= 5 + \frac{5}{9} = 5\frac{5}{9}$$

Since the denominators are the same, subtract the whole-number parts and then subtract the fraction parts.

Vertically:

$$\begin{array}{r} 8\frac{7}{9} \\ -\ 3\frac{2}{9} \\ \hline 5\frac{5}{9} \end{array}$$

The process is similar to that for adding mixed numbers.

It is sometimes necessary to "borrow" from the whole number in order to subtract the fractions. For example

$$8\frac{2}{5} - 3\frac{3}{4} = ?$$

First, write in columns and build each fraction to the common denominator, 20.

$$\begin{array}{rcl} 8\frac{2}{5} & = & 8\frac{8}{20} \\ -\ 3\frac{3}{4} & = & 3\frac{15}{20} \\ \hline \end{array}$$

Since we cannot subtract $\frac{15}{20}$ from $\frac{8}{20}$, we need to "borrow." So we rename $8\frac{8}{20}$ by "borrowing" 1 from the 8.

$$8\frac{8}{20} = 7 + 1\frac{8}{20}$$

Borrow 1 from the 8 and add it to the fraction part.

$$= 7 + \frac{28}{20}$$

Change the mixed number, $1\frac{8}{20}$, to an improper fraction.

$$= 7\frac{28}{20}$$

Write as a mixed number.

CAUTION

Do not write $\frac{18}{20}$. If we "borrow" 1 from 8, we must add 1 $\left(\frac{20}{20}\right)$ to $\frac{8}{20}$.

The example can now be completed.

$$\begin{array}{rrrl} 8\frac{2}{5} = & 8\frac{8}{20} = & 7\frac{28}{20} & \textbf{Rename by "borrowing" 1.} \\ -\ 3\frac{3}{4} = & 3\frac{15}{20} = & 3\frac{15}{20} & \\ \hline & & 4\frac{13}{20} & \textbf{Subtract.} \end{array}$$

To subtract mixed numbers

1. Build the fractions so they have a common denominator.
2. Subtract the fractions. If the fractions cannot be subtracted, rename the first mixed number by "borrowing" 1 from the whole-number part to add to the fraction part. Then subtract the fractions.
3. Subtract the whole numbers.
4. Simplify.

Warm Ups A–H

A. Subtract: $48\frac{5}{9} - 22\frac{2}{5}$

B. Subtract: $21\frac{5}{8} - 17$

Examples A–H

Directions: Subtract. Write as a mixed number.

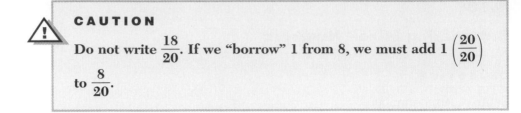

Strategy: Subtract the fractions and subtract the whole numbers. If necessary borrow. Simplify.

A. Subtract: $47\frac{5}{8} - 36\frac{3}{7}$

Strategy. Write the mixed numbers in columns to group the whole numbers and to group the fractions.

$$\begin{array}{rrl} 47\frac{5}{8} = & 47\frac{35}{56} & \textbf{Build the fractions to the common denominator, 56.} \\ -\ 36\frac{3}{7} = & 36\frac{24}{56} & \\ \hline & 11\frac{11}{56} & \textbf{Subtract.} \end{array}$$

B. Subtract: $13\frac{2}{3} - 8$

$13\frac{2}{3} - 8 = 5\frac{2}{3}$ **Subtract the whole numbers.**

Answers to Warm Ups A. $26\frac{7}{45}$ B. $4\frac{5}{8}$

C. Subtract: $19\frac{11}{30} - 12\frac{1}{5}$

$$\begin{array}{rl} 19\frac{11}{30} = 19\frac{11}{30} \\ -\ 12\frac{1}{5} = 12\frac{6}{30} \\ \hline 7\frac{5}{30} = 7\frac{1}{6} \end{array}$$

Write in columns and build $\frac{1}{5}$ to a denominator of 30.

Subtract and simplify.

C. Subtract: $32\frac{17}{40} - 27\frac{1}{8}$

D. Subtract: $15\frac{5}{12} - 7\frac{7}{12}$

Strategy: Since $\frac{7}{12}$ cannot be subtracted from $\frac{5}{12}$, we will need to borrow.

$$\begin{array}{rl} 15\frac{5}{12} = 14 + 1\frac{5}{12} = 14\frac{17}{12} \\ -\ 7\frac{7}{12} \qquad = \ \ 7\frac{7}{12} \\ \hline 7\frac{10}{12} \\ = 7\frac{5}{6} \end{array}$$

Borrow 1 from 15 and then change the mixed number to an improper fraction.

Subtract.

Simplify.

D. Subtract: $18\frac{11}{15} - 8\frac{12}{15}$

E. Subtract: $43\frac{7}{12} - 21\frac{11}{15}$

$$\begin{array}{rl} 43\frac{7}{12} = 43\frac{35}{60} = 42 + 1\frac{35}{60} = 42\frac{95}{60} \\ -\ 21\frac{11}{15} = 21\frac{44}{60} \qquad\qquad = 21\frac{44}{60} \\ \hline 21\frac{51}{60} \\ = 21\frac{17}{20} \end{array}$$

The LCM of 12 and 15 is 60. Borrow 1 from 43 and then change the mixed number to an improper fraction.

Subtract.

Simplify.

E. Subtract: $47\frac{3}{8} - 32\frac{14}{15}$

F. Subtract: $11 - 2\frac{5}{9}$

Strategy: We may think of 11 as $11\frac{0}{9}$ in order to get a common denominator for the improper fraction. Or think: $11 = 10 + 1 = 10 + \frac{9}{9} = 10\frac{9}{9}$.

$$\begin{array}{rl} 11 = 10 + 1\frac{0}{9} = 10\frac{9}{9} \\ -\ 2\frac{5}{9} \qquad = \ \ 2\frac{5}{9} \\ \hline = 8\frac{4}{9} \end{array}$$

F. Subtract: $33 - 11\frac{5}{7}$

Answers to Warm Ups C. $5\frac{3}{10}$ D. $9\frac{14}{15}$ E. $14\frac{53}{120}$ F. $21\frac{2}{7}$

CAUTION

Do not just bring the fraction $\frac{5}{9}$ down and subtract the whole numbers; $11 - 2\frac{5}{9} \neq 9\frac{5}{9}$. The fraction must also be subtracted.

Calculator Example

G. Subtract: $72\frac{7}{9} - 25\frac{13}{15}$

G. Subtract: $31\frac{1}{3} - 18\frac{3}{4}$

$$31\frac{1}{3} - 18\frac{3}{4} = \frac{151}{12} = 12\frac{7}{12}$$

On a calculator with a fraction key it is not necessary to find the common denominator. The calculator is programmed to subtract and simplify. Some calculators may not change the improper difference to a mixed number.

H. Jamie weighs $138\frac{1}{2}$ lb and decides to lose some weight. She loses a total of $5\frac{3}{4}$ lb in one week. What is her weight after the loss?

H. Shawn McCord brings a roast for Sunday dinner that weighs 7 lb. He cuts off some fat and takes out a bone. The meat left weighs $4\frac{1}{3}$ lb. How many pounds of bone and fat does he trim off?

Strategy: Subtract the weight of the remaining meat from the original weight of the roast.

$$\begin{array}{rl} 7 & = 6 + 1\frac{0}{3} = 6\frac{3}{3} \\ -\ 4\frac{1}{3} & = 4\frac{1}{3} \\ \hline & \quad 2\frac{2}{3} \end{array}$$

Borrow 1 from 7 and rename it as an improper fraction.

Shawn trims off $2\frac{2}{3}$ lb of fat and bone.

Answers to Warm Ups G. $46\frac{41}{45}$ H. Her weight is $132\frac{3}{4}$ lb.

NAME ____________ CLASS ____________ DATE ____________

Exercises 4.10

OBJECTIVE: *Subtract mixed numbers.*

A.

Subtract. Write the results as mixed numbers where possible. Simplify.

1. $16\frac{6}{7} - 9\frac{4}{7}$

2. $31\frac{5}{9} - 20\frac{2}{9}$

3. $212\frac{37}{80} - 109\frac{21}{80}$

4. $205\frac{6}{11} - 112\frac{4}{11}$

5. $10\frac{7}{8} - 5\frac{3}{4}$

6. $11\frac{4}{5} - 7\frac{3}{10}$

7. $15 - 8\frac{3}{7}$

8. $32 - 14\frac{1}{6}$

9. $145\frac{2}{3} - 27\frac{1}{2}$

10. $6\frac{5}{6} - 3\frac{3}{10}$

11. $5\frac{1}{4} - 2\frac{3}{4}$

12. $7\frac{3}{8} - 4\frac{5}{8}$

13. $26\frac{1}{10} - 10\frac{9}{10}$

14. $19\frac{3}{8} - 8\frac{5}{8}$

15. $212\frac{1}{9} - 57$

16. $400\frac{3}{8} - 62$

17. $7\frac{1}{4} - 5\frac{5}{8}$

18. $9\frac{9}{16} - 2\frac{5}{6}$

B.

19. $21\frac{15}{16} - 12\frac{7}{16}$

20. $24\frac{11}{12} - 13\frac{7}{12}$

21. $310\frac{23}{24} - 254\frac{5}{8}$

22. $118\frac{7}{12} - 93\frac{1}{4}$

23. $76\frac{7}{15} - 50\frac{1}{12}$

24. $9\frac{9}{16} - 3\frac{5}{12}$

25. $37\frac{2}{3} - 15\frac{11}{12}$

26. $28\frac{1}{3} - 15\frac{7}{9}$

27. $30\frac{7}{16} - 22\frac{5}{6}$

28. $33\frac{17}{30} - 25\frac{7}{9}$

29. $45 - 16\frac{2}{3}$

30. $76 - 26\frac{2}{3}$

31. $5\frac{31}{32} - \frac{3}{16}$

32. $8\frac{11}{20} - \frac{9}{10}$

33. $7\frac{13}{18} - \frac{5}{12}$

34. $9\frac{1}{6} - \frac{5}{8}$

35. $43\frac{5}{6} - 21$

36. $48\frac{6}{11} - 17$

37. $64\frac{13}{15} - 28\frac{29}{40}$

38. $70\frac{11}{18} - 58\frac{41}{45}$

39. $75\frac{7}{12} - 47\frac{13}{18}$

40. $82\frac{4}{15} - 56\frac{7}{12}$

C.

Subtract. Write the results as mixed numbers where possible.

41. $18\frac{5}{24} - 11\frac{3}{40}$

42. $25\frac{7}{12} - 14\frac{11}{15}$

43. $16 - 14\frac{17}{20}$

44. $41 - 29\frac{17}{36}$

45. $3\frac{11}{15} - 2\frac{9}{10}$

46. $6\frac{1}{3} - 5\frac{5}{6}$

47. Find the error(s) in the statement: $16 - 13\frac{1}{4} = 3\frac{3}{4}$. Correct the statement. Explain how you would avoid this error.

48. Find the error(s) in the statement: $5\frac{1}{2} - 2\frac{3}{4} = 3\frac{1}{4}$. Correct the statement. Explain how you would avoid this error.

49. Find the difference of $6\frac{23}{25}$ and $4\frac{14}{15}$.

50. Find the difference of $10\frac{19}{32}$ and $8\frac{43}{48}$.

51. Han Kwong trims bone and fat from a $6\frac{3}{4}$ lb roast. The meat left weighs $5\frac{1}{4}$ lb. How many pounds does she trim off?

52. Patti has a piece of lumber that measures $10\frac{7}{12}$ ft, and it is to be used in a spot that calls for a length of $8\frac{5}{12}$ ft. How much of the board must be cut off?

53. Dick harvests $30\frac{3}{4}$ tons of wheat. He sells $12\frac{3}{10}$ tons to the Cartwright Flour Mill. How many tons of wheat does he have left?

54. A $14\frac{3}{4}$ in. casting shrinks $\frac{3}{16}$ in. on cooling. Find the size when the casting is cold.

NAME ______________________________ CLASS ______________ DATE ______________

55. Larry and Greg set out to hike 42 miles in two days. At the end of the first day they have covered $24\frac{3}{10}$ miles. How many miles do they have yet to go?

56. Frank pours $8\frac{3}{10}$ yd of cement for a fountain. Another fountain takes $5\frac{3}{8}$ yd. How much more cement is needed for the larger fountain than for the smaller one?

57. The Williams family recycles an average of $134\frac{2}{5}$ lb of material per month. The Madera family recycles an average of $102\frac{5}{8}$ lb of material per month. How many more pounds of material are recycled by the Williams family in a year?

58. The town of Fredonia averages $35\frac{1}{5}$ in. of rain per year. The town of Wheatland averages $27\frac{13}{15}$ in. of rain per year. Over a 10-yr period, how much more rain falls on Fredonia than on Wheatland?

59. The graph displays the average yearly rainfall for five cities.

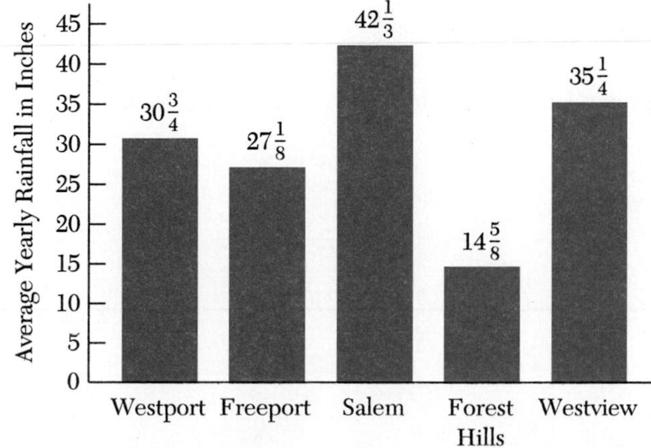

a. How much more rain falls in Westport during a year than in Freeport?

b. In a 10-yr period, how much more rain falls in Salem than in Forest Hills?

c. If the average rainfall in Westview doubles, how much more rain would it receive than Salem?

60. The town of Hartview averages $35\frac{1}{5}$ in. of rain per year. The town of Sunrise averages $27\frac{13}{15}$ in. of rain per year. Over a 5-yr period, how much more rain falls on Hartview than on Sunrise?

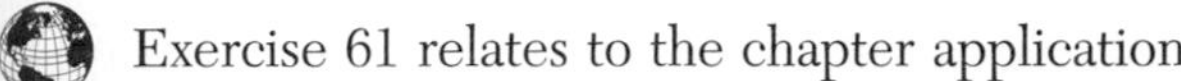

Exercise 61 relates to the chapter application.

61. Some recipes call for regular milk and vinegar as a substitute for buttermilk. A recipe calls for 1 c of regular milk less 1 tbs, and then 1 tbs of vinegar is added for a total of 1 c of "buttermilk replacement." How much regular milk is needed for this recipe?

STATE YOUR UNDERSTANDING

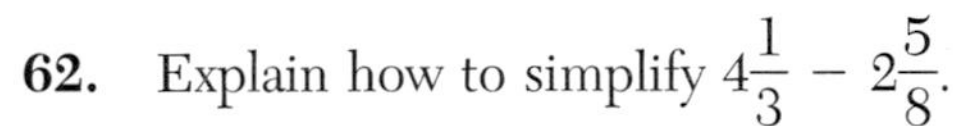

62. Explain how to simplify $4\frac{1}{3} - 2\frac{5}{8}$.

63. When you "borrow" 1 to subtract mixed numbers, explain the fraction form it is written in and explain why.

CHALLENGE

64. Does $4\left(5\frac{5}{6}\right) - 3\left(2\frac{7}{12}\right) = 6\left(5\frac{1}{2}\right) - 5\frac{3}{4}$?

65. Does $3\left(7\frac{7}{8}\right) - 13\frac{1}{10} = 6\left(3\frac{3}{16}\right) - 7\left(1\frac{8}{35}\right)$?

66. A snail climbs $4\frac{2}{3}$ ft in a day and slips back $1\frac{5}{16}$ ft at night. What is the snail's net distance in 24 hr? How many days will it take the snail to make a net gain of over 20 ft?

GROUP ACTIVITY

67. Have each member of the group create an application involving subtraction of mixed numbers. Trade them around and have other members solve them. Share the best two with the rest of the class.

MAINTAIN YOUR SKILLS (SECTIONS 2.1, 4.1, 4.2, 4.3, 4.8)

Simplify.

68. $\frac{208}{272}$

69. $\frac{286}{429}$

70. Change $\frac{145}{24}$ to a mixed number.

71. Change $31\frac{7}{8}$ to an improper fraction.

Multiply.

72. $\frac{15}{24} \cdot \frac{9}{25} \cdot \frac{16}{27}$

73. $\frac{16}{25} \cdot \frac{25}{28} \cdot \frac{14}{15}$

74. $\frac{1}{2} \cdot \frac{5}{8} \cdot \frac{3}{5} \cdot 24$

75. $\frac{2}{3} \cdot \frac{6}{11} \cdot \frac{1}{4} \cdot 22$

76. Last week when Karla filled the tank of her car with gasoline, the odometer read 57,832 miles. Yesterday when she filled the tank with 18 gallons of gasoline, the odometer read 58,336 miles. How many miles to the gallon was the car averaging?

77. A pet-food canning company packs Feelein Cat Food in cans, each containing $7\frac{3}{4}$ oz of cat food. Each empty can weighs $1\frac{1}{2}$ oz. Twenty-four cans are packed in a case that weighs 10 oz empty. What is the shipping weight of five cases of the cat food?

Getting Ready for Algebra

OBJECTIVE

Solve equations of the form $x + \frac{a}{b} = \frac{c}{d}$ where a, b, c, and d are whole numbers.

HOW AND WHY

We have solved equations in which whole numbers were either added to or subtracted from a variable. Now we solve equations in which fractions or mixed numbers are either added to or subtracted from the variable. We use the same procedure as with whole numbers.

Warm Ups A–C

A. Solve: $x - \frac{5}{8} = 3\frac{7}{8}$

Examples A–C

Directions: Solve.

Strategy: Add or subtract the same number from each side of the equation to isolate the variable.

A. Solve: $x - \frac{4}{5} = 3\frac{1}{2}$

$$x - \frac{4}{5} = 3\frac{1}{2}$$

$$x - \frac{4}{5} + \frac{4}{5} = 3\frac{1}{2} + \frac{4}{5}$$ **Eliminate the subtraction by adding $\frac{4}{5}$ to each side of the equation.**

$$x = 3\frac{1}{2} + \frac{4}{5}$$

$$x = 3\frac{5}{10} + \frac{8}{10}$$ **Build each fraction to the denominator, 10.**

$$x = 3\frac{13}{10}$$ **Change the improper fraction to a mixed number and add.**

$$x = 4\frac{3}{10}$$

CHECK:

$$4\frac{3}{10} - \frac{4}{5} = 3\frac{1}{2}$$ **Substitute $4\frac{3}{10}$ for x in the original equation.**

$$3\frac{1}{2} = 3\frac{1}{2}$$

The solution is $x = 4\frac{3}{10}$.

Answer to Warm Up A. $x = 4\frac{1}{2}$

B. Solve: $4\frac{1}{2} = x + 2\frac{2}{3}$

$$4\frac{1}{2} = x + 2\frac{2}{3}$$

$$4\frac{1}{2} - 2\frac{2}{3} = x + 2\frac{2}{3} - 2\frac{2}{3}$$ **Eliminate the addition by subtracting $2\frac{2}{3}$ from both sides of the equation.**

$$4\frac{1}{2} - 2\frac{2}{3} = x$$

$$1\frac{5}{6} = x$$

CHECK:

$$4\frac{1}{2} = 1\frac{5}{6} + 2\frac{2}{3}$$ **Substitute $1\frac{5}{6}$ for x in the original equation.**

$$4\frac{1}{2} = 4\frac{1}{2}$$

The solution is $x = 1\frac{5}{6}$.

B. Solve: $6\frac{1}{2} = a + 5\frac{5}{8}$

C. On Tuesday $2\frac{3}{8}$ in. of rain fell on Kansas City. This brought the total for the last five consecutive days to $14\frac{1}{2}$ in. What was the rainfall for the first 4 days?

Strategy: First write the English version of the equation.

$$\begin{pmatrix}\text{rain on first}\\ \text{four days}\end{pmatrix} + \begin{pmatrix}\text{rain on}\\ \text{Tuesday}\end{pmatrix} = \text{total rain}$$

Let x represent the inches of rain on the first four days.

$$x + 2\frac{3}{8} = 14\frac{1}{2}$$ **Translate to algebra.**

$$-2\frac{3}{8} \quad -2\frac{3}{8}$$ **Subtract $2\frac{3}{8}$ from each side.**

$$x = 12\frac{1}{8}$$

Since $12\frac{1}{8} + 2\frac{3}{8} = 14\frac{1}{2}$, $12\frac{1}{8}$ in. of rain fell during the first 4 days.

C. When $3\frac{7}{8}$ miles of new freeway opened last month it brought the total length to $21\frac{3}{5}$ miles. What was the original length of the freeway?

Answers to Warm Ups B. $a = \frac{7}{8}$ C. The original freeway was $17\frac{29}{40}$ miles long.

NAME ______________________ CLASS __________ DATE __________

Exercises

Solve.

1. $a + \frac{3}{8} = \frac{7}{8}$

2. $y - \frac{1}{8} = \frac{5}{8}$

3. $c - \frac{3}{16} = \frac{7}{16}$

4. $w + \frac{5}{12} = \frac{11}{12}$

5. $x + \frac{5}{8} = \frac{7}{9}$

6. $x - \frac{7}{8} = \frac{3}{2}$

7. $y - \frac{5}{7} = \frac{8}{9}$

8. $y + \frac{5}{9} = \frac{9}{10}$

9. $a + \frac{9}{8} = \frac{12}{5}$

10. $a - \frac{7}{4} = \frac{5}{8}$

11. $c - 1\frac{1}{2} = 2\frac{2}{3}$

12. $c + 2\frac{4}{5} = 3\frac{5}{8}$

13. $x + 4\frac{3}{4} = 7\frac{8}{9}$

14. $x - 7\frac{8}{9} = 5\frac{7}{8}$

15. $12 = w + 8\frac{5}{6}$

16. $25 = m + 15\frac{5}{8}$

17. $a - 13\frac{5}{6} = 22\frac{11}{18}$

18. $b + 23\frac{11}{12} = 34\frac{1}{3}$

19. $c + 44\frac{13}{21} = 65\frac{5}{7}$

20. $x - 27\frac{5}{8} = 48\frac{2}{3}$

21. The stock of the Acme Corporation rises $3\frac{7}{8}$ points on Friday to hit a new high of $95\frac{1}{2}$. What was the price of a share of Acme stock on Thursday?

22. Pele brought in $35\frac{3}{4}$ lb of tin to be recycled last week. This brings his total for the month to $112\frac{1}{2}$ lb. How many pounds had he already brought in this month?

23. Freeda bought a supply of nails for her construction project. She has used $18\frac{2}{3}$ lb and has $27\frac{1}{3}$ lb left. How many pounds of nails did she buy?

24. For cross-country race practice Althea has run $10\frac{7}{10}$ miles. She needs to run an additional $13\frac{3}{10}$ miles to meet the quota set by her coach. How many miles does the coach want her to run?

4.11

Order of Operations; Average; Formulas Involving π

OBJECTIVES

1. Do any combinations of operations with fractions.
2. Find the average of a set of fractions.
3. Evaluate geometric formulas that involve the number π.

VOCABULARY

The **circumference** of a circle is the distance around the circle.

The **radius** of a circle is the distance from the center to any point on the circle.

The **diameter** of a circle is twice the radius.

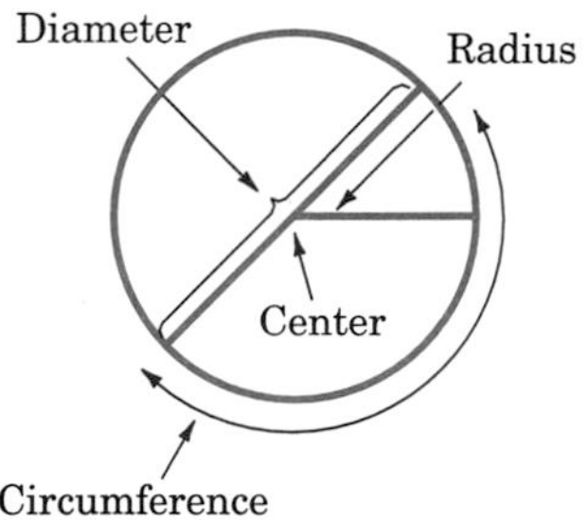

HOW AND WHY

Objective 1

Do any combinations of operations with fractions.

The order of operations for fractions is the same as for whole numbers.

To evaluate an expression with more than one operation

1. Parentheses—Do the operations within grouping symbols first (parentheses, fraction bar, etc.) in the order given in steps 2, 3, and 4.
2. Exponents—Do the operations indicated by exponents.
3. Multiply and divide—Do only multiplication and division as they appear from left to right.
4. Add and subtract—Do addition and subtraction as they appear from left to right.

Table 4.3 summarizes some of the processes that need to be remembered when working with fractions.

TABLE 4.3 OPERATIONS WITH FRACTIONS

Operation	Find the LCM and Build	Change Mixed Numbers to Improper Fractions	Invert Divisor and Multiply	Simplify Answer
Add	Yes	No	No	Yes
Subtract	Yes	No	No	Yes
Multiply	No	Yes	No	Yes
Divide	No	Yes	Yes	Yes

Warm Ups A–D

Examples A–D

Directions: Perform the indicated operations.

Strategy: Follow the order of operations that are used for whole numbers.

A. Simplify: $\frac{7}{8} - \frac{3}{4} \cdot \frac{8}{9}$

A. Simplify: $\frac{5}{6} - \frac{1}{2} \cdot \frac{2}{3}$

$\frac{5}{6} - \frac{1}{2} \cdot \frac{2}{3} = \frac{5}{6} - \frac{1}{3}$ **Multiplication is performed first.**

$= \frac{5}{6} - \frac{2}{6}$ **Build $\frac{1}{3}$ to a denominator of 6.**

$= \frac{3}{6} = \frac{1}{2}$ **Subtract and simplify.**

B. Simplify: $\frac{5}{8} \div \frac{9}{16} \cdot \frac{7}{5}$

B. Simplify: $\frac{1}{2} \div \frac{2}{3} \cdot \frac{1}{4}$

$\frac{1}{2} \div \frac{2}{3} \cdot \frac{1}{4} = \frac{1}{2} \cdot \frac{3}{2} \cdot \frac{1}{4}$ **Division is performed first, as it appears first from left to right.**

$= \frac{3}{16}$ **Multiply from left to right.**

C. Simplify: $\left(\frac{5}{8}\right)^2 \cdot \frac{8}{15} - \frac{1}{8}$

C. Simplify: $\left(\frac{3}{4}\right)^2 \cdot \frac{2}{5} - \frac{1}{5}$

$\left(\frac{3}{4}\right)^2 \cdot \frac{2}{5} - \frac{1}{5} = \frac{9}{\cancel{16}_{8}} \cdot \frac{\cancel{2}^{1}}{5} - \frac{1}{5}$ **Exponentiation is done first then simplify.**

$= \frac{9}{40} - \frac{1}{5}$ **Multiply.**

$= \frac{9}{40} - \frac{8}{40}$

$= \frac{1}{40}$ **Subtract.**

D. Jill, Jenny, and Joan have equal shares in a gift shop. Jill sells her share. She sells $\frac{1}{4}$ to Jenny, and the rest to Joan. What share of the gift shop is now owned by Joan?

D. Gwen, Sam, Carlos, and Sari have equal shares in a florist shop. Sam sells his share. He sells $\frac{3}{8}$ of his share to Gwen, $\frac{1}{2}$ to Carlos, and the rest to Sari. What share of the florist shop is now owned by Gwen?

Strategy: Since each of the four had equal shares, each of them owned $\frac{1}{4}$ of the business. To find the share Gwen now owns, add her original share, $\frac{1}{4}$, to $\frac{3}{8}$ of Sam's share, which was $\frac{1}{4}$.

Answers to Warm Ups A. $\frac{5}{24}$ B. $\frac{14}{9}$ or $1\frac{5}{9}$ C. $\frac{1}{12}$ D. Joan owns $\frac{7}{12}$ of the gift shop.

$\frac{1}{4} + \frac{3}{8} \cdot \frac{1}{4} = \frac{1}{4} + \frac{3}{32}$ **Multiply first.**

$= \frac{8}{32} + \frac{3}{32}$ **Build $\frac{1}{4}$ to the denominator 32.**

$= \frac{11}{32}$

Gwen now owns $\frac{11}{32}$ of the florist shop.

HOW AND WHY

Objective 2

Find the average of a set of fractions.

To find the average of a set of fractions, divide the sum of the fractions by the number of fractions. The procedure is the same for all types of numbers.

Examples E–G

Directions: Find the average.

Strategy: Find the sum of the set of numbers and then divide by the number of numbers.

E. Find the average: $\frac{1}{2}$, $\frac{1}{3}$, and $\frac{3}{4}$

$\frac{1}{2} + \frac{1}{3} + \frac{3}{4} = \frac{6}{12} + \frac{4}{12} + \frac{9}{12}$ **Add the three numbers in the set.**

$= \frac{19}{12}$

$\frac{19}{12} \div 3 = \frac{19}{12} \cdot \frac{1}{3}$ **Divide the sum by 3, the number of numbers in the set.**

$= \frac{19}{36}$

The average is $\frac{19}{36}$.

Warm Ups E–G

E. Find the average: $\frac{5}{6}$, $\frac{7}{8}$, and $\frac{3}{4}$

Answer to Warm Up E. $\frac{59}{72}$

F. A class of 12 students takes a 20-problem test. The results are listed in the table:

Number of Students	Fraction of Problems Correct
1	$\frac{20}{20}$
2	$\frac{19}{20}$
4	$\frac{16}{20}$
5	$\frac{14}{20}$

What is the average?

F. A class of 10 students takes a 12-problem test. The results are listed in the table:

Number of Students	Fraction of Problems Correct
1	$\frac{12}{12}$
2	$\frac{11}{12}$
3	$\frac{10}{12}$
4	$\frac{9}{12}$

What is the average?

Strategy: To find the class average, add all the grades together and divide by 10. There were two scores of $\frac{11}{12}$, three scores of $\frac{10}{12}$, and four scores of $\frac{9}{12}$ in addition to one perfect score of $\frac{12}{12}$.

$$\frac{12}{12} + 2\left(\frac{11}{12}\right) + 3\left(\frac{10}{12}\right) + 4\left(\frac{9}{12}\right)$$

Find the sum of the scores.

$$\frac{12}{12} + \frac{22}{12} + \frac{30}{12} + \frac{36}{12} = \frac{100}{12}$$

$$\frac{100}{12} \div 10 = \frac{100}{12} \cdot \frac{1}{10}$$

$$= \frac{10}{12}$$

Divide the sum by the number of students in the class.

CAUTION

Do not simplify the answer, since the test scores are based on 12.

The class average is $\frac{10}{12}$ correct.

Answer to Warm Up F. The class average is $\frac{16}{20}$.

G. Find the average: $7\frac{1}{2}$, $6\frac{1}{4}$, and $4\frac{5}{8}$

$7\frac{1}{2} + 6\frac{1}{4} + 4\frac{5}{8} = 18\frac{3}{8}$ **Find the sum of the numbers.**

$18\frac{3}{8} \div 3 = \frac{147}{8} \div 3$ **Divide by the number of numbers in the set. Change the mixed number to an improper fraction.**

$= \frac{147}{8} \cdot \frac{1}{3}$

$= \frac{49}{8} = 6\frac{1}{8}$ **Multiply, simplify, and change to a mixed number.**

The average is $6\frac{1}{8}$.

G. Find the average: $3\frac{5}{6}$, $4\frac{1}{2}$, and $2\frac{2}{3}$

HOW AND WHY

Objective 3

Evaluate geometric formulas that involve the number π.

Formulas for geometric figures that involve circles contain the number called pi (π). The number π is the quotient of the *circumference* of (distance around) the circle and its diameter. This number is the same for *every* circle no matter how large or small. This remarkable fact was discovered over a long period of time historically and during that time a large number of approximations have been used. Here are some of the approximations:

$3 \quad 3\frac{1}{8} \quad 3\frac{1}{7}\left(\text{or } \frac{22}{7}\right) \quad \frac{355}{113}$

Because $\frac{22}{7}$ is an approximate value, when we use this fraction we write the symbol for approximately equal to, $\pi \approx \frac{22}{7}$. There are also decimal approximations for π (see Chapter 5).

Circle Formulas

If C is the circumference, d is the diameter, and r is the radius of a circle, then

$C = \pi d$ or $C = 2\pi r$

If A if the area, then

$A = \pi r^2$

Answer to Warm Up G. $3\frac{2}{3}$.

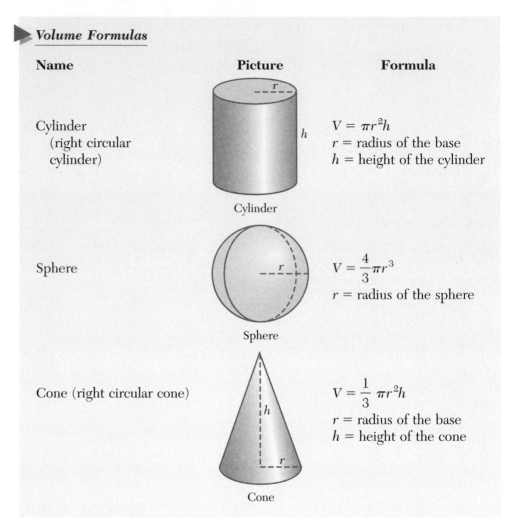

Because cylinders, spheres, and cones contain circles, their volume formulas also contain the number π.

The circumference C is measured in units of length (such as inches, feet, meters). The area is measured in square units (such as in^2, ft^2, m^2). Volume is measured in cubic units (such as in^3, ft^3, m^3).

Volume Formulas

Name	Picture	Formula
Cylinder (right circular cylinder)	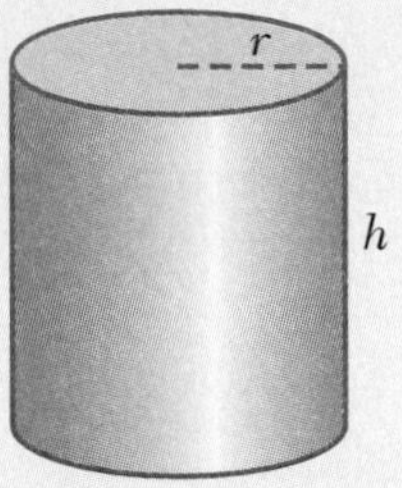 Cylinder	$V = \pi r^2 h$ r = radius of the base h = height of the cylinder
Sphere	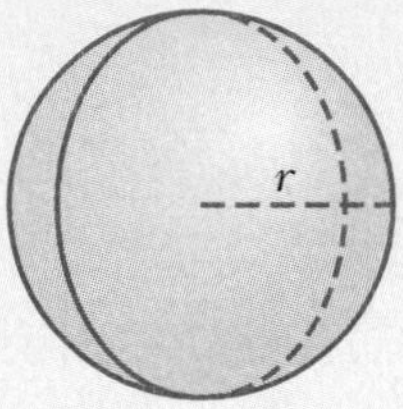 Sphere	$V = \frac{4}{3}\pi r^3$ r = radius of the sphere
Cone (right circular cone)	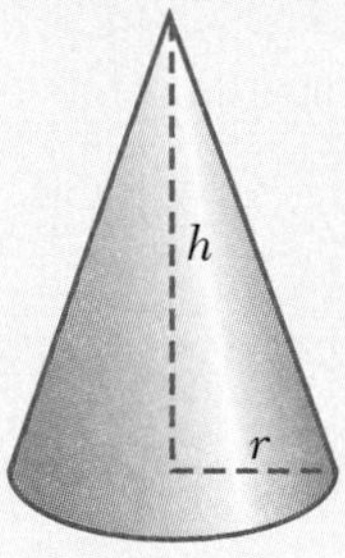 Cone	$V = \frac{1}{3}\pi r^2 h$ r = radius of the base h = height of the cone

Warm Ups H–K

Examples H–K

Directions: Evaluate the formula.

Strategy: Substitute the values of the measurements into the appropriate formula and evaluate. Use $\pi \approx \frac{22}{7}$.

H. Find the circumference of the circle.

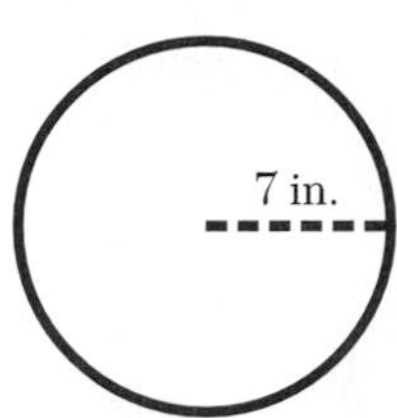

$C = 2\pi r$ **Formula for circumference.**

$C \approx 2\left(\frac{22}{7}\right)(7 \text{ in.})$ **Substitute.**

$C \approx 44$ in. **Multiply.**

The circumference is approximately 44 in.

H. Find the circumference of the circle.

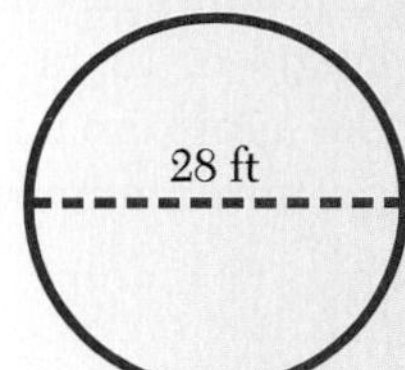

I. Find the area of a circle with a radius of $\frac{3}{4}$ inch.

$A = \pi r^2$ **Formula for area.**

$A \approx \left(\frac{22}{7}\right)\left(\frac{3}{4} \text{ in.}\right)^2$ **Substitute.**

$A \approx \left(\frac{22}{7}\right)\left(\frac{9}{16} \text{ in}^2\right)$ **Exponents are done first.**

$A \approx \frac{99}{56} \text{ in}^2$ **Multiply.**

$A \approx 1\frac{43}{56} \text{ in}^2$ **Change to a mixed number.**

The area is approximately $1\frac{43}{56}$ in^2.

I. Find the area of a circle with radius 14 cm.

Answers to Warm Ups H. The circumference is approximately 88 ft. I. The area is approximately 616 cm^2.

J. Find the volume of a cylinder that has a circular base with radius 12 cm and a height of 13 cm. Write the approximate answer as a mixed number.

J. Find the volume of the cylinder.

$V = \pi r^2 h$ **Formula for cylinder volume.**

$V \approx \frac{22}{7}(2\text{ ft})^2(6\text{ ft})$ **Substitute.**

$V \approx \frac{528}{7}\text{ ft}^3$

$V \approx 75\frac{3}{7}\text{ ft}^3$

The volume is approximately $75\frac{3}{7}$ ft^3.

K. Find the volume of a right circular cone that has a base diameter of 3 ft and a height of 7 ft. Write the approximate answer as a mixed number.

K. Find the volume of the cone.

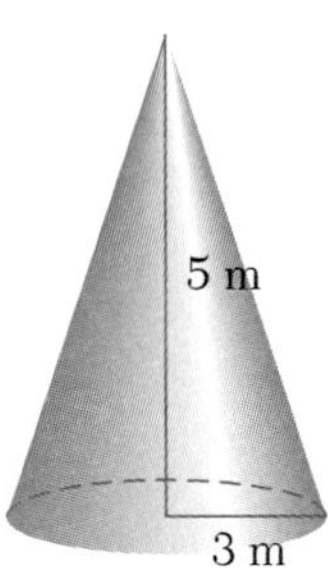

$V = \frac{1}{3}\pi r^2 h$ **Formula for cone volume.**

$V \approx \frac{1}{3}\left(\frac{22}{7}\right)(3\text{ m})^2(5\text{ m})$ **Substitute.**

$V \approx \frac{330}{7}\text{ m}^3$

$V \approx 47\frac{1}{7}\text{ m}^3$

The volume is approximately $47\frac{1}{7}$ m^3.

Answers to Warm Ups J. The volume is approximately $5883\frac{3}{7}$ cm^3. K. The volume is approximately $16\frac{1}{2}$ ft^3.

NAME ______________________ CLASS __________ DATE __________

Exercises 4.11

OBJECTIVE 1: *Do any combinations of operations with fractions.*

A.

Perform the indicated operations.

1. $\frac{5}{9} - \frac{2}{9} - \frac{1}{9}$

2. $\frac{5}{9} + \frac{3}{9} - \frac{1}{9}$

3. $\frac{1}{2} \cdot \left(\frac{3}{7} - \frac{1}{7}\right)$

4. $\frac{1}{3} \div \frac{1}{2} \cdot \frac{1}{3}$

5. $\frac{5}{6} - \frac{1}{2} \cdot \frac{2}{3}$

6. $\frac{1}{6} + \frac{1}{2} \div \frac{3}{2}$

7. $\frac{1}{4} + \frac{3}{8} \div \frac{1}{2}$

8. $\frac{1}{4} \div \frac{3}{8} + \frac{1}{2}$

9. $\frac{3}{4} \cdot \frac{1}{2} - \frac{3}{8}$

10. $\frac{5}{6} \cdot \frac{1}{3} - \frac{5}{18}$

11. $\frac{5}{8} \div \frac{3}{4} + \frac{3}{4}$

12. $\frac{1}{3} \div \frac{1}{6} + \frac{4}{9}$

13. $\frac{2}{3} + \left(\frac{1}{2}\right)^2$

14. $\frac{3}{4} - \frac{2}{3}\left(\frac{1}{2}\right)^2$

B.

15. $\frac{3}{4} \div \frac{1}{3} \cdot \frac{1}{6}$

16. $\frac{3}{4} \div \left(\frac{1}{3} \cdot \frac{1}{6}\right)$

17. $\frac{4}{5} - \frac{2}{5} \cdot \frac{1}{2}$

18. $\frac{4}{5} - \frac{2}{5} + \frac{1}{2}$

19. $\frac{5}{6} - \frac{3}{4} \div \frac{3}{2} + \frac{1}{2}$

20. $\frac{5}{8} - \frac{5}{6} \div \frac{4}{3} + \frac{1}{6}$

21. $\frac{5}{12} \cdot \frac{2}{5} + \frac{1}{3} \div \frac{1}{2} - \frac{1}{6}$

22. $\frac{4}{9} \div \frac{2}{3} - \frac{3}{4} \cdot \frac{5}{12} + \frac{1}{6}$

23. $\frac{7}{12} + \left(\frac{3}{4}\right)^2 - \frac{5}{9} \cdot \frac{33}{16}$

24. $\frac{3}{4} \cdot \frac{4}{5} - \frac{3}{0} + \left(\frac{2}{3}\right)^2$

25. $\frac{15}{16} - \left(\frac{5}{8}\right)^2 + \frac{7}{8} \div \frac{4}{3}$

26. $\frac{19}{25} - \frac{2}{5} + \left(\frac{3}{5}\right)^2 \div \frac{2}{3}$

27. $\frac{3}{8} - \left(\frac{2}{3} \div \frac{4}{5} - \frac{1}{2}\right)$

28. $\frac{3}{4} + \left(\frac{4}{5} \cdot \frac{5}{8} + \frac{2}{3}\right)$

OBJECTIVE 2: *Find the average of a set of fractions.*

A.

Find the average.

29. $\frac{1}{9}$ and $\frac{7}{9}$

30. $\frac{1}{5}, \frac{2}{5}$, and $\frac{3}{5}$

31. $\frac{2}{7}, \frac{3}{7}$, and $\frac{4}{7}$

32. $\frac{3}{11}, \frac{6}{11}$, and $\frac{9}{11}$

33. $\frac{2}{7}, \frac{3}{7}$, and $\frac{5}{7}$

34. $\frac{7}{15}, \frac{2}{15}$, and $\frac{4}{15}$

35. $\frac{2}{3}, \frac{3}{4}$, and $\frac{3}{2}$

36. $\frac{3}{4}, 1\frac{3}{4}$, and $3\frac{3}{4}$

37. $2, 3\frac{1}{2}$, and $6\frac{1}{2}$

38. $1\frac{2}{3}, 3\frac{2}{3}$, and $6\frac{2}{3}$

B.

39. $\frac{3}{4}, \frac{4}{3}, \frac{3}{2}$, and $\frac{2}{3}$

40. $\frac{5}{6}, \frac{7}{12}, \frac{5}{4}$, and $\frac{3}{2}$

41. $\frac{3}{8}, \frac{1}{4}, \frac{1}{2}$, and $\frac{3}{4}$

42. $\frac{2}{3}, \frac{5}{12}, \frac{1}{2}, \frac{3}{4}$, and $\frac{5}{6}$

43. $3\frac{2}{3}, 4\frac{5}{6}$, and $2\frac{5}{9}$

44. $6\frac{7}{8}, 8\frac{3}{4}$, and $8\frac{1}{2}$

45. $5\frac{1}{3}, 6\frac{2}{5}$, and $9\frac{13}{15}$

46. $\frac{1}{2}, 1\frac{1}{6}, \frac{7}{9}$, and $5\frac{1}{3}$

OBJECTIVE 3: *Evaluate geometric formulas that involve the number π.*

A.

47. Find the circumference of a circle with radius 7 cm.

48. Find the area of a circle with radius 7 cm.

49. Find the circumference.

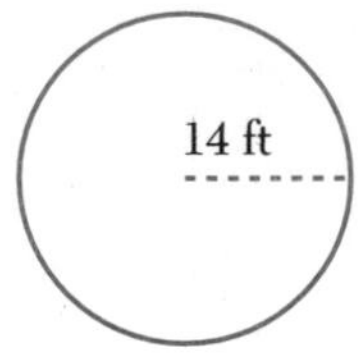

50. Find the area of the circle in Exercise 49.

51. Find the circumference.

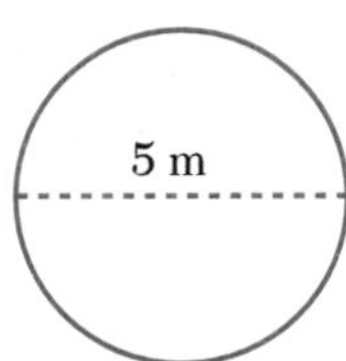

52. Find the area of the circle in Exercise 51.

53. Find the volume.

54. Find the volume.

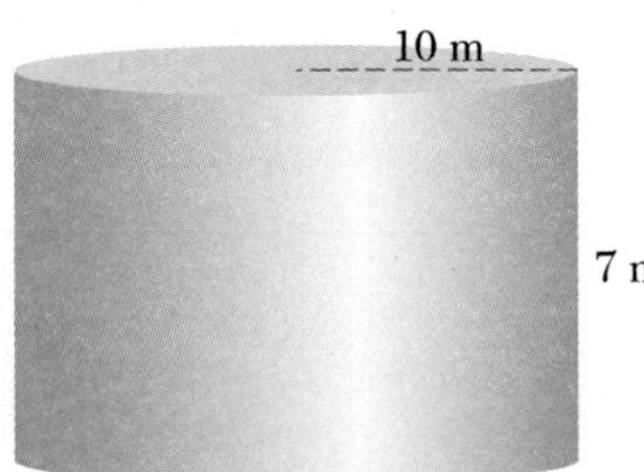

B.

55. Find the perimeter.

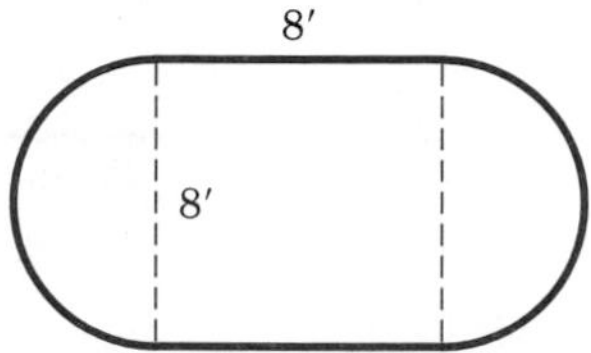

56. Find the perimeter.

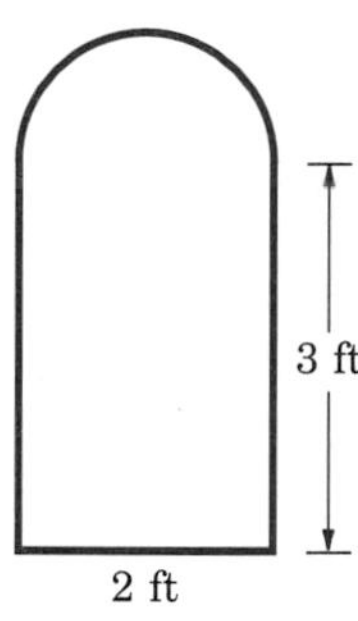

57. Find the area.

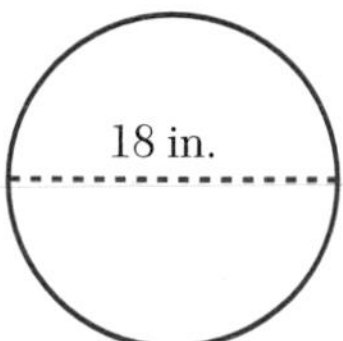

58. Find the area.

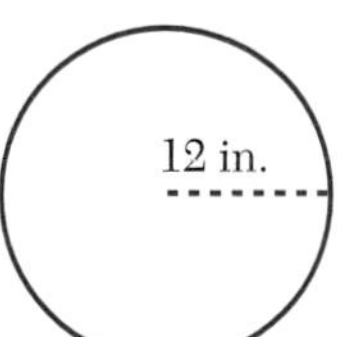

59. Find the area.

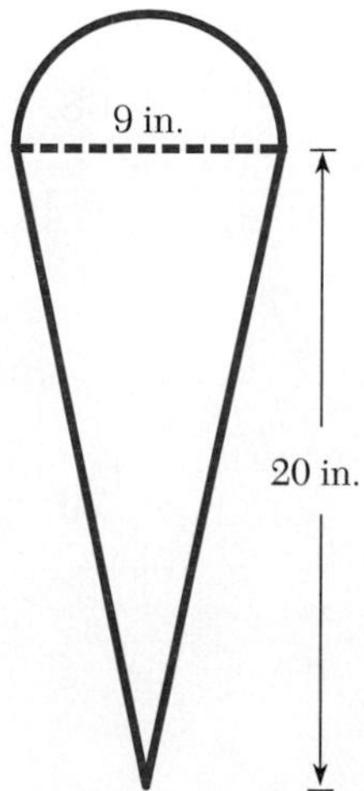

60. Find the area.

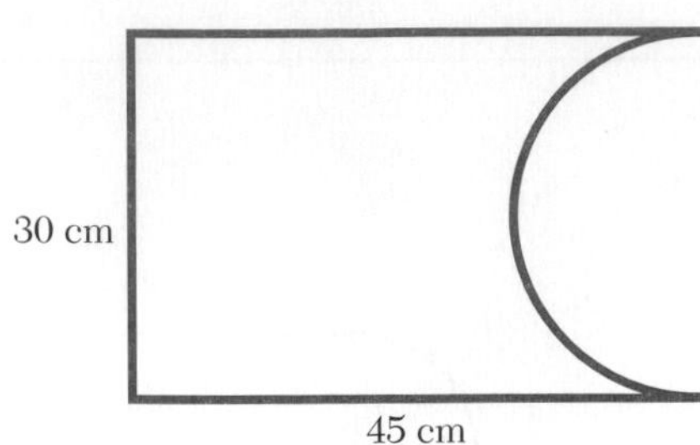

61. Find the volume of a cylinder with radius $6\frac{1}{2}$ in. and height 4 ft.

62. Find the volume of a sphere with diameter 4 cm.

C.

Perform the indicated operations.

63. $\frac{17}{9} - \left(\frac{1}{6} + \frac{1}{2} \div \frac{5}{2}\right) \cdot \frac{5}{3}$

64. $\frac{5}{2} \cdot \frac{2}{15} \div 3 + \frac{3}{4} \div \frac{3}{2}$

65. $\left(\frac{5}{8} - \frac{1}{2} \cdot \frac{3}{4}\right) \div \frac{1}{2} + \frac{1}{2}$

66. $\left(\frac{29}{12} - \frac{5}{4} \cdot \frac{3}{2}\right) \div \frac{5}{6} + \frac{3}{4}$

Find the average.

67. $3\frac{2}{3}$, $5\frac{1}{6}$, and $4\frac{2}{9}$

68. $8\frac{1}{4}$, $2\frac{4}{5}$, and $9\frac{9}{10}$

69. $\frac{7}{15}$, 9, $\frac{2}{3}$, and 4

70. 7, $\frac{7}{12}$, 13, and $\frac{7}{8}$

71. Find the area of a circle if the radius is $\frac{7}{11}$ in. Let $\pi \approx \frac{22}{7}$.

72. Find the area of a circle if the diameter is $\frac{7}{4}$ cm. Let $\pi \approx \frac{22}{7}$.

73. Wayne catches six salmon. The salmon measure $23\frac{1}{4}$ in., $31\frac{5}{8}$ in., $42\frac{3}{4}$ in., $28\frac{5}{8}$ in., $35\frac{3}{4}$ in., and 40 in. in length. What is the average length of the salmon?

74. Nurse Karla weighs five new babies at General Hospital. They weigh $6\frac{1}{2}$ lb, $7\frac{3}{4}$ lb, $9\frac{3}{8}$ lb, $7\frac{1}{2}$ lb, and $8\frac{7}{8}$ lb. What is the average weight of the babies?

75. A class of 15 students took a 10-problem quiz. Their results were as follows:

Number of Students	Fraction of Problems Correct
1	$\frac{10}{10}$ (all correct)
2	$\frac{9}{10}$
3	$\frac{8}{10}$
5	$\frac{7}{10}$
1	$\frac{6}{10}$
2	$\frac{5}{10}$
1	$\frac{2}{10}$

What was the class average?

76. On the second quiz, the class in Exercise 75 scored as follows:

Number of Students	Fraction of Problems Correct
2	$\frac{10}{10}$ (all correct)
1	$\frac{9}{10}$
2	$\frac{8}{10}$
2	$\frac{7}{10}$
4	$\frac{5}{10}$
3	$\frac{3}{10}$
1	$\frac{2}{10}$

What was the class average?

77. What is the average of the five highest scores in Exercises 75?

78. What is the average of the five highest scores in Exercise 76?

79. Kohough Inc. packs a variety carton of canned seafood. Each carton contains three $3\frac{1}{2}$-oz cans of smoked sturgeon, five $6\frac{3}{4}$-oz cans of tuna, four $5\frac{1}{2}$-oz cans of salmon, and four $10\frac{1}{2}$-oz cans of sardines. How many ounces of seafood are in the carton? If the carton sells for \$52, to the nearest cent, what is the average cost per ounce?

80. In a walk for charity, seven people walk $2\frac{7}{8}$ miles, six people walk $3\frac{4}{5}$ miles, nine people walk $4\frac{1}{4}$ miles, and five people walk $5\frac{3}{4}$ miles. What is the total number of miles walked? If the charity raises $2355, what is the average amount raised per mile, rounded to the nearest dollar?

81. Find the volume. Let $\pi \approx \frac{22}{7}$.

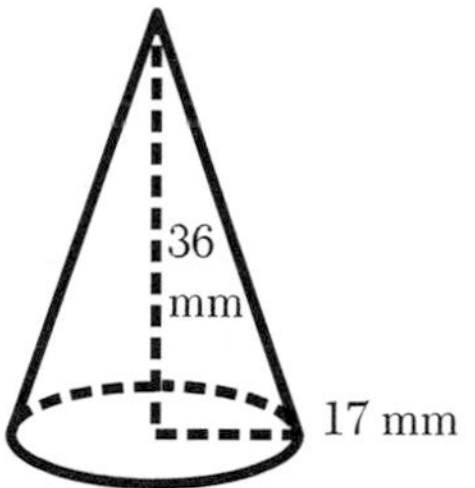

82. Find the volume of a sphere with a diameter of 2 ft 4 in. in cubic inches. Let $\pi \approx \frac{22}{7}$.

Exercises 83–84 relate to the chapter application.

83. Sandy and Hannah can fill four Christmas sampler boxes with one recipe each of nutballs, thumbprints, and lemon squares.

a. Nutballs require $1\frac{1}{2}$ c of flour, thumbprints require $2\frac{3}{4}$ c of flour, and lemon squares require 2 c of flour. What is the average amount of flour needed for the three recipes?

b. Nutballs require $\frac{3}{4}$ c of sugar, thumbprints require $1\frac{1}{2}$ c, and lemon squares require $1\frac{1}{4}$ c. What is the average amount of sugar needed for the three recipes?

84. Sandy's pecan chocolate chip cookie recipe calls for $2\frac{1}{4}$ c flour and $1\frac{1}{2}$ c sugar. Hannah's dipped butterfinger recipe calls for $1\frac{1}{2}$ c flour and $1\frac{1}{4}$ c sugar. The women make three pecan chocolate chip recipes and four dipped butterfinger recipes.

a. What is the total amount of flour needed?

b. What is the total amount of sugar needed?

STATE YOUR UNDERSTANDING

85. Write out the order of operations for fractions. How is it different from the order of operations for whole numbers?

86. Must the average of a group of numbers be larger than the smallest number and smaller than the largest number? Why?

CHALLENGE

Perform the indicated operations.

87. $2\frac{5}{8}\left(4\frac{1}{5} - 3\frac{5}{6}\right) \div 2\frac{1}{2}\left(3\frac{1}{7} + 2\frac{1}{5}\right)$

88. $1\frac{2}{5}\left(5\frac{1}{5} - 4\frac{3}{4}\right)^2 \div 4\frac{1}{2}\left(3\frac{1}{7} - 2\frac{1}{3}\right)^2$

89. The Acme Fish Company pays $1500 per ton for crab. Jerry catches $3\frac{2}{5}$ tons, his brother Joshua catches $1\frac{1}{2}$ times as many as Jerry. Their sister, Salicita, catches $\frac{7}{8}$ the amount that Joshua does. What is the total amount paid to the three people by Acme Fish Company (to the nearest dollar)?

GROUP ACTIVITY

90. A satellite travels in a circular orbit around the Earth once every $2\frac{1}{2}$ hr. The satellite is orbiting 2900 km above the surface of the Earth. The diameter of the Earth is approximately 12,800 km. Calculate the speed of the satellite and explain your strategy.

NAME ______________________ CLASS ____________ DATE ____________

MAINTAIN YOUR SKILLS (SECTIONS 3.5, 4.3, 4.7)

Divide.

91. $\frac{7}{9} \div \frac{16}{3}$

92. $\frac{7}{3} \div \frac{14}{5}$

93. $\frac{25}{32} \div \frac{15}{36}$

94. $\frac{25}{36} \div \frac{15}{32}$

Multiply.

95. $\frac{15}{28} \cdot \frac{21}{45} \cdot \frac{20}{35}$

96. $\frac{9}{15} \cdot \frac{21}{28} \cdot \frac{35}{6}$

97. Prime factor 650.

98. Prime factor 975.

99. A coffee table is made of a piece of maple that is $\frac{3}{4}$ in. thick, a piece of chipboard that is $\frac{3}{8}$ in. thick, and a veneer that is $\frac{1}{8}$ in. thick. How thick is the table top?

100. A woman works a five-day week for the following hours: $6\frac{3}{4}$ hr, $7\frac{1}{3}$ hr, $6\frac{2}{3}$ hr, $9\frac{3}{4}$ hr, and $7\frac{1}{2}$ hr. How many hours does she work for the week? What is her pay if the rate is \$$5\frac{1}{2}$ per hour?

NAME ______________________ CLASS __________ DATE __________

CHAPTER 4

Group Project *(2 weeks)*

OPTIONAL

One of the major applications of statistics is in predicting future occurrences. Before the future can be predicted, statisticians study what has happened in the past and look for patterns. If a pattern can be detected, and it is reasonable to assume that nothing will happen to interrupt the pattern, then it is a relatively easy matter to predict the future simply by continuing the pattern. Automobile insurance companies, for instance, study the occurrences of traffic accidents among various groups of people. Once they have identified a pattern, they use this to predict future accident rates, which in turn are used to set insurance rates. When a group, such as teenage boys, is identified as having a higher incidence of accidents, their insurance rates are set higher.

Dice

While predicting accident rates is a very complicated endeavor, there are other activities in which the patterns are relatively easy to find. Take, for instance, the act of rolling a die. The die has six sides, marked 1 to 6. Theoretically, each side has an equal chance of ending in the up position after a roll. Fill in the following table by rolling a die 120 times.

Side up	1	2	3	4	5	6
Times rolled						

Theoretically, each side will be rolled the same amount as the others. Since you rolled the die 120 times and there were 6 possible outcomes, each side should come up $120 \div 6 = 20$ times. How close to 20 are your outcomes in the table? What do you suppose are reasons for not getting a perfectly distributed table?

Mathematicians are likely to express the relationships in this situation using the concept of *probability*, which is a measure of the likelihood of a particular event occurring. We describe the probability of an event with a fraction. The numerator of the fraction is the number of different ways the desired event can occur, and the denominator of the fraction is the total number of possible outcomes. So the probability of rolling a 2 on the die is $\frac{1}{6}$ because there is only one way to roll a 2 but there are 6 possible outcomes when rolling a die. What is the probability of rolling a 5? What is the probability of rolling a 6? Nonmathematicians are more likely to express this relationship using the concept of *odds*. They would say that the odds of rolling a 2 are 1 in 6. This means that for every 6 times you roll a die, you can expect one of them to result in a 2.

Coin Toss

Suppose you and a friend each flip a coin. What are all the possible joint outcomes? What is the probability of getting two heads? What is the probability of getting two tails? What is the probability of getting one head and one tail?

What does it mean if the probability of an event is $\frac{3}{3}$? Is it possible for the probability of an event to be $\frac{5}{4}$? Explain.

Cards

Suppose you pick a card at random out of a deck of playing cards. What is the probability that the card will be the queen of hearts? What is the probability that the card will be a queen? What is the probability that the card will be a heart? Fill out the table below and try to discover the relationship among these three probabilities.

Probability of a Queen	Probability of a Heart	Probability of the Queen of Hearts

For a card to be the queen of hearts, two conditions must hold true at the same time. The card must be a queen *and* the card must be a heart. Make a guess about the relationship of the probabilities when two conditions must occur simultaneously. Test your guess by considering the probability of drawing a black seven. What are the two conditions that must be true in order for the card to be a black seven? What are their individual probabilities? Was your guess correct?

What two conditions must be true when you draw a red face card? What is the probability of drawing a red face card?

Suppose you pick a card at random out of a deck of playing cards. What is the probability that the card will be a three or a four? What is the probability that the card will be a three? A four? Fill in the table below to try to discover the relationship between these probabilities.

Probability of a 3	Probability of a 4	Probability of a 3 or a 4

A card is a three *or* a four if either condition holds. Make a guess about the relationship of the probabilities when either of two conditions must be true. Test your guess by calculating the probability that a card will be a heart or a club. Was your guess correct?

Sometimes a complicated probability is easier to calculate using a back-door approach. For instance, suppose you needed to calculate the probability that a card drawn is an ace or a two or a three or a four or a five or a six or a seven or an eight or a nine or a ten or a jack or a queen. You can certainly add the individual probabilities (what do you get?). However, another way to look at the situation is to ask what is the probability of not getting a king. We reason that if you do not get a king then you do get one of the desired cards. We calculate this by subtracting the probability of getting a king from 1. This is because 1 must be the sum of all the probabilities that totally define the set (in this case, the sum of the probabilities of getting a king, and the probability of getting one of the other cards). Verify that you get the same probability using both methods.

NAME ______________________ CLASS ______________ DATE ______________

CHAPTER 4

True–False Concept Review

Check your understanding of the language of basic mathematics. Tell whether each of the following statements is True (always true) or False (not always true). For each statement you judge to be false, revise it to make a statement that is true.

ANSWERS

1. It is not possible to picture an improper fraction using unit regions. **1.** ______

2. The fraction $\frac{2}{3}$ written as a mixed number is $0\frac{2}{3}$. **2.** ______

3. The whole number 1 can also be written as a proper fraction. **3.** ______

4. A fraction is another way of writing a division problem. **4.** ______

5. When a fraction is completely simplified, its value remains the same. **5.** ______

6. Every improper fraction can be simplified. **6.** ______

7. There are some fractions with large numerators and denominators that cannot be further simplified. **7.** ______

8. Two mixed numbers can be subtracted without first changing them to improper fractions. **8.** ______

9. The reciprocal of a mixed number greater than 1 is a proper fraction. **9.** ______

10. The quotient of two nonzero fractions can always be found by multiplication. **10.** ______

11. Building fractions is the opposite of simplifying fractions. **11.** ______

12. The primary reason for building fractions is so that they will have a common denominator. **12.** ______

13. Unlike fractions have different numerators. **13.** ______

14. Mixed numbers must be changed to improper fractions before adding them. **14.** ____________

15. It is sometimes necessary to use "borrowing" to subtract mixed numbers as we do when subtracting some whole numbers. **15.** ____________

16. The order of operations for fractions is the same as the order of operations for whole numbers. **16.** ____________

17. The average of three different fractions is larger than at least one of the fractions. **17.** ____________

18. The product of two fractions is always larger than the two fractions. **18.** ____________

NAME ______________________ CLASS ______________ DATE ______________

CHAPTER 4

Test

ANSWERS

1. Change $\frac{61}{16}$ to a mixed number. **1.** ______________

2. Add: $\frac{7}{8} + \frac{5}{12}$ **2.** ______________

3. Change $8\frac{7}{9}$ to an improper fraction. **3.** ______________

4. List these fractions from the smallest to the largest: $\frac{2}{5}, \frac{3}{10}, \frac{3}{8}$ **4.** ______________

5. Change 17 to an improper fraction. **5.** ______________

6. Find the missing numerator: $\frac{5}{8} = \frac{?}{56}$ **6.** ______________

7. Add: **7.** ______________

$$\begin{array}{r} 5\frac{3}{10} \\ +\ 3\frac{5}{6} \\ \hline \end{array}$$

8. Multiply. Write the result as a mixed number. $\left(4\frac{5}{8}\right)\left(3\frac{4}{5}\right)$ **8.** ______________

9. Perform the indicated operations: $\frac{3}{8} \div \frac{3}{4} + \frac{1}{4}$ 9. ________

10. Simplify $\frac{64}{96}$ completely. 10. ________

11. Subtract: $13\frac{4}{5} - 7$ 11. ________

12. Multiply: $\frac{3}{4} \cdot \frac{8}{9} \cdot \frac{15}{16}$ 12. ________

13. Subtract: $\frac{4}{5} - \frac{3}{10}$ 13. ________

14. Divide: $1\frac{1}{3} \div 3\frac{5}{9}$ 14. ________

15. Multiply: $\frac{3}{7} \cdot \frac{4}{5}$ 15. ________

16. Subtract: 16. ________

$$\begin{array}{r} 11\frac{7}{12} \\ -\ \ 4\frac{14}{15} \\ \hline \end{array}$$

17. Simplify $\frac{132}{352}$ completely. 17. ________

18. Add: $\frac{3}{35} + \frac{3}{14} + \frac{1}{10}$ 18. ________

19. What is the reciprocal of $2\frac{5}{8}$? 19. ________

20. What is the reciprocal of $\frac{5}{16}$? **20.** ____________

21. Which of these fractions are proper? $\frac{7}{8}, \frac{8}{8}, \frac{9}{8}, \frac{7}{9}, \frac{9}{7}, \frac{8}{9}, \frac{9}{9}$ **21.** ____________

22. Divide: $\frac{10}{3} \div \frac{8}{9}$ **22.** ____________

23. Subtract: **23.** ____________

$$\begin{array}{r} 11\frac{7}{10} \\ -\ 3\frac{3}{8} \\ \hline \end{array}$$

24. ____________

24. Write the fraction for the shaded part of this figure.

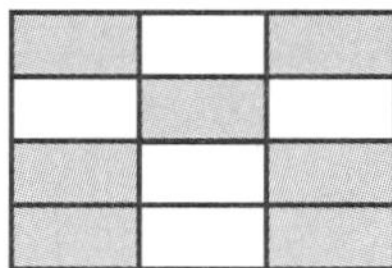

25. Subtract: $13 - 7\frac{7}{12}$ **25.** ____________

26. Add: $\frac{5}{14} + \frac{5}{14}$ **26.** ____________

27. Find the average of $1\frac{3}{4}, \frac{1}{2}, 3\frac{1}{8}$, and $\frac{1}{8}$. **27.** ____________

28. Multiply: $\left(\frac{8}{25}\right)\left(\frac{9}{16}\right)$ **28.** ____________

29. True or false? $\frac{4}{7} < \frac{11}{16}$ **29.** ____________

30. Which of the fractions represent the number 1?

$\frac{6}{5}, \frac{5}{5}, \frac{7}{6}, \frac{6}{6}, \frac{7}{7}, \frac{6}{7}, \frac{5}{7}$

30. ____________________

31. A rail car contains $126\frac{1}{2}$ tons of baled hay. A truck that is being used to unload the hay can haul $5\frac{3}{4}$ tons in one load. How many truckloads of hay are in the rail car?

31. ____________________

32. Jill wants to make up 20 bags of homemade candy for the local bazaar. Each bag will contain $1\frac{1}{4}$ lb of candy. How many pounds of candy must she make?

32. ____________________

Good Advice for Studying

Learning to Learn Math

Learning mathematics is a building process. For example, if you have not mastered fractions, rational expressions are difficult to learn because they require an understanding of the rules for fractions. Therefore, if you are having difficulty with the current topic, you may not have mastered a previous skill that you need. It will be necessary to go back and learn/relearn this skill before you can continue.

Learning math also means learning not just skills, but how and where and when to *apply* the skills. For example: if it takes 16 gallons of gas to travel 320 miles, how many miles to the gallon are you getting? What skill would you use to solve this problem? (Answer: dividing) Reading the application problems and thinking of situations where you have used or could use these concepts helps integrate the concept into your experience.

Learning mathematics is learning something basic to daily life and to virtually every field of science and business. The examples in the book state the problem and a *strategy* for solving the problem. This strategy applies to several related problems. Learning mathematics is learning strategies to solve related problems.

The more you begin to appreciate mathematics as relevant to your life, the more you will see mathematics as worthy of your time, and the more committed you will feel to studying mathematics. Here is an activity that may give you fresh opportunities to see how much mathematics relates to your life. First, create a simple "web" or "map" with "math" at the center and spokes out from this center naming areas in life where math comes up—areas where math is useful. Capture as many areas as you can in a five-minute period.

Second, turn the page over and construct a second "web," but this time choose one of your "math areas" for the center and create spokes that capture subtopics or subheadings in this particular math area. Take another five minutes for this second web. Next, create a problem that you believe to be solvable, from your own experience, and that might be enticing for someone else to solve.

If you are working with a partner, trade problems and see if you can solve each others' problems. Talk about how you might approach the problems, and whether they are stated clearly and believably. Here are some criteria for a good problem:

- Enough data and information to solve the problem
- Clear statement of what you need to find
- Not too many questions included
- Appropriate reading level and clearly written
- Appears solvable and not too scary
- Makes the reader care about wanting to solve the problem

Going through this activity may help you become more aware of mathematics in your daily life, and give you a greater understanding of problems and problem-solving.

Decimals

APPLICATION

Sports hold a universal attraction. People all over the world enjoy a good game. For some sports it is relatively easy to determine which athlete is the best. In track and swimming for instance, each contestant races against the clock and the fastest time wins. In team sports, it is easy to tell which team wins, but sometimes difficult to determine how the individual athletes compare with one another. In order to make comparisons more objective, we often use sports statistics.

The easiest kind of statistic is simply to count how many times an athlete performs a particular feat in a single game. In basketball, for instance, it is usual to count the number of points scored, the number of rebounds made, and the number of assists for each player.

Consider the following statistics of members of the Houston Rockets in their four-game sweep of Orlando in the 94–95 National Basketball Association Championship.

Player	Points Scored	Rebounds	Assists
Olajuwon	131	46	22
Drexler	86	38	27
Horry	71	40	15
Elie	65	17	13
Cassell	57	7	12
Smith	30	7	16
Brown	12	11	0

1. Which player was the best in the championship series? Why?
2. Which two players have the best statistics that are closest to each other?
3. Which is more important in basketball, rebounds or assists?

5.1

Decimals: Reading, Writing, and Rounding

OBJECTIVES

1. Write word names from place value names and place value names from word names.
2. Round a given decimal.

VOCABULARY

Decimal numbers, more commonly referred to as **decimals,** are another way of writing fractions and mixed numbers. The digits used to write whole numbers and a period called a **decimal point** are used to write place value names for these numbers.

The **number of decimal places** is the number of digits to the right of the decimal point. **Exact decimals** are decimals that show exact values. **Approximate decimals** are rounded values.

HOW AND WHY

Objective 1

Write word names from place value names and place value names from word names.

Decimals are written by using a standard place value which is an extension of the way we write whole numbers. Numbers such as 12.65, 0.45, 0.795, 1306.94, and 19.36956 are examples of decimals.

In general, the place value for decimals is

1. The same as whole numbers for digits to the left of the decimal point, and
2. A fraction whose denominator is 10, 100, 1000, and so on, for digits to the right of the decimal point.

The digits to the right of the decimal point have place values of

$$\frac{1}{10^1} = \frac{1}{10} = 0.1$$

$$\frac{1}{10^2} = \frac{1}{10 \cdot 10} = \frac{1}{100} = 0.01$$

$$\frac{1}{10^3} = \frac{1}{10 \cdot 10 \cdot 10} = \frac{1}{1000} = 0.001$$

$$\frac{1}{10^4} = \frac{1}{10 \cdot 10 \cdot 10 \cdot 10} = \frac{1}{10{,}000} = 0.0001$$

and so on, in that order from left to right.

Using the ones place as the central position, the place values of a decimal are:

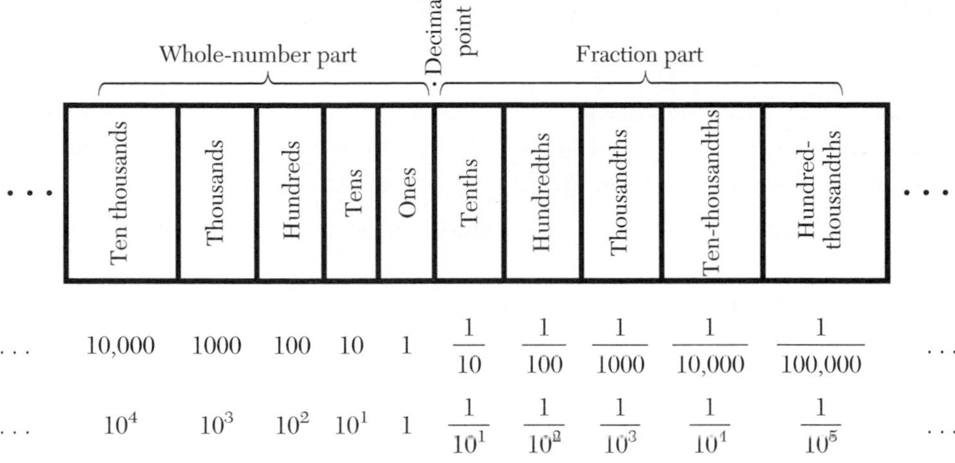

Note that the names of all the fraction parts end in "ths" and that the decimal point separates the whole-number part from the fraction part.

$$26.573 = \underbrace{26}_{\text{Whole-number part}} + \underbrace{.573}_{\text{Fraction part}}$$

If the decimal point is not written, as in the case of a whole number, the decimal point is understood to follow the ones place thus,

$23 = 23. \qquad 9 = 9. \qquad 612 = 612.$

We can write an expanded form of the decimal using fractions with denominators that are powers of 10. So

$$95.632 = 90 + 5 + \frac{6}{10} + \frac{3}{100} + \frac{2}{1000}$$

The expanded form can also be written:

9 tens + 5 ones + 6 tenths + 3 hundredths + 2 thousandths

Table 5.1 shows how to write the word names for 226.35 and 0.127.

TABLE 5.1 WORD NAMES FOR DECIMALS

	Number to Left of Decimal Point	Decimal Point	Number to Right of Decimal Point	Place Value of Last Digit
Place value name	226	.	35	$\frac{1}{100}$
Word name of each	Two hundred twenty-six	and	thirty-five	hundredths
Word name of decimal	Two hundred twenty-six and thirty-five hundredths			
Place value name	0	.	127	$\frac{1}{1000}$
Word name of each	Omit	Omit	One hundred twenty-seven	Thousandths
Word name of decimal	One hundred twenty-seven thousandths			

For numbers greater than zero and less than one (such as 0.127) the digit 0 is written in the ones place.

To write the word name for a decimal

1. Write the name for the whole number to the left of the decimal point.
2. Write the word "and" for the decimal point.
3. Write the whole-number name for the number to the right of the decimal point.
4. Write the place value of the digit farthest to the right.

If the decimal has only zero or no digit to the left of the decimal point, omit steps 1 and 2.

Some other numbers and their corresponding word names are:

Number	Word Name
10.21	Ten and twenty-one hundredths
0.723	Seven hundred twenty-three thousandths
0.00045	Forty-five hundred-thousandths
6.006	Six and six thousandths

To write the place value name for a decimal

1. Write the whole number (the number before the word "and").
2. Write a decimal point for the word "and."
3. Ignoring the place value name, write the name for the number following the word "and." Insert zeros, if necessary, between the decimal point and the digits following it to ensure that the place on the far right has the given place value.

So the place value name for three hundred ten and sixty-five thousandths is

310 **First write the whole number to the left of the word "and."**

310. **Write a decimal point for the word "and."**

310.065 **The number after the word "and" is 65. A zero is inserted to place the "5" in the thousandths place.**

Warm Ups A–D

Examples A–D

Direction: Write the word name.

Strategy: Write the word name for the whole number to the left of the decimal point. Then write "and" for the decimal point. Finally, write the word name for the number to the right of the decimal point followed by the place value of the digit farthest to the right.

A. Write the word name for 0.58.

Fifty-eight — **Write the word name for the number right of the decimal point.**

Fifty-eight hundredths — **Next, write the place value of the digit 8. The word name for 0 in the ones place may be written or omitted. "Zero and fifty-eight hundredths" is correct but unnecessary.**

A. Write the word name for 0.31.

B. Write the word name for 0.0034.

Thirty-four ten-thousandths

B. Write the word name for 0.0089.

C. Write the word name for 13.65.

Thirteen — **Write the word name for the whole number left of the decimal point.**

Thirteen and — **Write "and" for the decimal point.**

Thirteen and sixty-five — **Write the word name for the number right of the decimal point.**

Thirteen and sixty-five hundredths — **Write the place value of the digit 5.**

C. Write the word name for 35.97.

D. Janet called an employee to find the measurement of the outside diameter of a new wall clock the company is manufacturing. She asked the employee to check the plans. What is the word name the employee will read to her? The clock is shown below.

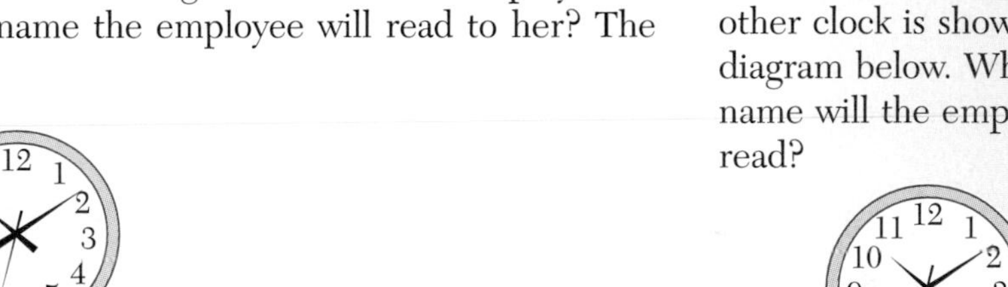

The employee will read "Nine and two hundred twenty-five thousandths inches."

D. The measurement of the outside diameter of another clock is shown in the diagram below. What word name will the employee read?

Examples E–F

Directions: Write the place value name.

Strategy: Write the digit symbols for the corresponding words. Replace the word "and" with a decimal point.

E. Write the place value name for fifteen hundred-thousandths.

15 — **First, write the number for fifteen.**

.00015 — **The place value "hundred-thousandths" indicates five decimal places, so write three zeros before the numeral fifteen and then a decimal point. This puts the numeral 5 in the hundred-thousandths place.**

0.00015 — **Since the number is between zero and one, we write a "0" in the ones place.**

Warm Ups E–F

E. Write the place value name for twenty-nine thousandths.

Answers to Warm Ups A. Thirty-one hundredths B. Eighty-nine ten-thousandths C. Thirty-five and ninety-seven hundredths D. The employee will read "Eleven and three hundred seventy-five thousandths inches." E. 0.029

F. Write the place value name for "seven hundred three and three hundred seven ten-thousandths."

F. Write the place value name for "four hundred five and four hundred five ten-thousandths."

405	**The whole number part is 405.**
405.	**Write the decimal point for "and."**
405.0405	**The "number" after "and" is 405. A zero is inserted so the numeral 5 is in the ten-thousandths place.**

HOW AND WHY

Objective 2

Round a given decimal.

Decimals can be either *exact* or *approximate*. For example, decimals that count money are exact. The figure $56.35 shows an exact amount. Most decimals that describe measurements are approximations. For example, 6.1 ft shows a person's height to the nearest tenth of a foot, and 1.9 m shows the height to the nearest tenth of a meter, but neither is an exact measure.

Decimals are rounded using the same procedure as for whole numbers. Using a ruler, we round 2.563.

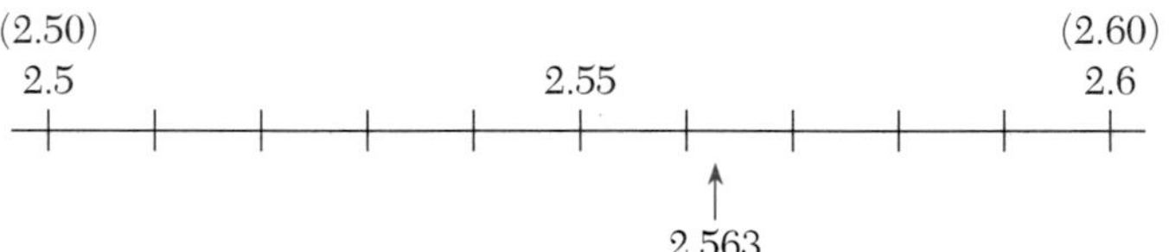

To the nearest tenth, 2.563 is rounded to 2.6, because it is closer to 2.6 than to 2.5. Rounded to the nearest hundredth, 2.563 is rounded to 2.56, because it is closer to 2.56 than to 2.57.

To round 7.4765 to the nearest hundredth, without drawing a number line, draw an arrow under the hundredths place to identify the round-off place.

7.4765

We must choose between 7.47 and 7.48. Since the digit to the right of the round-off position is 6, the number is more than halfway to 7.48. So we choose the larger number.

$7.4765 \approx 7.48$

CAUTION

Do not replace the dropped digits with zeros if the round-off place is to the right of the decimal point. $7.4765 \approx 7.4800$ indicates a round-off position of ten-thousandths.

Answer to Warm Up F. 703.0307

To round a decimal number to a given place value

1. Draw an arrow under the given place value. (After enough practice, you will be able to round mentally and will not need the arrow.)
2. If the digit to the right of the arrow is 5, 6, 7, 8, or 9, add 1 to the digit above the arrow. That is, round to the larger number.
3. If the digit to the right of the arrow is 0, 1, 2, 3, or 4, keep the digit above the arrow. That is, round to the smaller number.
4. Write whatever zeros are necessary after the arrow so that the number above the arrow has the same place value as the original. See Example H.

This method is sometimes called the "four-five" rule. Although this rounding procedure is the most commonly used, it is not the only way to round. Many government agencies round by *truncation;* that is, by dropping the digit after the decimal point. Thus, $87.32 ≈ $87. It is common for retail stores to round up for any amounts smaller than one cent. Thus, $3.553 ≈ $3.56. There is also a rule for rounding numbers in science, which is sometimes referred to as the "even-odd" rule. You might learn and use a different round-off rule depending on what kind of work you are doing.

Examples G–J

Directions: Round as indicated.

Strategy: Draw an arrow under the given place value. Examine the digit to the right of the arrow to determine whether to round up or down.

G. Round 0.5928 to the nearest hundredth.

0.5928 ≈ 0.59 **The digit to the right of the round-off place is 2, so round down.**
↑

H. Round 5582.9 to the nearest thousand.

5582.9 ≈ 6000 **Three zeros must be written after the 6 to keep it in the thousands place.**
↑

I. Round 389.6 to the nearest unit.

389.6 ≈ 390 **The number to the right of the round-off place is 6, so we round up by adding 389 + 1 = 390.**
↑

J. Round 45.3737 and 8.9973 to the nearest unit, the nearest tenth, the nearest hundredth, and the nearest thousandth.

	Unit		Tenth		Hundredth		Thousandth
45.3737 ≈	45	≈	45.4	≈	45.37	≈	45.374
8.9973 ≈	9	≈	9.0	≈	9.00	≈	8.997

CAUTION

The zeros following the decimal in 9.0 and 9.00 are necessary to show that the original was rounded to the nearest tenth and hundredth, respectively.

Warm Ups G–J

G. Round 0.69288 to the nearest thousandth.

H. Round 663.89 to the nearest ten.

I. Round 4519.8 to the nearest unit.

J. Round 12.8947 to the nearest unit, to the nearest tenth, to the nearest hundredth, and the nearest thousandth.

Answers to Warm Ups G. 0.693 H. 660 I. 4520 J. 13; 12.9; 12.89; 12.895

NAME ______________________ CLASS ______________ DATE ______________

Exercises 5.1

OBJECTIVE 1: *Write word names from place value names and place value names from word names.*

A.

Write the word name.

1. 0.12

2. 0.34

3. 0.267

4. 0.712

5. 6.0004

6. 5.3002

7. 4.67

8. 13.09

Write the place value name.

9. Eleven hundredths

10. Forty-five hundredths

11. One hundred eleven thousandths

12. Five hundred fourteen thousandths

13. Two and nineteen thousandths

14. One and six hundredths

15. Twenty-one ten-thousandths

16. Fifteen ten-thousandths

B.

Write the word name.

17. 0.504

18. 5.04

19. 50.04

20. 5.004

21. 18.0205

22. 45.0051

23. 30.008

24. 300.080

Write the place value name.

25. Twelve thousandths

26. Twelve thousand

27. Seven hundred and ninety-six thousandths

28. Six hundred and seven thousandths

29. Five hundred five and five thousandths

30. Five and five hundred five thousandths

31. Fourteen and fourteen ten-thousandths

32. One hundred four and one hundred four thousandths

OBJECTIVE 2: *Round a given decimal.*

A.

Round to the nearest unit, tenth, and hundredth.

	Unit	Tenth	Hundredth
33. 15.888			
34. 51.666			
35. 477.774			
36. 344.333			
37. 0.7392			
38. 0.92937			

Round to the nearest cent.

39. \$33.5374

40. \$84.4167

41. \$246.4936

42. \$368.1625

B.

Round to the nearest ten, hundredth, and thousandth.

	Ten	Hundredth	Thousandth
43. 12.5532			
44. 21.3578			
45. 245.2454			
46. 118.0752			
47. 0.5536			
48. 0.9695			

NAME ______________________ CLASS ____________ DATE ____________

Round to the nearest dollar.

49. $10.78

50. $15.49

51. $1129.38

52. $3178.48

C.

Use the graph to answer Exercises 53–56. The graph shows the precipitation for four weeks in a southern city.

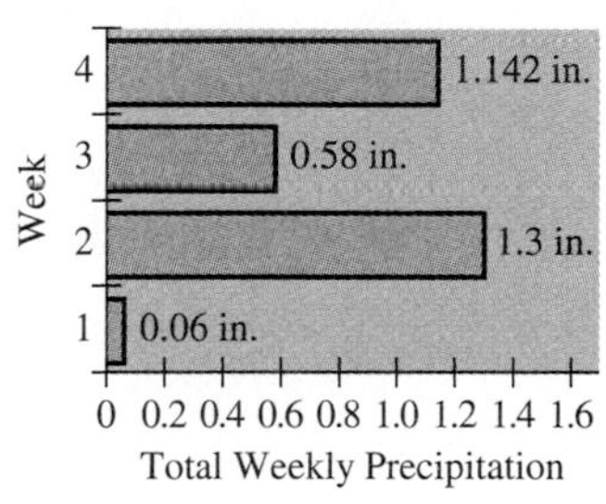

53. Write the word name for the number of inches of precipitation during the second week.

54. Write the word name for the number of inches of precipitation during the fourth week.

55. Round the amount of precipitation in week four to the nearest tenth of an inch.

56. Round the amount of precipitation in week one to the nearest tenth of an inch.

57. Dan buys a deep fryer that has a marked price $64.79. What word name does he write on the check?

58. Fari buys a truckload of organic fertilizer for her yard. The price of the load is $106.75. What word name does she write on the check?

Exercises 59–62 refer to the figure below.

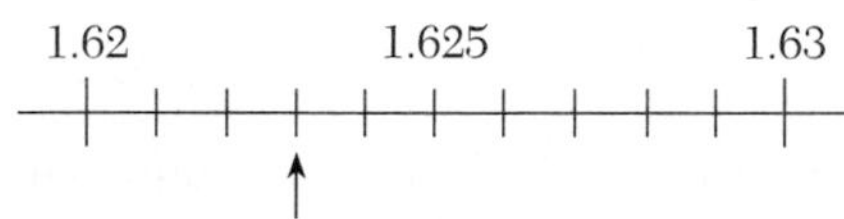

59. What is the position of the arrow to the nearest hundredth?

60. What is the position of the arrow to the nearest thousandth?

61. What is the position of the arrow to the nearest tenth?

62. What is the position of the arrow to the nearest unit?

Write the word name.

63. 567.9023

64. 1001.1001

65. The computer at Grant's savings company shows that his account, including the interest he has earned, has a value of $1617.37099921. Round the value of the account to the nearest cent.

66. In doing her homework, Catherine's calculator shows the answer to a division exercise is 34.78250012. If she is to round the answer to the nearest thousandth, what answer does she report?

Write the place value name.

67. Two hundred thirteen and one thousand one hundred one ten-thousandths

68. Four thousand five and four hundred fifteen hundred-thousandths

Round to the indicated place value.

	Hundred	Hundredth	Ten-thousandth
69. 2567.90347			
70. 982.678456			
71. 5790.350024			
72. 86025.47782			

Exercises 73–75 relate to the chapter application.

73. In January 1906 a Stanley car with a steam engine set a one-mile speed record by going 127.659 miles per hour. Round this rate to the nearest tenth of a mile per hour.

74. In March 1927 a Sunbeam set a one-mile speed record by going 203.790 mph. What place value was this rate rounded to?

75. In October 1970 a Blue Flame set a one-mile speed record by going 622.407 mph. Explain why it is incorrect to round the rate to 622.5 mph.

STATE YOUR UNDERSTANDING

76. Explain the difference between an exact decimal value and an approximate decimal value. Give an example of each.

77. Explain in words, the meaning of the value of the 4s in the numerals 43.29 and 18.64. Include some comment on how and why the values of the digit 4 are alike and how and why they are different.

78. Consider the decimal represented by abc.defg. Explain how to round this number to the nearest hundredth.

CHALLENGE

79. Round 8.28282828 to the nearest thousandth. Is the rounded value less than or greater than the original value? Write an inequality to illustrate your answer.

80. Round 7.7777777 to the nearest unit. Is the rounded value less than or greater than the original value? Write an inequality to illustrate your answer.

81. Write the place value name for "Two hundred fifteen ten-millionths."

82. Write the word name for 40715.300519.

GROUP ACTIVITY

83. Discuss with the members of the group cases in which you think rounding by "truncating" is the best way to round. Rounding by truncating means to drop all digit values to the right of the rounding position. For example 2.77 ≈ 2, $19.33 ≈ $19, 34,999 ≈ 34,000. Can each of the groups think of a situation in which such rounding is actually used?

MAINTAIN YOUR SKILLS (SECTIONS 4.1, 4.3, 4.4, 4.7)

Change to a mixed number.

84. $\frac{58}{17}$

85. $\frac{124}{15}$

Change to an improper fraction.

86. $14\frac{10}{11}$

87. $27\frac{3}{5}$

Divide.

88. $\dfrac{15}{4} \div \dfrac{5}{8}$

89. $\dfrac{14}{15} \div \dfrac{35}{40}$

Add and simplify.

90. $\dfrac{23}{48} + \dfrac{13}{48}$

91. $\dfrac{18}{55} + \dfrac{8}{55} + \dfrac{9}{55}$

92. A spring has 15 coils. Each coil requires $2\frac{1}{5}$ in. of wire. How many inches of wire does it take to make one spring?

93. How many springs like those in Exercise 92 can be made from 200 in. of wire?

5.2

Changing Decimals to Fractions, Listing in Order

OBJECTIVES

1. Change a decimal to a fraction.
2. List a set of decimals from smallest to largest.

HOW AND WHY

Objective 1

Change a decimal to a fraction.

The word name of a decimal is also the word name of a fraction.

Consider 0.575.

READ: five hundred seventy-five thousandths

WRITE: $\frac{575}{1000}$

So, $0.575 = \frac{575}{1000} = \frac{23}{40}$

Some other examples are shown in the table.

Place Value Name	Word Name	Fraction
0.21	Twenty-one hundredths	$\frac{21}{100}$
0.3	Three tenths	$\frac{3}{10}$
0.125	One hundred twenty-five thousandths	$\frac{125}{1000} = \frac{1}{8}$
0.5	Five tenths	$\frac{5}{10} = \frac{1}{2}$

To change a decimal to a fraction

1. Read the decimal word name.
2. Write the fraction that has the same value.
3. Simplify.

Notice that because of place value, the number of decimal places in a decimal tells us the number of zeros in the denominator of the fraction. This fact can be used as another way to write the fraction or to check that the fraction is correct:

$$\underbrace{2.78}_{\text{Two decimal places}} = 2\frac{78}{\underbrace{100}_{\text{Two zeros}}} = 2\frac{39}{50} \quad \text{or} \quad 2.78 = \frac{278}{100} = 2\frac{39}{50}$$

Warm Ups A–C

A. Change 0.51 to a fraction.

B. Write 0.625 as a fraction.

C. Write as a mixed number: 112.35

Examples A–C

Directions: Change the decimal to a fraction or mixed number.

Strategy: Say the word name to yourself and write the fraction or mixed number that is equivalent. Simplify.

A. Change 0.83 to a fraction.

Eighty-three hundredths **Word name.**

$\frac{83}{100}$ **Write as a fraction.**

B. Write 0.475 as a fraction.

Four hundred seventy-five thousandths **Word name.**

$\frac{475}{1000} = \frac{19}{40}$ **Write as a fraction and simplify.**

C. Write as a mixed number: 45.06

Forty-five and six hundredths **Word name.**

$45\frac{6}{100} = 45\frac{3}{50}$ **Write as a mixed number and simplify.**

HOW AND WHY

Objective 2

List a set of decimals from smallest to largest.

Fractions can be listed in order, when they have a common denominator, by ordering the numerators. This idea can be extended to decimals when they have the same number of decimal places. For instance, $0.26 = \frac{26}{100}$ and $0.37 = \frac{37}{100}$ have a common denominator when written in fraction form. So 0.26 is less than 0.37; or $0.26 < 0.37$.

We can see that $1.5 < 3.6$ because 1.5 is to the left of 3.6 on the number line.

The decimals 0.3 and 0.15 have a common denominator when a zero is placed after the 3. Thus

$$0.3 = \frac{3}{10} = \frac{3}{10} \cdot \frac{10}{10} = \frac{30}{100} = 0.30$$

Answers to Warm Ups A. $\frac{51}{100}$ B. $\frac{5}{8}$ C. $112\frac{7}{20}$

so that

$$0.3 = \frac{30}{100} \quad \text{and} \quad 0.15 = \frac{15}{100}$$

Then, since $\frac{15}{100} < \frac{30}{100}$, we conclude that $0.15 < 0.3$.

There are many forms for decimal numbers that are equivalent. For example,

$6.3 = 6.30 = 6.300 = 6.3000 = 6.30000$

$0.85 = 0.850 = 0.8500 = 0.85000 = 0.850000$

$45.982 = 45.9820 = 45.98200 = 45.982000 = 45.9820000$

The zeros to the right of the decimal point following the last nonzero digit do not change the value of the decimal. Usually these extra zeros are not written, but they are useful when operating with decimals.

To list a set of decimals from smallest to largest

1. Make sure that all numbers have the same number of decimal places to the right of the decimal point by writing zeros to the right of the last digit when necessary.
2. Write the numbers in order as if they were whole numbers.
3. Remove the extra zeros.

Examples D–F

Directions: Is the statement true or false?

Strategy: Write each numeral with the same number of decimal places. Compare the values without regard to the decimal point.

D. True or false: $0.55 > 0.52$

$55 > 52$ is true. **Compare the numbers without regard to the decimal point.**

So, $0.55 > 0.52$ is true.

E. True or false: $0.812 < 0.81$

$0.812 < 0.810$ **Write with the same number of decimal places.**

$812 < 810$ is false. **Without regard to the decimal point, 812 is larger.**

So, $0.812 < 0.81$ is false.

F. True or false: $12.007 > 12.03$

$12.007 > 12.030$ **Write with the same number of decimal places.**

$12007 > 12030$ is false.

So, $12.007 > 12.03$ is false.

Warm Ups D–F

D. True or false: $0.23 < 0.22$

E. True or false: $0.61 < 0.6089$

F. True or false: $33.7 > 33.698$

Answers to Warm Ups D. False E. False F. True

Warm Ups G–I

G. List 0.65, 0.592, 0.648, and 0.632 from smallest to largest.

H. List 1.03, 1.0033, 1.0333, and 1.0303 from smallest to largest.

I. Roberto and Rachel measure the same coin. Roberto measures 0.879 and Rachel measures 0.7969. Whose measure is larger?

Examples G–I

Directions: List the decimals from smallest to largest.

Strategy: Write zeros on the right so that all numbers have the same number of decimal places. Compare the numbers as if they were whole numbers and then remove the extra zeros.

G. List 0.52, 0.537, 0.5139, and 0.521 from smallest to largest.

0.5200 0.5370 0.5139 0.5210	**First, write all numbers with the same number of decimal places by inserting zeros on the right.**
0.5139, 0.5200, 0.5210, 0.5370	**Second, write the numbers in order as if they were whole numbers.**
0.5139, 0.52, 0.521, 0.537	**Third, remove the extra zeros.**

H. List 9.357, 9.361, 9.3534, and 9.358 from smallest to largest.

9.357 = 9.3570 9.361 = 9.3610 9.3534 = 9.3534 9.358 = 9.3580	**Step 1.**
9.3534, 9.3570, 9.3580, 9.3610	**Step 2.**
9.3534, 9.357, 9.358, 9.361	**Step 3.**

I. Using a micrometer to measure the diameter of a foreign coin, Mike measures 0.8841 and Mildred measures 0.883. Whose measure is larger?

0.8841, 0.8830 **Write with the same number of decimal places and compare as whole numbers.**

Mike's measure is larger because 8841 > 8830.

Answers to Warm Ups G. 0.592, 0.632, 0.648, 0.65 H. 1.0033, 1.03, 1.0303, 1.0333 I. Roberto's measure is larger.

NAME ______________________ CLASS ____________ DATE ____________

Exercises 5.2

OBJECTIVE 1: *Change a decimal to a fraction.*

A.

Change each decimal to a fraction and simplify if possible.

1. 0.33 **2.** 0.97 **3.** 0.75 **4.** 0.8

5. One hundred eleven thousandths **6.** Five hundred thirteen thousandths

7. 0.34 **8.** 0.98 **9.** 0.48 **10.** 0.55

B.

Change the decimal to a fraction or mixed number and simplify.

11. 2.73 **12.** 16.47 **13.** 0.875 **14.** 0.375

15. 3.64 **16.** 17.88 **17.** 3.564 **18.** 7.450

19. Two hundred thousandths **20.** Five hundred-thousandths

OBJECTIVE 2: *List a set of decimals from smallest to largest.*

A.

List the set of decimals from smallest to largest.

21. 0.6, 0.7, 0.1 **22.** 0.07, 0.03, 0.025

23. 0.05, 0.6, 0.07 **24.** 0.04, 0.1, 0.01

25. 4.16, 4.161, 4.159

26. 7.18, 7.183, 7.179

Is the statement true or false?

27. $0.61 < 0.6$

28. $0.44 < 0.04$

29. $9.54 > 9.45$

30. $3.41 > 3.14$

B.

List the set of decimals from smallest to largest.

31. 0.0729, 0.073001, 0.072, 0.073, 0.073015

32. 3.009, 0.301, 0.3008, 0.30101

33. 0.888, 0.88799, 0.8881, 0.88579

34. 8.36, 8.2975, 8.3599, 8.3401

35. 20.004, 20.04, 20.039, 20.093

36. 71.4506, 71.0456, 71.0546, 71.6405

NAME ______________________ CLASS ______________ DATE ______________

Is the statement true or false?

37. 3.1231 < 3.1213

38. 4.1243 > 4.124

39. 13.1204 < 13.2014

40. 53.1023 > 53.1203

C.

41. The probability that a flipped coin will come up heads three times in a row is 0.125. Write this as a reduced fraction.

42. The probability that a flipped coin will come up heads twice and tails once out of three flips is 0.375. Write this as a reduced fraction.

Exercises 43–45 refer to the following graph:

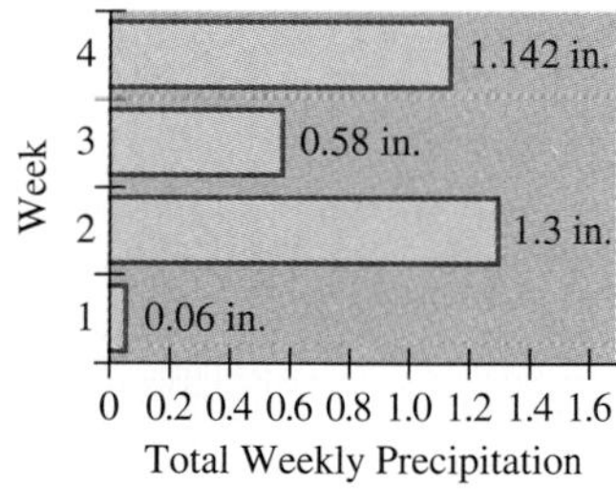

43. Write a simplified fraction to show the number of inches of precipitation in the first week.

44. Write a mixed number to show the number of inches of precipitation in the fourth week.

45. Which week has the least precipitation? The most precipitation?

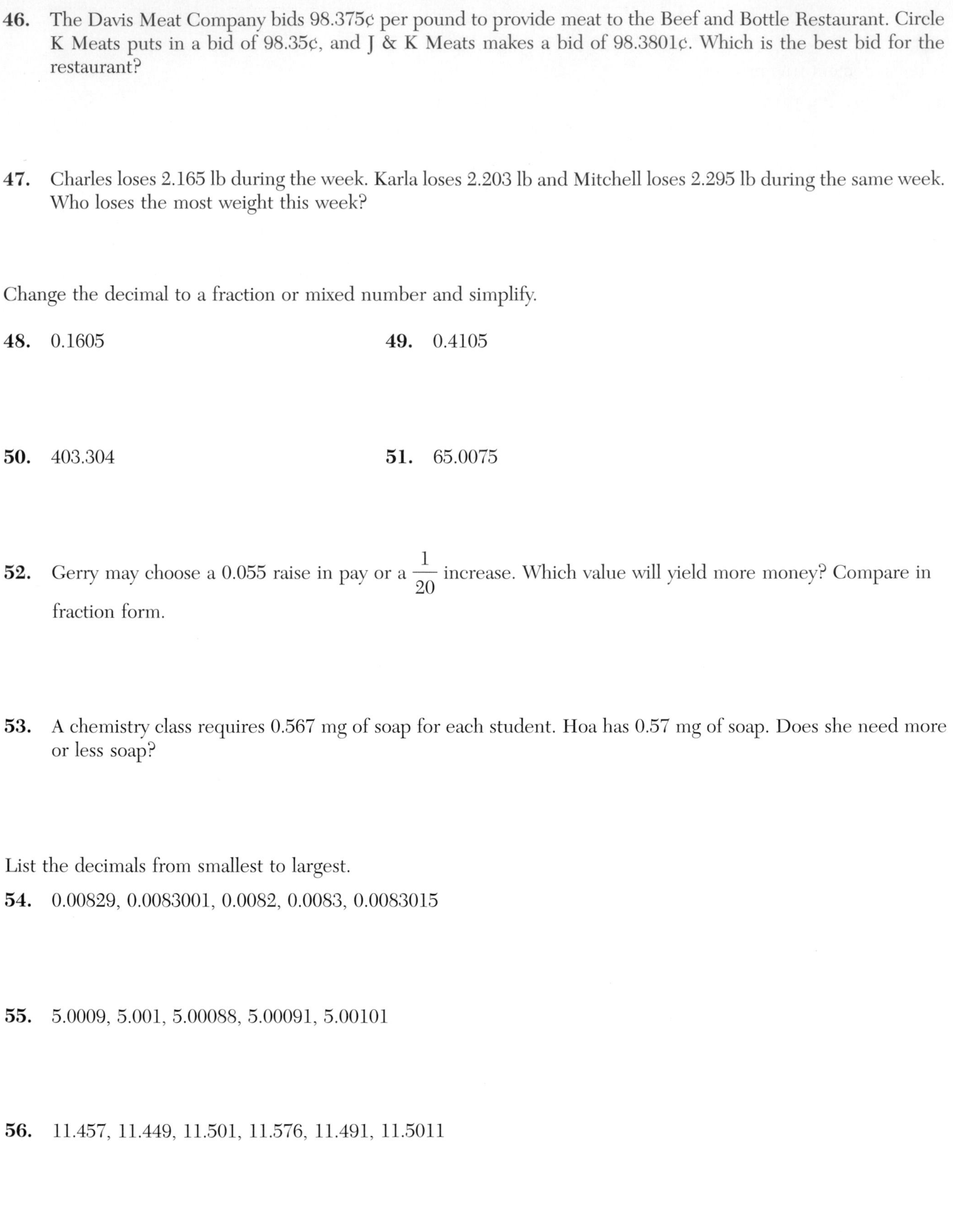

46. The Davis Meat Company bids 98.375¢ per pound to provide meat to the Beef and Bottle Restaurant. Circle K Meats puts in a bid of 98.35¢, and J & K Meats makes a bid of 98.3801¢. Which is the best bid for the restaurant?

47. Charles loses 2.165 lb during the week. Karla loses 2.203 lb and Mitchell loses 2.295 lb during the same week. Who loses the most weight this week?

Change the decimal to a fraction or mixed number and simplify.

48. 0.1605

49. 0.4105

50. 403.304

51. 65.0075

52. Gerry may choose a 0.055 raise in pay or a $\frac{1}{20}$ increase. Which value will yield more money? Compare in fraction form.

53. A chemistry class requires 0.567 mg of soap for each student. Hoa has 0.57 mg of soap. Does she need more or less soap?

List the decimals from smallest to largest.

54. 0.00829, 0.0083001, 0.0082, 0.0083, 0.0083015

55. 5.0009, 5.001, 5.00088, 5.00091, 5.00101

56. 11.457, 11.449, 11.501, 11.576, 11.491, 11.5011

57. 23.98, 22.01, 24.99, 23.67, 22.89, 23.86, 22.81, 23.76

58. Maria is using a pattern to make her wedding dress. The pattern calls for 9.4375 yd of material. Change the measure to a mixed number and simplify.

59. For a bridesmaid's bow Maria may choose 0.825 yd or $\frac{6}{7}$ yd for the same price. Which should she choose to get the most fabric? Compare in fraction form.

Exercises 60–61 relate to the chapter application.

60. At one point in the 1997–98 NBA season the listed teams had won the given decimal fraction of their games: Seattle, 0.789; L.A. Lakers, 0.757; San Antonio, 0.684; Utah, 0.667; Miami, 0.667; New York, 0.556; Indiana, 0.694; Chicago, 0.692. Rank these teams from best record to the worst record.

61. At one point in the 1997–98 NBA season Tim Hardaway had hit 0.835 of his free throws. Express his record as a fraction. What fraction of his free throws did he miss?

62. Betty Crocker cake mixes, when prepared as directed, have the following decimal fraction of the calories per slice from fat: Apple Cinnamon, 0.36; Butter Pecan, 0.4; Butter Recipe/Chocolate, 0.43; Chocolate Chip, 0.42; Spice, 0.38; Golden Vanilla, 0.45. If each slice contains 280 calories, which cake has the most calories from fat? Least calories from fat?

63. Hash brown potatoes have the following number of fat grams per serving: frozen plain, 8.95 g, frozen with butter sauce, 8.9 g; and homemade with vegetable oil, 10.85 g. Write the fat grams as mixed numbers and simplify. Which serving of hash browns has the least amount of fat?

STATE YOUR UNDERSTANDING

64. Explain how the number line can be a good visual aid for determining which of two numerals has the larger value.

CHALLENGE

65. Change 0.44, 0.404, and 0.04044 to fractions and simplify.

66. Determine whether each statement is true or false.

a. $7.44 < 7\frac{7}{18}$ b. $8.6 > 8\frac{5}{9}$ c. $3\frac{2}{7} < 3.285$ d. $9\frac{3}{11} > 9.271$

GROUP ACTIVITY

67. Find a pattern, a vehicle manual, and/or a parts list whose measurements are given in decimal form. Change the measurements to fraction form.

68. Have each member of the group write one fraction and one decimal each with values between 3 and 4. Then as a group, list all of the fractions and decimal values from smallest to largest.

MAINTAIN YOUR SKILLS (SECTIONS 4.3, 4.4)

Perform the indicated operations.

69. $\frac{25}{27} \cdot \frac{18}{35} \cdot \frac{7}{15}$

70. $\frac{36}{75} \cdot \frac{15}{16} \cdot \frac{40}{27}$

71. $4\frac{2}{5} \cdot 2\frac{4}{5}$

72. $4\frac{2}{5} \div 2\frac{4}{5}$

73. $\left(3\frac{1}{2}\right)\left(6\frac{3}{4}\right)$

74. $\left(7\frac{2}{3}\right)\left(8\frac{3}{4}\right)$

75. $\left(16\frac{2}{3}\right) \div \left(4\frac{1}{6}\right)$

76. $\left(16\frac{1}{6}\right) \div \left(4\frac{2}{3}\right)$

77. $\left(12\frac{1}{2}\right) \div \left(1\frac{1}{4}\right)$

78. $\left(12\frac{1}{4}\right) \div \left(1\frac{1}{2}\right)$

5.3

Adding and Subtracting Decimals

OBJECTIVES

1. Add decimals.
2. Subtract decimals.
3. Estimate the sum or difference of decimals.

HOW AND WHY

Objective 1

Add decimals.

What is the sum of 6.3 + 2.5? We make use of the expanded form of the decimal to explain addition.

$$\begin{array}{rl} 6.3 = & 6 \text{ ones} + 3 \text{ tenths} \\ + \underline{2.5} = & \underline{2 \text{ ones} + 5 \text{ tenths}} \\ & 8 \text{ ones} + 8 \text{ tenths} = 8.8 \end{array}$$

We use the same principle for adding decimals that we use for whole numbers. That is, we add like place values. The vertical form gives us a natural grouping of the ones and tenths. By inserting zeros so all the numbers have the same number of decimal places, the addition 2.8 + 13.4 + 6.22 is written 2.80 + 13.40 + 6.22.

$$\begin{array}{r} 2.80 \\ 13.40 \\ +\ \underline{6.22} \\ 22.42 \end{array}$$

To add decimals

1. Write in columns with the decimal points aligned. Insert extra zeros to help align the place values.
2. Add the decimals as if they were whole numbers.
3. Align the decimal point in the sum with those above.

Examples A–C

Directions: Add.

Strategy: Write each numeral with the same number of decimal places, align the decimal points, and add.

A. Add: 1.3 + 21.41 + 32 + 0.05

$$\begin{array}{r} 1.30 \\ 21.41 \\ 32.00 \\ +\ \underline{0.05} \\ 54.76 \end{array}$$

Write each numeral with two decimal places. The extra zeros help line up the place values.

Warm Ups A–C

A. Add: 2.4 + 37.52 + 19 + 0.08

Answer to Warm Up A. 59

B. Add: 8.4068 + 0.0229 + 4.56 + 34.843

Calculator Example

B. Add: 6.3975 + 0.0116 + 3.41 + 18.624

Strategy: The extra zeros do not need to be inserted. The calculator will automatically align the like place values when adding.
The sum is 28.4431.

C. What is the total cost of a pair of emerald earrings if the retail price is $103.95, the federal tax is $4.16, the state sales tax is $8.32, and the city sales tax is $0.83?

C. What is the total cost of an automobile tire if the retail price is $67.95, the federal excise tax is $2.72, the state sales tax is $5.44, and the local sales tax is $0.68?

Strategy: Add the retail price and the taxes.

$$\begin{array}{r} \$67.95 \\ 2.72 \\ 5.44 \\ +\ \underline{0.68} \\ \$76.79 \end{array}$$

The tire costs $76.79.

HOW AND WHY

Objective 2

Subtract decimals.

What is the difference of 6.59 − 2.34? To find the difference we write the numbers in column form aligning the decimal points. Now subtract as if they were whole numbers.

$$\begin{array}{r} 6.59 \\ -\underline{2.34} \\ 4.25 \end{array}$$

The decimal point in the difference is aligned with those above.

When necessary, we can regroup, or "borrow," as with whole numbers. What is the difference 6.271 − 3.845?

$$\begin{array}{r} 6.271 \\ -\underline{3.845} \end{array}$$

We need to borrow one from the hundredths column (1 hundredth = 10 thousandths) and we need to borrow one from the ones column (1 one = 10 tenths).

$$\begin{array}{r} {\scriptstyle 5\ \ 12\ 6\ 11} \\ \not{6}.\ \not{2}\ \not{7}\ \not{1} \\ -\underline{3.\ 8\ 4\ 5} \\ 2.\ 4\ 2\ 6 \end{array}$$

So the difference is 2.426.

Sometimes it is necessary to write zeros on the right so the numbers have the same number of decimal places. See Example E.

To subtract decimals

1. Write the decimals in columns with the decimal points aligned. Insert extra zeros to align the place values.
2. Subtract the decimals as if they were whole numbers.
3. Align the decimal point in the difference with those above.

Answers to Warm Ups B. 47.8327 C. The earrings cost $117.26.

Examples D–H

Directions: Subtract.

Strategy: Write each numeral with the same number of decimal places, align the decimal points, and subtract.

D. Subtract: 5.831 − 0.287

$$\begin{array}{r} 5.831 \\ -0.287 \\ \hline \end{array}$$

Line up the decimal points so the place values are aligned.

$$\begin{array}{r} 5.\,8\,\overset{2}{\not 3}\,\overset{11}{\not 1} \\ -0.\,2\,8\,7 \\ \hline \end{array}$$

Borrow 1 hundredth from the 3 in the hundredths place. (1 hundredth = 10 thousandths)

$$\begin{array}{r} 5.\,\overset{7}{\not 8}\,\overset{12}{\not 2}\,\overset{11}{\not 1} \\ -0.\,2\,8\,7 \\ \hline 5.\,5\,4\,4 \end{array}$$

Borrow 1 tenth from the 8 in the tenths place. (1 tenth = 10 hundredths)

CHECK:

$$\begin{array}{r} 0.287 \\ +5.544 \\ \hline 5.831 \end{array}$$

Check by adding.

The difference is 5.544.

E. Subtract 2.94 from 6.

$$\begin{array}{r} 6.\,0\,0 \\ -2.\,9\,4 \\ \hline \end{array}$$

We write 6 as 6.00 so that both numerals will have the same number of decimal places.

$$\begin{array}{r} \overset{5}{\not 6}.\,\overset{10}{\not 0}\,0 \\ -2.\,9\,4 \\ \hline \end{array}$$

We need to borrow to subtract in the hundredths place. Since there is a 0 in the tenths place, we start by borrowing 1 from the ones place. (1 one = 10 tenths)

$$\begin{array}{r} \overset{5}{\not 6}.\,\overset{9}{\not 10}\,\overset{10}{\not 0} \\ -2.\,9\,4 \\ \hline 3.\,0\,6 \end{array}$$

Now borrow 1 tenth to add to the hundredths place. (1 tenth = 10 hundredths) Subtract.

CHECK:

$$\begin{array}{r} 2.94 \\ +3.06 \\ \hline 6.00 \end{array}$$

The difference is 3.06.

Warm Ups D–H

D. Subtract: 7.946 − 0.378

E. Subtract 5.736 from 7.

Answers to Warm Ups D. 7.568 E. 1.264

F. Find the difference of 9.382 and 5.736. Round to the nearest tenth.

F. Find the difference of 6.271 and 3.845. Round to the nearest tenth.

$$\begin{array}{r} \scriptstyle 5\ \ 12\,6\,11 \\ \not{6}.\not{2}\,\not{7}\,\not{1} \\ -3.\,8\,4\,5 \\ \hline 2.\,4\,2\,6 \end{array}$$

The check is left for the student.

The difference is 2.4 to the nearest tenth.

CAUTION

Do not round before subtracting. Note the difference if we do:

6.3 − 3.8 = 2.5

Calculator Example

G. Subtract: 540.7445 − 445.895

G. Subtract: 345.9673 − 298.893

Strategy: The calculator automatically lines up the decimal points.

The difference is 47.0743.

H. Mickey buys a video for $18.69. She gives the clerk a $20 bill to pay for the cassette. How much change does she get?

H. Marta purchases a small radio for $33.89. She gives the clerk two $20 bills to pay for the radio. How much change does she get?

Strategy: Since two $20 bills are worth $40, subtract the cost of the radio from $40.

$$\begin{array}{r} \$40.00 \\ -\ \ 33.89 \\ \hline \$\ \ 6.11 \end{array}$$

Marta gets $6.11 in change.

Clerks sometimes make change by counting backwards, that is, by adding to $33.89 the amount necessary to equal $40.

$33.89 + a penny = $33.90
$33.90 + a dime = $34.00
$34.00 + 1 dollar = $35.00
$35.00 + 5 dollars = $40.00

So the change is $0.01 + $0.10 + $1 + $5 = $6.11.

Answers to Warm Ups F. 3.6 G. 94.8495 H. Mickey gets $1.31 in change.

HOW AND WHY

Objective 3

Estimate the sum or difference of decimals.

The sum or difference of decimals can be estimated by rounding each to the largest nonzero place value of the numbers and then adding or subtracting the rounded values. For instance,

0.756	0.8	**The largest place value is tenths, so round each number to the nearest tenth.**
0.092	0.1	
0.0072	0.0	
+0.0205	+0.0	
	0.9	

The estimate of the sum is 0.9. One use of the estimate is to see if the sum of the group of numbers is correct. If the calculated sum is not close to the estimated sum, 0.9, you should check the addition by re-adding. In this case the calculated sum, 0.8757, is close to the estimate.

Example I

Directions: Estimate the difference. Then subtract and compare.

Strategy: Round each number to the largest nonzero place value of the numbers. Then subtract and compare.

I. 0.9569 − 0.875

1.0	**The largest nonzero place value is tenths.**
−0.9	
0.1	

Now subtract and compare.

8 15	
0. 9 5 6 9	
−0. 8 7 5 0	**Insert a zero so we have the same number of decimal places.**
0. 0 8 1 9	

The difference is 0.0819 and is close to the estimate, 0.1.

Warm Up I

I. 0.00782 − 0.00298

Answer to Warm Up I. 0.005; 0.00484

NAME ______________________ CLASS ____________ DATE ____________

Exercises 5.3

OBJECTIVE 1: *Add decimals.*

A.

Add.

1. 0.4 + 0.3

2. 0.8 + 0.3

3. 2.5 + 1.3

4. 6.7 + 2.1

5. 1.4 + 2.1 + 4.2

6. 3.2 + 1.1 + 2.4

7. 23.3 + 4.13

8. 17.7 + 2.28

9. To add 4.5, 6.78, 9.342, and 23 first rewrite each with _________ decimal places.

10. The sum of 6.7, 8.93, 5.4321, and 45.72 has _________ decimal places.

B.

11.
$$\begin{array}{r} 8.3 \\ +5.541 \\ \hline \end{array}$$

12.
$$\begin{array}{r} 7.6 \\ +6.44 \\ \hline \end{array}$$

13.
$$\begin{array}{r} 8.28 \\ 0.28 \\ 12.3 \\ +\ 2.54 \\ \hline \end{array}$$

14.
$$\begin{array}{r} 9.06 \\ 0.82 \\ 11.5 \\ +\ 4.35 \\ \hline \end{array}$$

15. $0.438 + 0.834 + 1.483$

16. $1.254 + 1.425 + 0.524$

17. $0.0017 + 1.007 + 7 + 1.071$

18. $1.0304 + 1.4003 + 1.34 + 0.403$

19. $37.008 + 38.007 + 3.87 + 3.708$

20. $82.005 + 8.25 + 2.085 + 28.55$

21.
$$\begin{array}{r} 7.5 \\ 14.378 \\ +\underline{33.6583} \end{array}$$

22.
$$\begin{array}{r} 10.03 \\ 223.231 \\ +\underline{5603.3056} \end{array}$$

23.
$$\begin{array}{r} 43.524 \\ 12.8 \\ +\underline{774.943} \end{array}$$

24.
$$\begin{array}{r} 314.143 \\ 712.217 \\ +\underline{333.444} \end{array}$$

25. Find the sum of 9.76, 9.6, 0.581, and 7.04.

26. Find the sum of 0.9855, 4.913, 6.72, and 3.648.

OBJECTIVE 2: *Subtract decimals.*

A.

Subtract.

27. $0.9 - 0.6$

28. $2.7 - 2.2$

29. $5.7 - 2.3$

30. $0.25 - 0.12$

31. $\begin{array}{r} 8.31 \\ -\underline{3.21} \end{array}$

32. $\begin{array}{r} 17.48 \\ -\underline{6.23} \end{array}$

33. $\begin{array}{r} 19.05 \\ -\underline{12.64} \end{array}$

34. $\begin{array}{r} 6.28 \\ -\underline{1.19} \end{array}$

35. Subtract 8.11 from 16.20.

36. Find the difference of 18.477 and 9.2.

B.

Subtract.

37. $\begin{array}{r} 0.612 \\ -\underline{0.155} \end{array}$

38. $\begin{array}{r} 3.457 \\ -\underline{2.509} \end{array}$

39. $\begin{array}{r} 2.712 \\ -\underline{1.148} \end{array}$

40. $\begin{array}{r} 7.303 \\ -\underline{0.178} \end{array}$

41. 5.678 − 3.069

42. 4.823 − 1.167

43. 134.98 − 67.936

44. 405.4057 − 316.316

45. Subtract 9.34 from 12.1.

46. Subtract 3.576 from 8.4.

47. Find the difference of 8.642 and 8.573.

48. Find the difference of 71.505 and 69.948.

OBJECTIVE 3: *Estimate the sum or difference of decimals.*

A.

Estimate the sum or difference.

49. 0.09346 + 0.0371 + 0.0444

50. 0.00567 + 0.003211 + 0.00123

51. 0.678 − 0.351

52. 0.074 − 0.056

53. 3.895 + 2.045 + 0.34 + 0.045

54. 0.45 + 0.0021 + 0.0043 + 0.1001

55. 0.075 − 0.0023

56. 0.00675 − 0.000984

57.
$$\begin{array}{r} 0.542 \\ 0.125 \\ 0.32974 \\ +\underline{0.7421} \end{array}$$

58.
$$\begin{array}{r} 0.0346 \\ 0.067 \\ 0.02389 \\ +\underline{0.06002} \end{array}$$

B.

Estimate the sum or difference, then do the addition or subtraction.

59.
$$\begin{array}{r} 0.0764 \\ -\underline{0.03621} \end{array}$$

60.
$$\begin{array}{r} 0.0056982 \\ -\underline{0.003781} \end{array}$$

61. 0.0342 + 0.00687 + 0.057294 + 0.00843

62. 2.005 + 0.8741 + 0.006723 + 5.0555

63. 38.9867 − 27.3074

64. 6.7382 − 0.88914

65. 0.067 + 0.456 + 0.0964 + 0.5321 + 0.112

66. 4.005 + 0.875 + 3.96 + 7.832 + 4.009

67. 0.0456 + 0.7834 + 0.456 − (0.3097 + 0.5067)

68. 3.079 + 7.935 − (0.983 + 3.115 + 6.22)

C.

69. On a vacation trip, Manuel stops for gas four times. The first time he bought 9.2 gallons. At the second station he bought 11.9 gallons, and at the third he bought 15.4 gallons. At the last stop he bought 12.6 gallons. How much gas did he buy on the trip?

70. Heather wrote five checks in the amounts of \$45.78, \$23.90, \$129.55, \$7.75, and \$85. She has \$295.67 in her checking account. Does she have enough money to cover the five checks?

71. Find the sum of 45.984, 134.6, 98.992, 89.56, and 102.774. Round the sum to the nearest tenth.

72. Find the sum of 235.98, 785.932, 6.94432, 11.116, and 8.0034. Round the sum to the nearest thousandth.

Exercises 73–77 relate to the chapter application.

73. Muthoni runs a race in 12.16 sec, whereas Sera runs the same race in 11.382 sec. How much faster is Sera?

74. A skier posts a race time of 1.257 min. A second skier posts a time of 1.32 min. The third skier completes the race in 1.2378 min. Find the difference between the fastest and the slowest times.

75. A college men's 4 × 100 m relay track team has runners with individual times of 9.35 sec, 9.91 sec, 10.04 sec, and 9.65 sec. What is the time for the relay?

76. A high school girl's swim team has a 200 yd freestyle relay whose members have times of 21.79 sec, 22.64 sec, 22.38 sec, and 23.13 sec. What is the time for the relay?

77. A high school women's track coach knows that the rival school's team in the 4 × 100 m relay has a time of 52.78 sec. If the coach knows that her top three sprinters have times of 12.83, 13.22, and 13.56 sec, how fast does the fourth sprinter need to be in order to beat the rival school's relay team?

Use the table for Exercises 78–80. The table shows the annual annuity sales in billions of dollars.

Annuity Sales in Billions of Dollars

1985	$4.5	1990	$12
1986	$8.1	1991	$17.3
1987	$9.3	1992	$28.5
1988	$7.2	1993	$46.6
1989	$9.8	1994	$50.4

78. Find the total sales for the ten years.

79. How many more dollars were invested in 1992 than in 1989?

80. How many more dollars were invested in the 1990s than in the 1980s?

81. Doris makes a gross salary (before deductions) of $2796 per month. She has the following monthly deductions: federal income tax, $254.87; state income tax; $152.32; Social Security, $155.40; Medicare, $35.61; retirement contribution, $82.45; union dues, $35; and health insurance, $134.45. Find her actual take-home (net) pay.

82. Jack goes shopping with \$75 in cash. He pays \$14.99 for a T-shirt, \$9.50 for a CD, and \$25.75 for a sweater. On the way home he buys \$18.55 worth of gas. How much money does he have left?

83. In 1996 the average interest on a 30-year home mortgage dropped from 8.23 percent to 7.88 percent in one week. What was the drop in interest rate?

84. What is the total cost of a cart of groceries that contains bread for \$1.88, bananas for \$2.12, cheese for \$5.87, cereal for \$3.57, coffee for \$7.82, and meat for \$7.89?

85. How high from the ground level is the top of the smokestack in the drawing below? Round to the nearest foot.

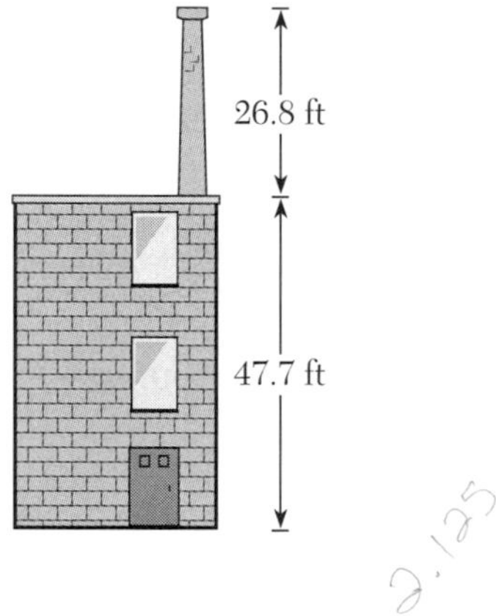

2.125

86. Find the length of the piston skirt (A) shown in the following drawing below if the other dimensions are as follows: $B = 0.3125$ in., $C = 0.250$ in., $D = 0.3125$ in., $E = 0.250$ in., $F = 0.3125$ in., $G = 0.375$ in., $H = 0.3125$.

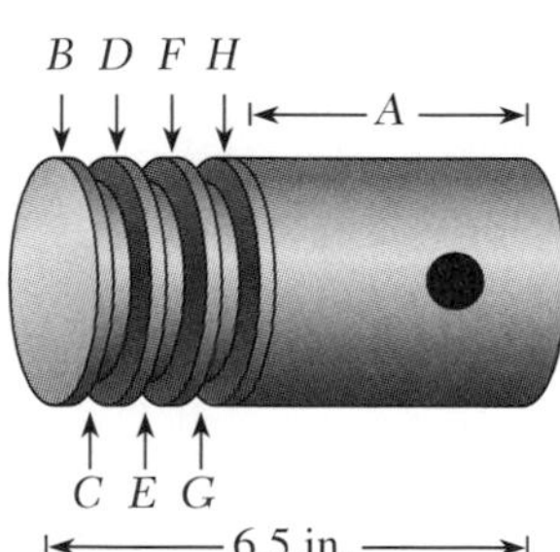

87. What is the center-to-center distance, A, between the holes in the diagram below?

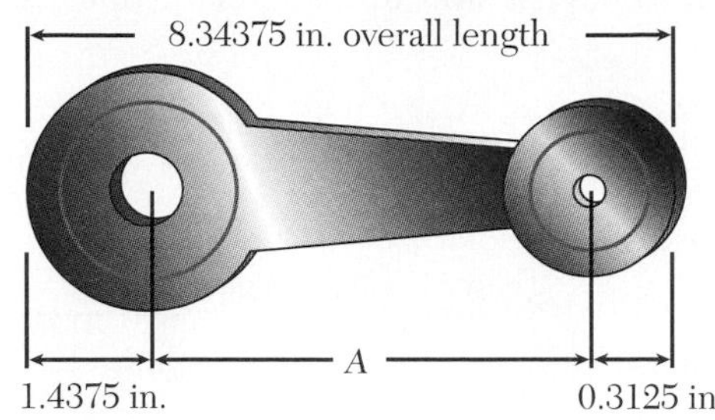

88. What is the total length of the connecting bar shown below?

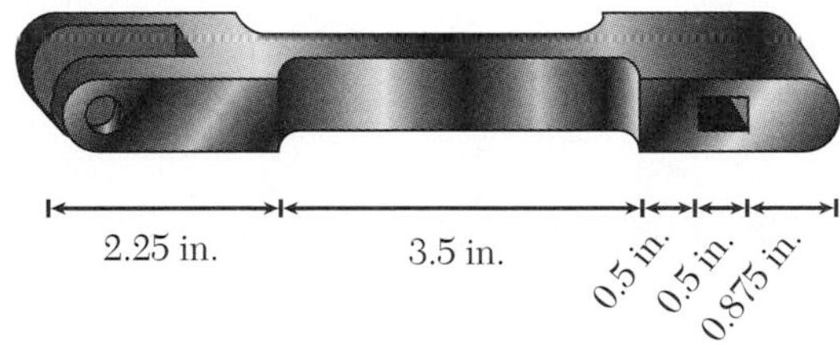

STATE YOUR UNDERSTANDING

89. Explain the procedure for adding 2.005, 8.2, 0.00004, and 3.

90. Explain the similarities between subtracting decimals and subtracting fractions.

91. Copy the table and fill it in.

Operation on Decimals	Procedure	Example
Addition		
Subtraction		

CHALLENGE

92. How many 8.75s must be added to have a sum that is greater than 200?

93. Find the missing number in the sequence: 0.4, 0.8, 1.3, __________, 2.6, 3.4, 4.3, 5.3.

94. Find the missing number in the sequence: 0.2, 0.19, 0.188, __________, 0.18766, 0.187655.

95. Which number in the following group is 11.1 less than 989.989: 999.999, 989.999, 988.889, 979.889, or 978.889?

96. Write the difference between $6\frac{9}{16}$ and 4.99 in decimal form.

97. Round the sum of 9.8989, 8.9898, 7.987, and 6.866 to the nearest tenth.

GROUP ACTIVITY

98. As a group, review the multiplication of fractions. Have each member make up a pair of fractions whose denominators are in the list: 10, 100, 1000, and 10,000. Find the product of each pair and change it to decimal form. In group discussion, make up a rule for multiplying decimals.

MAINTAIN YOUR SKILLS (SECTIONS 4.4, 4.7, 4.8, 4.9, 4.10)

Add.

99. $\frac{1}{2} + \frac{3}{4} + \frac{1}{8}$

100. $\frac{1}{3} + \frac{7}{12} + \frac{5}{6}$

101. $3\frac{4}{5} + 5\frac{1}{3} + 6\frac{2}{15} + 2\frac{7}{30}$

102. $7\frac{5}{8} + 8\frac{11}{16} + 3\frac{3}{4}$

Subtract.

103. $\frac{31}{32} - \frac{7}{8}$

104. $\frac{19}{20} - \frac{5}{12}$

105. $17\frac{7}{12} - 8\frac{5}{6}$

106. $23\frac{8}{9} - 5\frac{5}{7}$

107. A board is $4\frac{7}{8}$ in. thick. If $\frac{1}{16}$ in. is sanded from each side, how thick will the board be?

108. A new spool of wire contains $23\frac{1}{4}$ lb of wire. The spool costs \$465. What is the cost per pound?

Getting Ready for Algebra

OBJECTIVE

Solve equations that involve addition and subtraction of decimals.

HOW AND WHY

We solve equations that involve addition and subtraction of decimals in the same way as equations with whole numbers and fractions.

To solve an equation using addition or subtraction

1. Add the same number to both sides of the equation to isolate the variable, or
2. Subtract the same number from both sides of the equation to isolate the variable.

Examples A–D

Directions: Solve.

Strategy: Isolate the variable by adding or subtracting the same number to or from each side.

A. $8.6 = x + 3.5$

$8.6 = x + 3.5$

$8.6 - 3.5 = x + 3.5 - 3.5$ **Eliminate the addition by subtracting 3.5 from both sides of the equation. Since subtraction is the inverse of addition, the variable will be isolated.**

$5.1 = x$ **Simplify.**

CHECK:

$8.6 = 5.1 + 3.5$ **Substitute 5.1 for x in the original equation and simplify.**

$8.6 = 8.6$

The solution is $x = 5.1$.

B. $z - 12.6 = 27.7$

$$\begin{aligned} z - 12.6 &= 27.7 \\ +12.6 &= +12.6 \\ \hline z &= 40.3 \end{aligned}$$

Eliminate the subtraction by adding 12.6 to both sides of the equation. Since subtraction is the inverse of addition, the variable will be isolated.

CHECK:

$40.3 - 12.6 = 27.7$ **Substitute 40.3 for z in the original equation and simplify.**

$27.7 = 27.7$

The solution is $z = 40.3$.

Warm Ups A–D

A. $15.4 = p + 2.9$

B. $t - 9.37 = 6.5$

Answers to Warm Ups A. $p = 12.5$ B. $t = 15.87$

C. $c + 45.349 = 47$

C. $b + 12.875 = 22.4$

$$b + 12.875 = 22.4$$

$$b + 12.875 - 12.875 = 22.4 - 12.875$$ **Subtract 12.875 from both sides and simplify.**

$$b = 9.525$$

CHECK:

$9.525 + 12.875 = 22.4$ **Substitute 9.525 for b in the original equation and simplify.**

$$22.4 = 22.4$$

The solution is $b = 9.525$.

D. $w - 34.87 = 19.6641$

D. $y - 3.947 = 7.0721$

$$\begin{aligned} y - 3.947 &= \ 7.0721 \\ \underline{+\ 3.947} &= +\ \underline{3.9470} \\ y &= \ 11.0191 \end{aligned}$$ **Add 3.947 to both sides and simplify.**

CHECK:

$11.0191 - 3.947 = 7.0721$ **Substitute 11.0191 for y in the original equation and simplify.**

$$7.0721 = 7.0721$$

The solution is $y = 11.0191$.

E. A farmer practicing "sustainability" farming reduced his soil erosion by 1.7 tons in one year. If he lost 3.65 tons of topsoil this year to erosion, how many tons did he lose last year?

E. The price of a graphing calculator decreased by \$18.30 over the past year. What was the price a year ago if the calculator now sells for \$79.95?

First write the English version of the equation.

(cost last year) − (decrease in cost) = cost this year

Let x represent the cost last year.

$$x - 18.30 = 79.95$$ **Translate to algebra.**

$$x - 18.30 + 18.30 = 79.95 + 18.30$$ **Add 18.30 to both sides and simplify.**

$$x = 98.25$$

Since $98.25 - 18.30 = 79.95$, the cost of the calculator last year was \$98.25.

Answers to Warm Ups C. $c = 1.651$ D. $w = 54.5341$ E. The farmer lost 5.35 tons to erosion.

NAME ______________________ CLASS ____________ DATE ____________

Exercises

Solve.

1. $13.8 = x + 2.3$

2. $2.408 = x + 1.7$

3. $y - 0.7 = 14.28$

4. $w - 0.03 = 0.378$

5. $t + 0.05 = 0.123$

6. $x + 11.6 = 367.72$

7. $x - 2.6 = 9.4$

8. $y - 8.1 = 0.33$

9. $2.66 = w + 0.04$

10. $12 = x + 5.3$

11. $t - 7.14 = 0.09$

12. $w - 0.06 = 0.235$

13. $1.56 = a + 0.78$

14. $24.8 = w - 0.65$

15. $2.2 = x - 4.36$

16. $4 = 2.3 + x$

17. $a + 78.4 = 100$

18. $b + 29.76 = 45$

19. $s - 2.4 = 3.889$

20. $r - 3.8 = 5.6231$

21. $c + 567.8 = 1043.82$

22. $d + 234.87 = 505.1$

23. The price of an energy-efficient hot-water heater decreased by \$46.98 over the past 2 yr. What was the price two years ago if the heater now sells for \$359.99?

24. In one state the use of household biodegradable cleaners increased by 2444.67 lb per month because of state laws banning phosphates. How many pounds of these cleaners were used before the new laws if the average use now is 5780.5 lb?

25. The selling price of a personal computer is $995.95. If the cost is $855.29, what is the markup?

26. The selling price of a new tire is $58.95. If the markup on the tire is $12.84, what is the cost (to the store) of the tire?

27. A shopper needs to buy a bus pass and some groceries. The shopper has $43 with which to make both purchases. If the bus pass costs $19, write and solve an equation that represents the shopper's situation. How much can the shopper spend on groceries?

28. In a math class the final grade is determined by adding the test scores and the homework scores. If a student has a homework score of 18 and it takes a total of 90 to receive a grade of A, what total test score must the student have to receive a grade of A? Write an equation and solve it to determine the answer.

5.4

Multiplying Decimals

OBJECTIVE

1. Multiply decimals.
2. Estimate the product of decimals.

HOW AND WHY

Objective 1

Multiply decimals.

The "multiplication table" for decimals is the same as for whole numbers. In fact, decimals are multiplied the same way as whole numbers, with one exception, the location of the decimal point in the product. To discover the rule for the location of the decimal point, we will use what we already know about multiplication of fractions. First, change the decimal form to fraction form to find the product. Next, change the product back to decimal form and observe the number of decimal places in the product. Consider the examples in Table 5.2. We see that the product in decimal form has the same number of decimal places as the total number of places in the decimal factors.

TABLE 5.2 MULTIPLICATION OF DECIMALS

Decimal Form	Fraction Form	Product of Fractions	Product as a Decimal	Number of Decimal Places in Product
0.3×0.8	$\frac{3}{10} \times \frac{8}{10}$	$\frac{24}{100}$	0.24	Two
11.2×0.07	$\frac{112}{10} \times \frac{7}{100}$	$\frac{784}{1000}$	0.784	Three
0.02×0.13	$\frac{2}{100} \times \frac{13}{100}$	$\frac{26}{10000}$	0.0026	Four

The shortcut is to multiply the numbers and insert the decimal point. If necessary, insert zeros so that there are enough decimal places. The product of 0.2×0.3 has two decimal places, because tenths multiplied by tenths yields hundredths.

$$0.2 \times 0.3 = 0.06 \quad \text{because} \quad \frac{2}{10} \times \frac{3}{10} = \frac{6}{100}$$

To multiply decimals

1. Multiply the numbers as if they were whole numbers.
2. Locate the decimal point by counting the number of decimal places (to the right of the decimal point) in both factors. The total of these two counts is the number of decimal places the product must have.
3. If necessary, zeros are inserted at the *left of the numeral* so there are enough decimal places (see Example D).

When multiplying decimals it is not necessary to align the decimal points in the decimals being multiplied.

Examples A–F

Directions: Multiply.

Strategy: First multiply the numbers, ignoring the decimal points. Place the decimal point in the product by counting the number of decimal places in the two factors. Insert zeros if necessary to produce the number of required places.

A. Multiply: (0.7)(11)

(0.7)(11) = 7.7 **Multiply 7 and 11. The total number of decimal places in both factors is one (1), so there is one decimal place in the product.**

So, (0.7)(11) = 7.7.

B. Find the product of 0.8 and 0.21.

(0.8)(0.21) = 0.168 **There are three decimal places in the product because the total number of places in the factors is three.**

So the product of 0.8 and 0.21 is 0.168.

C. Find the product of 5.67 and 3.8.

$$\begin{array}{r} 5.67 \\ \times \quad 3.8 \\ \hline 4536 \\ 1701 \\ \hline 21.546 \end{array}$$

Multiply the numbers as if they were whole numbers. There are three decimal places in the product.

So the product of 5.67 and 3.8 is 21.546.

D. Multiply 1.2 times 0.0004.

$$\begin{array}{r} 1.2 \\ \times \; 0.0004 \\ \hline 0.00048 \end{array}$$

Since 1.2 has one decimal place and 0.0004 has four decimal places, the product must have five decimal places. We must insert three zeros at the left so there are enough places in the product.

So 1.2 times 0.0004 is 0.00048.

Calculator Example

E. Find the product: (12.87)(64.862)

The product is 834.77394. **The calculator will automatically place the decimal point in the correct position.**

F. If exactly eight strips of metal, each 3.875 in. wide, are to be cut from a piece of sheet metal, what is the smallest (in width) piece of sheet metal that can be used?

Strategy: To find the width of the piece of sheet metal we multiply the width of one of the strips by the number of strips needed.

$$\begin{array}{r} 3.875 \\ \times \qquad 8 \\ \hline 31.000 \end{array}$$

The extra zeros can be dropped.

The piece must be 31 in. wide.

Warm Ups A–F

A. Multiply: (9)(0.6)

B. Find the product of 0.7 and 0.33.

C. Find the product of 16.9 and 0.34.

D. Multiply 0.05 times 0.015.

E. Find the product: 23.84(79.035)

F. If 12 strips, each 6.45 cm wide, are to be cut from a piece of sheet metal, what is the narrowest piece of sheet metal that can be used?

Answers to Warm Ups A. 5.4 B. 0.231 C. 5.746 D. 0.00075 E. 1884.1944 F. The sheet metal must be 77.4 cm wide.

HOW AND WHY

Objective 2

Estimate the product of decimals.

The product of two decimals can be estimated by rounding each number to the largest nonzero place in each number and then multiplying the rounded numbers. For instance,

0.0673	0.07	**Round to the nearest hundredth.**
× 0.79	× 0.8	**Round to the nearest tenth.**
	0.056	**Multiply.**

The estimate of the product is 0.056. One use of the estimate is to see if the product is correct. If the calculated product is not close to 0.056, you should check the multiplication. In this case the actual product is 0.053167, which is close to the estimate.

To estimate the product of two decimals

1. Round each number to its largest nonzero place.
2. Multiply the rounded numbers.

Example G

Directions: Estimate the product. Then find the product and compare.

Strategy: Round each number to its largest nonzero place. Multiply the rounded numbers. Multiply the original numbers.

G. 0.316 and 9.107

0.3	**Round to the nearest tenth.**
× 9	**Round to the nearest one.**
2.7	

```
     0.316
×    9.107
     2 212
    31 60
  2 844
  2.877812
```

The product is 2.877812 and is close to the estimated product of 2.7.

Warm Up G

G. 3.871 and 0.0051

Answer to Warm Up G. 0.02; 0.0197421

NAME ____________________ CLASS __________ DATE __________

Exercises 5.4

OBJECTIVE 1: *Multiply decimals.*

A.

Multiply.

1. $\begin{array}{r} 0.4 \\ \times\ \underline{8} \end{array}$

2. $\begin{array}{r} 0.6 \\ \times\ \underline{7} \end{array}$

3. $\begin{array}{r} 1.5 \\ \times\ \underline{6} \end{array}$

4. $\begin{array}{r} 2.1 \\ \times\ \underline{8} \end{array}$

5. 3×0.09

6. 0.07×8

7. 0.5×0.4

8. 0.4×0.2

9. 0.03×0.5

10. 0.7×0.008

11. 0.16×0.4

12. 1.2×0.05

13. The number of decimal places in the product of 9.456 and 4.23 is __________.

14. In the product $0.034 \times ? = 0.0408$, the number of decimal places in the missing factor is __________.

B.

Multiply.

15. $\begin{array}{r} 7.45 \\ \times\ \underline{0.002} \end{array}$

16. $\begin{array}{r} 1.13 \\ \times\ \underline{0.005} \end{array}$

17. $\begin{array}{r} 1.45 \\ \times\ \underline{4.6} \end{array}$

18. $\begin{array}{r} 4.23 \\ \times\ \underline{3.2} \end{array}$

19. $\begin{array}{r} 7.84 \\ \times 0.53 \\ \hline \end{array}$

20. $\begin{array}{r} 45.8 \\ \times 0.12 \\ \hline \end{array}$

21. $\begin{array}{r} 0.346 \\ \times \quad 7.8 \\ \hline \end{array}$

22. $\begin{array}{r} 0.073 \\ \times \quad 9.6 \\ \hline \end{array}$

23. Find the product of 8.52 and 3.54.

24. Find the product of 6.79 and 1.34.

25. Multiply: 6.5(0.6)(0.03)

26. Multiply: 3.6(1.6)(0.012)

OBJECTIVE 2: *Estimate the product of decimals.*

A.

Estimate the product.

27. 32(0.845)

28. 680(0.00231)

29. 0.045(0.0672)

30. 0.00389(0.0912)

31. (16.95)(0.0781)

32. (0.00439)(31.9)

33. $\begin{array}{r} 0.0875 \\ \times \ 0.021 \\ \hline \end{array}$

34. $\begin{array}{r} 0.2753 \\ \times 0.9631 \\ \hline \end{array}$

B.

Estimate the product and then multiply.

35. $\begin{array}{r} 23.5 \\ \times \ 0.47 \\ \hline \end{array}$

36. $\begin{array}{r} 18.6 \\ \times \ 0.32 \\ \hline \end{array}$

37. $\begin{array}{r} 0.356 \\ \times \ 0.067 \\ \hline \end{array}$

38. $\begin{array}{r} 0.832 \\ \times \ 0.041 \\ \hline \end{array}$

39. $\begin{array}{r} 0.0975 \\ \times \underline{3.92} \end{array}$

40. $\begin{array}{r} 0.00732 \\ \times \underline{7.05} \end{array}$

41. $\begin{array}{r} 0.825 \\ \times \underline{0.0054} \end{array}$

42. $\begin{array}{r} 0.575 \\ \times \underline{0.00378} \end{array}$

C.

Use the table for Exercises 43–47. The table shows the amount of gas purchased by Grant and the price he paid per gallon for five fill-ups.

Number of Gallons	Price per Gallon
20.7	$1.375
20.4	$1.405
19.3	$1.447
18.9	$1.393
18.4	$1.523

43. What is the total number of gallons of gas that Grant purchased?

44. To the nearest cent, how much did he pay for the second fill-up?

45. To the nearest cent, how much did he pay for the fifth fill-up?

46. To the nearest cent, what is the total amount he paid for the five fill-ups?

47. At which price per gallon did he pay the least for his fill-up?

Estimate the product and then multiply.

48. (223.6)(8.45)

49. (98.67)(3.52)

50. (343.17)(8.73)

51. (12.6)(760.02)

Multiply.

52. (9.58)(5.63)(23.22). Round to the nearest hundredth.

53. (9.86)(146.3)(14.83). Round to the nearest thousandth.

54. (2.15)(1.8)(0.54)(13.5). Round to the nearest tenth.

55. (5.7)(0.57)(5.07)(50.7). Round to the nearest hundredth.

56. Joe earns \$9.85 per hour. How much does he earn if he works 30.25 hr in one week? Round to the nearest cent.

NAME ______________________ CLASS ____________ DATE ____________

57. If upholstery fabric costs \$37.89 per yard, how much will Joanne pay for 14.75 yd? Round to the nearest cent.

Use the table for Exercises 58–61. The table shows the cost of renting a car from a local agency.

Type of Car	Cost per Day	Price per Mile Driven
Compact	\$25.95	\$0.25
Mid-size	\$39.72	\$0.32
Luxury	\$47.85	\$0.42

58. What does it cost to rent a compact car for 4 days if it is driven 324 miles?

59. What does it cost to rent a mid-size car for 3 days if it is driven 312 miles?

60. What does it cost to rent a luxury car for 6 days if it is driven 453 miles?

61. Which costs less, renting a mid-size car for 3 days or a compact car for 5 days if both are driven 345 miles? How much less does it cost?

62. Tiffany can choose any of the following ways to finance her new car. Which method is the least expensive in the long run?

\$650 down and \$305.54 per month for 5 years
\$350 down and \$343.57 per month for 54 months
\$400 down and \$386.42 per month for 4 years

63. A new freezer-refrigerator is advertised at three different stores as follows:

Store 1: \$75 down and \$85.95 for 18 months
Store 2: \$125 down and \$63.25 per month for 24 months
Store 3: \$300 down and \$109.55 per month for 12 months

Which store is selling the freezer-refrigerator for the least total cost?

64. An order of 31 bars of steel is delivered to a machine shop. Each bar is 19.625 ft long. Find the total linear feet of steel in the order.

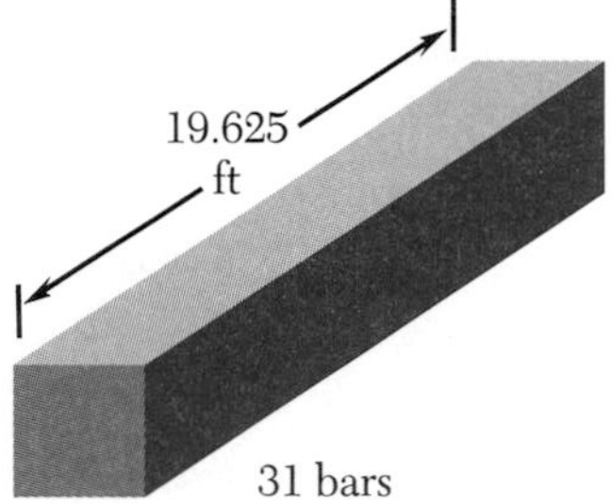

65. From a table in a machinist's handbook, it is determined that hexagon steel bars 1.125 in. across weigh 3.8 lb per running foot. Using this constant, find the weight of a 1.125 in. hexagon steel bar that is 18.875 ft long.

 Exercises 66–68 relate to the chapter application.

In Olympic diving, seven judges rate each dive using a whole or half number between 0 and 10. The high and low scores are thrown out and the remaining scores are added together. The sum is then multiplied by 0.6 and then by the difficulty factor of the dive to obtain the total points awarded.

66. A diver does a reverse $1\frac{1}{2}$ somersault with $2\frac{1}{2}$ twists, a dive with a difficulty factor of 2.9. She receives scores of 6.0, 6.5, 6.5, 7.0, 6.0, 7.5, and 7.0. What are the total points awarded for the dive?

67. Another diver also does a reverse $1\frac{1}{2}$ somersault with $2\frac{1}{2}$ twists. This diver receives scores of 7.5, 6.5, 7.5, 8.0, 8.0, 7.5, and 8.0. What are the total points awarded for the dive?

68. A cut-through reverse $1\frac{1}{2}$ somersault has a difficulty factor of 2.6. What is the highest number of points possible with this dive?

69. In 1970 the per capita consumption of red meat was 132 lb. In 1980 the consumption was 126.4 lb. In 1990, the amount consumed was 112.3 lb per person. Compute the total weight of red meat consumed by a family of four using the rates for each of these years. Discuss the reasons for the change in consumption.

70. The fat content in a 3-oz serving of common meats and fish is as follows: beef rib, 7.4 g; beef top round, 3.4 g; beef top sirloin, 4.8 g; dark meat chicken without skin, 8.3 g; white meat chicken without skin, 3.8 g; pink salmon, 3.8 g; and Atlantic cod, 0.7 g. Which contains the most fat grams, 3 servings of beef ribs, 6 servings of beef top round, 4 servings of beef top sirloin, 2 servings of dark meat chicken, 6 servings of white meat chicken, 5 servings of pink salmon, or 25 servings of Atlantic cod?

71. Older models of toilets use 5.5 gallons of water per flush. Models made in the 1970s use 3.5 gallons per flush. The new low-flow models use 1.55 gallons per flush. Assume each person flushes the toilet an average of five times per day. Determine the amount of water used in a town with a population of 34,782 in one day for each type of toilet. How much water is saved using the low-flow model as opposed to the pre-1970s model?

72. The annual property tax bills arrive in early November in some states. The McNamara house is assessed for $97,700. The annual tax rate is $0.199 per hundred dollars of assessment. Find what they owe in taxes.

73. Find the property tax on the Gregory Estate, which is assessed at $1,895,750. The tax rate in the area is $1.235 per hundred dollars of assessment. Round to the nearest dollar.

STATE YOUR UNDERSTANDING

74. Explain how to determine the number of decimal places needed in the product of two decimals.

75. Suppose you use a calculator to multiply (0.006)(3.2)(68) and get 13.056. Explain, using placement of the decimal point in a product, how you can tell that at least one of the numbers was entered incorrectly. How can such errors occur? How can you estimate the answers before using the calculator so you can avoid such errors?

CHALLENGE

76. What is the smallest whole number you can multiply by 0.66 to get a product that is greater than 55?

77. What is the largest whole number you can multiply by 0.78 to get a product that is less than 58.9?

78. Find the missing number in the following sequence: 1.8, 0.36, 0.108, 0.0432, ________.

79. Find the missing number in the following sequence: 3.1, 0.31, ________, 0.0000031, 0.0000000031.

GROUP ACTIVITY

80. Visit a grocery store or use newspaper ads to "purchase" the items listed. Have each member of your group use a different store or chain. Which members of your group "spent" the most? least? Which group "spent" the most? least?

Three 12-packs of Diet Pepsi
Eight gallons of 2% milk
Five pounds of hamburger
Four cans of the store-brand creamed corn
Five 4-roll packs of toilet paper
Seven pounds of butter
72 hamburger rolls
12 large boxes of Cheerios

MAINTAIN YOUR SKILLS (SECTION 1.5)

Multiply or divide as indicated.

81. 445(100)

82. 75(10,000)

83. 78,300 ÷ 100

84. 15,000,000 ÷ 100,000

85. 12(1,000,000)

86. 803×10^3

87. $67{,}000 \div 10^2$

88. $380{,}000 \div 10^4$

89. 23×10^7

90. $4{,}210{,}000{,}000 \div 10^5$

5.5

Multiplying and Dividing by Powers of 10; Scientific Notation

OBJECTIVES

1. Multiply or divide a number by a power of ten.
2. Write a number in scientific notation or change a number in scientific notation to its place value name.

VOCABULARY

Recall that a **power of 10** is the value obtained when ten is written with an exponent.

Scientific notation is a special way to write numbers as a product using a number between 1 and 10 and a power of 10.

HOW AND WHY

Objective 1

Multiply or divide a number by a power of 10.

The shortcut used in Section 1.5 for multiplying and dividing by 10 or a power of 10 works in a similar way with decimals. Consider the following products:

0.5	0.23	5.67
× 10	× 10	× 10
0	0	0
5 0	2 30	56 70
5.0 = 5	2.30 = 2.3	56.70 = 56.7

Note in each case that multiplying a decimal by 10 has the effect of moving the decimal point one place to the right.

Since $100 = 10 \cdot 10$, multiplying by 100 is the same as multiplying by 10 two times in succession. So, multiplying by 100 has the effect of moving the decimal point two places to the right. For instance

$(0.53)(100) = 0.53(10 \cdot 10) = (0.53 \cdot 10) \cdot 10 = 5.3 \cdot 10 = 53$

Since $1000 = 10 \cdot 10 \cdot 10$, the decimal point will move three places to the right when multiplying by 1000. Since $10{,}000 = 10 \cdot 10 \cdot 10 \cdot 10$, the decimal point will move four places to the right when multiplying by 10,000, and so on in the same pattern:

$(0.08321)(10{,}000) = 832.1$

Zeros may have to be placed on the right in order to move the correct number of decimal places:

$(2.3)(1000) = 2.300 = 2300$

In this problem, two zeros are placed on the right.

Since multiplying a decimal by 10 has the effect of moving the decimal point one place to the right, dividing a number by 10 must move the decimal point one place to the left. Again, we are using the fact that multiplication and division are inverse operations. Division by 100 will move the decimal point two places to the left, and so on. Thus,

$347.1 \div 100 = 347.1 = 3.471$

$0.763 \div 1000 = 0.000763$

Three zeros are placed on the left so that the decimal point may be moved three places to the left.

To multiply a number by a power of 10

Move the decimal point to the right. The number of places to move is shown by the number of zeros in the power of 10.

To divide a number by a power of 10

Move the decimal point to the left. The number of places to move is shown by the number of zeros in the power of 10.

Warm Ups A–G

A. Multiply: 45.89(10)

B. Multiply: 0.23(100)

C. Find the product of 23.7 and 10^4.

D. Divide: $445 \div 10$

Examples A–G

Directions: Multiply or divide as indicated.

Strategy: To multiply by a power of 10, move the decimal point to the right, inserting zeros as needed. To divide by a power of 10, move the decimal point to the left. The exponent of 10 specifies the number of places to move the decimal point.

A. Multiply: 4.572(10)

4.572(10) = 45.72 **Multiplying by 10 or 10^1 moves the decimal point one place to the right.**

So, 4.572(10) = 45.72.

B. Multiply: 0.618(100)

0.618(100) = 61.8 **Multiplying by 100 or 10^2 moves the decimal point two places to the right.**

So, 0.618(100) = 61.8.

C. Find the product of 90.5 and 10^3.

$90.5(10^3) = 90{,}500$ **Multiplying by 10^3 moves the decimal point three places to the right. Two zeros must be inserted on the right to make the move.**

So the product of 90.5 and 10^3 is 90,500.

D. Divide: $34.6 \div 10$

$34.6 \div 10 = 3.46$ **Dividing by 10 or 10^1 moves the decimal point one place to the left.**

So, $34.6 \div 10 = 3.46$.

Answers to Warm Ups A. 458.9 B. 23 C. 237,000 D. 44.5

E. Divide: 452.3 ÷ 100

452.3 ÷ 100 = 4.523 **Dividing by 100 or 10^2 moves the decimal point two places to the left.**

So, 452.3 ÷ 100 = 4.523.

E. Divide: 786.7 ÷ 100

F. Find the quotient: 71.8 ÷ 10^4

71.8 ÷ 10^4 = 0.00718 **Move the decimal point four places to the left. Zeros are inserted on the left to make the move.**

So, 71.8 ÷ 10^4 = 0.00718.

F. Find the quotient; 86.395 ÷ 10^4

G. A stack of sheet metal contains 100 sheets. The stack is 8.75 in. high. How thick is each sheet of metal?

8.75 ÷ 100 = 0.0875 **To find the thickness of each sheet, divide the height by the number of sheets.**

Each sheet is 0.0875 in. thick.

G. One hundred sheets of clear plastic is 0.05 in. thick. How thick is each sheet? (This is the thickness of some household plastic wrap.)

HOW AND WHY

Objective 2

Write a number in scientific notation or change a number in scientific notation to its place value name.

Scientific notation is widely used in science, technology, and industry to write large and small numbers. Every "scientific calculator" has a key for entering numbers in scientific notation. This notation makes it possible for a calculator or computer to deal with much larger or smaller numbers than those that take up 8, 9, or 10 spaces on the display.

For example, see Table 5.3.

SCIENTIFIC NOTATION

A number in scientific notation is written as the product of two numbers. The first number is between 1 and 10 (including 1 but not 10) and the second number is a power of 10. You must use a "×" to show the multiplication.

TABLE 5.3 SCIENTIFIC NOTATION

Word Name	Place Value Name	Scientific Notation	Calculator or Computer Display
One million	1,000,000	1×10^6	1. 06 or 1 E 6
Five billion	5,000,000,000	5×10^9	5. 09 or 5 E 9
One trillion three billion	1,003,000,000,000	1.003×10^{12}	1.003 12 or 1.003 E 12

Small numbers are shown by writing the power of 10 using a negative exponent. (You will learn more about this when you take a course in algebra.) For now, remember that multiplying by a negative power of 10 is the same as *dividing* by a power of 10, which means you will be moving the decimal point to the left.

Answers to Warm Ups E. 7.867 F. 0.0086395 G. Each sheet of plastic is 0.0005 in. thick.

TABLE 5.4 SCIENTIFIC NOTATION

Word Name	Place Value Name	Scientific Notation	Calculator or Computer Display
Seven thousandths	0.007	7×10^{-3}	7. −03 or 7 E −3
Six ten-millionths	0.0000006	6×10^{-7}	6. −07 or 6 E −7
Fourteen hundred-billionths	0.00000000014	1.4×10^{-10}	1.4 −10 or 1.4 E −10

The shortcut for multiplying by a power of 10 is to move the decimal to the right, and the shortcut for dividing by a power of 10 is to move the decimal to the left.

To write a number in scientific notation

1. Move the decimal point right or left so that only one digit remains to the left of the decimal point. The result will be a number between 1 and 10. If the choice is 1 or 10 itself, use 1.
2. Multiply the decimal found in step 1 by a power of 10. The exponent of 10 to use is one that will make the new product equal to the original number.
   a. If you had to move the decimal to the left, multiply by the same number of 10s as the number of places moved.
   b. If you had to move the decimal to the right, divide (by writing a negative exponent) by the same number of 10s as the number of places moved.

To change from scientific notation to place value name

1. If the exponent of 10 is positive, multiply by as many 10s (move the decimal to the right as many places) as the exponent shows.
2. If the exponent of 10 is negative, divide by as many 10s (move the decimal to the left as many places) as the exponent shows.

For numbers larger than 1:

Place-Value Name:	12,000	3,400,000	12,300,000,000,000	
Number Between 1 and 10:	1.2	3.4	1.23	**Move the decimal (which is after the units place) to the left until the number is between 1 and 10 (one digit to the left of the decimal).**
Scientific Notation:	1.2×10^4	3.4×10^6	1.23×10^{13}	**Multiply each by a power of ten that shows how many places left the decimal moved, or how many places you would have to move to the right to recover the original number.**

For numbers smaller than 1:

Place-Value Name:	0.000033	0.00000007	0.0000000000345	
Number between 1 and 10:	3.3	7.	3.45	**Move the decimal to the right until the number is between 1 and 10.**
Scientific Notation:	3.3×10^{-5}	7×10^{-8}	3.45×10^{-11}	**Divide each by the power of ten that shows how many places right the decimal moved. Show this division by a negative power of 10.**

It is important to note that scientific notation is not rounding. The scientific notation has exactly the same value as the original name.

Examples H–J

Directions: Write in scientific notation.

Strategy: Move the decimal point so that there is one digit to the left. Multiply or divide this number by the appropriate power of 10 so the value is the same as the original number.

H. 782,000,000

7.82 is between 1 and 10 — **Move the decimal eight places to the left.**

$7.82 \times 100,000,000$ is 782,000,000 — **Moving the decimal left is equivalent to dividing by 10 for each place.**

$782,000,000 = 7.82 \times 10^8$ — **To keep the values the same, we multiply by 10 eight times.**

I. 0.0000000092

9.2 is between 1 and 10 — **Move the decimal nine places to the right.**

$9.2 \div 1,000,000,000$ is 0.0000000092 — **Moving the decimal right is equivalent to multiplying by 10 for each place.**

$0.0000000092 = 9.2 \times 10^{-9}$ — **To keep the values the same, we divide by 10 nine times.**

J. Approximately 12,000,000 people in the United States have type II diabetes. Write this number in scientific notation.

1.2 is between 1 and 10.

$1.2 \times 10,000,000 = 12,000,000$

$12,000,000 = 1.2 \times 10^7$

In scientific notation the number of people with type II diabetes is 1.2×10^7.

Warm Ups H–J

H. 13,000,000

I. 0.00000774

J. The age of a 22-yr-old student is approximately 694,000,000 sec. Write this number in scientific notation.

Answers to Warm Ups H. 1.3×10^7 I. 7.74×10^{-6} J. The age of a 22-yr-old student is approximately 6.94×10^8 sec.

Warm Ups K–L

K. Write the place value name for 3.45×10^{-7}.

L. Write the place value name for 9.1×10^{8}.

Examples K–L

Directions: Write the place value name.

Strategy: If the exponent is positive, move the decimal point to the right as many places as shown in the exponent. If the exponent is negative, move the decimal point to the left as many places as shown by the exponent.

K. Write the place value name for 6.7×10^{-5}.

$6.7 \times 10^{-5} = 0.000067$ **The exponent is negative, so move the decimal point 5 places to the left. That is, divide by 10 five times.**

L. Write the place value name for 4.033×10^{10}.

$4.033 \times 10^{10} = 40{,}330{,}000{,}000$ **The exponent is positive, so move the decimal point 10 places to the right. That is, multiply by 10 ten times.**

Answers to Warm Ups K. 0.000000345 L. 910,000,000

NAME ______________________ CLASS __________ DATE __________

Exercises 5.5

OBJECTIVE 1: *Multiply or divide a number by a power of 10.*

A.

Multiply or divide.

1. $4.25 \div 10$

2. $56.98 \div 10$

3. $(3.67)(100)$

4. $28.9(100)$

5. $(0.62833)(1000)$

6. $(34.6211)(1000)$

7. $\dfrac{569.2}{1000}$

8. $\dfrac{9568.3}{1000}$

9. $\dfrac{5645}{100}$

10. $\dfrac{3459}{1000}$

11. 0.87×10^4

12. 4.3×10^5

13. To multiply 4.56 by 10^5 move the decimal point five places to the __________.

14. To divide 4.56 by 10^3 move the decimal point three places to the __________.

B.

Multiply or divide.

15. $(6.274)(1000)$

16. $8.75(100)$

17. $1.85 \div 10$

18. $912.5 \div 1000$

19. $36.9(1000)$

20. $0.6783(10)$

21. $\dfrac{6895.3}{10{,}000}$

22. $\dfrac{213.775}{100{,}000}$

23. $14.78(100{,}000)$

24. $5.732(1{,}000{,}000)$

25. $1367.94 \div 100$

26. $78.94 \div 1000$

27. $45.8 \div 100{,}000$

28. $2.789 \div 1{,}000{,}000$

OBJECTIVE 2: *Write a number in scientific notation or change a number in scientific notation to its place value name.*

A.

Write in scientific notation.

29. 230,000

30. 4700

31. 0.00035

32. 0.0000521

33. 467.95

34. 1245.6

Write the place value name.

35. 6×10^4

36. 7×10^2

37. 8×10^{-3}

38. 2×10^{-5}

39. 4.78×10^3

40. 9.02×10^5

B.

Write in scientific notation.

41. 780,000

42. 4,520,000

43. 0.0000345

44. 0.0007432

45. 0.0000000000821

46. 0.00000002977

47. 3567.003

48. 56.8004

Write the place value name.

49. 1.345×10^{-6}

50. 8.031×10^{-5}

51. 7.11×10^9

52. 8.032×10^8

53. 4.44×10^{-7}

54. 3.9×10^{-10}

55. 5.6723×10^2

56. 7.892111×10^5

C.

57. Ken's Shoe Store buys 100 pairs of shoes that cost \$22.29 per pair. What is the total cost of the shoes?

58. If Mae's Shoe Store buys 100 pairs of shoes for a total cost of \$4897, what is the cost of each pair of shoes?

59. Ms. James buys 100 acres of land at a cost of \$985 per acre. What is the total cost of her land?

60. If 1000 bricks weigh 5900 lb, how much does each brick weigh?

61. The total land area of the Earth is approximately 52,000,000 square miles. What is the total area written in scientific notation?

62. A local computer store offers a small computer with 1152K (1,152,000) bytes of memory. Write the number of bytes in scientific notation.

Exercises 63–65 relate to the chapter application.

In baseball, a hitter's batting average is calculated by dividing the number of hits by the number of times at bat. Mathematically, this number is always between zero and one.

63. In 1988 Wade Boggs led the American League with a batting average of 0.366. However, players and fans would say that Boggs has an average of "three hundred sixty-six." Mathematically, what are they doing to the actual number?

64. Explain why the highest possible batting average is 1.0.

65. The major league player with the highest season batting average in this century is Roger Hornsby of St. Louis. In 1924 he batted 424. Change this to the mathematically calculated number of his batting average.

66. The wave length of a red light ray is 0.000000072 cm. Write this length in scientific notation.

67. The time it takes light to travel 1 km is approximately 0.0000033 sec. Write this time in scientific notation.

68. The speed of light is approximately 1.116×10^7 miles per minute. Write the place value name of this speed.

69. Earth is approximately 1.5×10^8 km from the sun. Write the place value name of this distance.

70. The shortest wavelength of visible light is approximately 4×10^{-5} cm. Write the place value name of this length.

71. A sheet of paper is approximately 1.3×10^{-3} in. thick. Write the place value name of the thickness.

72. A family in the Northeast used 3.276×10^8 BTUs of energy during 1989. A family in the Midwest used 3.312×10^8 BTUs in the same year. A family in the South used 3.933×10^8 BTUs and a family in the West used 1.935×10^8 BTUs. Write the place value name for the total energy usage for the four families.

73. In 1990 the per capita consumption of fish was 15.5 lb. In the same year the per capita consumption of poultry was 63.6 lb and that of red meat was 112.3 lb. Write the total amount of each category consumed by 100,000 people in scientific notation.

74. The population of Cabot Cove was approximately 10,000 in 1995. During the year, the community consumed a total of 276,000 gallons of milk. What was the per capita consumption of milk in Cabot Cove in 1995?

75. In 1980, $24,744,000,000 was spent on air pollution abatement. Ten years later, $26,326,000,000 was spent. In scientific notation, how much more money was spent in 1990 than in 1980? What is the average amount of increase per year during the period?

STATE YOUR UNDERSTANDING

76. Find a pair of numbers whose product is larger than ten trillion. Explain how scientific notation makes it possible to multiply these factors on a calculator. Why is it not possible without scientific notation?

CHALLENGE

77. A parsec is a unit of measure used to determine distance between stars. One parsec is approximately 206,265 times the average distance of the Earth from the sun. If the average distance from the Earth to the sun is approximately 93,000,000 miles, find the approximate length of one parsec. Write the length in scientific notation. Round the number in scientific notation to the nearest hundredth.

78. Light will travel approximately 5,866,000,000,000 miles in one year. Approximately how far will light travel in eight years? Write the distance in scientific notation. Round the number in scientific notation to the nearest thousandth.

Simplify.

79. $\dfrac{(3.25 \times 10^{-3})(2.4 \times 10^{3})}{(4.8 \times 10^{-4})(2.5 \times 10^{-3})}$

80. $\dfrac{(3.25 \times 10^{-7})(2.4 \times 10^{6})}{(4.8 \times 10^{4})(2.5 \times 10^{-3})}$

GROUP ACTIVITY

81. Find the 1990 population for the 10 largest and the 5 smallest cities in your state. Round these numbers to the nearest thousand. Find the total number of pounds of fruit, at the rate of 92.3 lb per person, and the total number of pounds of vegetables, at the rate of 11.2 lb per person, consumed in each of these 15 cities.

MAINTAIN YOUR SKILLS (SECTIONS 1.4, 2.2, 2.3)

Divide.

82. $38\overline{)18{,}050}$

83. $76\overline{)8208}$

84. $103\overline{)21{,}527}$

85. $\dfrac{312}{11}$

86. $\dfrac{3{,}467}{23}$

87. Find the quotient of 297,168 and 123.

88. Find the quotient of 231,876 and 203.

89. Find the quotient of 2,529,948 and 525. Round to the nearest 10.

90. Find the perimeter of a rectangular field that is 645 ft long and 317 ft wide.

91. Find the area of a rectangle that is 3 ft long and 18 in. wide. Find the area in square inches.

5.6

Dividing Decimals

OBJECTIVES

1. Divide decimals.

2. Estimate the quotient of decimals.

HOW AND WHY

Objective 1

Divide decimals.

Division of decimals is the same as division of whole numbers, with one exception. The exception is the location of the decimal point in the quotient.

As with multiplication, we examine the fraction form of division to discover the method of placing the decimal point in the quotient. First, change the decimal form to fraction form to find the quotient. Next, change the quotient to decimal form. Consider the information in Table 5.5.

TABLE 5.5 DIVISION BY A WHOLE NUMBER

Decimal Form	Fraction Form	Division Fraction Form	Division Decimal Form
$3\overline{)0.36}$	$\frac{36}{100} \div 3$	$\frac{36}{100} \div 3 = \frac{\overset{12}{\cancel{36}}}{100} \cdot \frac{1}{\underset{1}{\cancel{3}}} = \frac{12}{100}$	$\overset{0.12}{3\overline{)0.36}}$
$8\overline{)0.72}$	$\frac{72}{100} \div 8$	$\frac{72}{100} \div 8 = \frac{\overset{9}{\cancel{72}}}{100} \cdot \frac{1}{\underset{1}{\cancel{8}}} = \frac{9}{100}$	$\overset{0.09}{8\overline{)0.72}}$
$5\overline{)0.3}$	$\frac{3}{10} \div 5$	$\frac{3}{10} \div 5 = \frac{3}{10} \cdot \frac{1}{5} = \frac{3}{50} = \frac{6}{100}$	$\overset{0.06}{5\overline{)0.3}}$

We can see from Table 5.5 that the decimal point for the quotient of a decimal and a whole number is written directly above the decimal point in the dividend. It may be necessary to insert zeros to do the division. See Example B.

When a decimal is divided by 7, the division process may not have a remainder of zero at any step:

```
   0.97
 7)6.85
   6 3
    55
    49
     6
```

At this step we can write zeros to the right of the digit 5, since 6.85 = 6.850 = 6.8500 = 6.85000 = 6.850000.

```
   0.97857
7)6.85000
  6 3
    55
    49
     60
     56
      40
      35
       50
       49
        1
```

It appears that we might go on inserting zeros and continue endlessly. This is indeed what happens. Such decimals are called "nonterminating, repeating decimals." For example, the quotient of this division is sometimes written

$0.97857142857142\ldots$ or $0.97\overline{857142}$

The bar written above the sequence of digits, 857142, indicates that these digits are repeated endlessly.

In practical applications we stop the division process one place value beyond the accuracy required by the situation and then round. Therefore,

```
   0.97
7)6.85
  6 3
    55
    49
     6     Stop
```

```
   0.9785
7)6.8500
  6 3
    55
    49
     60
     56
      40
      35
       5     Stop
```

$6.85 \div 7 \approx 1.0$ **Rounded to the nearest tenth.**

$6.85 \div 7 \approx 0.979$ **Rounded to the nearest thousandth.**

Now let's examine division when the divisor is also a decimal. We use what we already know about division with a whole number divisor.

TABLE 5.6 DIVISION BY A DECIMAL

Decimal Form	Conversion to a Whole Number Divisor	Decimal Form of the Division
$0.3\overline{)0.36}$	$\frac{0.36}{0.3} \cdot \frac{10}{10} = \frac{3.6}{3} = 1.2$	$0.3\overline{)0.3\,6}$ = 1.2
$0.4\overline{)1.52}$	$\frac{1.52}{0.4} \cdot \frac{10}{10} = \frac{15.2}{4} = 3.8$	$0.4\overline{)1.5\,2}$ = 3.8
$0.08\overline{)0.72}$	$\frac{0.72}{0.08} \cdot \frac{100}{100} = \frac{72}{8} = 9$	$0.08\overline{)0.72}$ = 9.
$0.25\overline{)0.3}$	$\frac{0.3}{0.25} \cdot \frac{100}{100} = \frac{30}{25} = 1.2$	$0.25\overline{)0.30\,0}$ = 1.2
$0.006\overline{)4.8}$	$\frac{4.8}{0.006} \cdot \frac{1000}{1000} = \frac{4800}{6} = 800$	$0.006\overline{)4.800}$ = 800.

We see from Table 5.6 that we moved the decimal point in both the divisor and the dividend the number of places to make the divisor a whole number. Then divide as before.

To divide two numbers

1. If the divisor is not a whole number, move the decimal point in both the divisor and dividend to the right the number of places necessary to make the divisor a whole number.
2. Place the decimal point in the quotient above the decimal point in the dividend.
3. Divide as if both numbers are whole numbers.
4. Round to the given place value. (If no round-off place is given, divide until the remainder is zero or round as appropriate in the problem. For instance, in problems with money, round to the nearest cent.)

Examples A–G

Directions: Divide. Round as indicated.

Strategy: If the divisor is not a whole number, move the decimal point in both the divisor and the dividend to the right the number of places necessary to make the divisor a whole number. The decimal point in the quotient is found by writing it directly above the decimal (as moved) in the dividend.

A. Divide: 15)15.795

```
    1.053
15)15.795
   15
    07
    00
     79
     75
      45
      45
       0
```

The numerals in the quotient are lined up in columns that have the same place value as those in the dividend.

CHECK:

```
   1.053
×     15
   5 265
  10 53
  15.795
```

So the quotient is 1.053.

CAUTION

Write the decimal point for the quotient directly above the decimal point in the dividend.

Warm Ups A–G

A. Divide: 13)14.365

Answer to Warm Up A. 1.105

B. Find the quotient of 2.16 and 16.

B. Find the quotient of 1.88 and 8.

Strategy: Recall that the quotient of a and b is written $a \div b$ or $b\overline{)a}$.

$$\begin{array}{r} 0.23 \\ 8\overline{)1.88} \\ \underline{1\,6} \\ 28 \\ \underline{24} \\ 4 \end{array}$$

Here the remainder is not zero, so the division is not complete. We write a zero on the right (1.880) without changing the value of the dividend and continue dividing.

$$\begin{array}{r} 0.235 \\ 8\overline{)1.880} \\ \underline{1\,6} \\ 28 \\ \underline{24} \\ 40 \\ \underline{40} \\ 0 \end{array}$$

Both the quotient (0.235) and the rewritten dividend (1.880) have three decimal places. Check by multiplying 8 × 0.235.

$$\begin{array}{r} 0.235 \\ \times \quad\;\; 8 \\ \hline 1.880 \end{array}$$

The quotient is 0.235.

C. Divide 357.3 by 23; round the quotient to the nearest hundredth.

C. Divide 589.5 by 21; round the quotient to the nearest hundredth.

$$\begin{array}{r} 28.071 \\ 21\overline{)589.500} \\ \underline{42} \\ 169 \\ \underline{168} \\ 1\,5 \\ \underline{0\,0} \\ 1\,50 \\ \underline{1\,47} \\ 30 \\ \underline{21} \\ 9 \end{array}$$

It is necessary to place two zeros on the right in order to round to the hundredths place, since the division must be carried out one place past the place to which you wish to round.

The quotient is approximately 28.07.

D. Divide: 2.48 ÷ 0.7; round to the nearest hundredth.

D. Divide: 1.32 ÷ 0.7; round to the nearest hundredth.

$0.7\overline{)1.32}$

First, move both decimal points one place to the right so the divisor is the whole number, 7. The same result is obtained by multiplying both divisor and dividend by 10.

$$\begin{array}{r} 1.8 \\ 7\overline{)13.2} \\ \underline{7} \\ 6\,2 \\ \underline{5\,6} \\ 6 \end{array}$$

$$\frac{1.32}{0.7} \times \frac{10}{10} = \frac{13.2}{7}$$

Answers to Warm Ups B. 0.135 C. 15.53 D. 3.54

```
     1.885
  7)13.200
    7
    6 2
    5 6
      60
      56
       40
       35
        5
```

The number of zeros you place on the right depends on either the directions for rounding or your own choice of the number of places. Here we find the approximate quotient rounded to the nearest hundredth.

The quotient is approximately 1.89.

E. Divide 0.57395 by 0.067; round to the nearest thousandth.

0.067)0.57395

```
      8.5664
  67)573.9500
     536
      37 9
      33 5
       4 45
       4 02
         430
         402
          280
          268
           12
```

Move both decimals three places to the right. It is necessary to insert two zeros on the right in order to round to the thousandths place.

The quotient is approximately 8.566.

E. Divide 0.85697 by 0.083; round to the nearest thousandth.

Calculator Example

F. Find the quotient of 92.1936 and 6.705 and round to the nearest thousandth.

$92.1936 \div 6.705 \approx 13.74997763$ **The calculator will automatically place the decimal point in the correct position.**

The quotient is 13.750, to the nearest thousandth.

F. Find the quotient of 453.843 and 7.098 and round to the nearest thousandth.

G. What is the cost per ounce of a 16-oz bottle of iced tea that costs 70¢? This is called the "unit price" and is used for comparing prices. Many stores are required to show this price for the food they sell.

Strategy: To find the unit price (cost per ounce), we divide the cost by the number of ounces.

```
     4.375
  16)70.000
     64
      6 0
      4 8
      1 20
      1 12
         80
         80
          0
```

The iced tea costs 4.375¢ per ounce.

G. What is the unit price of corn chips if a 13.5-oz bag costs $2.97?

Answers to Warm Ups E. 10.325 F. 63.940 G. The unit price of corn chips is $0.22 or 22¢.

HOW AND WHY

Objective 2

Estimate the quotient of decimals.

The quotient of two decimals can be estimated by rounding each number to its largest nonzero place value and then dividing the rounded numbers. For instance,

$0.27\overline{)0.006345}$ $\quad$ $0.3\overline{)0.006}$ $\quad$ **Round each to its largest nonzero place.**

$$\begin{array}{r} 0.02 \\ 3\overline{)0.06} \\ \underline{6} \\ 0 \end{array}$$

Move the decimal points one place to the right so the divisor is a whole number. Divide.

The estimate of the quotient is 0.02.

One use of the estimate is to see if the quotient is correct. If the calculated quotient is not close to 0.02, you should check the division. In this case the actual quotient is 0.0235, which is close to the estimate.

The estimated quotient will not always come out even. When this happens, round the first partial quotient and use it for the first nonzero entry in the quotient. For instance,

$7.41\overline{)34.54542}$ $\quad$ $7\overline{)30}$ $\quad$ **Round each number to its largest nonzero place.**

$$\begin{array}{r} 4 \\ 7\overline{)30} \end{array}$$

Divide 30 by 7 for the first partial quotient. Since 4(7) = 28 and 5(7) = 35, we choose the closer value, 4.

The estimated quotient is 4. The exact quotient is 4.662.

As you get more proficient at estimating, you may want to round the dividend to two nonzero digits to get a closer estimate. In this case we would have

$$\begin{array}{r} 5 \\ 7\overline{)35} \end{array}$$

Round 34.54542 to 35, two nonzero digits.

We now have an estimate of 5, which is closer to the actual quotient 4.662.

To estimate the quotient of decimals

1. Round each decimal to its largest nonzero place.
2. Divide the rounded numbers.
3. If the first partial quotient has a remainder, choose the digit that gives the closer value when multiplied by the divisor.

Example H

Directions: Estimate the quotient and find the quotient to the nearest ten-thousandth.

Strategy: Round each number to its largest nonzero place and then divide the rounded numbers. If the first partial quotient has a remainder, choose the digit that gives the closer value when multiplied by the divisor.

H. Divide 0.05682 by 0.908.

$0.908\overline{)0.05682}$ $\quad 0.9\overline{)0.06}$ **Round. Divide and choose between 0.06 and 0.07.**

$$\begin{array}{r} 0.07 \\ 9\overline{)0.60} \end{array}$$

The estimate is 0.07.

$$\begin{array}{r} 0.06257 \\ 908\overline{)56.82} \\ \underline{54\,48} \\ 2\,340 \\ \underline{1\,816} \\ 5240 \\ \underline{4540} \\ 7000 \\ \underline{6356} \\ 644 \end{array}$$

The estimated quotient is 0.07 and the quotient is 0.0626 to the nearest ten-thousandth.

Warm Up H

H. Divide 0.0039 by 0.723.

Answer to Warm Up H. 0.006; 0.0054

NAME ____________________ CLASS __________ DATE __________

Exercises 5.6

OBJECTIVE 1: *Divide decimals.*

A.

Divide.

1. $7\overline{)3.5}$

2. $6\overline{)4.8}$

3. $2\overline{)19.6}$

4. $3\overline{)19.2}$

5. $0.1\overline{)18.31}$

6. $0.1\overline{)2.72}$

7. $242.4 \div 0.12$

8. $337.7 \div 0.11$

9. To divide 2.65 by 0.05 we first multiply both the dividend and the divisor by 100 so we are dividing by a __________.

10. To divide 0.4763 by 0.287 we first multiply both the dividend and the divisor by __________.

B.

Divide.

11. $80\overline{)1008}$

12. $80\overline{)104.8}$

13. $16.64 \div 32$

14. $3.936 \div (32)$

Divide and round to the nearest tenth.

15. $7\overline{)8.96}$

16. $8\overline{)0.912}$

17. $1.3\overline{)11.778}$

18. $6.7\overline{)34.562}$

Divide and round to the nearest thousandth.

19. $2.2\overline{)34.22}$

20. $24.9\overline{)60.363}$

OBJECTIVE 2: *Estimate the quotient.*

A.

Estimate the quotient.

21. $4\overline{)0.0782}$

22. $7\overline{)0.0359}$

23. $0.2\overline{)1.67}$

24. $0.3\overline{)5.823}$

25. $0.468 \div 0.523$

26. $0.2489 \div 0.1943$

NAME ______________________ CLASS __________ DATE __________

27. $34 \div 0.0756$

28. $76 \div 0.08659$

B.

Estimate the quotient. Divide and round to the nearest hundredth.

29. $64\overline{)6211.84}$

30. $32\overline{)201.824}$

31. $2.97\overline{)0.2267}$

31. $3.46\overline{)0.5699}$

33. $0.12 \div 0.007$

34. $0.15 \div 0.0083$

C.

35. Find the quotient of 9.19 and 0.11, and round to the nearest hundredth.

36. Find the quotient of 0.481 and 94, and round to the nearest hundredth.

Use the table below for Exercises 37–43. The table shows some prices from a grocery store.

Item	Quantity	Price	Item	Quantity	Price
Oranges	3 lb	$4.42	Potatoes	15 lb	$1.16
Strawberries	6 pints	$4.77	Rib Steak	3.24 lb	$9.62
Syrup	36 oz	$2.77	Ham	4.6 lb	$16.05

37. Find the unit price (price per pound) of oranges. Round to the nearest tenth of a cent.

38. Find the unit price (price per ounce) of syrup. Round to the nearest tenth of a cent.

39. Find the unit price of rib steak. Round to the nearest tenth of a cent.

40. Find the unit price of potatoes. Round to the nearest tenth of a cent.

41. Using the unit price find the cost of 7 lb of ham. Round to the nearest cent.

42. Using the unit price find the cost of 11 pints of strawberries. Round to the nearest cent.

NAME ________________ CLASS ________ DATE ________

43. Using unit pricing, find the total cost of 6 lb of rib steak, 2 pints of strawberries, 4 lb of potatoes, and 5 lb of oranges.

44. Ninety-seven alumni of Tech U. donated $7635 to the university. To the nearest cent, what was the average donation?

Estimate the quotient and divide. Round the actual quotient to the nearest hundredth.

45. 83.568 ÷ 0.393

46. 97.342 ÷ 0.3367

47. 0.3491 ÷ 0.0562

48. 0.67321 ÷ 0.03599

49. Vern bought a pair of green socks. The socks are on sale at three pairs for $12.45. How much does he pay for the socks?

50. June drove 354.6 miles on 12.3 gallons of gas. What is her mileage (miles per gallon)? Round to the nearest mile per gallon.

51. A 65-gallon drum of cleaning solvent in an auto repair shop is being used at the rate of 1.94 gallons per day. At this rate, how many days will the drum last? Round to the nearest day.

52. The Williams Construction Company uses cable that weighs 3.5 lb per foot. A partly filled spool of the cable is weighed. The cable itself weighs 813 lb after taking off for the weight of the spool. To the nearest foot, how many feet of cable are on the spool?

53. A plumber connects four buildings' sewers to the public sewer line. The total bill for the job is $6355.48. What is the average cost for each connection?

54. A carpenter works 195.5 hr in one month. How many hours did she work each day if the month contained 23 work days? (Assume that she worked the same number of hours each day.)

55. A 1 ft I beam weighs 32.7 lb. What is the length of a beam weighing 630.6 lb? Find the length to the nearest tenth of a foot.

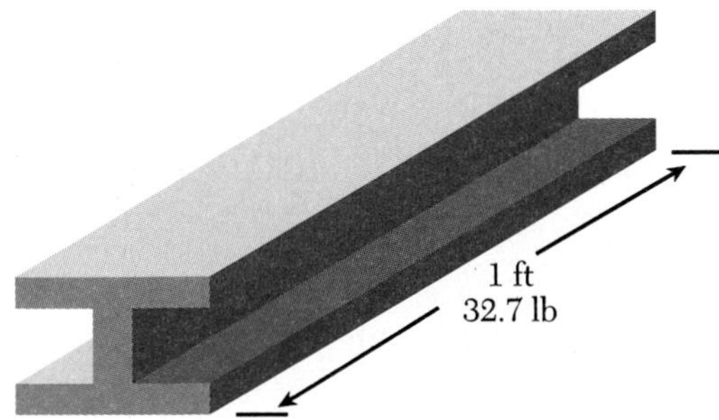

56. A contractor estimates the labor cost of pouring 75 yd^3 of concrete to be \$4050. How much does he allow for each cubic yard of concrete?

57. What is the cost of the illustrated clay flue lining if 56 ft cost \$312.60? Round to the nearest cent.

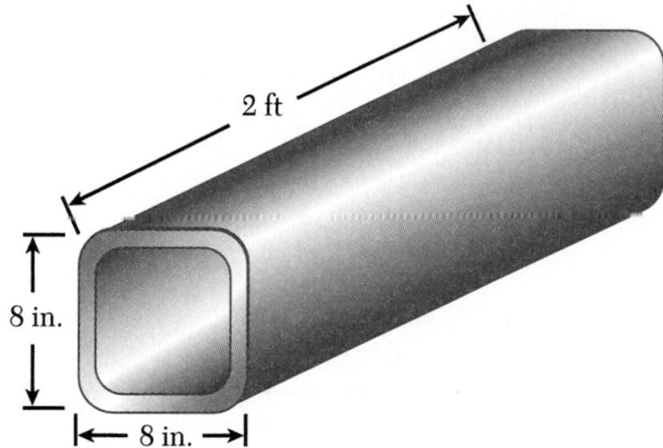

58. Allowing 0.125 in. of waste for each cut, how many bushings, which are 1.25 in. in length can be cut from an 8-in. length of bronze? What is the length of the piece that is left?

Exercises 59–62 relate to the chapter application.

In baseball, a pitcher's earned run average (ERA) is calculated by dividing the number of earned runs by the quotient of the number of innings pitched and 9. The lower a pitcher's ERA the better.

59. Suppose a pitcher allowed 26 earned runs in 80 innings of play. Calculate his ERA and round to the nearest hundredth.

60. A pitcher allows 15 earned runs in 100 innings. Calculate his ERA, rounding to the nearest hundredth.

61. A runner's stolen base average is the quotient of the number of bases stolen and the number of attempts. As with the batting averages, this number is usually rounded to the nearest thousandth. Calculate the stolen base average of a runner who stole 18 bases in 29 attempts.

62. A good stolen base average is 0.700 or higher. Express this as a fraction and say in words what the fraction represents.

STATE YOUR UNDERSTANDING

63. Describe a procedure for determining the placement of the decimal in a quotient. Include an explanation for the justification of the procedure.

64. Explain how to find the quotient of 4.1448 ÷ 0.0012.

65. Copy the table below and fill it in.

Operation on Decimals	Procedure	Example
Division		

NAME ______________________ CLASS __________ DATE __________

CHALLENGE

66. What will be the value of $3000 invested at 0.06 interest compounded quarterly at the end of one year? (Compounded quarterly means that the interest earned for the quarter, the annual interest divided by four, is added to the principal and then earns interest for the next quarter.) How much more is earned by compounding quarterly instead of annually?

GROUP ACTIVITY

67. Determine the distance each member of your group travels to school each day. Find the average distance to the nearest hundredth of a mile for your group. Compare these results with the class. Find the average distance for the entire class. Recalculate the class average after throwing out the longest and the shortest distances. Are the averages different? Why?

68. In diving, a back $3\frac{1}{2}$ somersault, tuck position has a difficulty factor of 3.3. What do the judges need to score this dive in order for it to be worth the same number of points as a perfect cut-through reverse $1\frac{1}{2}$ somersault that has a difficulty factor of 2.6? (See Exercises 66–68 in section 5.4.)

MAINTAIN YOUR SKILLS (SECTIONS 2.1, 4.10, 4.11)

Perform the indicated operations.

69. $2\frac{1}{4} - \frac{3}{4} + 1\frac{5}{8}$

70. $3 + 2\frac{3}{4} - 1\frac{5}{16}$

71. $17 - 5\frac{7}{9}$

72. $23 - 19\frac{15}{16}$

73. Mr. Lewis buys 350 books for $60 at an auction. He sells two fifths of them for $25. He also sells 25 books at $1.50 each, 45 books at $1 each, and gives away the rest. How many books does he give away? What is his total profit if his handling cost is $15?

74. Find the average of $\frac{2}{3}$, $\frac{5}{8}$, and $\frac{2}{9}$.

Find the sum.

75. 3 lb + 8 lb + 11 lb + 9 lb

76. 14 m + 54 m + 67 m + 101 m

77. If a 3-ft 8-in. board is cut into four pieces of equal length, what is the length of each piece in inches? (Assume no waste in cutting the board.)

78. Mary uses 3 quarts and 1 pint from a container containing 4 gallons of soup. How much soup is left in the container?

Getting Ready for Algebra

OBJECTIVE

Solve equations that involve multiplication and division of decimals.

HOW AND WHY

We solve equations that involve multiplication and division of decimals in the same way as equations with whole numbers and fractions.

To solve an equation using multiplication or division

1. Divide both sides of the equation by the same number to isolate the variable, or
2. Multiply both sides of the equation by the same number to isolate the variable.

Examples A–E

Directions: Solve.

Strategy: Isolate the variable by multiplying or dividing each side by the same number.

A. $1.5x = 43.5$

$\frac{1.5x}{1.5} = \frac{43.5}{1.5}$ **Since x is multiplied by 1.5, we divide both sides by 1.5. The division is usually written in fraction form. Since division is the inverse of multiplication, the variable is isolated.**

$x = 29$ **Simplify.**

CHECK:

$1.5(29) = 43.5$ **Substitute 29 for x in the original equation and simplify.**

$43.5 = 43.5$

The solution is $x = 29$.

B. $18.2 = \frac{a}{12.7}$

$12.7(18.2) = (12.7)\frac{a}{12.7}$ **Since a is divided by 12.7, we multiply both sides by 12.7. Since multiplication is the inverse of division, the variable is isolated.**

$231.14 = a$

Warm Ups A–E

A. $1.2t = 84$

B. $10.9 = \frac{r}{6.8}$

Answers to Warm Ups A. $t = 70$ B. $74.12 = r$

CHECK:

$18.2 = \frac{231.14}{12.7}$ **Substitute 231.14 for *a* in the original equation and simplify.**

$18.2 = 18.2$

The solution is $a = 231.14$.

C. $0.075a = 1.065$

C. $22.3y = 62.44$

$\frac{22.3y}{22.3} = \frac{62.44}{22.3}$ **Divide both sides by 22.3 to eliminate the multiplication and simplify.**

$y = 2.8$

CHECK:

$22.3(2.8) = 62.44$ **Substitute 2.8 for *y* in the original equation and simplify.**

$62.44 = 62.44$

The solution is $y = 2.8$.

D. $\frac{x}{0.508} = 3.5$

D. $\frac{c}{0.234} = 1.2$

$(0.234)\frac{c}{0.234} = (0.234)(1.2)$ **Multiply both sides by 0.234 and simplify.**

$c = 0.2808$

CHECK:

$\frac{0.2808}{0.234} = 1.2$ **Substitute 0.2808 for *c* in the original equation and simplify.**

$1.2 = 1.2$

The solution is $c = 0.2808$.

E. Use the formula in Example E to find the number of calories per serving if there is a total of 4039 calories in 35 servings.

E. The total number of calories T is given by the formula $T = sC$, where s represents the number of servings and C represents the number of calories per serving. Find the number of calories per serving in 7.5 servings if the total number of calories is 948.

First substitute the known values into the formula:

$T = sC$

$948 = 7.5C$ **Substitute $T = 948$ and $s = 7.5$.**

$\frac{948}{7.5} = \frac{7.5C}{7.5}$ **Divide both sides by 7.5 to eliminate the multiplication.**

$126.4 = C$

Since $7.5(126.4) = 948$, the number of calories per serving is 126.4.

Answers to Warm Ups C. $a = 14.2$ D. $x = 1.778$ E. There are 115.4 calories per serving.

NAME ______________________ CLASS ____________ DATE ____________

Exercises

Solve.

1. $2.3x = 13.8$

2. $1.7x = 0.408$

3. $0.07y = 14.28$

4. $0.03w = 0.378$

5. $0.123 = 2.05t$

6. $367.72 = 11.6x$

7. $1.1m = 0.044$

8. $0.004p = 8$

9. $0.016q = 7$

10. $6 = 0.004w$

11. $8 = 0.016h$

12. $\frac{x}{2.6} = 9.4$

13. $\frac{y}{8.1} = 0.33$

14. $0.03 = \frac{b}{0.23}$

15. $0.215 = \frac{c}{0.48}$

16. $\frac{w}{0.04} = 2.66$

17. $0.0425 = \frac{x}{23}$

18. $0.09 = \frac{t}{7.14}$

19. $\frac{w}{0.06} = 0.235$

20. $\frac{y}{12.3} = 1.07$

21. $\frac{z}{14.5} = 2.08$

22. $\dfrac{c}{9.07} = 1.003$

23. The total number of calories T is given by the formula $T = sC$, where s represents the number of servings and C represents the number of calories per serving. Find the number of servings if the total number of calories is 3885 and there are 155.4 calories per serving.

24. Use the formula in Exercise 23 to find the number of servings if the total number of calories is 9883.8 and there are 115.6 calories per serving.

25. Ohm's law is given by the formula $E = IR$, where E is the voltage (number of volts), I is the current (number of amperes), and R is the resistance (number of ohms). What is the current in a circuit if the resistance is 15 ohms and the voltage is 126 volts?

26. Use the formula in Exercise 25 to find the current in a circuit if the resistance is 16 ohms and the voltage is 134 volts.

27. Find the length of a rectangle that has a width of 12.2 ft and an area of 262.3 ft^2.

28. Find the width of a rectangular plot of ground that has an area of 5987.64 m^2 and a length of 123 m.

29. Each student in a certain instructor's math classes hands in 20 homework assignments. During the term, the instructor has graded a total of 3500 homework assignments. How many students does this instructor have in all her classes? Write and solve an equation to determine the answer.

30. Twenty-four plastic soda bottles were recycled and made into one shirt. At this rate, how many shirts can be made from 800 soda bottles? Write and solve an equation to determine the answer.

5.7

Another Look at Conversion of Units

OBJECTIVE

Convert units of measure.

HOW AND WHY

Objective

Convert units of measure.

In this section we convert units of measure that require decimal representation. Most of these conversions are approximate, so we round to an indicated place value. For instance, to convert 4.1 ft to yards,

$$4.1 \text{ ft} = \frac{4.1 \cancel{\text{ft}}}{1} \cdot \frac{1 \text{ yd}}{3 \cancel{\text{ft}}} = \frac{4.1 \times 1}{1 \times 3} \text{ yd} = \frac{4.1}{3} \text{ yd} \approx 1.37 \text{ yd}$$ **(to the nearest hundredth)**

Decimal representation provides us with an alternative method for conversion within the metric system. This method utilizes the fact that the metric system has a base-ten place value system. The following conversion chart, based on the prefixes and the base unit, will help. We include in the chart three seldom used prefixes, h for hecto, 100; da for deka, 10; and d for deci, 0.1. For example, 1 hectogram = 100 grams.

k	h	da	base unit	d	c	m

To convert from one metric measure to another, move the decimal point the same number of places and in the same direction as you do to go from the original prefix to the new one on the chart. This is the same process we used to divide or multiply by powers of 10. For instance,

4.5 cg = ? kg

k	h	da	base unit	d	c	m
0	0	0	0	0	4.	5

The "k" prefix is five places to the left of the "c" prefix, so move the decimal point five places to the left.

4.5 cg = 0.000045 kg

Also,

56 kℓ = ? mℓ
56 kℓ = 56,000,000 mℓ

The "m" prefix is six places to the right of the "k" prefix, so move the decimal place six places to the right.

Three other widely used prefixes are

mega -one million -1,000,000
giga -one billion -1,000,000,000
nano -one-billionth -0.000000001

For example, 80 megabytes means 80(1,000,000) bytes or 80,000,000 bytes and 15 nanoseconds means 15(0.000000001) seconds or 0.000000015 seconds. These large and small numbers are used in the sciences and world of computers.

The world is almost entirely metric and the United States is moving in that direction. We are seeing comparisons of the two systems by double listing. This is apparent on marked packages of food products (ounces and grams) and on most speedometers (mph and kph).

We perform conversions between the systems (English and metric) using the basic units. Since the systems were developed independently, there are no exact comparisons. All conversion units are approximate.

With Table 5.7 we can convert between the two systems.

TABLE 5.7 ENGLISH AND METRIC CONVERSIONS

English–Metric Conversions	Metric–English Conversions
1 inch ≈ 2.54 centimeters	1 centimeter ≈ 0.3937 inch
1 foot ≈ 0.3048 meter	1 meter ≈ 3.281 feet
1 yard ≈ 0.9144 meter	1 meter ≈ 1.094 yards
1 mile ≈ 1.609 kilometers	1 kilometer ≈ 0.6214 mile
1 quart ≈ 0.946 liter	1 liter ≈ 1.057 quarts
1 gallon ≈ 3.785 liters	1 liter ≈ 0.2642 gallon
1 ounce ≈ 28.35 grams	1 gram ≈ 0.0353 ounce
1 pound ≈ 453.59 grams	1 gram ≈ 0.0022 pound

For example, convert 45 inches to centimeters.

$$45 \text{ in.} \approx \frac{45 \cancel{\text{in.}}}{1} \cdot \frac{2.54 \text{ cm}}{1 \cancel{\text{in.}}}$$

Multiply by the conversion factor $\frac{2.54 \text{ cm}}{1 \text{ in.}}$ from the table.

$$\approx \frac{45(2.54)}{(1)(1)} \text{ cm}$$

$$\approx 114.3 \text{ cm}$$

So 45 in. is approximately 114.3 cm.

CAUTION

1. Because the conversions in the chart are all rounded to the nearest hundredth, thousandth, or ten-thousandth, we cannot expect closer accuracy when using them.
2. Since there are two or more conversion factors that can be used for each conversion, it is possible for answers to vary slightly, depending on which factors are chosen.

Consider the conversion of 5 m to inches. First we use the conversion factor 1.094 yd ≈ 1 m.

$$\frac{5\text{ m}}{1} \approx \frac{5\text{ m}}{1} \cdot \frac{1.094\text{ yd}}{1\text{ m}} \cdot \frac{3\text{ ft}}{1\text{ yd}} \cdot \frac{12\text{ in.}}{1\text{ ft}}$$
$$\approx 196.92\text{ in.}$$

Now use the conversion factor, 3.281 ft ≈ 1 m.

$$\frac{5\text{ m}}{1} \approx \frac{5\text{ m}}{1} \cdot \frac{3.281\text{ ft}}{1\text{ m}} \cdot \frac{12\text{ in.}}{1\text{ ft}}$$
$$\approx 196.86\text{ in.}$$

The different conversion factors gave different answers. In practice the difference in the results may or may not be significant. If it is, then a more accurate table can be used.

All conversions between the systems in the examples and exercises are rounded to the nearest tenth. Since these conversions are approximate, the results may not always be accurate.

For conversions involving area, Table 5.8 is useful.

TABLE 5.8 AREA CONVERSIONS

$1\text{ ft}^2 = 144\text{ in}^2$	$1\text{ cm}^2 = 100\text{ mm}^2$
$1\text{ yd}^2 = 9\text{ ft}^2$	$1\text{ m}^2 = 10{,}000\text{ cm}^2$
$1\text{ mi}^2 = 3{,}097{,}600\text{ yd}^2$	$1\text{ km}^2 = 1{,}000{,}000\text{ m}^2$
$1\text{ in}^2 \approx 6.4516\text{ cm}^2$	$1\text{ cm}^2 \approx 0.155\text{ in}^2$
$1\text{ ft}^2 \approx 0.0929\text{ m}^2$	$1\text{ m}^2 \approx 10.764\text{ ft}^2$
$1\text{ mi}^2 \approx 2.590\text{ km}^2$	$1\text{ km}^2 \approx 0.3861\text{ mi}^2$

Other conversion tables can be found inside the back cover.

Examples A–J

Directions: Convert the units of measure. If the conversion is not exact, round to the nearest tenth.

Strategy: Multiply the given unit(s) of measure by fractions which show equivalent measures to get the desired unit(s). Simplify.

A. Convert 6.85 pints to gallons.

$$6.85\text{ pt} = \frac{6.85\text{ pt}}{1} \cdot \frac{1\text{ qt}}{2\text{ pt}} \cdot \frac{1\text{ gal}}{4\text{ qt}}$$ **Multiply by the conversion factors.**

$$= \frac{6.85(1)(1)}{1(2)(4)}\text{ gal}$$
$$= 0.85625\text{ gal}$$

So 6.85 pints is 0.85625 gallons.

Warm Ups A–J

A. Convert 0.32 miles to feet.

Answer to Warm Up A. 1689.6 ft

B. Convert 0.43 grams to centigrams.

B. Convert 34.8 meters to kilometers.

Strategy: Use the conversion chart to move the decimal point. This is an exact conversion so we will not round.

34.8 m = 0.0348 km **The "k" unit is three places to the left of the base unit "m." Move the decimal point three places left.**

So 34.8 m = 0.0348 km.

C. Convert 12 ounces to grams.

C. Convert 3 pints to liters.

Strategy: First change pints to quarts, as we have a conversion from quarts to liters.

$$3 \text{ pt} \approx \frac{3 \cancel{\text{pt}}}{1} \cdot \frac{1 \cancel{\text{qt}}}{2 \cancel{\text{pt}}} \cdot \frac{0.946\ \ell}{1 \cancel{\text{qt}}}$$ **Multiply by the conversion factors.**

$$\approx \frac{3(1)(0.946)}{1(2)(1)}\ \ell$$ **Divide. Then round to the nearest tenth.**

$$\approx 1.419\ \ell$$

So 3 pints are approximately 1.4 liters.

D. Convert 64 feet per minute to meters per minute.

D. Convert 55 miles per hour to kilometers per hour.

$$\frac{55 \text{ mi}}{\text{hr}} \approx \frac{55 \cancel{\text{mi}}}{\text{hr}} \cdot \frac{1 \text{ km}}{0.6214 \cancel{\text{mi}}}$$ **Multiply by the conversion factor.**

$$\approx \frac{55(1)}{1(0.6214)} \frac{\text{km}}{\text{hr}}$$

$$\approx 88.5 \frac{\text{km}}{\text{hr}}$$

So 55 miles per hour is approximately 88.5 kph.

E. Convert 2 pounds to kilograms.

E. Convert 2 kilograms to pounds.

$$2 \text{ kg} \approx \frac{2 \cancel{\text{kg}}}{1} \cdot \frac{1000 \cancel{\text{g}}}{1 \cancel{\text{kg}}} \cdot \frac{0.0022 \text{ lb}}{1 \cancel{\text{g}}}$$ **Multiply by the conversion factors.**

$$\approx \frac{2(1000)(0.0022)}{1(1)(1)} \text{ lb}$$

$$\approx 4.4 \text{ lb}$$

So 2 kg is approximately 4.4 lb.

F. Convert 8.95 kiloliters to gallons.

F. Convert 8.3 liters to quarts.

$$8.3\ \ell \approx \frac{8.3 \cancel{\ell}}{1} \cdot \frac{1.057 \text{ qt}}{1 \cancel{\ell}}$$ **Multiply by the conversion factor.**

$$\approx \frac{8.3(1.057)}{1(1)} \text{ qt}$$

$$\approx 8.7731 \text{ qt}$$

So 8.3 liters is approximately 8.8 quarts.

Answers to Warm Ups B. 43 cg C. 340.2 g D. 19.5 meters per minute E. 0.9 kg F. 2364.6 gallons

G. Convert 25 centigrams per centimeter to ounces per inch. Round to the nearest hundredth.

$$\frac{25\text{ cg}}{1\text{ cm}} \approx \frac{25\not{c}\not{g}}{1\not{c}\not{m}} \cdot \frac{1\not{g}}{100\not{c}\not{g}} \cdot \frac{0.0353\text{ oz}}{1\not{g}} \cdot \frac{1\not{c}\not{m}}{0.3937\text{ in.}}$$

$$\approx \frac{25(1)(0.0353)(1)}{1(100)(1)(0.3937)} \frac{\text{oz}}{\text{in.}}$$

$$\approx 0.02\frac{\text{oz}}{\text{in.}}$$

So 25 cg per centimeter is approximately 0.02 oz per inch.

G. Convert 1.3 pounds per foot to grams per meter. Round to the nearest hundredth.

H. Mary is driving an old car on a road in Canada where the speed limit is posted as 80 kilometers per hour. The speedometer is registering 45 miles per hour. Is she driving within the speed limit?

Strategy: Change the posted speed limit to mi/hr and compare with her speed.

$$\frac{80\text{ km}}{1\text{ hr}} \approx \frac{80\not{k}\not{m}}{1\text{ hr}} \cdot \frac{1\text{ mi}}{1.609\not{k}\not{m}}$$

$$\approx \frac{(80)(1)}{1(1.609)} \frac{\text{mi}}{\text{hr}}$$

$$\approx 49.72\frac{\text{mi}}{\text{hr}}$$

The speed limit of 80 kph is approximately 49.7 mph, so Mary is within the limit.

H. Mary is traveling on another road in Canada where the speed limit is posted as 115 kilometers per hour. Can she drive 70 miles per hour and be within the speed limit?

I. Convert 504.8 in^2 to yd^2.

$$504.8\text{ in}^2 = \frac{504.8\not{i}\not{n}^2}{1} \cdot \frac{1\not{f}\not{t}^2}{144\not{i}\not{n}^2} \cdot \frac{1\text{ yd}^2}{9\not{f}\not{t}^2}$$

Multiply by the conversion factors.

$$= \frac{504.8(1)(1)}{1(144)(9)}\text{ yd}^2$$

$$\approx 0.4\text{ yd}^2$$

So 504.8 in^2 is approximately 0.4 yd^2.

I. Convert 0.45 yd^2 to in^2.

J. Convert 3.6 mi^2 to km^2.

$$3.6\text{ mi}^2 \approx \frac{3.6\not{m}\not{i}^2}{1} \cdot \frac{2.590\text{ km}^2}{1\not{m}\not{i}^2}$$

$$\approx \frac{3.6(2.590)}{1(1)}\text{ km}^2$$

$$\approx 9.3\text{ km}^2$$

So 3.6 mi^2 is approximately 9.3 km^2.

J. Convert 7.8 m^2 to ft^2.

Answers to Warm Ups G. 1934.6 grams per meter H. Yes, Mary can travel at 70 mph and be within the speed limit. I. 583.2 in^2 J. 84.0 ft^2

NAME ______________________ CLASS ____________ DATE ____________

Exercises 5.7

OBJECTIVE: *Convert units of measure.*

A.

Convert the units as shown.

1. 8 oz = ? pounds

2. 30 quarts = ? gallons

3. 15 mm = ? cm

4. 600 g = ? kilograms

5. 550 mℓ = ? ℓ

6. 456 mm = ? km

7. 6.9 ft = ? yd

8. 35 min = ? hour

9. 3.02 ft = ? in.

10. 0.7 pint = ? cup

B.

11. 1.83 m = ? mm

12. 4.756 cg = ? mg

13. 4.56 mi = ? ft

14. 3.62 hr = ? minutes

Convert and round to the nearest hundredth. (Answers may vary depending on the conversion factors used.)

15. 16 in. = ? cm

16. 15 lb = ? kg

17. 4.5 ℓ = ? qt

18. 20 yd = ? m

19. 37.9 cm = ? in.

20. 0.85 ℓ = ? qt

21. 9.5 in^2 = ? ft^2

22. 16.85 ft^2 = ? yd^2

23. 53,400 cm^2 = ? m^2

24. 67,930 m^2 = ? km^2

C.

Convert and round to the nearest tenth.

25. 14.8 ft = ? m

26. 19.8 yd = ? m

27. 3.2 kg = ? lb

28. 67.4 km = ? mi

29. 8235 m = ? mi

30. 9.67 mi = ? m

31. 7.5 cups = ? cℓ

32. 0.65 kℓ = ? pints

33. $14.5 \text{ ft}^2 = ? \text{ m}^2$

34. $78.5 \text{ cm}^2 = ? \text{ in}^2$

Convert and round to the nearest hundredth.

35. $\frac{1.5 \text{ lb}}{\text{ft}} = \frac{? \text{ g}}{\text{m}}$

36. $\frac{55 \text{ miles}}{\text{hour}} = \frac{? \text{ meters}}{\text{second}}$

37. $\frac{5.2 \text{ lb}}{\text{ft}^2} = \frac{? \text{ g}}{\text{cm}^2}$

38. $\frac{45 \text{ ft}}{\text{sec}} = \frac{? \text{ m}}{\text{min}}$

39. $\frac{525 \text{ g}}{\ell} = \frac{? \text{ lb}}{\text{qt}}$

40. $\frac{\$1.52}{\text{lb}} = \frac{? \ \$}{\text{kg}}$

41. A box of Wheat Bran Flakes weighs 18.9 oz. To the nearest tenth, how many grams does it weigh?

42. A farmer's harvest of strawberries averages 10 tons per acre. To the nearest tenth, express the harvest in kilograms per acre.

43. The Heatherton Corporation reimburses its employees 30¢ per mile when they use their private car on company business. The company is opening a plant in Canada. What reimbursement per kilometer should the company pay, to the nearest cent per kilometer?

44. The Williams Company advertises that a water pump will deliver water at 35 gallons per minute. Nyen needs to put an ad in a Mexican paper. How many liters per minute should he put in the ad, to the nearest liter per minute?

45. A newborn baby elephant at the Oxnard Zoo weighs 295 lb. To the nearest tenth, express the weight in kilograms.

46. The Japanese fishing fleet has a weekly catch of 110,000 kg of whiting. To the nearest ton, express the catch in tons.

47. The Georgia Pacific Corporation grows trees for poles. The length of a typical pole is 95 ft. To the nearest tenth, express the length in meters.

48. The average length of a salmon returning to the Oakridge Hatchery is 37 in. To the nearest tenth, express the length in centimeters.

49. The Coffee Company packages 3 kg of coffee that sells in Canada for \$18.95. The company also packages 6 lb of coffee that sells for \$17.95 in the United States. Assuming both dollar amounts are in American currency, which is the better buy?

50. Laura drives 67 miles to work each day. Her cousin Mabel, who lives in England, drives 115 km to work each day. Who has the shorter drive?

51. The average rainfall in Freeport, Florida, is 26.83 in. Express the average to the nearest tenth of a centimeter.

52. The Toyo Tire Company recommends that its premium tire be inflated to 1.5 kg per square centimeter. In preparation to sell the tire in America, Mishi must convert the pressure to pounds per square inch. What measure, to the nearest pound per square inch, must she use?

 Exercises 53–55 relate to the chapter application.

In swimming in the United States, there is a short-course season that takes place in 25-yd pools, and a long-course season that takes place in 50-m pools. Consequently, swimmers gather both yard and meter times in their events. Occasionally, for purposes of qualification for major meets, it is necessary to convert a yard time into its equivalent meter time.

53. Doug can swim the 100-yd breaststroke in 1 min 10 sec. Express this rate as seconds per yard. Then convert to seconds per meter, to the nearest ten-thousandth.

54. Using your results from Exercise 53, what should Doug's time be for the 100-m breaststroke? Round to the the nearest hundredth.

55. In doing yard-to-meter swimming conversions, there is an extra consideration, the number of turns in the race. In a 100-yd race, there are four lengths of the pool and 3 turns. By contrast in a 100-m race, there are two lengths of the pool and only one turn. When converting times, some allowance must be made for the extra turns in the yard pool. Since turning is faster than straight swimming, it is usual to add 1 sec to the yard time for each extra turn and then convert to meters. Following this rule, what is Doug's expected time for the 100-m breaststroke?

56. In 1995 Miguel Indurain won the Tour de France, completing the 3635-km course in a total time of 92 hr 44 min 59 sec. Calculate his average speed in kilometers per hour and in miles per hour. Round to the nearest hundredth.

57. A ski resort in Canada reports a snow pack of 275 cm. How many feet of snow pack should be reported to the American press? Round to the nearest tenth of a foot.

58. A ski resort in Idaho reports a snow pack of 156 in. How many centimeters of snow pack should be reported to the Canadian press? Round to the nearest centimeter.

59. How many tablets, each containing 235 mg of aspirin, does it take to make 1 oz of aspirin?

60. How many tablets of lisinopril can be made from 1 oz if each tablet contains 5 mg?

STATE YOUR UNDERSTANDING

61. Explain why converting units within the same system results in exact answers while converting units between systems results in approximate answers.

62. Explain why you might get different answers when converting 25 m to feet using conversion units from the chart in this section.

63. Explain how to convert 40 mph to kilometers per minute.

CHALLENGE

64. Convert $\frac{110 \text{ lb}}{\text{ft}^2}$ to $\frac{\text{kg}}{\text{m}^2}$.

65. Mary has 6 gallons 3 quarts 1 pint of blackberry juice. To the nearest tenth, how many liters does she have?

66. The Candy Basket receives an order of candy from Europe that costs \$21 per kilogram. The store plans to sell the candy for \$14.50 per pound. To the nearest dollar, what is the profit on the sale of 100 lb of candy?

GROUP ACTIVITY

67. Write a brief history of the metric system, including the controversy over the change in Canada, Great Britain, and the United States. Present your findings as an oral report in class.

MAINTAIN YOUR SKILLS (SECTIONS 2.2, 2.3, 4.1, 4.2, 4.3)

Simplify.

68. $\frac{95}{114}$

69. $\frac{168}{216}$

Write as an improper fraction.

70. $4\frac{8}{11}$

71. $18\frac{5}{7}$

Write as a mixed number.

72. $\frac{215}{12}$

73. $\frac{459}{25}$

Perform the indicated operation. Simplify.

74. $\left(\frac{10}{63}\right)\left(\frac{15}{16}\right)\left(\frac{24}{25}\right)$

75. $\frac{25}{36} \div \frac{75}{96}$

Use the diagram for exercises 76–77.

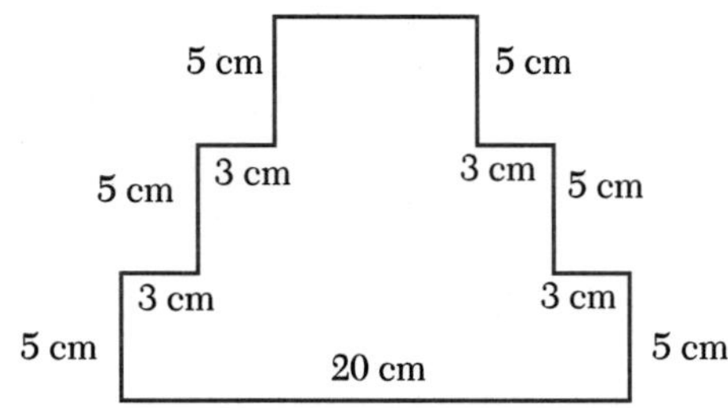

76. Find the perimeter of the figure.

77. Find the area of the figure.

5.8

Changing Fractions to Decimals

OBJECTIVE

Change fractions to decimals.

HOW AND WHY

Objective

Change fractions to decimals.

Every decimal can be written as a whole number times the place value of the last digit on the right:

$$0.73 = 73 \times \frac{1}{100} = \frac{73}{100}$$

The fraction has a power of 10 for the denominator. Any fraction that has only prime factors of 2 and 5 in the denominator can be written as a decimal by building the denominator to a power of 10.

$$\frac{3}{5} = \frac{3}{5} \cdot \frac{2}{2} = \frac{6}{10} = 0.6$$

$$\frac{7}{20} = \frac{7}{20} \cdot \frac{5}{5} = \frac{35}{100} = 0.35$$

Every fraction can be thought of as a division problem $\left(\frac{3}{5} = 3 \div 5\right)$. Therefore, another method for changing fractions to decimals is division. As you discovered in a previous section, many division problems with decimals do not have a zero remainder at any point. If the denominator of the fraction has prime factors other than 2 or 5, the quotient could be a nonterminating decimal. The fraction $\frac{1}{6}$ is an example:

$$\frac{1}{6} = 0.166666666\ldots \quad \text{or} \quad 0.1\overline{6}$$

The bar over the 6 indicates that the decimal repeats the number 6 forever. In the exercises for this section, round the division to the indicated decimal place or use the repeat bar as directed.

CAUTION

Be careful to use an equal sign (=) when your conversion is exact and an approximately equal sign (≈) when you have rounded.

To change a fraction to a decimal

Divide the numerator by the denominator.

To change a mixed number to a decimal

Change the fraction part to a decimal and add to the whole number part.

Warm Ups A–G

A. Change $\frac{13}{20}$ to a decimal.

B. Change $6\frac{21}{25}$ to a decimal.

Examples A–G

Directions: Change the fraction or mixed number to a decimal.

Strategy: Divide the numerator by the denominator. Round as indicated. If the number is a mixed number, add the decimal to the whole number.

A. Change $\frac{9}{40}$ to a decimal.

$$\begin{array}{r} 0.225 \\ 40\overline{)9.000} \\ \underline{8\,0} \\ 1\,00 \\ \underline{80} \\ 200 \\ \underline{200} \\ 0 \end{array}$$

Divide the numerator 9 by the denominator 40.

Therefore, $\frac{9}{40} = 0.225$.

B. Change $9\frac{13}{50}$ to a decimal.

$$\begin{array}{r} 0.26 \\ 50\overline{)13.00} \\ \underline{10\,0} \\ 3\,00 \\ \underline{3\,00} \\ 0 \end{array}$$

Divide the numerator 13 by the denominator 50.

$9\frac{13}{50} = 9.26$ **Add the decimal to the whole number.**

or

$\frac{13}{50} = \frac{13}{50} \cdot \frac{2}{2} = \frac{26}{100} = 0.26$ **A fraction with a denominator that has only 2 or 5 for prime factors can be changed to a decimal by building.**

So, $9\frac{13}{50} = 9 + 0.26 = 9.26$.

Answers to Warm Ups A. 0.65 B. 6.84

C. Change $\frac{13}{16}$ to a decimal.

$$\begin{array}{r} 0.8125 \\ 16\overline{)13.0000} \\ \underline{12\,8} \\ 20 \\ \underline{16} \\ 40 \\ \underline{32} \\ 80 \\ \underline{80} \\ 0 \end{array}$$

This fraction can be changed by building to a denominator of 10,000 but the factor is not easily recognized, unless we use a calculator.

$$\frac{13}{16} = \frac{13}{16} \cdot \frac{625}{625} = \frac{8125}{10{,}000} = 0.8125$$

So, $\frac{13}{16} = 0.8125$.

C. Change $\frac{33}{40}$ to a decimal.

CAUTION

Most fractions cannot be changed to terminating decimals because the denominators contain factors other than 2 and 5. In these cases we round to the indicated place value or use a repeat bar.

D. Change $\frac{11}{12}$ to a decimal rounded to the nearest hundredth.

$$\begin{array}{r} 0.916 \\ 12\overline{)11.000} \\ \underline{10\,8} \\ 20 \\ \underline{12} \\ 80 \\ \underline{72} \\ 8 \end{array}$$

Divide 11 by 12. Carry out the division to three decimal places and round to the nearest hundredth.

So, $\frac{11}{12} \approx 0.92$.

D. Change $\frac{5}{7}$ to a decimal rounded to the nearest hundredth.

E. Change $\frac{5}{9}$ to a decimal. Do not round.

$$\begin{array}{r} .555 \\ 9\overline{)5.000} \\ \underline{4\,5} \\ 50 \\ \underline{45} \\ 50 \\ \underline{45} \\ 5 \end{array}$$

We see that the division will not have a zero remainder. So we use the repeat bar to show the quotient.

So $\frac{5}{9} = 0.5\overline{5}$ or $0.\overline{5}$.

E. Change $\frac{3}{11}$ to a decimal. Do not round.

Answers to Warm Ups C. 0.825 D. 0.71 E. $0.\overline{27}$

Calculator Example

F. Change $\frac{121}{147}$ to a decimal rounded to the nearest thousandth.

F. Change $\frac{87}{123}$ to a decimal rounded to the nearest ten-thousandth.

$87 \div 123 \approx 0.7073170732$

So $\frac{87}{123} \approx 0.7073$ to the nearest ten-thousandth.

G. Change the measurements on the figure to the nearest tenth for use with a ruler marked in tenths.

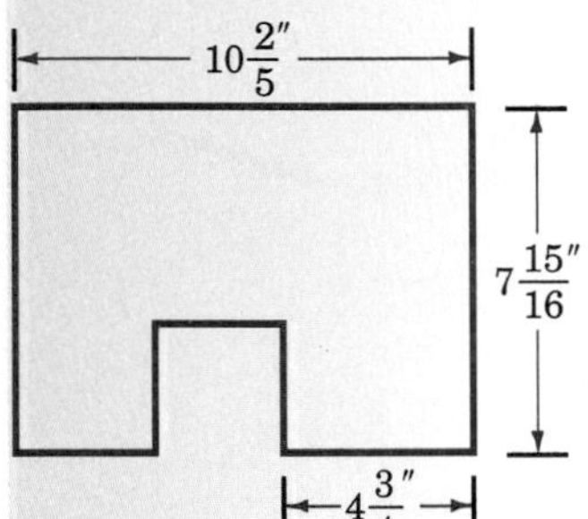

G. Jan needs to make a pattern of the shape shown below. Her ruler is marked in tenths. Change all the measurements to tenths so she can make an accurate pattern.

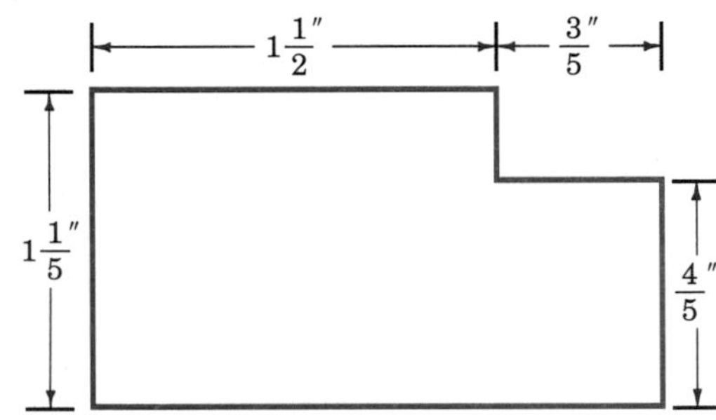

So that Jan can use her ruler for more accurate measure, each fraction is changed to a decimal rounded to the nearest tenth.

$1\frac{1}{2} = 1\frac{5}{10} = 1.5$

$\frac{3}{5} = \frac{6}{10} = 0.6$

$1\frac{1}{5} = 1\frac{2}{10} = 1.2$

$\frac{4}{5} = \frac{8}{10} = 0.8$

Each fraction and mixed number can be changed by either building each to a denominator of 10 as shown or by dividing the numerator by the denominator. The measurements on the drawing can be labeled:

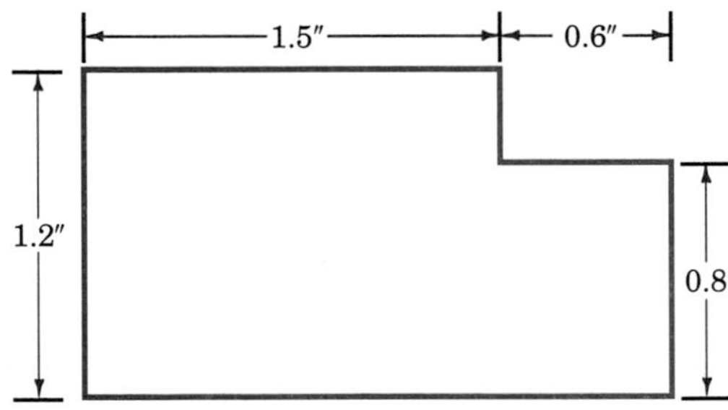

Answers to Warm Ups F. 0.823 G. The decimal measures are $10\frac{2}{5}$ in. = 10.4 in.; $7\frac{15}{16}$ in. ≈ 7.9 in.; and $4\frac{3}{4}$ in. ≈ 4.8 in.

NAME ______________________ CLASS ____________ DATE ____________

Exercises 5.8

OBJECTIVE: *Change fractions to decimals.*

A.

Change the fraction or mixed number to a decimal.

1. $\frac{3}{4}$ **2.** $\frac{7}{10}$ **3.** $\frac{3}{8}$ **4.** $\frac{5}{8}$

5. $\frac{11}{16}$ **6.** $\frac{1}{32}$ **7.** $3\frac{11}{20}$ **8.** $6\frac{13}{20}$

9. $11\frac{3}{125}$ **10.** $12\frac{7}{50}$

B.

Change to a decimal rounded to the indicated place value.

	Tenth	Hundredth		Tenth	Hundredth
11. $\frac{3}{7}$			**16.** $\frac{3}{14}$		
12. $\frac{2}{9}$			**17.** $\frac{8}{15}$		
13. $\frac{5}{11}$			**18.** $\frac{11}{19}$		
14. $\frac{5}{6}$			**19.** $5\frac{17}{18}$		
15. $\frac{2}{13}$			**20.** $11\frac{9}{17}$		

Change each of the following fractions to decimals. Use the repeat bar.

21. $\frac{5}{11}$ **22.** $\frac{7}{9}$ **23.** $\frac{5}{12}$ **24.** $\frac{8}{15}$

C.

Change each of the following fractions to decimals to the nearest indicated place value.

	Hundredth	Thousandth
25. $\frac{15}{43}$		
26. $\frac{12}{31}$		
27. $\frac{28}{57}$		
28. $\frac{85}{91}$		

Change to a decimal. Use the repeat bar.

29. $\frac{5}{13}$ **30.** $\frac{11}{12}$ **31.** $\frac{6}{7}$ **32.** $\frac{23}{26}$

33. A piece of blank metal stock is $1\frac{5}{8}$ in. in diameter. A micrometer measures in decimal units. If the stock is measured with the micrometer, what will the reading be?

34. A wrist pin is $\frac{15}{16}$ in. in diameter. What would the micrometer reading be?

35. Convert the measurements in the following drawing to decimals:

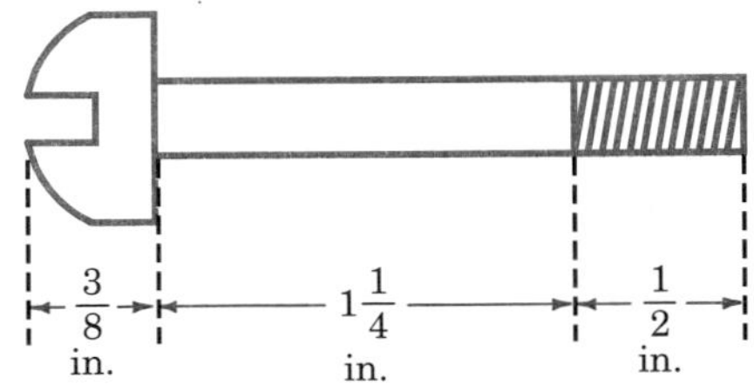

NAME ______________________ CLASS ____________ DATE ____________

36. Stephen needs $\frac{27}{40}$ in. of rope. What is the decimal equivalent?

Change to a decimal. Round as indicated.

	Hundredths	Thousandths	Ten-thousandths
37. $\frac{31}{33}$			
38. $21\frac{42}{57}$			
39. $7\frac{81}{133}$			
40. $\frac{467}{1533}$			

41. A remnant of material $1\frac{1}{4}$ yd long costs \$6.08. Find the cost per yard of the fabric using fractions. Recalculate the same cost using decimals. Which is easier? Why?

42. A telephone lobbyist works $37\frac{3}{4}$ hr during one week. If he is paid \$45.40 per hour, compute his gross wages for the week. Did you use decimals or fractions to do the calculation? Why?

43. In a pole vault meet conducted by telephone, the highest jump in Iowa is $23\frac{7}{16}$ ft. The highest jump in Texas is 23.439 ft. Which state has the winning jump?

STATE YOUR UNDERSTANDING

44. Write a short paragraph on the uses of decimals and of fractions. Include examples in which fractions are more useful and examples in which decimals are more fitting.

CHALLENGE

45. Which is larger, 0.0036 or $\frac{2}{625}$?

46. Which is larger, 2.5×10^{-4} or $\frac{3}{2000}$?

47. First decide whether the fraction $\frac{1.23}{80}$ is more or less than 0.1. Then change the fraction to a decimal. Were you correct in your estimate?

48. First decide whether the fraction $\frac{62}{0.125}$ is more or less than 100. Then change the fraction to a decimal. Were you correct in your estimate?

GROUP ACTIVITY

49. Select 10 stocks from the NYSE and assume you have 1000 shares of each. From today's paper, calculate the current value of your holdings. Using the reported changes, calculate the value of your holdings yesterday. How much money did you make or lose?

MAINTAIN YOUR SKILLS (SECTIONS 5.3, 5.4)

Add.

50. $0.32 + 0.024 + 3.1 + 7.214$

51. $375.3 + 2.8041 + 1984.7$

52. $3475 + 3.475 + 34.75 + 377.05$

53. $314.2 + 2.314 + 42.13 + 4312$

Subtract.

54. $48 - 27.83$

55. $49.721 - 36$

56. $17.66 - 16$

57. $17 - 16.66$

58. A flight of stairs has five risers. What is the height of the flight of stairs if each riser is 7.5 in. high?

59. Katya made the following transactions in her checking account for June. With a starting balance of $65.87, she made a deposit of $562.34, wrote checks for $255, $34.22, and $75.29, and made another deposit of $281.17. What is the balance of her account now?

5.9

Order of Operations; Average

OBJECTIVES

1. Do any combination of operations with decimals.
2. Find the average of a set of decimals.

HOW AND WHY

Objective 1

Do any combination of operations with decimals.

The order of operations for decimals is the same as that for whole numbers and fractions.

Order of operations

To simplify an expression with more than one operation follow these steps:

1. Parentheses — Do the operations within grouping symbols first (parentheses, fraction bar, etc.), in the order given in steps 2, 3, and 4.
2. Exponents — Do the operations indicated by exponents.
3. Multiply and divide — Do only multiplication and division as they appear from left to right.
4. Add and subtract — Do addition and subtraction as they appear from left to right.

Many formulas involve the number pi, represented by the symbol π. (See Section 4.11.) Pi is found by dividing the circumference of a circle by its diameter. Pi has no exact decimal representation. We have already used $\frac{22}{7}$ to approximate pi.

Using decimals we can get a better approximation, $\pi \approx 3.14159$. For the purposes of this text we will round the value of pi to the nearest hundredth, so we will let $\pi \approx 3.14$. Scientific calculators have a key, $\boxed{\pi}$, that gives π to eight or ten decimal places. When using your calculator, use the π key for greater accuracy.

Examples A–F

Directions: Perform the indicated operations.

Strategy: Use the same order of operations as for whole numbers and fractions.

A. Simplify: 0.63 − 0.26(0.25)

0.63 − 0.26(0.25) = 0.63 − 0.065	**Multiplication is done first.**
= 0.565	**Subtract.**

So 0.63 − 0.26(0.25) = 0.565.

Warm Ups A–F

A. Simplify: 0.82 − 0.19(0.3)

Answer to Warm Up A. 0.763

B. Simplify: $0.016 \div 0.5(3.17)$

B. Simplify: $0.88 \div 0.11(3.02)$

$0.88 \div 0.11(3.02) = 8(3.02)$ **Division is done first, since it occurs first.**

$= 24.16$ **Multiply.**

So $0.88 \div 0.11(3.02) = 24.16$.

C. Simplify: $(2.1)^2 - (0.3)^4$

C. Simplify: $(5.1)^2 - (1.2)^3$

$(5.1)^2 - (1.2)^3 = 26.01 - 1.728$ **Exponents are done first.**

$= 24.282$ **Subtract.**

So $(5.1)^2 - (1.2)^3 = 24.282$.

Calculator Example

D. Simplify: $7.634 \div 2.2 + (0.34)(12) - 5.6$

D. Simplify: $4.94 \div 1.9 + (3.4)(6.23) + 4.81$

Strategy: All but the least expensive calculators have algebraic logic. The operations can be entered in the same order as the exercise.

So $4.94 \div 1.9 + (3.4)(6.23) + 4.81 = 28.592$.

E. A second trough in a feed lot has a radius r of 3.75 ft and a length of 17.82 ft. Using the formula in Example E, find the volume of this trough. Round to the nearest tenth.

E. The formula for the volume of one half a cylindrical tank is $V = \frac{1}{2}\pi r^2 h$. These tanks are often used for watering cattle or for feed troughs. Find the volume if $\pi \approx 3.14$, $r = 2.25$ ft, and $h = 12.35$ ft. Round to the nearest tenth.

$V = \frac{1}{2}\pi r^2 h$

$V \approx 0.5(3.14)(2.25)^2(12.35)$ **Substitute; $\frac{1}{2} = 0.5$.**

≈ 98.159343

So the volume is approximately 98.2 ft^3.

F. A mall is to have a central area in the form of a circle. It will be constructed with decorative bricks with a concrete base. If the circle is to have a diameter of 92 feet, how many square feet of bricks will the central area have? Round to the nearest square foot.

F. A shopping mall is to have a central patio in the form of a circle. It will be constructed with decorative bricks with a concrete base. If the circle is to have a diameter of 75 feet, how many square feet of bricks will the patio have? Round to the nearest square foot.

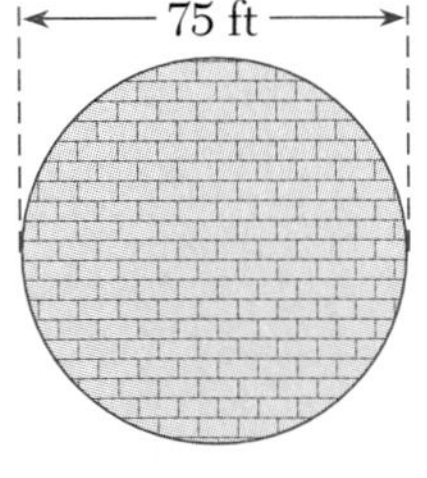

Formula:

$A = \pi r^2$ **Area of a circle.**

$\approx 3.14\left(\frac{75}{2}\right)^2$ **Substitute; r is one half the diameter. Let $\pi \approx 3.14$.**

$\approx 3.14(37.5)^2$

$\approx 3.14(1406.25)$

≈ 4415.625

The mall patio will have about 4416 ft^2 of brick.

Answers to Warm Ups B. 0.10144 C. 4.4019 D. 1.95 E. The volume is approximately 393.4 ft^3. F. The central area of the mall will have about 6644 ft^2 of brick.

HOW AND WHY

Objective 2

Find the average of a set of decimals.

The method for finding the average of a set of decimals is the same as that for whole numbers and fractions.

To find the average of a set of decimals

1. Add the numbers.

2. Divide the sum by the number of numbers in the set.

Examples G–H

Directions: Find the average.

Strategy: Use the same procedure as for whole numbers and fractions.

G. Find the average of 0.35, 0.65, 1.76, and 0.08.

$0.35 + 0.65 + 1.76 + 0.08 = 2.84$ **First add the numbers.**

$2.84 \div 4 = 0.71$ **Second, divide by four, the number of numbers.**

So the average of 0.35, 0.65, 1.76, and 0.08 is 0.71.

H. Pedro's grocery bills for the past five weeks were:

Week 1: $134.86

Week 2: $210.54

Week 3: $178.10

Week 4: $143.92

Week 5: $207.63

What is Pedro's average cost of groceries per week for the five weeks?

$$\begin{array}{r} 134.86 \\ 210.54 \\ 178.10 \\ 143.92 \\ +\underline{207.63} \\ 875.05 \end{array}$$

Add the weekly totals and divide by 5, the number of weeks.

$875.05 \div 5 = 175.01$

Pedro's average weekly cost of groceries is $175.01.

Warm Ups G–H

G. Find the average of 7.8, 0.55, 6.7, 1.01, and 0.04.

H. Mary's weekly car expenses, including parking, for the past six weeks were:

Week 1: $45.85

Week 2: $38.42

Week 3: $49.93

Week 4: $55.86

Week 5: $73.65

Week 6: $35.75

What is Mary's average weekly car expense for the six weeks?

Answers to Warm Ups G. 3.22 H. Mary's average weekly car expense is $49.91.

NAME ______________________ CLASS ______________ DATE ______________

Exercises 5.9

OBJECTIVE 1: *Do any combination of operations with decimals.*

A.

Perform the indicated operations.

1. $0.7 + 0.5 - 0.3$

2. $0.8 - 0.5 + 0.6$

3. $0.24 \div 8 - 0.01$

4. $0.32 \div 4 + 0.07$

5. $2.4 - 3(0.7)$

6. $3.6 + 3(0.2)$

7. $2(1.2) + 3(2.1)$

8. $3(1.5) - 0.3(10)$

9. $0.16 + (0.2)^2$

10. $0.52 - (0.4)^2$

B.

11. $5.13 - 3.75 + 0.85 - 1.71$

12. $0.09 + 3.65 - 2.17 + 0.89$

13. $4.2 \div 3.5(2.7)$

14. $36.04 \div 6.8(0.15)$

15. $(15.6)(2.5) \div (0.3)$

16. $(7.5)(3.42) \div 0.15$

17. $(7.3)^2 - 8.3(2.1)$

18. $(5.3)^2 + 4.8 \div 0.25$

19. $(6.7)(1.4)^3 \div 0.7$

20. $(3.1)^3 - (0.8)^2 + 4.5$

OBJECTIVE 2: *Find the average of a set of decimals.*

A.

Find the average.

21. 4.5, 6.5

22. 3.6, 5.2

23. 8.2, 9.6

24. 7.6, 5.4

25. 4.5, 3.3, 1.2

26. 1.2, 3.6, 1.8

27. 4.1, 4.3, 4.2

28. 2.7, 3.4, 1.4

29. 3.1, 2.1, 5.1, 2.1

30. 4.2, 5.2, 6.2, 9.2

B.

31. 4.5, 8.12, 7.3, 5.1

32. 7.32, 9.45, 2.4, 6.1

33. 16.4, 8.72, 11.5, 3.23

34. 3.4, 8.2, 5.43, 7.2

35. 5.5, 17.3, 4.7, 23.5, 9.6

36. 45.3, 67.2, 30.2, 12.5, 27.1

37. 0.234, 0.451, 0.673, 0.924

38. 0.321, 0.843, 0.553, 0.213

39. 0.4523, 0.5421, 0.8674, 0.9234, 0.4535

40. 1.234, 4.678, 14.89, 0.872, 8.775

C.

41. Elmer goes shopping and buys 3 cans of cream-style corn at 69¢ per can, 4 cans of tomato soup at 89¢ per can, 2 bags of corn chips at $2.19 per bag, and 6 candy bars at 55¢ each. How much did Elmer spend?

42. Christie buys school supplies for her children. She buys 6 pads of paper at $1.10 each, 5 pens at 89¢ each, 4 erasers at 39¢ each, and 4 boxes of crayons at $1.49 each. How much does she spend?

43. The common stock of Oracle Corporation closed at \$49.75, \$51.375, \$52.75, \$48.875, and \$50.5 during one week in 1997. What is the average closing price of the stock?

44. A consumer watchdog group priced a can of a certain brand of soup at six different grocery stores. They found the following prices: \$2.19, \$2.25, \$1.95, \$1.89, \$2.05, and \$2.40. What is the average selling price of a can of the soup? Round to the nearest cent.

Perform the indicated operations.

45. $(5.4)(9.1) - (3.3)(6.7) + (3.2)^2$

46. $8.9 + 3.45(2.1) - 5.6(0.21) + 11$

47. $10.075 - [(3.4)(2.56) + 1.01]$

48. $6.3 - [(1.3)^2 - 0.95]$

49. $8.5(4.35 + 2.37 - 2.21) - 5.62(0.15)$

50. $7.8(5.61 - 3.46 + 4.44) - 3.4(2.25)$

51. The Adams family had the following natural gas bills for last year:

January	$145.75	July	$12.75
February	132.83	August	8.93
March	84.87	September	14.62
April	69.85	October	29.91
May	55.43	November	34.72
June	25.67	December	83.45

The gas company will allow them to make equal payments this year equal to the monthly average of last year. How much will the payment be? Round to the nearest cent.

52. Crescent City police report the following number of automobile accidents for a week:

Monday	27	Tuesday	43
Wednesday	17	Thursday	38
Friday	24	Saturday	26
Sunday	9		

To the nearest tenth, what is the average number of accidents reported per day?

53. Tim Raines, the fullback for the East All-Stars, gained the following yards in six carries: 5.6 yd, 3.5 yd, 7.8 yd, 12 yd, 10.5 yd, and 1 yd. What was the average gain per carry? Round to the nearest tenth of a yard.

54. The price per gallon of the same grade of gasoline at eight different service stations is 1.485, 1.399, 1.409, 1.385, 1.415, 1.395, 1.425, and 1.379. What is the average price per gallon at the eight stations? Round to the nearest thousandth.

55. Pens cost $1.19 for two, while pencils cost 45¢ for four. What is the total cost of 20 of each?

56. The wholesale cost of shampoo is \$1.11 per bottle, while the wholesale cost of conditioner is \$0.89. The Fancy Hair Beauty Salon sells the shampoo for \$8.49 a bottle and the conditioner for \$8.19 a bottle. What is the net income on the sale of a case, 24 bottles, of each product?

57. The stage at an outdoor amphitheater is in the shape of a semicircle of radius 125 feet. Find the area (A) of the stage, $A = \dfrac{\pi r^2}{2}$ and $\pi \approx 3.14$.

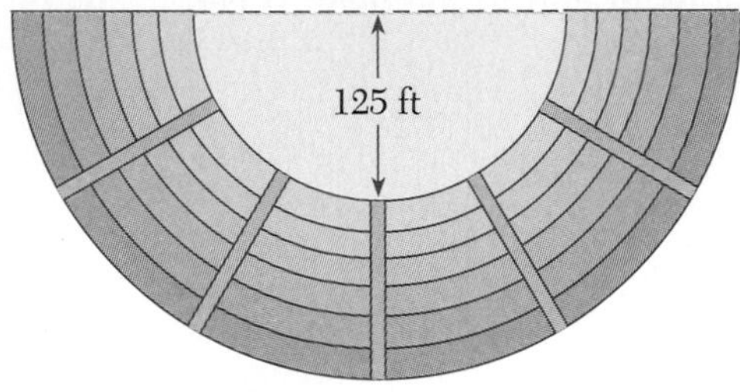

58. The plaza at the new mall in downtown Seattle is in the shape of a square with a semicircle at one end. If the side of the square is 60 feet, how many square feet are in its area? $A = s^2 + \dfrac{\pi r^2}{2}$, $\pi \approx 3.14$, $s = 60$, and $r = 30$.

59. Find the volume of a cylinder that has a diameter of 3 ft and a height of 15 ft. Use the formula $V = \pi r^2 h$. Find the number of gallons, to the nearest gallon, of water the tank will hold. Use the fact that 1 ft$^3 \approx 7.48$ gal.

60. A painter needs to find the area of the surface of a cylindrical drum. If the drum has radius $r = 7.85$ feet and a height of 10.5 feet, what is the area of the drum, not counting the top and bottom? If a gallon of industrial paint covers 125 ft^2, how many gallons of paint must he buy? Formula: $A = 2\pi rh$, let $\pi \approx 3.14$.

NAME ______________________ CLASS ____________ DATE ____________

STATE YOUR UNDERSTANDING

61. Explain the difference between evaluating $0.3(5.1)^2 + 8.3 \div 5$ and $[0.3(5.1)^2 + 8.3] \div 5$. How do the symbols indicate the order of the operations?

CHALLENGE

Insert grouping symbols to make each statement true.

62. $2 \cdot 8.1 \div 5 - 1 = 4.05$

63. $3.62 \div 0.02 + 72.3 \cdot 0.2 = 0.25$

64. $3.62 \div 0.02 + 8.6 \cdot 0.51 = 96.696$

65. $1.4^2 - 0.8^2 = 1.3456$

66. The average of 4.56, 8.23, 16.5 and a missing number is 8.2975. Find the missing number.

GROUP ACTIVITY

67. Go to the library and use the daily newspaper to provide the data for you to calculate the following: (a) average daily high temperature, (b) average daily low temperature, (c) average daily rainfall, and (d) average daily minutes of daylight. Use data from your town over the past 7 days.

MAINTAIN YOUR SKILLS (SECTIONS 2.4, 5.2, 5.3, 5.8)

Change to a decimal.

68. $\frac{15}{16}$

69. $\frac{25}{32}$

70. $\frac{17}{80}$

Change to a fraction and simplify.

71. 0.92

72. 0.728

73. 0.045

74. The sale price of a hand-held calculator is \$22.75. If the sale price was marked down \$7.28 from the original price, what was the original price?

75. The price of a Tomaya VCR is \$348.95. The store is going to put it on sale at a discount of \$59.50. What price should the clerk put on the VCR for the sale?

76. Find the volume of a rectangular solid that has dimensions of 7 ft by 3 ft by 4 ft.

77. Find the volume of a cube that is 11 cm on an edge.

Getting Ready for Algebra

OBJECTIVE

Solve equations that require more than one operation.

HOW AND WHY

We solve equations that require more than one operation in the same way as equations with whole numbers and fractions.

To solve an equation that requires more than one operation

1. Eliminate the addition or subtraction by performing the inverse operation.
2. Eliminate the multiplication by dividing both sides by the same number, that is, perform the inverse operation.

Examples A–C

Directions: Solve.

Strategy: Isolate the variable by performing the inverse operations.

A. $2.5x - 3.7 = 18.8$

$2.5x - 3.7 + 3.7 = 18.8 + 3.7$ **Eliminate the subtraction by adding 3.7 to both sides.**

$2.5x = 22.5$

$\frac{2.5x}{2.5} = \frac{22.5}{2.5}$ **Eliminate the multiplication by dividing both sides by 2.5.**

$x = 9$

CHECK:

$2.5(9) - 3.7 = 18.8$

$22.5 - 3.7 = 18.8$

$18.8 = 18.8$

The solution is $x = 9$.

B. $6.3 = 1.75x + 1.05$

$6.30 = 1.75x + 1.05$

$-1.05 = \quad - 1.05$ **Subtract 1.05 from both sides.**

$5.25 = 1.75x$

$\frac{5.25}{1.75} = \frac{1.75x}{1.75}$ **Divide both sides by 1.75.**

$3 = x$

Warm Ups A–C

A. $0.06y - 2.8 = 0.5$

B. $2.9494 = 2.49t + 2.8$

Answers to Warm Ups A. $y = 55$ B. $0.06 = t$

CHECK:

$6.3 = 1.75(3) + 1.05$

$6.3 = 5.25 + 1.05$

$6.3 = 6.3$

The solution is $x = 3$.

C. Use the formula in Example C to find the Celsius temperature that corresponds to 104.9°F.

C. The formula relating temperatures measured in degrees Fahrenheit and degrees Celsius is $F = 1.8C + 32$. Find the Celsius temperature that corresponds to 63.14°F.

First substitute the known values into the formula:

$$F = 1.8C + 32$$

$$63.14 = 1.8C + 32 \quad \textbf{Substitute } F = 63.14.$$

$$63.14 - 32 = 1.8C + 32 - 32 \quad \textbf{Subtract 32 from each side.}$$

$$31.14 = 1.8C$$

$$\frac{31.14}{1.8} = \frac{1.8C}{1.8} \quad \textbf{Divide each side by 1.8.}$$

$$17.3 = C$$

Since $1.8(17.3) + 32 = 63.14$, the temperature is 17.3°C.

Answer to Warm Up C. The temperature is 40.5°C.

Exercises

Solve.

1. $2.5x - 8.9 = 13.6$

2. $0.25x - 2.2 = 0.47$

3. $2.4x + 2.6 = 4.04$

4. $14w + 0.004 = 43.404$

5. $4.115 = 2.15t + 3.9$

6. $10.175 = 1.25y + 9.3$

7. $0.03x - 13.5 = 2.22$

8. $0.07r - 2.35 = 61.7$

9. $7x + 9.06 = 11.3$

10. $13x + 14.66 = 15.7$

11. $3.45m - 122 = 109.15$

12. $11.6t - 398 = 193.6$

13. $2000 = 92y + 482$

14. $1500 = 48w + 516$

15. $50p - 149 = 1.1$

16. $14.4 = 0.44y + 5.6$

17. $7.5 = 2.3 + 0.13x$

18. $7 = 0.25w - 3.6$

19. $9 = 1.25h - 0.1$

20. $1000 = 90y + 415$

21. $1250 = 80c - 130$

22. The formula relating temperatures measured in degrees Fahrenheit and degrees Celsius is $F = 1.8C + 32$. Find the Celsius temperature that corresponds to 221°F.

23. Use the formula in Exercise 22 to find the Celsius temperature that corresponds to 35.6°F.

24. The formula for the balance of a loan D is $D + NP = B$, where P represents the monthly payment, N represents the number of payments, and B represents the amount of money borrowed. Find the amount of the monthly payment Gina must make if she borrows $1300 for 2 years with a remaining balance of $400. Round to the nearest cent.

25. Use the formula in Exercise 24 to find the amount of the monthly payment Morales must make if he borrows $7500 for 5 years with a remaining balance of $400. Round to the nearest cent.

26. Catherine is an auto mechanic. She charges $22 per hour for her labor. The cost of any parts needed is in addition to her labor charge. How many hours of labor result from a repair job in which the total bill (including $125 for parts) is $301? Write and solve an equation to determine the answer.

27. A car rental agency charges $23 per day plus $0.22 per mile to rent one of their cars. Determine how many miles were driven by a customer after a 3-day rental that cost $344. Write and solve an equation to determine the answer.

NAME ______________________ CLASS ____________ DATE ____________

CHAPTER 5

Group Project *(3–4 weeks)*

OPTIONAL

a. Each person in the group selects three stocks and tracks their progress for two full weeks. Record the beginning price, the daily change, and closing price.

b. For each stock, calculate the net change for the 2 weeks and the average daily change.

c. The entire group has $10,000 to invest for 1 week, with the objective being to make as much money as possible. The group must buy at least two different stocks but not more than four different ones. All stocks purchased must be from the ones tracked by the members of the group. Once the decision is made, the group's stocks are tracked for 1 week.

d. The group issues a final report. The report includes the data gathered, analysis of the data, a clear rationale for the investment decision, and the final results.

NAME ______________________ CLASS __________ DATE __________

CHAPTER 5

True–False Concept Review

Check your understanding of the language of basic mathematics. Tell whether each of the following statements is True (always true) or False (not always true). For each statement you judge to be false, revise it to make a statement that is true.

ANSWERS

1. The word name for 0.502 is "five hundred and two thousandths."

1. ______________

2. .502 and 0.502 name the same number.

2. ______________

3. To write 0.75 in expanded form we write $\frac{75}{100}$.

3. ______________

4. Since 0.145 is read "one hundred forty-five thousandths," we write $\frac{145}{1000}$ and reduce to change the decimal to a fraction.

4. ______________

5. True or false: $0.821597 > 0.84$.

5. ______________

6. Since $2.3 > 1.5$ is true, 1.5 is to the left of 2.3 on the number line.

6. ______________

7. To list a group of decimals in order, we need to write or think of all the numbers as having the same number of decimal places.

7. ______________

8. Decimals are either exact or approximate.

8. ______________

9. To round 123.3477 to the nearest tenth, we write 123.4 since the 4 in the hundredths place rounds up to 5 since it is followed by a 7.

9. ______________

10. The sum of 0.5 and 0.25 is 0.75.

10. ______________

11. $8.5 - 0.2 = 6.5$. 11. ______

12. The final answer of a multiplication problem will always contain the same number of decimal places as the total number of places in the two numbers being multiplied. 12. ______

13. To multiply a number by a positive power of ten, move the decimal point in the number to the right the same number of places as the number of zeros in the power of 10. 13. ______

14. To divide a number by a positive power of 10 that is written in exponent form, move the decimal the same number of places to the right as the exponent indicates. 14. ______

15. To change 1.56×10^{-5} to place value form, move the decimal five places to the right. 15. ______

16. To divide a number by a decimal, first change the decimal to a whole number by moving the decimal point to the right. 16. ______

17. All fractions can be changed to exact terminating decimals. 17. ______

18. The decimal 0.5649 rounds to 0.57 to the nearest hundredth. 18. ______

19. The order of operations for decimals is the same as for whole numbers. 19. ______

20. To find the average of a group of decimals, find their sum and divide by the number of decimals in the group. 20. ______

NAME ______________________ CLASS ______________ DATE ______________

CHAPTER 5

Test

ANSWERS

1. Divide. Round the answer to the nearest thousandth: $0.87\overline{)2.4678}$ — 1. ______

2. List the following decimals from the smallest to the largest: 0.728, 0.731, 0.7279, 0.7299, 0.7308. — 2. ______

3. Write the word name for 32.032. — 3. ______

4. Multiply: 6.78(8.3) — 4. ______

5. Write as a decimal: $\frac{7}{125}$ — 5. ______

6. Round to the nearest hundredth: 45.997 — 6. ______

7. Subtract: $34 - 13.736$ — 7. ______

8. Change to a mixed number with the fraction part simplified: 16.925 — 8. ______

9. Write in scientific notation: 0.00052 — 9. ______

10. Write as an approximate decimal to the nearest hundredth: $\frac{8}{13}$ — 10. ______

11. Round to the nearest hundred: 56,885.678 — 11. ______

12. Perform the indicated operations: $4.56 \div 0.6 \times 1.03 + 7.5$ — 12. ______

13. Subtract: $\begin{array}{r} 17.356 \\ \underline{14.448} \end{array}$ — 13. ______

14. Change to place value form: 2.66×10^{-3} — 14. ______

15. Write the place value name for "three hundred twelve and fifty-eight thousandths." **15.** ______

16. Multiply: 0.00342(100,000) **16.** ______

17. Write in scientific notation: 55,810 **17.** ______

18. Add: 2.34 + 0.543 + 13.56 + 6.7 **18.** ______

19. Multiply: 91.4(0.0032) **19.** ______

20. Divide: $35\overline{)2.38}$ **20.** ______

21. Convert 56 mph to the nearest whole kilometer per hour. (1 mile ≈ 1.609 km) **21.** ______

22. For each of the four Sundays of February, the offering at the Chapel on the Hill was \$1258.50, \$2067.25, \$1850.50, and \$3962.75. What was the average Sunday offering? **22.** ______

23. Add: **23.** ______

$$\begin{array}{r} 345.78 \\ 23.678 \\ 1002.005 \\ 3.45 \\ 19.67 \\ +\ 345.231 \\ \hline \end{array}$$

24. During a canned vegetable sale Grant buys 14 cans of various vegetables. If the sale price is 4 cans for \$1.79, how much does Grant pay for the canned vegetables? **24.** ______

25. At the beginning of the 1995–96 NBA season, Michael Jordan was the all-time leader in points scored per game, with 32.2. How many games had he played if he scored a total of 21,998 points (to the nearest game)? **25.** ______

26. In baseball, the slugging percentage is calculated by dividing the number of total bases (a double is worth two bases) by the number of times at bat and then multiplying by 1000. What is the slugging percentage of a player who has 195 bases in 290 times at bat? Round to the nearest whole number. **26.** ______

27. What is the perimeter of a rectangle that has a length of 14.5 m and a width of 9.34 m? **27.** ______

28. Harold and Jerry go on diets. Initially Harold weighed 245.7 lb and Jerry weighed 213.8 lb. After one month of the diet, Harold weighed 229.84 lb and Jerry weighed 198.6 lb. Who lost the most weight and by how much? **28.** ______

Good Advice for Studying

Preparing for Tests

Testing usually causes the most anxiety for students. By studying more effectively, you can eliminate many of the causes of anxiety. But there are also other ways to prepare that will help relieve your fears.

If you are math anxious, you actually may study too much out of fear of failure and not allow enough time for resting and nurturing yourself. Every day, allow yourself some time to focus on your concerns, feelings, problems, or anything that might distract you when you try to study. Then, when these thoughts distract you, say to yourself, "I will not think about this now. I will later at _____ o'clock. Now I have to focus on math." If problems become unmanageable, make an appointment with a college counselor.

Nurturing is any activity that will help you recharge your energy. Choose an activity that makes you feel good such as going for a walk, daydreaming, reading a favorite book, doing yard work, taking a bubble bath, or playing basketball.

Other ways to keep your body functioning effectively under stress are diet and exercise. Exercise is one of the most beneficial means of relieving stress. Try to eat healthy foods and drink plenty of water. Avoid caffeine, nicotine, drugs, alcohol, and "junk food."

Plan to have all your assignments finished two days before the test, if possible. The day before the test should be completely dedicated to reviewing and practicing for the test.

Many students can do the problems, but cannot understand the instructions and vocabulary, so they do not know where to begin. Review any concepts that you have missed or any that you were unsure of or "guessed at."

When you feel comfortable with all of the concepts, you are ready to take the practice test at the end of the chapter. You should simulate the actual testing situation as much as possible. Have at hand all the tools that you will use on the real test: sharpened pencils, eraser, and calculator (if your instructor allows). Give yourself the same amount of time as you'll be given on the actual test. Plan a time for your practice test when you can be sure there will be no interruptions. Work each problem slowly and carefully. Remember, if you make a mistake by rushing through a problem and have to do it over, it will take more time than doing the problem carefully in the first place.

After taking the test, go back and study topics referenced with the answers you missed or that you feel you do not understand. You should now know if you are ready for the test. If you have been studying effectively and did well on the practice test, you should be ready for the real test. You are prepared!

6

Ratio and Proportion

APPLICATION

One of the ways we relate to our surroundings is by judging the size of various objects relative to our bodies. Have you ever gone up a flight of stairs in which the stairs were taller than normal? You are acutely aware that something is not right, because your body is expecting a certain amount of space between the stairs. We tend to trip on the first few stairs, because we do not lift our feet high enough. But then our body makes an adjustment and we successfully negotiate the rest of the stairs but with more effort than is usually required. Have you ever observed a toddler climbing stairs? Try to imagine how difficult it would be to climb stairs that were as large compared to our bodies as normal stairs are compared to a toddler's body. This concept of relative size is referred to both mathematically and in the real world as "proportionality."

Complete the following table:

	Adult	Toddler
Height of stair in inches	8	8
Body height in inches		

Write a fraction using the numbers in the Adult column. Use the stair measurement as the numerator and the body height measurement as the denominator. State what it means in the context of this situation. Do the same for the numbers in the Toddler column. Which fraction is larger? Explain what this means in context.

Many movies stem from the premise that someone or something is proportionally larger or smaller than normal. Early treatments include the King Kong movies. More recent examples include "Honey I Shrunk The Kids" and its sequels. Such movies are special challenges to the set designers, who must ensure that all the surroundings are proportionally correct so that the premise is visually believable. In this chapter we will investigate some of the proportions used in the movie "Honey I Shrunk The Kids."

6.1

Ratio and Rate

OBJECTIVES

1. Write a fraction that shows a ratio comparison of two like measurements.
2. Write a fraction that shows a rate comparison of two unlike measurements.
3. Write a unit rate.

VOCABULARY

A **ratio** is a comparison of two quantities by division.

Like measurements have the same unit of measure.

Unlike measurements have different units of measure.

A **rate** is a comparison of two unlike quantities by division.

A **unit rate** is a rate with a denominator of one unit.

HOW AND WHY

Objective 1

Write a fraction that shows a ratio comparison of two like measurements.

Two numbers can be compared by subtraction or by division. If we compare 12 and 3, since $12 - 3 = 9$, we could say that 12 is 9 more than 3.

And since $12 \div 3 = 4$, we say that 12 is 4 times larger than 3.

The indicated division, $12 \div 3$, is called a *ratio.* These are common ways to write the ratio to compare 12 and 3:

12:3 $\quad 12 \div 3 \quad$ 12 to 3 $\quad \dfrac{12}{3}$

Since we are comparing 12 to 3, 12 is written first or placed in the numerator of the fraction.

Here we write ratios as fractions. Since a ratio is a fraction, it can be simplified. The ratio $\frac{4}{6}$ is simplified to $\frac{2}{3}$. If the ratio contains two like measurements, it can be simplified in the same way as a fraction.

$$\frac{\$12}{\$25} = \frac{12}{25}$$ **The units, $, are dropped since they are the same.**

$$\frac{15 \text{ miles}}{25 \text{ miles}} = \frac{3}{5}$$ **The common units are dropped and the fraction simplified.**

Examples A–C

Directions: Write a ratio in simplified form.

Strategy: Write the ratio as a simplified fraction.

A. Write the ratio of 84 to 105.

$\frac{84}{105} = \frac{4}{5}$ **Write 84 in the numerator and simplify.**

The ratio of 84 to 105 is $\frac{4}{5}$.

B. Write the ratio of the length of a room to its width if the room is 24 ft by 18 ft.

$\frac{24 \text{ ft}}{18 \text{ ft}} = \frac{24}{18} = \frac{4}{3}$ **Simplify the fraction.**

The ratio of the length to the width is $\frac{4}{3}$.

C. Write the ratio of 2 dimes to 5 quarters. (Compare in cents.)

$\frac{2 \text{ dimes}}{5 \text{ quarters}} = \frac{20 \text{ cents}}{125 \text{ cents}} = \frac{20}{125} = \frac{4}{25}$

The ratio of 2 dimes to 5 quarters is $\frac{4}{25}$.

Warm Ups A–C

A. Write the ratio of 16 to 20.

B. Write the ratio of the length of a room to its width if the room is 36 ft by 28 ft.

C. Write the ratio of 45 mm to 2 m. (Compare in millimeters.)

HOW AND WHY

Objective 2

Write a fraction that shows a rate comparison of two unlike measurements.

Fractions are also used to compare unlike measurements. The rate of $\frac{31 \text{ children}}{10 \text{ families}}$ compares the unlike measurements "31 children" and "10 families." A common application of a rate is computing gas mileage. For example, if a car runs 208 miles on 8 gallons of gas, we compare miles to gallons by writing $\frac{208 \text{ miles}}{8 \text{ gallons}}$. This rate can be simplified as long as the units are stated, not dropped.

$$\frac{208 \text{ miles}}{8 \text{ gallons}} = \frac{104 \text{ miles}}{4 \text{ gallons}} = \frac{26 \text{ miles}}{1 \text{ gallon}} = 26 \text{ miles per gallon} = 26 \text{ mpg}$$

CAUTION

When units are different, they are *not* dropped.

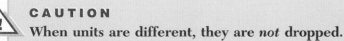

Answers to Warm Ups A. $\frac{4}{5}$ B. $\frac{9}{7}$ C. $\frac{9}{400}$

Warm Ups D–F

D. Write the rate of 15 people to 8 tables.

E. Write the rate of 12 TVs to 8 homes.

F. The following spring the committee repeated the tree program. This time they sold 770 oak trees and 440 birch trees.

1. What is the rate of oak trees to birch trees sold?

2. What is the rate of birch trees to the total number of trees?

Examples D–F

Directions: Write a rate in simplified form.

Strategy: Write the simplified fraction and retain the unlike units.

D. Write the rate of 10 chairs to 11 people.

$$\frac{10 \text{ chairs}}{11 \text{ people}}$$ **The units must be kept since they are different.**

E. Write the rate of 8 cars to 6 homes.

$$\frac{8 \text{ cars}}{6 \text{ homes}} = \frac{4 \text{ cars}}{3 \text{ homes}}$$

F. An urban environmental committee urged the local citizens to plant deciduous trees around their homes as a means of conserving energy. (The leaves provide shade in the summer and the fallen leaves allow the sun to shine through the limbs in the winter.) The committee provided the trees to the citizens at cost. After the program had been completed, they determined that 825 oak trees and 675 birch trees had been sold.

1. What is the rate of the number of oak trees to the number of birch trees sold?

2. What is the rate of the number of oak trees to the total number of trees sold?

1. Strategy: Write the first unit, 825 oak trees, in the numerator, and the second unit, 675 birch trees, in the denominator.

$$\frac{825 \text{ oak trees}}{675 \text{ birch trees}} = \frac{11 \text{ oak trees}}{9 \text{ birch trees}}$$ **Simplify.**

The rate is $\frac{11 \text{ oak trees}}{9 \text{ birch trees}}$, that is, 11 oak trees were sold for every 9 birch trees sold.

2. Strategy: Write the first unit, 825 oak trees, in the numerator, and the second unit, total number of trees, in the denominator.

$$\frac{825 \text{ oak trees}}{1500 \text{ trees total}} = \frac{11 \text{ oak trees}}{20 \text{ trees total}}$$ **Simplify.**

The rate is $\frac{11 \text{ oak trees}}{20 \text{ trees total}}$, that is, 11 out of every 20 trees sold were oak trees.

HOW AND WHY

Objective 3

Write a unit rate.

When a rate is simplified so that the denominator is one unit, then we have a *unit* rate. For example,

$$\frac{208 \text{ miles}}{8 \text{ gallons}} = \frac{26 \text{ miles}}{1 \text{ gallon}}$$ **Read "26 miles per gallon."**

Answers to Warm Ups D. $\frac{15 \text{ people}}{8 \text{ tables}}$ E. $\frac{3 \text{ TVs}}{2 \text{ homes}}$ F. 1. The rate is $\frac{7 \text{ oak trees}}{4 \text{ birch trees}}$. 2. The rate is $\frac{4 \text{ birch trees}}{11 \text{ trees total}}$.

Simplifying rates can lead to statements such as "There are 3.1 children to a family," since

$$\frac{31 \text{ children}}{10 \text{ families}} = \frac{3.1 \text{ children}}{1 \text{ family}}$$

The last rate is a comparison, not a fact, since no family has 3.1 children.

To write a unit rate given a rate

1. Do the indicated division.
2. Retain the units.

Examples G–I

Directions: Write as a unit rate.

Strategy: Do the indicated division so that the denominator is one unit.

G. Write the unit rate for $\frac{\$2.34}{3 \text{ cans of peas}}$.

Strategy: Do the division by 3. Retain the units.

$$\frac{\$2.34}{3 \text{ cans of peas}} = \frac{\$0.78}{1 \text{ can of peas}}$$

The unit rate is 78¢ per can.

H. Write the unit rate for $\frac{270 \text{ miles}}{12.5 \text{ gallons}}$.

$$\frac{270 \text{ miles}}{12.5 \text{ gallons}} = \frac{21.6 \text{ miles}}{1 \text{ gallon}}$$ **Divide numerator and denominator by 12.5.**

The unit rate is 21.6 mpg.

Calculator Example

I. The population density of a region is a unit rate. The rate is the number of people per one square mile of area. Find the population density of Stone County if the population is 13,550 and the area of the county is 1700 square miles. Round your answer to the nearest tenth.

Strategy: Write the rate and divide the numerator by the denominator using your calculator.

$$\text{Density} = \frac{13{,}550 \text{ people}}{1700 \text{ square miles}}$$

$$= \frac{7.970588235 \text{ people}}{1 \text{ square mile}}$$ **Divide.**

$$\approx \frac{8.0 \text{ people}}{1 \text{ square mile}}$$ **Round to the nearest tenth.**

The density is 8.0 people per square mile, to the nearest tenth.

Warm Ups G–I

G. Write the unit rate for $\frac{492 \text{ lb}}{12 \text{ in}^2}$.

H. Write the unit rate for $\frac{301 \text{ miles}}{14 \text{ gallons}}$.

I. What was the estimated population density of Los Angeles in 1995 if the population was estimated at 10,414,000 and the area is 1110 square miles? Round your answer to the nearest whole number.

Answers to Warm Ups G. The unit rate is 41 lb per in^2. H. The unit rate is 21.5 mpg. I. The population density was 9382 people per square mile.

NAME ______________________ CLASS ______________ DATE ______________

Exercises 6.1

OBJECTIVE 1: *Write a fraction that shows a ratio comparison of two like measurements.*

A.

Write as a ratio in simplified form.

1. 7 to 35

2. 9 to 54

3. 12 m to 10 m

4. 12 ft to 9 ft

5. 40 cents to 45 cents

6. 30 dimes to 54 dimes

B.

7. 1 dime to 4 nickels (compare in cents)

8. 3 quarters to 5 dimes (compare in cents)

9. 16 in. to 2 ft (compare in inches)

10. 3 ft to 3 yd (compare in feet)

11. 200 cm to 3 km (compare in centimeters)

12. 200 yd to 5 miles (compare in yards)

OBJECTIVE 2: *Write a fraction that shows a rate comparison of two unlike measurements.*

A.

Write a rate and simplify.

13. 8 people to 11 chairs

14. 6 families to 18 children

15. 110 miles in 2 hours

16. 264 km in 3 hr

17. 63 miles to 3 gallons

18. 100 km to 4 gallons

19. 88 lb to 33 ft

20. 36 buttons to 24 bows

B.

21. 18 trees to 63 ft

22. 91 TVs to 52 houses

23. 38 books to 95 students

24. 750 people for 3000 tickets

25. 765 people to 27 rooms

26. 8780 households to 6 cable companies

27. 345 pies to 46 sales

28. $17.68 per 34 lb of apples

OBJECTIVE 3: *Write a unit rate.*

A.

29. 50 miles to 2 hr

30. 60 miles to 4 min

31. 36 ft to 9 sec

32. 75 m to 3 min

33. 132 yd to 2 billboards

34. 60 yd to 15 posts

35. 90¢ per 10 lb of potatoes

36. 117¢ per 3 lb of broccoli

NAME ______________________ CLASS ______________ DATE ______________

B.

Write a unit rate. Round to the nearest tenth.

37. 825 miles per 22 gallons

38. 13,266 km per 220 gallons

39. 1000 ft to 12 sec

40. 1000 yd to 15 min

41. 12,095 lb to 45 square miles

42. 5486 kg to 290 cm^2

43. 225 gallons per 14 min

44. 850 ℓ per 14 min

C.

45. The parking lot in the lower level of the Senter Building has 18 spaces for compact cars and 24 spaces for larger cars.

a. What is the ratio of compact spaces to larger spaces?

b. What is the ratio of compact spaces to the total number of spaces?

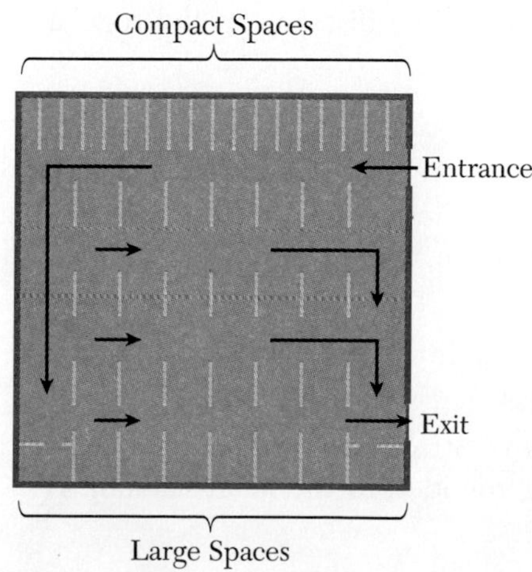

46. The Reliable Auto Repair Service building has eight stalls for repairing automobiles and four stalls for repairing small trucks.

a. What is the ratio of the number of stalls for small trucks to the number of stalls for automobiles?

b. What is the ratio of the number of stalls for small trucks to the total?

47. One section of the country has 3500 TV sets per 1000 houses. Another section has 500 TV sets per 150 houses. Are the rates of the TV sets to the number of houses the same in both parts of the country?

48. In Oakville there are 5000 automobiles per 3750 households. In Firland there are 6400 automobiles per 4800 households. Are the rates of the number of automobiles to the number of households the same?

49. A store bought a sofa for a cost of $175 and sold it for $300. What is the ratio of the cost to the selling price?

50. In Exercise 49, what is the ratio of the markup to the cost?

51. A coat is regularly priced at $99.99, but during a sale its price is $66.66. What is the ratio of the sale price to the regular price?

52. In Exercise 51, what is the ratio of the discount to the regular price?

53. What is the population density of the city of Dryton if there are 22,450 people and the area is 230 square miles? Simplify to a unit comparison, rounded to the nearest tenth.

54. What is the population density of Struvaria if 950,000 people live there and the area is 18,000 square miles? Simplify to a unit comparison, rounded to the nearest tenth.

55. What was the population density of your city in 1997?

56. What was the population density of your state in 1997?

57. In the United States, four people use an average of 250 gallons of water per day. One hundred gallons are used to flush the toilet, 80 gallons in baths/showers, 35 gallons doing laundry, 15 gallons washing dishes, 12 gallons for cooking and drinking, and 8 gallons in the bathroom sink.

a. Write the ratio of laundry use to toilet use.

b. Write the ratio of bathing/showering use to dishwashing use.

58. Using Exercise 57,

a. Write the ratio of cooking/drinking use to dishwashing use.

b. Write the ratio of laundry use per person.

59. Drinking water is considered to be polluted when a pollution index of 0.05 mg of lead per liter is reached. At that rate, how many milligrams of lead are enough to pollute 25 ℓ of drinking water?

60. Data indicate that 3 of every 20 rivers in the United States showed an increase in water pollution from 1974 to 1983. Determine how many rivers are in your state. At the same rate, determine how many of those rivers had an increased pollution level during the same period.

It is often difficult to compare the price of various food items, many times because of the packaging. Is a 14-oz can of pears for $0.89 a better buy than a 16-oz can of pears for $1.00? To help consumers compare, *unit pricing* is often posted. Mathematically we write the information as a rate and simplify to a one-unit comparison.

61. Write a ratio for a 14-oz can of pears that sells for $0.89 and simplify it to a unit price (price per 1 oz of pears). Do the same with the 16-oz can of pears for $1.00. Which is the better buy?

62. Which is the best buy: a 15-oz box of Cheerios for $2.49, a 20-oz box for $3.29, or a 2-lb 3-oz box for $5.39?

63. Which is the better buy: 5 lb of granulated sugar on sale for $4.95 or 25 lb of sugar for $24.90?

Some food items have the same unit price regardless of the quantity purchased. Other food items have a decreasing unit price as the size of the container increases. In order to determine which category a food falls in, find the unit price for each item.

64. Is the unit price of hamburger the same if 2.35 lb costs $4.44 and 3.52 lb costs $6.65?

65. Is the unit price of frozen orange juice the same if a 12-oz can costs $1.09 and a 16-oz can costs $1.30?

66. List five items that usually have the same unit price regardless of quantity purchased, and five that do not. What circumstances could cause an item to change categories?

Exercises 67–68 relate to the chapter application.

67. In the movie "Honey I Shrunk The Kids," two teenagers and their little brothers are accidentally reduced to $\frac{1}{4}$-in. tall. Assume that the average teenager in this country is 5′6″. Write a ratio that describes this relationship.

NAME ______________ CLASS ______________ DATE ______________

68. The kids encounter an ant, and one of them rides on the ant's back like riding a horse. Estimate the length of a horse and the length of an ant. Write a ratio between these two lengths, which should have the same value as the ratio you wrote in Exercise 67. Reduce each of your ratios to decimal form to decide if they are the same (or at least reasonably close). Is it consistent mathematically that the shrunken kids could ride an ant like riding a horse? Explain.

STATE YOUR UNDERSTANDING

69. Write a short paragraph explaining why ratios are useful ways to compare measurements.

 70. Explain the difference between a ratio, a rate, and a unit rate. Give an example of each.

CHALLENGE

71. Give an example of a ratio that is not a rate. Give an example of a rate that is not a ratio.

72. The ratio of noses to persons is $\frac{1}{1}$ or one-to-one. Find three examples of two-to-one ratios and three examples of three-to-one ratios.

73. Each gram of fat contains 9 calories. Chicken sandwiches at various fast-food places contain the following total calories and grams of fat.

		Calories	Grams of Fat
a.	RB's Light Roast Chicken Sandwich	276	7
b.	KB's Broiler Chicken Sandwich	267	8
c.	Hard B's Chicken Filet	370	13
d.	LJS's Baked Chicken Sandwich	130	4
e.	The Major's Chicken Sandwich	482	27
f.	Mickey's Chicken	415	19
g.	Tampico's Soft Chicken Taco	213	10
h.	Winston's Grilled Chicken Sandwich	290	7

Find the ratio of fat calories to total calories for each sandwich.

GROUP ACTIVITY

74. Have each member of your group select a country other than the United States. Each member is to use the library or other resources to find the population and area of the country selected. Calculate the population density for each country and compare your findings. Which country has the greatest population density? The least?

75. The golden ratio of 1.618 to 1 has been determined by artists to be very pleasing aesthetically. The ratio has been discovered to occur in nature in many places, including the human body. In particular, the ratio applies to successive segments of the fingers.

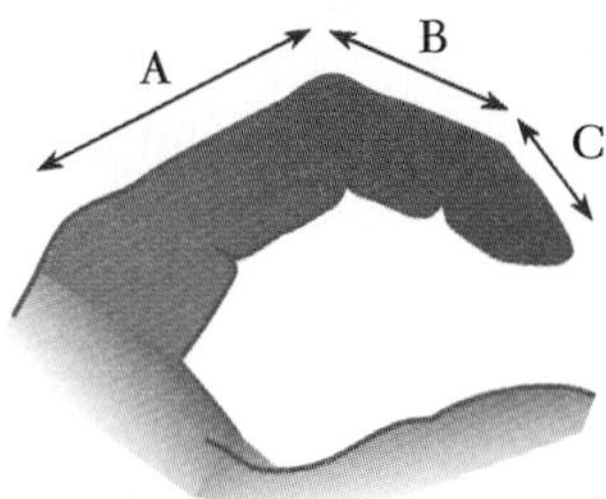

Measure as accurately as possible at least three fingers of everyone in the group. Calculate the ratio of successive segments, and fill in the table.

Name	Finger	A	B	C	$\frac{A}{B}$	$\frac{B}{C}$

Whose fingers come closest to the golden ratio? Can you find other body measures that have this ratio?

MAINTAIN YOUR SKILLS (SECTIONS 4.3, 4.5, 4.8, 4.10, 5.6, 5.7, 5.8)

Simplify.

76. $\frac{3}{8} \cdot \frac{4}{21}$

77. $\frac{5}{12} \div \frac{15}{28}$

78. $6.5(.03)$

79. $12.85 \div 2.5$

80. $\frac{7}{8}(0.6)$

81. $1\frac{3}{8} + 4\frac{3}{10}$

82. $9 - 4\frac{7}{15}$

83. $31.2 \div 1000$

Which is larger?

84. $\frac{3}{4}, \frac{2}{3}$

85. $\frac{7}{10}, \frac{18}{25}$

6.2

Solving Proportions

OBJECTIVES

1. Determine whether a proportion is true or false.

2. Solve a proportion.

VOCABULARY

A **proportion** is a statement that two ratios are equal.

In a proportion, **cross multiplication** means multiplying the numerator of each ratio times the denominator of the other.

Cross products are the products obtained from cross multiplication.

Solving a proportion means finding a missing number, usually represented as a letter or **variable,** that will make a proportion true.

HOW AND WHY

Objective 1

Determine whether a proportion is true or false.

The statement $\frac{14}{8} = \frac{35}{20}$ is a proportion. A proportion states that two rates or ratios are equal. To check whether the proportion is true or false we use "cross multiplication."

The proportion $\frac{14}{8} = \frac{35}{20}$ is true if the cross products are equal.

$\frac{14}{8} \stackrel{?}{=} \frac{35}{20}$

$14(20) \stackrel{?}{=} 8(35)$ **Find the cross products.**

$280 = 280$ **The cross products are equal.**

The proportion is true.

This test is actually a shortcut for converting both fractions to equivalent fractions with common denominators and checking that the numerators match. Let us examine the same proportion using the formal method.

$\frac{14}{8} \stackrel{?}{=} \frac{35}{20}$

$\frac{20}{20} \cdot \frac{14}{8} \stackrel{?}{=} \frac{35}{20} \cdot \frac{8}{8}$ **A common denominator is $8 \cdot 20 = 160$.**

$\frac{280}{160} = \frac{280}{160}$ **Multiply. The numerators are the same.**

The proportion is true.

The cross products in the first method are the numerators in the second method. This is the reason that checking the cross products is a valid procedure for determining the truth of a proportion.

To check whether a proportion is true or false

1. Check that the ratios or rates have the same units.
2. Cross multiply.
3. If the cross products are equal, the proportion is true.

Warm Ups A–C

A. Is $\frac{7}{8} = \frac{42}{48}$ true or false?

B. Is $\frac{5.6}{6.3} = \frac{5}{6}$ true or false?

C. Is $\frac{1 \text{ dollar}}{2 \text{ quarters}} = \frac{16 \text{ nickels}}{4 \text{ dimes}}$ true or false?

Examples A–C

Directions: Determine whether a proportion is true or false.

Strategy: Check the cross products. If they are equal the proportion is true.

A. Is $\frac{6}{5} = \frac{72}{60}$ true or false?

$\frac{6}{5} \times \frac{72}{60}$ **Find the cross products.**

$6(60) = 5(72)$

$360 = 360$ **True.**

The proportion is true.

B. Is $\frac{4.1}{7.1} = \frac{4}{7}$ true or false?

$\frac{4.1}{7.1} \times \frac{4}{7}$ **Find the cross products.**

$4.1(7) = 7.1(4)$

$28.7 = 28.4$ **False.**

The proportion is false.

C. Is $\frac{1 \text{ dollar}}{3 \text{ quarters}} = \frac{8 \text{ dimes}}{12 \text{ nickels}}$ true or false?

Strategy: The units in the rates are not the same. We change all units to cents and simplify.

$\frac{1 \text{ dollar}}{3 \text{ quarters}} = \frac{8 \text{ dimes}}{12 \text{ nickels}}$

$\frac{100 \text{ cents}}{75 \text{ cents}} = \frac{80 \text{ cents}}{60 \text{ cents}}$

$\frac{100}{75} \times \frac{80}{60}$ **Like units may be dropped.**

$100(60) = 75(80)$

$6000 = 6000$ **True.**

The proportion is true.

Answers to Warm Ups A. True B. False C. True

HOW AND WHY

Objective 2

Solve a proportion.

Proportions are used to solve many problems in science, technology, and business. There are four numbers or measures in a proportion. If three of the numbers are known, we can find the missing number.

For example,

$$\frac{x}{5} = \frac{15}{25}$$

$25x = 5(15)$ **Cross multiply.**

$25x = 75$

$x = 3$ **Divide each side by 25.**

Every multiplication fact can be written as a related division fact. The product divided by one factor gives the other factor. So $25 \cdot x = 75$ can be written as $x = 75 \div 25$.

$x = 75 \div 25$ **Rewrite as division.**

$x = 3$

The missing number is 3.

To solve a proportion:

1. Cross multiply.
2. Do the related division problem to find the missing number.

Examples D–G

Directions: Solve the proportion.

Strategy: Cross multiply, then write the related division and simplify.

D. Solve: $\frac{4}{9} = \frac{8}{x}$

$4x = 9(8)$ **Cross multiply.**

$4x = 72$ **Simplify.**

$x = 72 \div 4$ **Rewrite as division.**

$x = 18$ **Simplify.**

The missing number is 18.

Warm Ups D–G

D. Solve: $\frac{5}{9} = \frac{10}{y}$

Answer to Warm Up D. $y = 18$

E. Solve: $\frac{0.5}{c} = \frac{1.5}{0.75}$

E. Solve: $\frac{0.6}{t} = \frac{1.2}{0.84}$

$0.6(0.84) = 1.2t$ **Cross multiply.**

$0.504 = 1.2t$ **Simplify.**

$0.504 \div 1.2 = t$ **Rewrite as division.**

$0.42 = t$ **Simplify.**

The missing number is 0.42.

F. Solve: $\frac{\frac{3}{4}}{\frac{5}{8}} = \frac{\frac{1}{2}}{w}$

F. Solve: $\frac{\frac{3}{4}}{1\frac{2}{3}} = \frac{\frac{1}{2}}{x}$

$\frac{3}{4}x = \left(1\frac{2}{3}\right)\left(\frac{1}{2}\right)$ **Cross multiply.**

$\frac{3}{4}x = \left(\frac{5}{3}\right)\left(\frac{1}{2}\right)$

$\frac{3}{4} \cdot x = \frac{5}{6}$ **Simplify.**

$x = \frac{5}{6} \div \frac{3}{4}$ **Rewrite as division.**

$x = \frac{5}{6} \cdot \frac{4}{3}$ **Invert the divisor.**

$x = \frac{10}{9}$ **Simplify.**

The missing number is $\frac{10}{9}$ or $1\frac{1}{9}$.

G. Solve $\frac{8}{y} = \frac{1.82}{21.24}$ and round to the nearest hundredth.

Calculator Example

G. Solve $\frac{3}{z} = \frac{9.6}{7.32}$ and round to the nearest hundredth.

$3(7.32) = 9.6z$ **Cross multiply.**

$3(7.32) \div 9.6 = z$ **Rewrite as division.**

$2.2875 = z$ **Simplify using a calculator.**

$2.29 \approx z$ **Round.**

The missing number is 2.29 to the nearest hundredth.

Answers to Warm Ups E. $c = 0.25$ F. $w = \frac{5}{12}$ G. $y \approx 93.36$

NAME ______________________ CLASS ____________ DATE ____________

Exercises 6.2

OBJECTIVE 1: *Determine whether a proportion is true or false.*

A.

True or false.

1. $\frac{6}{3} = \frac{16}{8}$

2. $\frac{2}{3} = \frac{10}{15}$

3. $\frac{6}{8} = \frac{9}{12}$

4. $\frac{6}{9} = \frac{8}{12}$

5. $\frac{3}{2} = \frac{9}{4}$

6. $\frac{3}{4} = \frac{9}{16}$

B.

7. $\frac{18}{15} = \frac{12}{10}$

8. $\frac{16}{24} = \frac{10}{15}$

9. $\frac{35}{22} = \frac{30}{20}$

10. $\frac{24}{32} = \frac{36}{38}$

11. $\frac{27}{45} = \frac{30}{60}$

12. $\frac{36}{45} = \frac{20}{25}$

13. $\frac{2.8125}{3} = \frac{15}{16}$

14. $\frac{9.375}{3} = \frac{25}{8}$

OBJECTIVE 2: *Solve a proportion.*

A.

Solve.

15. $\frac{1}{2} = \frac{a}{18}$

16. $\frac{1}{3} = \frac{b}{18}$

17. $\frac{2}{6} = \frac{c}{18}$

18. $\frac{2}{9} = \frac{x}{18}$

19. $\frac{2}{a} = \frac{5}{10}$

20. $\frac{8}{b} = \frac{2}{5}$

21. $\frac{14}{28} = \frac{5}{c}$

22. $\frac{8}{12} = \frac{6}{d}$

23. $\frac{2}{3} = \frac{12}{x}$

24. $\frac{3}{6} = \frac{8}{y}$

25. $\frac{w}{7} = \frac{2}{28}$

26. $\frac{z}{2} = \frac{2}{12}$

B.

27. $\frac{x}{7} = \frac{3}{2}$

28. $\frac{y}{5} = \frac{3}{4}$

29. $\frac{2}{z} = \frac{5}{11}$

30. $\frac{12}{x} = \frac{16}{3}$

31. $\frac{16}{24} = \frac{y}{16}$

32. $\frac{9}{11} = \frac{z}{15}$

33. $\frac{15}{16} = \frac{12}{a}$

34. $\frac{28}{7} = \frac{50}{b}$

35. $\frac{0.1}{c} = \frac{0.2}{1.2}$

36. $\frac{0.5}{a} = \frac{0.2}{0.6}$

37. $\frac{\frac{3}{5}}{b} = \frac{8}{5}$

38. $\frac{\frac{2}{3}}{c} = \frac{\frac{8}{9}}{1\frac{7}{9}}$

NAME ______________________ CLASS ____________ DATE ____________

39. $\frac{0.9}{4.5} = \frac{0.05}{x}$

40. $\frac{1.2}{2.7} = \frac{3.4}{y}$

41. $\frac{7}{42} = \frac{w}{3}$

42. $\frac{80}{30} = \frac{b}{2}$

43. $\frac{y}{3} = \frac{9}{\frac{1}{8}}$

44. $\frac{s}{40} = \frac{\frac{3}{4}}{5}$

45. $\frac{t}{24} = \frac{3\frac{1}{2}}{10\frac{1}{2}}$

46. $\frac{w}{4\frac{1}{4}} = \frac{3\frac{1}{3}}{2\frac{1}{2}}$

Solve. Round to the nearest tenth.

47. $\frac{3}{11} = \frac{w}{5}$

48. $\frac{3}{11} = \frac{x}{15}$

49. $\frac{9}{35} = \frac{14}{y}$

50. $\frac{9}{35} = \frac{24}{z}$

Solve. Round to the nearest hundredth.

51. $\frac{1.5}{5.5} = \frac{a}{0.8}$

52. $\frac{1.5}{5.5} = \frac{b}{2.8}$

53. $\frac{\frac{3}{7}}{c} = \frac{9}{20}$

54. $\frac{\frac{3}{7}}{d} = \frac{9}{32}$

C.

Fill in the boxes so the statement is true. Explain your answer.

55. If $\frac{\square}{70} = \frac{x}{7}$, then $x = 1$.

56. If $\frac{\square}{35} = \frac{y}{7}$, then $y = 4$.

57. Find the error(s) in the statement: If $\frac{2}{5} = \frac{x}{19}$, then $2x = 5(19)$. Correct the statement. Explain how you would avoid this error.

58. Find the error(s) in the statement: If $\frac{3}{x} = \frac{7}{9}$, then $3x = 7(9)$. Correct the statement. Explain how you would avoid this error.

59. A box of Tide that is sufficient for 18 loads costs \$3.65. What is the most that a store brand of detergent can cost if the box is sufficient for 25 loads and is cheaper to use than Tide? To find the cost solve the proportion $\frac{\$3.65}{18} = \frac{c}{25}$ where c represents the cost of the store brand.

60. Data show that it takes the use of 18,000,000 gasoline-powered lawn mowers to produce the same amount of air pollution as 3,000,000 new cars. Determine the number of gasoline-powered lawn mowers that will produce the same amount of air pollution as 50,000 new cars. To find the number of lawn mowers, solve the proportion $\frac{18{,}000{,}000}{3{,}000{,}000} = \frac{L}{50{,}000}$ where L represents the number of lawn mowers.

61. For every 10 people in the United States, it is believed that 7 suffer from some form of migraine headache.

a. Determine how many migraine sufferers would be in a group of 350 people. To determine the number of migraine sufferers, solve the proportion $\frac{7}{10} = \frac{N}{350}$ where N represents the number of migraine sufferers.

b. If three times as many women as men suffer from migraines, find the number of men that suffer from migraines if in a certain group it is known that 306 women are affected with migraines. To determine the number of men, solve the proportion $\frac{1}{3} = \frac{\mathrm{M}}{306}$ where M represents the number of men.

c. It is believed that headaches in 8 out of every 40 migraine sufferers are related to diet. At this rate, determine the number of migraine headaches that could be related to diet in a group of 350 migraine sufferers. To determine the number of headaches related to diet, solve the proportion $\frac{8}{40} = \frac{D}{350}$ where D represents the number of migraine headaches that could be related to diet.

Exercises 62–64 relate to the chapter application.

62. Early in the movie, the shrunken kids encounter a butterfly. One of them comments that the butterfly has a 42-ft wingspan. This measure, of course, is not physically true. It is the shrunken kid's perception relative to his new size. To check the accuracy of this statement, we complete the table below and use it to set up a proportion.

	Shrunken Kid	Butterfly
Actual Size	0.25 in.	3 in.
Perceived Size	5.5 ft	42 ft

From the table, we set up the proportion $\frac{0.25}{5.5} = \frac{3}{42}$. Is this proportion true or false?

63. Perhaps the proportion in Exercise 62 is false because of the assumption that a butterfly has a 3-in. wingspan. Experiment with butterflies of other sizes until you find a size that makes the proportion true. Do you think this part of the movie is realistic mathematically?

64. Later in the movie, the kids find a Lego and crawl into the holes in the back of it to sleep. The diameter of the holes appeared to be around 4 ft, relative to the shrunken kids. Measure the diameter of the hole of a real Lego to complete the following table.

	Shrunken Kid	**Lego (Hole Diameter)**
Actual Size	0.25 in.	
Perceived Size	5.5 ft	4 ft

Set up a proportion from the table as in Exercise 62. Is your proportion true or false? How realistic is the set design on this point?

STATE YOUR UNDERSTANDING

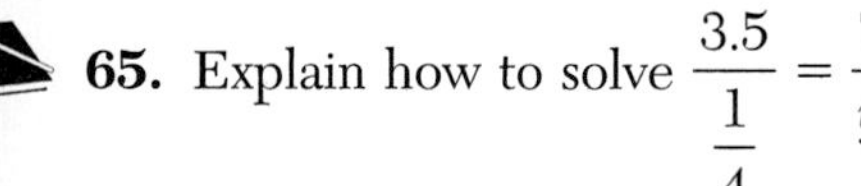

65. Explain how to solve $\dfrac{3.5}{\frac{1}{4}} = \dfrac{7}{y}$.

66. Look up the word "proportion" in the dictionary and write two definitions that differ from the mathematical definition of the word. Write three sentences, using the word "proportion," that illustrate each of the meanings.

CHALLENGE

Solve.

67. $\dfrac{9+3}{9+6} = \dfrac{8}{a}$

68. $\dfrac{5(9) - 2(5)}{8(6) - 3(2)} = \dfrac{8(5)}{b}$

Solve. Round to the nearest thousandth.

69. $\dfrac{7}{w} = \dfrac{18.92}{23.81}$

70. $\dfrac{7}{t} = \dfrac{18.81}{23.92}$

GROUP ACTIVITY

71. Five ounces of decaffeinated coffee contain approximately 3 mg of caffeine whereas 5 oz of regular coffee contain an average of 120 mg of caffeine. Five ounces of tea brewed for 1 min contain an average of 21 mg of caffeine. Twelve ounces of regular cola contain an average of 54 mg of caffeine. Six ounces of hot cocoa contain an average of 11 mg of caffeine. Twelve ounces of iced tea contain an average of 72 mg of caffeine. Determine the total amount of caffeine each member of your group consumed yesterday. Make a chart to illustrate this information. Combine this information with that of the other groups in your class to make a class amount. Make a class chart to illustrate this information. Determine the average amount of caffeine consumed by each member of the group and then by each member of the class. Compare these averages by making ratios. Discuss the similarities and the differences.

MAINTAIN YOUR SKILLS (SECTIONS 5.3, 5.4, 5.5, 5.6, 5.9)

72. Find the difference of 620.3 and 499.9781.

73. Find the average: 1.8, .006, 17, 8.5

74. Find the average: 6.45, 7.13, 5.11

75. Multiply: 4.835(10,000)

76. Divide: 4.835 ÷ 1000

77. Multiply: (0.875)(29)

78. Multiply: (12.75)(8.09)

79. Divide: $0.35\overline{)0.70035}$

80. Divide: $0.72\overline{)3.6}$

81. If gasoline is $1.249 per gallon, how much does Quan pay for 12.8 gallons? Round your answer to the nearest cent.

6.3

Applications of Proportions

OBJECTIVE

Solve word problems using proportions.

HOW AND WHY

Objective

Solve word problems using proportions.

If the ratio of two quantities is constant, the ratio can be used to find the missing part of a second ratio. For instance, if 2 lb of bananas cost \$0.48, what will 12 lb of bananas cost?

	Case I	Case II
Pounds of bananas	2	12
Cost in dollars	0.48	

In the table, the cost in Case II is missing. Call the missing value y.

	Case I	Case II
Pounds of bananas	2	12
Cost in dollars	0.48	y

Write the proportion using the ratios as shown in the chart.

$$\frac{2 \text{ lb of bananas}}{\$0.48} = \frac{12 \text{ lb of bananas}}{\$y}$$

Cross multiplying gives us

$(2 \text{ lb of bananas})(\$y) = (12 \text{ lb of bananas})(\$0.48)$

The units are the same on each side of the equation, so we can drop them and have

$2y = 12(0.48)$

$2y = 5.76$

$y = 5.76 \div 2$ Rewrite as division.

$y = 2.88$ Simplify.

So, 12 lb of bananas will cost \$2.88.

Using a table forces the units of a proportion to match. Therefore we usually do not write the units in the proportion itself. We always use the units in the answer.

 ***To solve word problems involving proportions***

1. Write the two ratios and form the proportion. (A table with two columns and two rows will help organize the data. The proportion will be shown in the boxes.)
2. Solve the proportion.
3. Write the solution including the appropriate units.

Warm Ups A–D

A. A sporting goods store advertises golf balls at 6 for \$7.25. At this rate, what will 3 dozen balls cost?

B. A house has a property tax of \$1260 and is valued at \$45,000. At the same rate, what will be the property tax on a house valued at \$56,000?

Examples A–D

Directions: Solve the following problems using proportions.

Strategy: Make a table with two columns and two rows. Label the columns Case I and Case II, and the rows with the units in the problem. Fill in the table with the quantities given, and assign a variable to the unknown quantity. Write the proportion contained in the table, and solve it. Write the solution including units.

A. If 3 cans of tuna fish sell for \$3.73, what is the cost of 48 cans of tuna fish?

	Case I	Case II
Cans	3	48
Cost	\$3.73	C

Make a table.

$$\frac{3}{3.73} = \frac{48}{C}$$ **Write the proportion.**

$$3C = (3.73)(48)$$ **Cross multiply.**

$$3C = 179.04$$

$$C = 59.68$$ **Divide.**

The cost of 48 cans of tuna fish is \$59.68.

B. Mary Alice pays \$1650 property tax on a house valued at \$55,000. At the same rate, what would be the property tax on a house valued at \$82,000?

	Case I	Case II
Tax	\$1650	T
Value	\$55,000	\$82,000

Make a table.

$$\frac{1650}{55{,}000} = \frac{T}{82{,}000}$$ **Write the proportion.**

$$1650(82{,}000) = 55{,}000T$$ **Cross multiply.**

$$135{,}300{,}000 = 55{,}000T$$

$$2460 = T$$ **Divide.**

The tax on the \$82,000 house is \$2460.

Answers to Warm Ups A. The cost of 3 dozen golf balls is \$43.50. B. The property tax will be \$1568.

C. On a road map of Texas, $\frac{1}{4}$ in. represents 50 miles. How many miles are represented by $1\frac{1}{2}$ in.?

	Case I	Case II
Inches	$\frac{1}{4}$	$1\frac{1}{2}$
Miles	50	N

Make a table.

$\dfrac{\frac{1}{4}}{50} = \dfrac{1\frac{1}{2}}{N}$ **Write the proportion.**

$\frac{1}{4}N = \left(1\frac{1}{2}\right)(50)$ **Cross multiply.**

$\frac{1}{4}N = \frac{3}{2}(50)$ **Change to an improper fraction.**

$\frac{1}{4}N = 75$ **Simplify.**

$N = 300$ **Divide.**

On the map $1\frac{1}{2}$ in. represent 300 miles.

C. On a road map of Jackson County, $\frac{1}{4}$ in. represents 25 miles. How many miles are represented by $2\frac{1}{2}$ in.?

D. The city fire code requires a school to have at least 50 square feet of floor space in a classroom for each three students that are in the class. What is the minimum number of square feet needed for 30 students?

	Case I	Case II
Students	3	30
Square Feet	50	S

$\frac{3}{50} = \frac{30}{S}$ **Write the proportion.**

$3S = (50)(30)$ **Cross multiply.**

$3S = 1500$

$S = 500$

The room must have at least 500 ft^2 for 30 students.

D. In another city the fire code requires a school to have at least 86 square feet for each 5 students. What is the minimum area needed for 30 students?

Answers to Warm Ups C. On the map, $2\frac{1}{2}$ in. represents 250 mi. D. The minimum area needed is 516 ft^2.

NAME ______ CLASS ______ DATE ______

Exercises 6.3

OBJECTIVE: *Solve word problems using proportions.*

A.

A photograph that measures 6 in. wide and 4 in. high is to be enlarged so that the width will be 15 in. What will be the height of the enlargement?

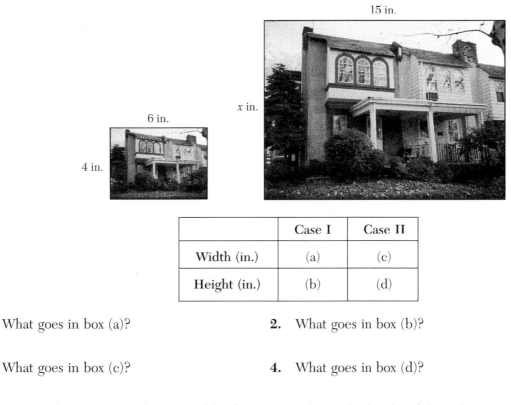

6 in.

4 in.

15 in.

x in.

	Case I	Case II
Width (in.)	(a)	(c)
Height (in.)	(b)	(d)

1. What goes in box (a)?

2. What goes in box (b)?

3. What goes in box (c)?

4. What goes in box (d)?

5. What is the proportion for the problem?

6. What is the height of the enlargement?

If a fir tree is 30 ft tall and casts a shadow of 18 ft, how tall is a tree that casts a shadow of 48 ft?

	First Tree	Second Tree
Height (ft)	(1)	(3)
Shadow (ft)	(2)	(4)

7. What goes in box (1)?

8. What goes in box (2)?

9. What goes in box (3)?

10. What goes in box (4)?

11. What is the proportion for the problem?

12. How tall is the second tree?

Jean and Jim are building a fence around their yard. From past experience they know that they are able to build 48 ft in 8 hr. If they work at the same rate, how many hours will it take them to complete the job if the perimeter of the yard is 288 ft?

	Case I	Case II
Time (hr)	(5)	(7)
Length of fence (ft)	(6)	(8)

13. What goes in box (5)?

14. What goes in box (6)?

15. What goes in box (7)?

16. What goes in box (8)?

17. What is the proportion for the problem?

18. How many hours will it take to build the fence?

B.

The Midvale Junior High School expects a fall enrollment of 910 students. The district assigns teachers at the rate of 3 teachers for every 65 students. The district currently has 38 teachers assigned to the school. How many additional teachers does the district need to assign to the school?

	Case I	Case II
Teachers	3	(e)
Students	65	(f)

19. What goes in box (e)?

20. What goes in box (f)?

21. What is the proportion for the problem?

22. How many teachers will be needed at the school next year?

23. How many additional teachers will need to be assigned?

The average restaurant in Universeville produces 30 lb of garbage in $1\frac{1}{2}$ days. How many pounds of garbage do they produce in 2 weeks (14 days)? (Use x for the missing number of pounds.)

	Case I	Case II
Days		
Garbage (lb)		

24. What goes in each of the four boxes?

25. What proportion should be used to solve this problem?

26. How many pounds of garbage do they have at the end of 2 weeks?

C.

27. Merle is knitting a sweater. The knitting gauge is eight rows to the inch. How many rows must she knit to complete $12\frac{1}{2}$ in. of the sweater?

	Case I	Case II
Rows		
Inches		

28. For every 2 hr a week that Helen is in class, she plans to spend 5 hr a week doing her homework. If she is in class 15 hr each week, how many hours will she plan to be studying each week?

29. If 16 lb of fertilizer will cover 1500 ft^2 of lawn, how much fertilizer is needed to cover 2500 ft^2?

30. If 30 lb of fertilizer covers 1500 ft^2 of lawn, how many square feet will 50 lb of fertilizer cover?

The Logan Community College basketball team won 12 of its first 15 games. At this rate how many games will they win if they play a 30-game schedule?

	Case I	Case II
Games won		
Games played		

31. What goes in each of the four boxes?

32. What is the proportion for the problem?

33. How many games should they win with a 30-game schedule?

34. John must do 25 hr of work to pay for the tuition for three college credits at the local university. If John is going to take 15 credits in the fall, how many hours will he need to work to pay for his tuition?

35. Using Exercise 34, if John works 40 hr per week, how many weeks will he need to work to pay for his tuition? (Any part of a week counts as a full week.)

36. Larry sells men's clothing at the University Men's Shop. If he sells $100 worth of clothing, he makes $15. How much does he make if he sells $340 worth of clothes?

37. Hazel sells automobiles at the Quality Used Car Company. If she sells an automobile for $1200 she is paid $60. If she sells an automobile for $2900, how much is she paid?

38. If gasoline sells for $1.229 per gallon, how many gallons can be purchased for $24.58?

39. If 44 oz of soap powder costs $4.84, how much does 20 oz cost?

40. In Jean's Vegetable Market, onions are priced at 2 lb for $0.63. If Mike buys 6 lb, what does he pay?

41. Twenty-five pounds of tomatoes cost $23.70 at the local market. At this rate, what is the cost of 10 lb?

42. A new car travels 369 miles in 8.2 hr. At the same rate, how long does it take to go 900 miles?

43. A brine solution is made by dissolving 1.5 lb of salt in one gallon of water. At this rate, how many gallons of water are needed when 9 lb of salt are used?

44. Celia earns a salary of $900 per month from which she saves $45 each month. Her salary is increased to $980 per month. How much must she save each month to save at the same rate?

45. Ginger and George have a room in their house that needs to be carpeted. It is determined that a total of 33 yd^2 of carpet are needed for the job. Hickson's Carpet Emporium will install the 33 yd^2 of carpet for $526.35. If Ginger and George decide to have a second room of their house carpeted and the room will need 22 yd^2 of carpet, at the same rate, how much will it cost to have the second room carpeted?

46. A 16-oz can of pears costs $0.98 and a 29-oz can costs $1.69. Is the price per ounce the same in both cases? If not, then what should be the price of the 29-oz can to equalize the price per ounce?

47. A doctor requires that Ida, the nurse, give 8 mg of a certain drug to a patient. The drug is in a solution that contains 20 mg in 1 cm^3. How many cubic centimeters should Ida use for the injection?

48. If a 24-ft beam of structural steel contracts 0.0036 in. for each drop of 5 degrees in temperature, then at the same rate, how much does a 50-ft beam of structural steel contract for a drop of 5 degrees in temperature?

49. If a package of gumdrops weighing 1.5 oz costs 45¢, at the same rate what is the cost of 1 lb (16 oz) of the gumdrops?

NAME ______________________ CLASS ______________ DATE ______________

50. The ratio of boys to girls taking math is 5 to 4. How many boys are in a math class of 81 students? (*Hint:* Fill in the rest of the table.)

	Case I	Case II
Number of boys		
Number of students	9	81

51. Betty prepares a mixture of nuts that has cashews and peanuts in a ratio of 3 to 7. How many pounds of each will she need to make 40 lb of the mixture?

52. The Local Health-Food Store is making a cereal mix that has nuts to cereal in a ratio of 2 to 7. If they want to make 126 oz of the mix, how many ounces of nuts will they need?

53. Debra is making green paint by using 3 quarts of blue paint for every 4 quarts of yellow paint. How much blue paint will she need to make 98 quarts of green paint?

54. A concrete mix takes 3 bags of cement for every 2 bags of sand and every 3 bags of gravel. How many bags of cement are necessary if 80 bags of the concrete mix are needed?

55. Mario makes meatballs for his famous spaghetti sauce by using 10 lb of ground round to 3 lb of additives. How many pounds of ground round should he buy for 91 lb of meatballs?

56. When \$1 is worth 210 drachma (Greek currency) and a used refrigerator costs \$247, what is the cost in drachmas?

57. When \$1 is worth £0.65 (British pound) and a computer costs \$2300, what is the cost in pounds?

58. When \$1 is worth 1455 lire (Italian currency), a pair of shoes costs 69,840 lire. What is the cost in dollars?

59. Auto batteries are sometimes priced proportionally to the number of years they are expected to last. If a \$35.85 battery is expected to last 36 months, what is the comparable price of a 60-month battery?

60. In 1960 only 6.7 lb of every 100 lb of waste was recovered. In 1970, this rose to 7.1 lb. By 1980, the amount was 9.7 lb. In 1990 the amount was up to 13.1 lb. Determine the amount of waste recovered from 56,000,000 lb of waste in each of these years.

61. The amount of ozone contained in 1 m^3 of air may not exceed 235 mg or the air is considered to be polluted. What is the greatest amount of ozone that can be contained in 12 m^3 of air and not be considered polluted?

62. A 5.5-oz can of Alpo cat food is priced at three cans for \$1.00. A 13-oz can is \$0.59. The store manager wants to put the smaller cans on sale so that they are the same unit price as the larger cans. What price should the smaller cans be marked?

63. A large box of brownie mix that makes four batches of brownies costs \$4.79 at a warehouse store. A box of brownie mix that makes one batch costs \$1.29 in a grocery store. By how much should the grocery store reduce each box so that their prices are competitive with the warehouse store?

 Exercises 64–66 relate to the chapter application.

64. You are the set designer for the movie "Honey I Shrunk The Kids." If the back yard was actually 60 ft long, how long is it in the perception of the shrunken kids? Use the following table.

	Shrunken Kid	Back Yard
Actual	.25 in.	
Perceived	5.5 ft	

65. One of the kids is nearly eaten when he falls in a bowl of Cheerios. How big does the set designer make a Cheerio in order to be mathematically accurate?

66. Along with the kids, a sofa was also shrunk. If the sofa was 6 ft long, how small did it become?

STATE YOUR UNDERSTANDING

67. What is a proportion? Write three examples of situations that are proportional.

68. Look on the label of any food package to find the number of calories in one serving. Use this information to create a problem that can be solved by a proportion. Write the solution of your problem in the same way as the examples in this section are written.

69. From a consumer's viewpoint, explain why it is not always an advantage for costs of goods and services to be proportional.

CHALLENGE

70. In 1982, approximately 25 California condors were alive. This low population was caused by losses from hunting, habitat loss, and poisoning. The U.S. Fish and Wildlife Service instituted a program that resulted in 73 condors alive in 1992. If this increase continues proportionally, predict how many condors will be alive in 2017.

71. The tachometer of a sports car shows the engine speed is 2800 revolutions per minute. The transmission ratio (engine speed to drive shaft speed) for the car is 2.5 to 1. Find the drive shaft speed.

72. Two families rented a mountain cabin for 19 days at a cost of $1905. The Santini family stayed for 8 days and the Nguyen family stayed for 11 days. How much did it cost each family? Round the rents to the nearest dollar.

NAME ______________________ CLASS ____________ DATE ____________

GROUP ACTIVITY

73. A $13\frac{1}{2}$-oz bag of Cheetos costs $2.69 and a 24-ounce bag costs $4.09. You have a coupon for $0.50 that the store will double. Divide the group into two teams. One team will formulate an argument that using the coupon on the smaller bag results in a better value. The other team will formulate an argument that using the coupon on the larger bag is better. Present your arguments to the whole group and select the one that is most convincing. Share your results with the rest of the class.

74. List all the types of recycling done by you and your group members. Determine how many people participate in each type of recycling. Determine ratios for each kind of recycling. Find the population of your city or county. Using your class ratios, determine how many people in your area are recycling each type of material. Make a chart to illustrate your findings. Contact your local recycling center to see how your ratios compare to their estimates. Explain the similarities and differences.

MAINTAIN YOUR SKILLS (SECTIONS 5.1, 5.2, 5.4, 5.7)

75. Round 37.4145 to the nearest thousandth.

76. Round 37.4145 to the nearest hundredth.

77. Compare the decimals 0.00872 and 0.011. Write the result as an inequality.

78. Compare the decimals 0.06 and 0.15. Write the result as an inequality.

79. What is the total cost of 11.9 gallons of gasoline that costs $1.379 per gallon? Round to the nearest cent.

80. A barrel of liquid weighs 429.5 lb. If the barrel weighs 22.5 lb and the liquid weighs 7.41 lb per gallon, how many gallons of liquid are in the barrel?

Change each decimal to a fraction.

81. 0.865

82. 0.01175

Change each fraction to a decimal rounded to the nearest thousandth.

83. $\frac{123}{225}$

84. $\frac{29}{350}$

NAME ______________________ CLASS __________ DATE __________

CHAPTER 6

Group Project *(2–3 weeks)*

OPTIONAL

The human body is the source of many common proportions. Artists have long studied the human figure in order to portray it accurately. Your group will be investigating how each member compares to the standard, and how various artists have used the standards.

Most adult bodies can be divided into eight equal portions. The first section is from the top of the head to the chin. Next is from the chin to the bottom of the sternum. The third section is from the sternum to the navel, and the fourth is from the navel to the bottom of the torso. The bottom of the torso to the bottom of the knee is two sections long, and the bottom of the knee to the bottom of the foot is the last two sections. (Actually, these last two sections are a little short. Most people agree that the body is actually closer to 7.5 sections but since this is hard to judge proportionally, we use eight sections and leave the bottom one short.)

Complete the following table for each group member.

Section	Length (cm)	Ratio of Section to Head (Actual)	Ratio of Section to Head (Expected)
Head			
Chin to sternum			
Sternum to navel			
Navel to torso bottom			
Torso bottom to knee			
Knee to foot			

Explain how your group arrived at the values in the last column. Which member of the group comes closest to the standards? Did you find any differences between the males and females in your group? Either draw a body using the standard proportions or get a copy of a figure from a painting and analyze how close the artist came to the standards.

A slightly different method of dividing the upper torso is to start at the bottom of the torso and divide into thirds at the waist and the shoulders. In this method, there is a pronounced difference between males and females. In females, the middle "third" between the waist and shoulders is actually shorter than the other two. In males, the bottom "third" from waist to bottom of the torso is shorter than the others. For each member of your group, fill out the following table.

Section	Length (cm)	Ratio of Section to Entire Upper Torso (Actual)	Ratio of Section to Entire Upper Torso (Expected)
Head to shoulders			
Shoulders to waist			
Waist to bottom of torso			

Explain how your group arrived at the values in the last column (these will depend on gender). Which member of your group comes closest to the standards? Either draw a body using the standard proportions or get a copy of a figure from a painting and analyze how close the artist came to the standards.

Children have different body proportions than adults, and these proportions change with the age of the child. Measure three children who are the same age. Use their head measurement as one unit and compute the ratio of head to entire body. How close are the three children's ratios to each other? Before the Renaissance, artists usually depicted children as miniature adults. This means that the proportions fit those in the first table rather than those you just discovered. Find a painting from before the Renaissance that contains a child. Calculate the child's proportions and comment on its proportions. Be sure to reference the painting you use.

NAME ______________________ CLASS ____________ DATE ____________

CHAPTER 6

True–False Concept Review

Check your understanding of the language of algebra and arithmetic. Tell whether each of the following statements is True (always true) or False (not always true). For each statement you judge to be false, revise it to make a statement that is true.

ANSWERS

1. A fraction can be regarded as a ratio. **1.** ______

2. A ratio is the comparison of two fractions. **2.** ______

3. $\frac{7 \text{ miles}}{1 \text{ gallon}} = \frac{14 \text{ miles}}{2 \text{ hours}}$ **3.** ______

4. To solve a proportion, we must know the values of three of the four numbers. **4.** ______

5. If $\frac{7}{3} = \frac{t}{4}$, then $t = \frac{3}{28}$. **5.** ______

6. In a proportion, two ratios are equal. **6.** ______

7. Twelve inches and one foot are like measures. **7.** ______

8. Ratios always compare like units. **8.** ______

9. To determine whether a proportion is true or false, the ratios must have the same units. **9.** ______

10. If a fir tree that is 18 ft tall casts a shadow of 17 ft, how tall is a tree that casts a shadow of 25 ft? The following table can be used to solve this problem. **10.** ______

	First Tree	Second Tree
Height	17	18
Shadow	x	25

NAME ______________________________ CLASS ______________ DATE ______________

CHAPTER 6

Test

ANSWERS

1. Write a ratio to compare 12 lb to 15 lb. — **1.** ________

2. On a test Ken answered 24 of 30 questions correctly. At the same rate, how many would he answer correctly if there were 100 questions on a test? — **2.** ________

3. Solve the proportion: $\frac{2.4}{8} = \frac{0.36}{w}$ — **3.** ________

4. Is the following proportion true or false? $\frac{16}{34} = \frac{24}{51}$ — **4.** ________

5. Solve the proportion: $\frac{13}{36} = \frac{y}{18}$ — **5.** ________

6. Is the following proportion true or false? $\frac{9\text{ in.}}{2\text{ ft}} = \frac{6\text{ in.}}{16\text{ in.}}$ — **6.** ________

7. If Mary is paid $49.14 for 7 hr of work, how much should she be paid for 12 hr of work? — **7.** ________

8. Write a ratio to compare 6 hr to 3 days (compare in hours). — **8.** ________

9. There is a canned food sale at the supermarket. A case of 24 cans of peas is priced at $19.68. What is the price of 10 cans? — **9.** ________

10. If 40 lb of beef contains 7 lb of bones, how many pounds of bones may be expected in 100 lb of beef? — **10.** ________

11. Solve the proportion: $\frac{0.3}{0.9} = \frac{0.8}{x}$ — **11.** ________

12. A charter fishing boat has been catching an average of 3 salmon for every 4 people they take fishing. At that rate, how many fish will they catch if over a period of time they take a total of 32 people fishing? — **12.** ________

13. On a trip home, Jennie used 12.5 gallons of gas. The trip odometer on her car registered 295 miles for the trip. She is planning a trip to see a friend who lives 236 miles away. How much gas will Jennie need for the trip?

13. ______________

14. Solve the proportion: $\frac{a}{6} = \frac{3.21}{3.6}$

14. ______________

15. Is the following a rate? $\frac{35 \text{ miles}}{2 \text{ hr}}$

15. ______________

16. If a 20-ft tree casts a 15-ft shadow, how long a shadow is cast by a 14-ft tree?

16. ______________

17. What is the population density of a town that is 40 square miles and has 5,270 people? Reduce to a one square mile comparison.

17. ______________

18. Solve the proportion and round your answer to the nearest hundredth: $\frac{4.78}{y} = \frac{32.5}{11.2}$

18. ______________

19. A landscape firm has a job that it takes a crew of three $4\frac{1}{2}$ hr to do. How many of these jobs could the crew of three do in 117 hrs?

19. ______________

20. The ratio of males to females in a literature class is 2 to 5. How many females are in a class of 49 students?

20. ______________

Good Advice for Studying

Low-Stress Tests

It's natural to be anxious before an exam. In fact, a little anxiety is actually good: it keeps you alert and on your toes. Obviously, too much stress over tests is not good. Here are some proven tips for taking low-stress tests.

1. Before going to the exam, find a place on campus where you can physically and mentally relax. Don't come into the classroom in a rush.
2. Arrive at the classroom in time to arrange all the tools you will need for the test: sharpened pencils, eraser, plenty of scratch paper, and a water bottle. Try to avoid talking with classmates about the test. Instead, concentrate on deep breathing and relaxation.
3. Before starting the test, on a separate piece of paper, write all the things you may forget while you are busy at work: formulas, rules, definitions, and reminders to yourself. Doing so relieves the load on your short-term memory.
4. Read all of the test problems and mark the easiest ones. Don't skip reading the directions. Note point values so that you don't spend too much time on problems that count only a little, at the expense of problems that count a lot.
5. Do the easiest problems first; do the rest of the problems in order of difficulty.
6. Estimate a reasonable answer before you make calculations. When you finish the problem, check to see that your answer agrees with your estimate.
7. If you get stuck on a problem, mark it and come back to it later.
8. When you have finished trying all the problems, go back to the problems you didn't finish and do what you can. Show all steps because you may get partial credit even if you cannot complete a problem.
9. When you are finished, go back over the test to see that all the problems are as complete as possible and that you have indicated your final answer. Use all of the time allowed, unless you are sure there is nothing more that you can do.
10. Turn in your test and be confident that you did the best job you could. Congratulate yourself on a low-stress test!

If you find yourself feeling anxious, try a 3×5 "calming" card. It may include the following: (a) a personal coping statement such as "I have studied hard and prepared well for this test, I will do fine."; (b) a brief description of your peaceful scene; and (c) a reminder to stop, breathe, and relax your tense muscles.

By now, you should be closer to taking control over math instead of allowing math to control you. You are avoiding learned helplessness (believing that other people or influences control your life). Perfectionism, procrastination, fear of failure, and blaming others are also ineffective attitudes that block your power of control. Take responsibility, and believe that you have the power within to control your life situation.

7 Percent

APPLICATION

The price we pay for everyday items such as food and clothing is theoretically simple. The manufacturer of the item sets the price based on how much it costs to produce and adds a small profit. The manufacturer then sells the item to a retail store, which in turn marks it up and sells it to you the consumer. But as you know, it is rarely as simple as that. The price you actually pay for an item also depends on the time of year, the availability of raw materials, the amount of competition between manufacturers of comparable items, the economic circumstances of the retailer, the geographic location of the retailer, and many other factors.

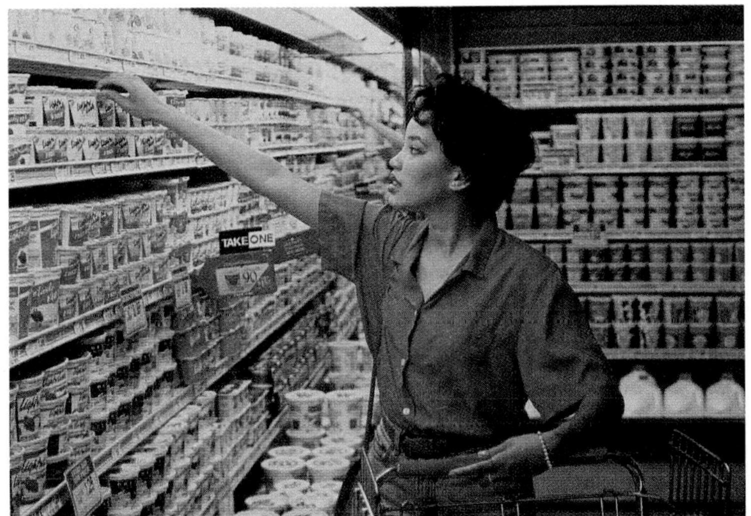

GROUP ACTIVITY

Select a common item whose price is affected by the factors listed below. Discuss how the factor varies and how the price of the item is affected. For each factor, make a plausible bar graph that shows the change in price as the factor varies. (You may estimate specific price levels.)

a. Time of year

b. Economic circumstances of the retailer

c. Competition of comparable products

7.1

The Meaning of Percent

OBJECTIVE

Write a percent to express a comparison of two numbers.

VOCABULARY

When ratios are used to compare numbers, the denominator is called the **base unit.** In comparing 80 to 100 $\left(\text{as the ratio } \frac{80}{100}\right)$, 100 is the base unit.

The **percent comparison,** or just the **percent,** is a ratio with a base unit of 100. The percent $\frac{80}{100} = (80)\left(\frac{1}{100}\right)$ is usually written 80%. The symbol % is read "percent," and $\% = \frac{1}{100} = 0.01$.

HOW AND WHY

Objective

Write a percent to express a comparison of two numbers.

The word "percent" means "by the hundred." It is from the Roman word *percentum.* In Rome, taxes were collected by the hundred. For example, if you had 100 cattle, the tax collector might take 14 of them to pay your taxes. Hence, 14 per one hundred, or 14 percent, would be the tax rate.

Look at Figure 7.1 to see an illustration of the concept of "by the hundred." The base unit is 100, and 24 of the 100 parts are shaded. The ratio of shaded parts to total parts is $\frac{24}{100} = 24\left(\frac{1}{100}\right) = 24\%$. We say that 24% of the unit is shaded.

Figure 7.1

Figure 7.1 also illustrates that if the numerator is smaller than the denominator, then not all of the base unit will be shaded, and hence the comparison will be less than 100%. If the numerator equals the denominator, the entire unit will be shaded and the comparison will be 100%. If the numerator is larger than the denominator, more than one entire unit will be shaded, and the comparison will be more than 100%.

The ratio of two numbers can be used to find the percent when the base unit is not 100. Compare 7 to 20. The ratio is $\frac{7}{20}$. Now find the equivalent ratio with a denominator of 100.

$$\frac{7}{20} = \frac{35}{100} = 35 \cdot \frac{1}{100} = 35\%$$

If the equivalent ratio with a denominator of 100 cannot be found easily, solve as a proportion. See Example F.

To find the percent comparison of two numbers

1. Write the ratio of the first number to the base number.
2. Find the equivalent ratio with denominator 100.
3. $\frac{\text{numerator}}{100} = \text{numerator} \cdot \frac{1}{100} = \text{numerator } \%$

Examples A–C

Directions: Write the percent of each region that is shaded.

Strategy: (1) Count the number of parts in each unit. (2) Count the number of parts that are shaded. (3) Write the ratio of these as a fraction and build the fraction to denominator of 100. (4) Write the percent using the numerator in step 3.

A. What percent of the unit is shaded?

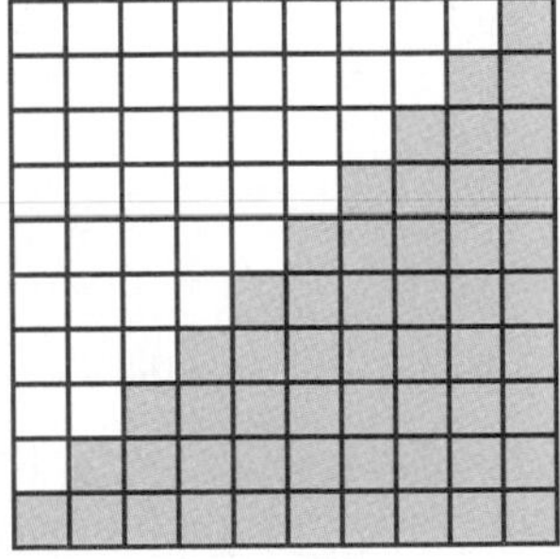

100 parts in the region

55 parts are shaded.

$\frac{55}{100}$

55%

So, 55% of the region is shaded.

Warm Ups A–C

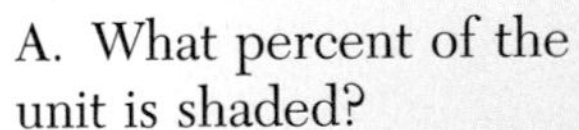

A. What percent of the unit is shaded?

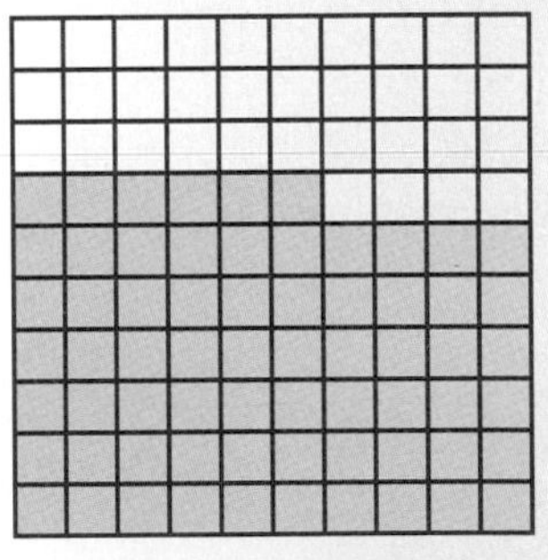

Answer to Warm Up A. 66%

B. What percent of the region is shaded?

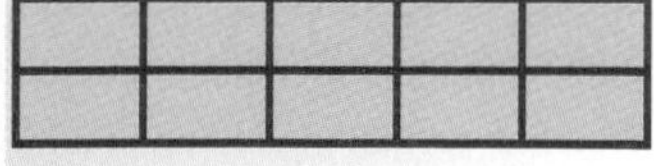

B. What percent of the region is shaded?

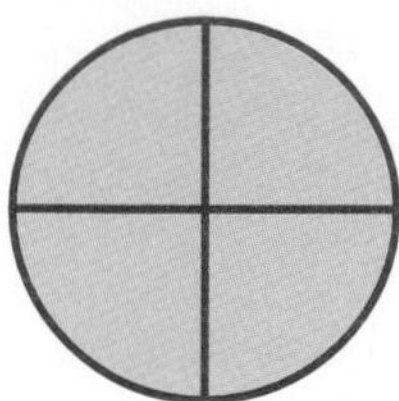

4 parts in the region

4 parts are shaded.

$\frac{4}{4} = \frac{100}{100}$

100%

So, 100% of the region is shaded.

C. What percent of the region is shaded?

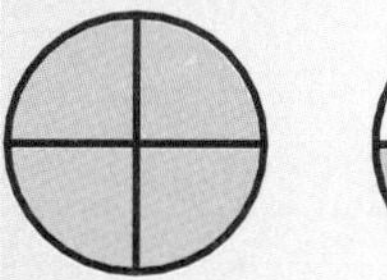

One unit   One unit

C. What percent of the region is shaded?

4 parts in each unit

5 parts are shaded

$\frac{5}{4} = \frac{125}{100}$ **Write as a fraction with a denominator of 100.**

125% **Write as a percent.**

So, 125% of a unit in the region is shaded.

Warm Ups D–H

Examples D–H

Directions: Write the percent for the comparison.

Strategy: Write the comparison in fraction form. Build the fraction to hundredths or write and solve a proportion and write the percent using the numerator.

D. At the last soccer match of the season, of the first 100 tickets sold, 73 were student tickets. What percent are student tickets?

D. At a football game, 27 women are among the first 100 fans to enter. What percent of the first 100 fans are women?

$\frac{27}{100} = 27 \cdot \frac{1}{100} = 27\%$ **The comparison of women to fans is 27 to 100. Write the fraction and change to a percent.**

So, 27% of the first 100 fans are women.

Answers to Warm Ups B. 100% C. 175% D. Of the 100 tickets, 73% were student tickets.

E. Write the ratio of 36 to 25 as a percent.

$\frac{36}{25} = \frac{144}{100}$ **Write the ratio and build to a denominator of 100.**

$= 144 \cdot \frac{1}{100}$

$= 144\%$ **Change to a percent.**

So, 36 is 144% of 25.

E. Write the ratio of 70 to 50 as a percent.

F. Write the ratio of 15 to 18 as a percent.

$\frac{15}{18} = \frac{R}{100}$ **Since we cannot easily write the fraction with a denominator of 100, we write a proportion to find the percent.**

$15(100) = 18R$ **Cross multiply.**

$1500 \div 18 = R$

$83\frac{1}{3} = R$

So,

$$\frac{15}{18} = \frac{83\frac{1}{3}}{100}$$

$$= 83\frac{1}{3} \cdot \frac{1}{100}$$

$$= 83\frac{1}{3}\%$$

So, 15 is $83\frac{1}{3}\%$ of 18.

F. Write the ratio of 8 to 12 as a percent.

Calculator Example

G. Compare 35 to 560 as a percent.

$\frac{35}{560} = \frac{R}{100}$ **Write as a proportion.**

$560R = 35(100)$ **Solve.**

$R = 35(100) \div 560$ **Evaluate using a calculator.**

$R = 6.25$

So, 35 is 6.25% of 560.

G. Compare 45 to 480 as a percent.

Answers to Warm Ups E. 140% F. $66\frac{2}{3}\%$ G. 9.375%

H. Of 300 fish caught in Olive Lake, 54 were tagged. What percent of the fish was tagged?

H. After a strong antismoking campaign in one state, 285 of 300 restaurants banned smoking. What percent of the restaurants banned smoking?

$$\frac{285}{300} = \frac{95}{100}$$ **Write the ratio comparison and simplify.**

$$= 95 \cdot \frac{1}{100}$$

$$= 95\%$$

So, 95% of the restaurants banned smoking.

Answer to Warm Up H. Of the 300 fish, 18% were tagged.

NAME ______________________ CLASS ____________ DATE ____________

Exercises 7.1

OBJECTIVE: *Write a percent to express a comparison of two numbers.*

A.

What percent of each of the following regions is shaded?

1.

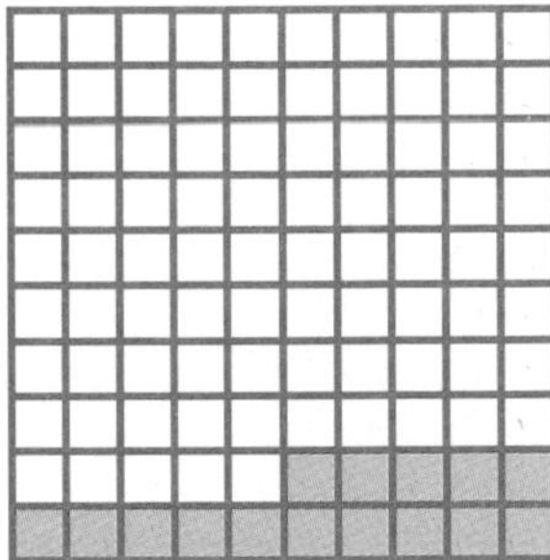

2.

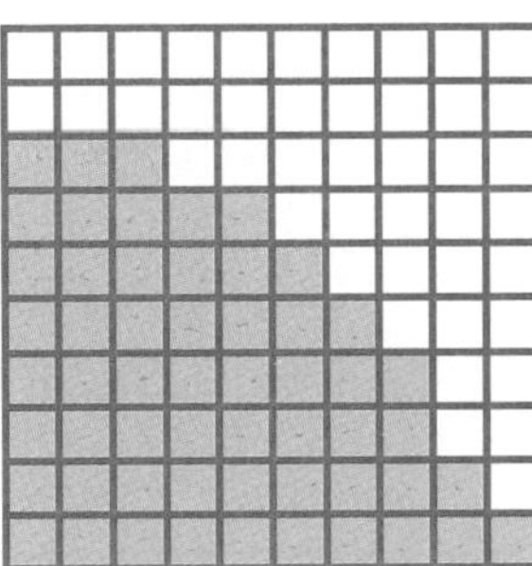

3.

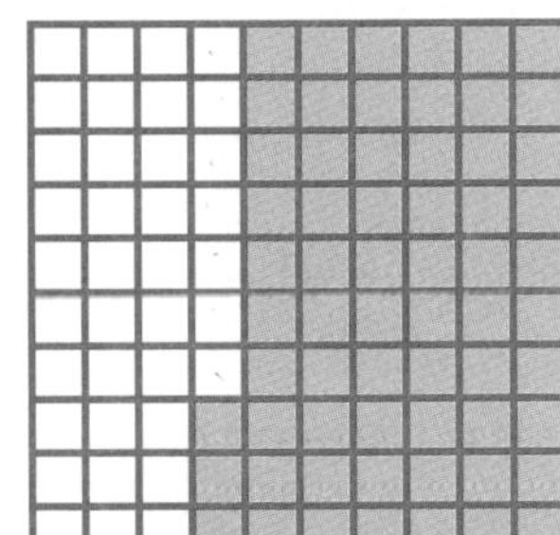

4.

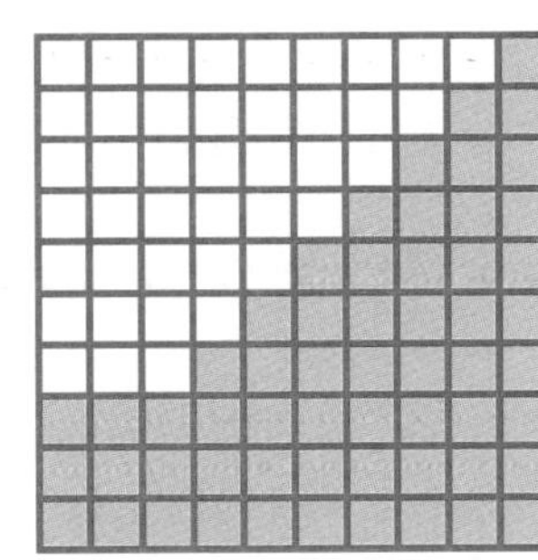

5.

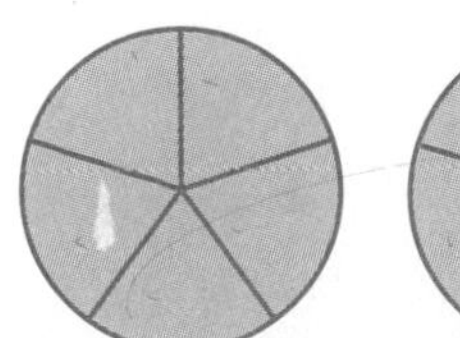
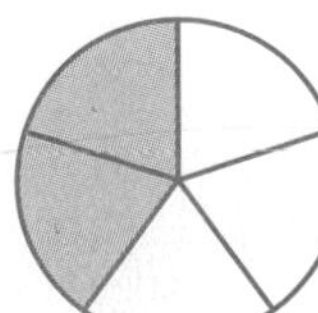

6.

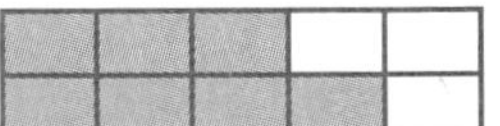

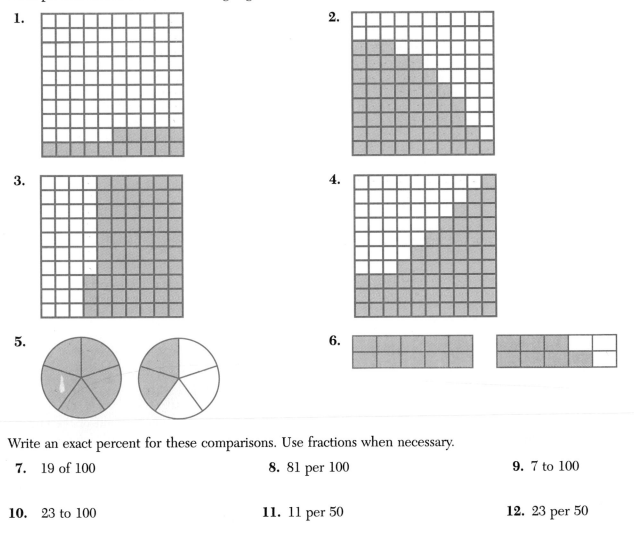

Write an exact percent for these comparisons. Use fractions when necessary.

7. 19 of 100

8. 81 per 100

9. 7 to 100

10. 23 to 100

11. 11 per 50

12. 23 per 50

13. 11 of 25

14. 21 to 25

15. 7 per 20

16. 13 per 20

B.

17. 17 to 10

18. 450 to 120

19. 170 of 170

20. 63 to 63

21. 24 to 16

22. 36 to 16

23. 75 to 200

24. 65 to 200

25. 9 per 15

26. 53 per 500

27. 75 per 80

28. 93 per 80

29. 28 to 42

30. 11 to 12

31. 95 to 114

32. 11 to 15

C.

33. The fact that 12% of all people are blonde indicates that __________ of 100 people are blonde.

34. In a recent election there was a 48% turn out of registered voters. This indicates that __________ of 100 registered voters turned out to vote.

35. In a recent mail-in election, 76 of every 100 eligible voters cast their ballots. What percent of the eligible voters exercised their right to vote?

36. Of the people who use Shiny toothpaste, 43 of 100 report fewer cavities. Of every 100 people who report, what percent do not report fewer cavities?

NAME ______________________ CLASS ____________ DATE ____________

Write an exact percent for these comparisons. Use fractions when necessary.

37. 133 to 152

38. 119 to 136

39. 105 to 25

40. 105 to 15

41. 319 to 600

42. 46 to 900

43. If a telephone tax is 8 cents per dollar, what percent is this?

44. For every $100 spent on gasoline, the state receives $7.50 tax. What percent of the price of gasoline is the state tax?

45. A bank pays $5.65 interest per year for every $100 in savings. What is the annual interest rate?

46. James has $500 in his savings account. Of that amount, $35 is interest that was paid to him. What percent of the total amount is the interest?

Exercises 47–50 are related to the chapter application.

47. Carol spends \$67 on a new outfit. If she has \$100, what percent of her money does she spend on the outfit?

48. A graphing calculator originally priced at \$100 is on sale for \$89. What is the percent of discount? (Discount is the difference between the original price and the sales price.)

49. Mickie bought a TV and makes monthly payments on it. Last year she paid a total of \$900. Of the total that she paid, \$135 was interest. What percent of the total was interest?

50. Jose buys a suit that was originally priced at \$100. He buys it for 35% off the original price. What does he pay for the suit?

51. Last year Mr. and Mrs. Johanson were informed that the property tax rate on their home was \$2.75 per \$100 of the house's assessed value. What percent is the tax rate?

STATE YOUR UNDERSTANDING

52. What is a percent? How is it related to fractions and decimals?

53. Explain the difference in meaning of the symbols 25% and 125%. In your explanation, use diagrams to illustrate the meanings. Contrast similarities and differences in the diagrams.

NAME ______________________ CLASS ____________ DATE ____________

CHALLENGE

54. Write the ratio of 101 to 500 as a fraction and as a percent.

55. Write the ratio of 503 to 800 as a fraction and as a percent.

GROUP ACTIVITY

56. Have the members of your group use the resource center to find some background about the percent symbol (%). Divide the task so that one member looks in a large dictionary, some look in different encyclopedias, and others look in other mathematics books. Together make a short report to the rest of the class on your findings.

MAINTAIN YOUR SKILLS (SECTIONS 5.5, 5.8, 5.9)

Change to a decimal.

57. $\frac{17}{32}$

58. $\frac{37}{64}$

Change to a decimal rounded to the nearest hundredths place.

59. $\frac{13}{18}$

60. $\frac{35}{24}$

Find the average.

61. 17.8, 14.3, 16.5, 20.8, 12.1

62. 210.8, 355.9, 175.6, 299.3

Divide.

63. $567.002 \div 10^5$

64. $33{,}000 \div 10^6$

65. Bill goes to the store with $15. He uses his calculator to keep track of the money he is spending. He decides that he could make the following purchases. Is he correct?

Article	*Cost*
2 loaves of bread	$0.99 each
5 cans of soup	$0.89 each
1 box of crackers	$1.29
2 lb hamburger	$1.59 per lb
6 cans of root beer	6 cans for $2.79

66. Ms. Henderson earns $14.85 per hour and works the following hours during one month. How much are her monthly earnings?

Week	*Hours*
1	35
2	30.25
3	25.5
4	36.75
5	6

7.2

Changing Decimals to Percents

OBJECTIVE

Write a given decimal as a percent.

HOW AND WHY

Objective

Write a given decimal as a percent.

In multiplication, where one factor is $\frac{1}{100}$, the indicated multiplication can be read as a percent. That is, $75\left(\frac{1}{100}\right) = 75\%$, $0.8\left(\frac{1}{100}\right) = 0.8\%$, and $\frac{3}{4}\left(\frac{1}{100}\right) = \frac{3}{4}\%$.

To write a number as a percent, multiply by $100 \cdot \frac{1}{100}$, a name for one. This is shown in Table 7.1.

TABLE 7.1 CHANGE A DECIMAL TO A PERCENT

Number	Multiply by 1 $100\left(\frac{1}{100}\right) = 1$	Multiply by 100	Percent
0.45	$0.45\ (100)\left(\frac{1}{100}\right)$	$45.\left(\frac{1}{100}\right)$	45%
0.2	$0.2\ (100)\left(\frac{1}{100}\right)$	$20.\left(\frac{1}{100}\right)$	20%
5	$5\ (100)\left(\frac{1}{100}\right)$	$500.\left(\frac{1}{100}\right)$	500%

In each case the decimal point is moved two places to the right and the percent symbol (%) is inserted.

To change a decimal to a percent

1. Move the decimal point two places to the right. (Write zeros on the right if necessary.)
2. Write the percent symbol (%) on the right.

Warm Ups A–G

A. Write 0.69 as a percent.

B. Change 0.09 to a percent.

C. Write 0.355 as a percent.

D. Change 0.002 to a percent.

E. Write 4 as a percent.

F. Change $0.44\overline{6}$ to a percent.

G. The tax code lists the tax rate on a zone 3 lot at 0.027. What is the tax rate expressed as a percent?

Examples A–G

Directions: Change the decimal to a percent.

Strategy: Move the decimal point two places to the right and write the percent sign on the right.

A. Write 0.47 as a percent.

0.47 = 47% **Move the decimal point two places to the right. Write the percent symbol on the right.**

B. Change 0.07 to a percent.

0.07 = 007% = 7% **Since the zeros are to the left of 7 we can drop them.**

C. Write 0.357 as a percent.

0.357 = 35.7%

D. Change 0.006 to a percent.

0.006 = 000.6% = 0.6% **This is six tenths of one percent.**

E. Write 8 as a percent.

8 = 8.00 = 800% **Insert two zeros on the right so we can move two decimal places. Eight hundred percent is 8 times 100%.**

F. Change $0.72\overline{3}$ to a percent.

$0.72\overline{3} = 72.\overline{3}\% = 72\frac{1}{3}\%$ **The repeating decimal $0.\overline{3} = \frac{1}{3}$.**

G. The tax rate on a building lot is given as 0.032. What is the tax rate expressed as a percent?

0.032 = 003.2% = 3.2%

So the tax rate expressed as a percent is 3.2%.

Answers to Warm Ups A. 69% B. 9% C. 35.5% D. 0.2% E. 400% F. $44.\overline{6}\%$ or $44\frac{2}{3}\%$ G. The tax rate is 2.7%.

NAME ______________________ CLASS __________ DATE __________

Exercises 7.2

OBJECTIVE: *Write a given decimal as a percent.*

A.

Write each decimal as a percent.

1. 0.47 **2.** 0.28 **3.** 6.51

4. 8.64 **5.** 0.07 **6.** 0.09

7. 1.17 **8.** 5.65 **9.** 17

10. 21 **11.** 0.006 **12.** 0.001

13. 0.732 **14.** 0.376 **15.** 0.45

16. 0.77 **17.** 3 **18.** 8

19. 0.416 **20.** 0.712

B.

21. 0.0956 **22.** 0.0487 **23.** 80

24. 45 **25.** 14.75 **26.** 6.091

27. 0.00034 **28.** 0.00236 **29.** 3.3

30. 1.95 **31.** 0.0345 **32.** 0.0213

33. 0.00011 **34.** 0.0021 **35.** 0.2054

36. 0.3608 **37.** $0.7416\overline{6}$ **38.** $0.033\overline{3}$

39. 0.1005 **40.** 0.30607

C.

41. If the tax rate on a person's income is 0.22, what is the rate expressed as a percent?

42. The completion rate in a certain math class is 0.73. What is the rate as a percent?

43. A merchant sold 0.52 of her sale items in the first day of the sale. What percent of the sale items were sold?

44. The sales tax in an eastern state is 0.063. Express this as a percent.

45. If 0.137 of the contestants in a race withdraw, what is the percent of withdrawals?

46. Mary Ellen measured the July rainfall. She found it was 0.345 of the year's total. Express this as a percent.

47. During the blizzard of 1997, the price of a snow blower increased by a factor of 1.73. Express this as a percent.

48. The price of the new 1999 Accord is 1.13 times the price of a 1998 model. Express this as a percent.

49. A swim team wins 0.685 of their meets. Write this as a percent.

50. An NBA basketball player makes 0.893 of his free throws. Express this as a percent.

NAME ______________________ CLASS __________ DATE __________

Exercises 51–52 relate to the chapter application.

51. The sale price of a can of beans is 0.89 of the original price. Express this as a percent. What "percent off" will the store advertise?

52. Mary spends 0.325 of her monthly income on groceries. What percent of her monthly income is spent on groceries?

53. Social Security paid by employees is found by multiplying the gross wages by 0.062. The Medicare payment is found by multiplying the gross wages by 0.0145. Express the sum of these amounts as a percent.

54. Cholesterol levels in Americans have dropped from 0.26 in 1981 to 0.2 in 1993. Express the difference as a percent.

STATE YOUR UNDERSTANDING

55. Explain how the decimal form and the percent form of a number are related. Give an example of each form.

CHALLENGE

Write as percents.

56. 0.0004507

57. 18,000

Write as percents without using repeating decimals.

58. 0.024 and $0.02\overline{4}$

59. 0.425 and $0.42\overline{5}$

GROUP ACTIVITY

60. Baseball batting averages are written as decimals. A batter with an average of 238 has hit an average of 238 times out of 1000 times at bat (0.238). Find the batting averages of the top five players in the American and National Leagues for the past five years. Express these averages as percents.

MAINTAIN YOUR SKILLS (SECTIONS 5.1, 5.3, 5.8, 5.9)

61. Round to the nearest thousandth: 3.87264

62. Round to the nearest ten-thousand: 345,891.62479

Perform the indicated operations.

63. $(0.37)(0.4) + 2.5 - (0.04)(0.02)$

64. $(0.75) - (0.5)(0.3) + (1.8)(0.2)$

65. $(0.18)(4.6) \div 36 - (0.15)(0.03)$

66. $0.6^2 - 0.3^3$

67. $[2.4^2 + 7.3(0.2)](0.3)^2 - 0.041$

68. $37 - 3.7^2(0.3)$

69. Marilyn needs to buy four textbooks for this semester. She goes to the bookstore and finds used texts that cost:

Text	*Cost*
Algebra	$39.75
Chemistry	$48.50
Psychology	$36.00
American History	$42.75

What is the total cost of the books that she needs? What is the average cost of the books?

70. Bill is making a table top. The two ends of the piece of wood he is using need to be equally trimmed and sanded. The piece of wood is 52.25 in. long. He requires a length of 48.5 in. The saw removes $\frac{3}{16}$ in. during each cut, and he wants to allow $\frac{1}{16}$ in. for finishing sanding on each end, how much should he cut from each end?

7.3

Changing Percents to Decimals

OBJECTIVE

Write a given percent as a decimal.

HOW AND WHY

Objective

Write a given percent as a decimal.

The percent symbol indicates multiplication by $\frac{1}{100}$, so

$$17\% = 17 \cdot \frac{1}{100} = \frac{17}{100} = 17 \div 100$$

As we learned in Section 5.5, dividing a number by 100 is done by moving the decimal point two places to the left.

$17\% = 17 \div 100 = 0.17$

To change a percent to a decimal

1. Move the decimal point two places to the left. (Write zeros on the left if necessary.)
2. Drop the percent symbol (%).

Examples A–F

Directions: Change the percent to a decimal.

Strategy: Move the decimal point two places to the left and drop the percent symbol.

A. Change 13.7% to a decimal.

13.7% = 0.137 **Move the decimal point two places left. Drop the percent symbol.**

B. Write 45% as a decimal.

45% = 0.45

C. Change 376% to a decimal.

376% = 3.76 **A value over 100% becomes a mixed number or whole number.**

Warm Ups A–F

A. Change 87.5% to a decimal.

B. Write 83% as a decimal.

C. Change 342% to a decimal.

Answers to Warm Ups A. 0.875 B. 0.83 C. 3.42

D. Write $34\frac{4}{5}\%$ as a decimal.

D. Write $65\frac{3}{10}\%$ as a decimal.

$65\frac{3}{10}\% = 65.3\%$ **Change the fraction to a decimal.**

$= 0.653$ **Change the percent to a decimal.**

So, $65\frac{3}{10}\% = 0.653$.

E. Change $61\frac{5}{6}\%$ to a decimal. Round to the nearest thousandth.

E. Change $7\frac{5}{9}\%$ to a decimal. Round to the nearest thousandth.

$7\frac{5}{9}\% = 7.5\overline{5}\%$ **By division, $\frac{5}{9} = 0.\overline{5}$.**

$= 0.075\overline{5}$ **Change to a decimal.**

≈ 0.076 **Round to the nearest thousandth.**

So, $7\frac{5}{9}\% \approx 0.076$.

F. When ordering cement, a contractor orders 2.8% more than is needed to allow for waste. What decimal will she enter into the computer to calculate the extra amount to be added to the order?

F. When ordering fresh vegetables, a grocer orders 8.6% more than is needed to allow for spoilage. What decimal is entered into the computer to calculate the amount of extra vegetables to be added to the order?

8.6% = 0.086 **Change the percent to a decimal.**

So the grocer will enter 0.086 in the computer.

Answers to Warm Ups D. 0.348 E. 0.618 F. The contractor will enter 0.028 in the computer.

NAME ______________________ CLASS ______________ DATE ______________

Exercises 7.3

OBJECTIVE: *Write a given percent as a decimal.*

A.

Write each of the following as a decimal.

1. 18% **2.** 69% **3.** 91% **4.** 37%

5. 73% **6.** 57% **7.** 3.46% **8.** 5.61%

9. 416% **10.** 715% **11.** 216.5% **12.** 156.7%

13. 0.03% **14.** 0.07% **15.** 213% **16.** 529%

17. 0.13% **18.** 0.467% **19.** 2340% **20.** 1511%

B.

21. 0.058% **22.** 0.0467% **23.** 100%

24. 300% **25.** 125% **26.** 345%

27. $\frac{1}{2}\%$ **28.** $\frac{1}{4}\%$ **29.** $\frac{5}{8}\%$

30. $\frac{4}{5}\%$ **31.** $29\frac{3}{4}\%$ **32.** $27\frac{2}{5}\%$

33. 0.00074% **34.** 0.000206% **35.** $\frac{7}{8}\%$

36. $\frac{5}{16}\%$ **37.** 67.443% **38.** 58.01%

39. 475.5%

40. 213.9%

C.

41. Employees just settled their new contract and got a 9.75% raise over the next two years. Express this as a decimal.

42. When bidding for a job, an estimator adds 8.75% to the bid to cover unexpected expenses. What decimal part is this?

43. Interest rates are expressed as percents. The Credit Union charges 10.75% interest on new auto loans. What decimal will they use to compute the interest?

44. What decimal is used to compute the interest on a mortgage that has an interest rate of 7.65%?

45. Unemployment is down 0.35%. Express this as a decimal.

46. The cost of living rose 0.72% during June. Express this as a decimal.

Change to a decimal rounded to the nearest thousandth.

47. $\frac{1}{6}\%$

48. $\frac{2}{3}\%$

49. $35\frac{1}{15}\%$

50. $72\frac{5}{7}\%$

51. $\frac{11}{6}\%$

52. $\frac{9}{8}\%$

53. In industrialized countries 60% of the river pollution is due to agricultural runoff. Change this to a decimal.

54. Recycling aluminum cans consumes 95% less energy than smelting new stocks of metal. Change this to a decimal.

55. In 1993, Americans had 4% lower blood cholesterol levels than in 1981. Change this to a decimal.

56. One serving of Quaker Toasted Oatmeal contains 8% of the daily recommended intake of sodium and 14% of the recommended intake of dietary fiber. Express these as decimals.

Exercises 57–58 relate to the chapter application.

57. The May Company advertises a sale at 45% off originally marked prices. What decimal will Carol use to calculate the savings on an item originally priced at $78.95?

58. Nordstom's advertises a sale of an additional 15% off sale prices. What decimal will Houng use to calculate the savings on a item sale priced at $55?

59. Find today's interest rates for home mortgages for 15 and 30 years. Express these as decimals.

60. Find the interest being paid in savings accounts at three different financial institutions in your area. Express these as decimals.

61. During the 1997 season, Randy Johnson of the Seattle Mariners, won 83.3% of the games in which he was involved in the decision. Express this as a decimal.

62. During the 1997 major league baseball season the Houston Astros won 52.2% of the games they played. Express this as a decimal.

STATE YOUR UNDERSTANDING

63. When changing a percent to a decimal, how can you tell when the decimal will be greater than one?

CHALLENGE

64. Change $11\frac{4}{17}\%$ to a decimal rounded to the nearest tenth and the nearest thousandth.

65. Change $56\frac{11}{12}\%$ to a decimal rounded to the nearest tenth and the nearest thousandth.

GROUP ACTIVITY

66. Research the major causes of the greenhouse effect. Find out which substances cause the greenhouse effect and the percent contributed by each. Write these percents in decimal form. In class discuss ways to reduce the greenhouse effect and write group reports on your findings.

MAINTAIN YOUR SKILLS (SECTIONS 5.2, 5.3, 5.8, 5.9)

Change to a decimal.

67. $\frac{7}{8}$

68. $\frac{9}{64}$

69. $\frac{19}{16}$

70. $\frac{24}{75}$

71. $\frac{27}{15}$

72. $\frac{117}{65}$

Change to a fraction.

73. 0.715

74. 0.1025

75. In one week Greg earns $245. His deductions (income tax, Social Security, and so on) total $38.45. What is his take-home pay?

76. The cost of gasoline is reduced from $0.695 per liter to $0.629 per liter. How much money is saved on an automobile trip that requires 340 liters?

7.4

Changing Fractions to Percents

OBJECTIVE

Change a fraction or mixed number to a percent.

HOW AND WHY

Objective

Change a fraction or mixed number to a percent.

We already know how to change fractions to decimals and decimals to percent. We combine the two ideas to change fractions to a percent.

To change a fraction or mixed number to a percent

1. Change to a decimal. The decimal is rounded or carried out as directed.
2. Change the decimal to percent.

Unless directed to round, the division is completed or else the quotient is written as a repeating decimal.

Examples A–H

Directions: Change the fraction or mixed number to a percent.

Strategy: Change the number to a decimal and then to a percent.

A. Change $\frac{2}{5}$ to a percent.

$\frac{2}{5} = 0.4$ **Divide 2 by 5 to change the fraction to a decimal.**

$= 40\%$ **Change the decimal to a percent.**

So, $\frac{2}{5} = 40\%$.

B. Write $\frac{9}{16}$ as a percent.

$\frac{9}{16} = 0.5625$ **Change to a decimal.**

$= 56.25\%$ **Change to a percent.**

So, $\frac{9}{16} = 56.25\%$.

Warm Ups A–H

A. Change $\frac{9}{10}$ to a percent.

B. Write $\frac{5}{8}$ as a percent.

Answers to Warm Ups A. 90% B. 62.5%

C. Change $\frac{5}{6}$ to a percent.

C. Change $\frac{1}{11}$ to a percent.

$\frac{1}{11} = 0.09\overline{09}$ **Write $\frac{1}{11}$ as a repeating decimal.**

$= 9.\overline{09}\%$ **Change to a percent.**

$= 9\frac{1}{11}\%$ **The repeating decimal $0.\overline{09} = \frac{1}{11}$.**

So, $\frac{1}{11} = 9.\overline{09}\%$ or $9\frac{1}{11}\%$.

D. Write $1\frac{1}{2}$ as a percent.

D. Write $2\frac{4}{5}$ as a percent.

$2\frac{4}{5} = 2.8$

$= 280\%$

So, $2\frac{4}{5} = 280\%$.

CAUTION

One tenth of a percent is a thousandth; that is,

$$\frac{1}{10} \text{ of } 1\% = \frac{1}{10} \text{ of } \frac{1}{100} = \frac{1}{1000} = 0.001.$$

E. Change $\frac{2}{7}$ to a percent. Round to the nearest tenth of a percent.

E. Change $\frac{5}{7}$ to a percent. Round to the nearest tenth of a percent.

Strategy: To write the percent rounded to the nearest tenth of a percent, we need to change the fraction to a decimal rounded to the nearest thousandth. (That is, we need three decimal places.)

$\frac{5}{7} \approx 0.714$ **Write as a decimal rounded to the nearest thousandth.**

$\approx 71.4\%$

So, $\frac{5}{7} \approx 71.4\%$.

F. Write $\frac{19}{160}$ as a decimal rounded to the nearest tenth of a percent.

F. Write $\frac{57}{320}$ as a percent rounded to the nearest tenth of a percent.

$\frac{57}{320} \approx 0.178$ **Write as a decimal rounded to the nearest thousandth.**

$\approx 17.8\%$

So, $\frac{57}{320} \approx 17.8\%$.

Answers to Warm Ups C. $83.\overline{3}\%$ or $83\frac{1}{3}\%$ D. 150% E. 28.6% F. 11.9%

Calculator Example

G. Change $2\frac{37}{40}$ to a percent.

Strategy: Change the mixed number to a fraction, then use the calculator to change to a decimal.

$2\frac{37}{40} = \frac{117}{40}$ **Change to an improper fraction.**

$= 2.925$

$= 292.5\%$

So, $2\frac{37}{40} = 292.5\%$.

G. Change $5\frac{21}{80}$ to a percent.

H. A motor that needs repair is only turning $\frac{7}{16}$ the number of revolutions per minute that is normal. What percent of the normal rate is this?

$\frac{7}{16} = 0.4375$

$= 43.75\%$

So, the motor is turning at 43.75% of its normal rate.

H. An eight-cylinder motor has only seven of its cylinders firing. What percent of the cylinders are firing?

Answers to Warm Ups G. 526.25% H. The percent of the cylinders that are firing is 87.5%.

Exercises 7.4

OBJECTIVE: *Change a fraction or mixed number to a percent.*

A.

Change each fraction to a percent.

1. $\frac{63}{100}$

2. $\frac{47}{100}$

3. $\frac{39}{50}$

4. $\frac{7}{10}$

5. $\frac{17}{20}$

6. $\frac{7}{25}$

7. $\frac{1}{2}$

8. $\frac{3}{5}$

9. $\frac{11}{25}$

10. $\frac{7}{50}$

11. $\frac{3}{20}$

12. $\frac{3}{25}$

13. $\frac{13}{10}$

14. $\frac{21}{20}$

15. $\frac{11}{8}$

16. $\frac{19}{16}$

17. $\frac{9}{1000}$

18. $\frac{13}{1000}$

19. $\frac{13}{25}$

20. $\frac{11}{20}$

B.

Change each fraction or mixed number to a percent.

21. $4\frac{3}{5}$

22. $6\frac{1}{4}$

23. $\frac{2}{3}$

24. $\frac{1}{6}$

25. $\frac{11}{6}$

26. $\frac{8}{3}$

27. $2\frac{5}{6}$

28. $4\frac{7}{12}$

29. $\frac{21}{400}$

30. $\frac{27}{200}$

Change each fraction or mixed number to a percent. Round to the nearest tenth of a percent.

31. $\frac{11}{15}$

32. $\frac{19}{28}$

33. $\frac{5}{9}$

34. $\frac{7}{9}$

35. $\frac{7}{11}$

36. $\frac{5}{13}$

37. $2\frac{8}{13}$

38. $4\frac{13}{15}$

39. $\frac{2}{21}$

40. $\frac{2}{23}$

C.

41. Mike Bibby made 17 of 20 free throw attempts in one game. What percent of the free throws did he make?

42. Maureen gets 19 problems correct on a 25-problem test. What percent are correct?

43. Five sevenths of the eligible voters turned out for a recent Hillsboro city election. What percent of the voters turned out, to the nearest tenth of a percent?

44. In a supermarket two eggs out of eleven dozen are lost because of cracks. What percent of the eggs must be discarded, to the nearest tenth of a percent?

45. The local soccer club won 9 of their first 17 games. What percent of the games did they win, rounded to the nearest tenth of a percent?

46. Joe Lucky won 17 of 19 hands at the blackjack table. What percent of the hands did he win, rounded to the nearest tenth of a percent?

Change each of the following fractions or mixed numbers to a percent rounded to the nearest hundredth of a percent.

47. $\frac{47}{400}$

48. $\frac{79}{8000}$

49. $\frac{93}{5000}$

50. $\frac{43}{1200}$

51. $4\frac{11}{13}$

52. $3\frac{2}{17}$

Exercises 53–54 relate to the chapter application.

53. Consumer reports indicate that the cost of food is $1\frac{1}{12}$ what it was one year ago. Express this as a percent. Round to the nearest tenth of a percent.

54. A department store advertises one-third off the regular price on Monday and an additional one-seventh off the original price on Tuesday. What percent is taken off the original price if the item is purchased on Tuesday? Round to the nearest whole percent.

55. A vitamin C tablet is listed as fulfilling $\frac{1}{8}$ the recommended daily allowance of vitamin C. Miguel takes 13 of these tablets per day to ward off a cold. What percent of the average recommended allowance is he taking?

56. A chicken sandwich contains 7 g of fat. Each gram of fat has 9 calories. If the entire sandwich contains 290 calories, what percent of the calories come from the fat content? Round to the nearest percent.

57. In 1995, one area of California had 211 smoggy days. What was the percent of smoggy days? In 1998 there were only 167 smoggy days. What was the percent that year? Compare these percents and discuss the possible reasons for this decline. Round the percents to the nearest tenth of a percent.

STATE YOUR UNDERSTANDING

58. What is special about those fractions that can be changed to a whole number percent?

CHALLENGE

59. Change $2\frac{4}{13}\%$ to the nearest tenth of a percent.

60. Change $\frac{6}{13}\%$ to the nearest hundredth of a percent.

NAME ______________________ CLASS __________ DATE __________

GROUP ACTIVITY

61. Keep a record of everything you eat for one entire day. Use exact amounts as much as possible. With a calorie and fat counter, compute the percent of fat in each item. Then find the percent of fat you consumed that day. The latest recommendations suggest that the fat content not exceed 30% per day. How did you do? Which foods have the highest and which the lowest fat content? Was this a typical day for you? Continue this exercise for one week and compare your daily percent of fat with others in your group. Compute a weekly average individually and as a group.

MAINTAIN YOUR SKILLS (SECTIONS 5.4, 5.6)

Multiply.

62. (8.003)(0.87)

63. (19)(0.0115)

64. (0.02)(0.2)(2.02)

65. (1.45)(4.05)(1.4)

Divide.

66. $0.38\overline{)5.738}$

67. $6.22\overline{)0.202772}$

68. Divide 48 by 6.2 and round to the nearest hundredth.

69. Divide 62 by 480 and round to the nearest thousandth.

70. If Abel works 37 hours and earns a total of $197.95, what is his hourly wage?

71. If Spencer loses 11.6 lb in two weeks, what is his rate of weight loss per day (that is, what is the average loss per day) to the nearest hundredth of a pound?

7.5

Changing Percents to Fractions

OBJECTIVE

Change percents to fractions or mixed numbers

HOW AND WHY

Objective

Change percents to fractions or mixed numbers.

The expression 45% is equal to $45 \cdot \frac{1}{100}$. This gives a very efficient method for changing a percent to a fraction. See Example A.

To change a percent to a fraction or a mixed number

1. Replace the percent symbol (%) with the fraction $\left(\frac{1}{100}\right)$.
2. If necessary rewrite the other factor as a fraction.
3. Multiply and simplify.

Examples A–E

Directions: Change the percent to a fraction or mixed number.

Strategy: Change the percent symbol to the fraction $\frac{1}{100}$ and multiply.

A. Change 45% to a fraction.

$45\% = 45 \cdot \frac{1}{100}$ **Replace the percent symbol (%) with $\frac{1}{100}$.**

$= \frac{45}{100}$ **Multiply.**

$= \frac{9}{20}$ **Simplify.**

CAUTION

You need to multiply by $\frac{1}{100}$, not just write it down.

So, $45\% = \frac{9}{20}$.

Warm Ups A–E

A. Change 65% to a fraction.

Answer to Warm Up A. $\frac{13}{20}$

B. Change 175% to a mixed number.

B. Change 318% to a mixed number.

$$318\% = 318 \cdot \frac{1}{100} \qquad \% = \frac{1}{100}$$

$$= \frac{318}{100} \qquad \textbf{Multiply.}$$

$$= \frac{159}{50} \qquad \textbf{Simplify.}$$

$$= 3\frac{9}{50} \qquad \textbf{Write as a mixed number.}$$

So, $318\% = 3\frac{9}{50}$.

C. Change 9.5% to a fraction.

C. Change 7.5% to a fraction.

$$7.5\% = 7.5 \cdot \frac{1}{100}$$

$$= 7\frac{1}{2} \cdot \frac{1}{100} \qquad \textbf{Change the decimal to a mixed number.}$$

$$= \frac{15}{2} \cdot \frac{1}{100} \qquad \textbf{Change the mixed number to a fraction.}$$

$$= \frac{15}{200} = \frac{3}{40} \qquad \textbf{Multiply and simplify.}$$

So, $7.5\% = \frac{3}{40}$.

D. Change $5\frac{5}{6}\%$ to a fraction.

D. Change $12\frac{2}{3}\%$ to a fraction.

$$12\frac{2}{3}\% = 12\frac{2}{3} \cdot \frac{1}{100}$$

$$= \frac{38}{3} \cdot \frac{1}{100}$$

$$= \frac{38}{300} = \frac{19}{150}$$

So, $12\frac{2}{3}\% = \frac{19}{150}$.

E. Greg scores 78% on a math test. What fraction of the questions does he get incorrect?

E. A biological study shows that spraying a forest for gypsy moths is 88% successful. What fraction of the moths survive the spraying?

Strategy: Subtract the 88% from 100% to find the percent of the moths that survived. Then change the percent that survive to a fraction.

$$100\% - 88\% = 12\%$$

$$12\% = 12 \cdot \frac{1}{100}$$

$$= \frac{12}{100} = \frac{3}{25}$$

So, $\frac{3}{25}$ or 3 out of 25 gypsy moths survived the spraying.

Answers to Warm Ups B. $1\frac{3}{4}$ C. $\frac{19}{200}$ D. $\frac{7}{120}$ E. Greg gets $\frac{11}{50}$ of the questions incorrect.

NAME ______________________ CLASS ____________ DATE ____________

Exercises 7.5

OBJECTIVE: *Change percents to fractions or mixed numbers.*

A.

Change each of the following percents to fractions or mixed numbers.

1. 6% **2.** 8% **3.** 45% **4.** 85%

5. 125% **6.** 175% **7.** 300% **8.** 700%

9. 80% **10.** 30% **11.** 64% **12.** 84%

13. 75% **14.** 90% **15.** 15% **16.** 95%

17. 150% **18.** 225% **19.** 83% **20.** 39%

B.

21. 28.5% **22.** 36.4% **23.** 6.8%

24. 3.5% **25.** 30.5% **26.** 40.8%

27. $\frac{1}{3}\%$ **28.** $\frac{2}{3}\%$ **29.** $7\frac{1}{2}\%$

30. $6\frac{1}{4}\%$

31. $\frac{3}{4}\%$

32. $\frac{3}{5}\%$

33. $52\frac{1}{2}\%$

34. $28\frac{3}{4}\%$

35. $18\frac{4}{7}\%$

36. $21\frac{1}{7}\%$

37. 3.75%

38. 1.25%

39. $116\frac{2}{3}\%$

40. $183\frac{1}{3}\%$

C.

41. George and Ethel paid 16% of their annual income in taxes last year. What fractional part of their income went to taxes?

42. One summer Judy earned 35% of her college expenses working at a local food-processing plant. What fractional part of her expenses does she earn that summer?

43. The Kilroys spend 36% of their monthly income on their mortgage payment. What fractional part goes toward the mortgage?

44. Jorge spends 42% of his salary on his car each month. What fractional part is spent on Jorge's car?

45. Grant Hill made 76% of his shots during an NBA game. What fractional part of his shots did Grant make?

46. The offensive team for the Kansas City Chiefs was on the field 56% of the time during a game with the Dallas Cowboys. What fractional part of the game were they on the field?

Write each of the following as a fraction.

47. $11\frac{1}{5}\%$

48. $16\frac{1}{4}\%$

49. $4\frac{4}{9}\%$

50. $1\frac{1}{9}\%$

51. 0.525%

52. 0.675%

53. The enrollment at City Community College this year is 116% of last year's enrollment. What fraction of last year's enrollment does this represent?

54. The city budget is 132% over last year's budget. What fraction does this represent?

Exercises 55–56 relate to the chapter application.

55. Juanita can save 28% on canned food if she buys in quantities of 12 or more cans of each type of food. If she buys in quantity, what fraction of the original price is she paying?

56. Tru buys light bulbs for 15% off, plus he sends in the coupon for a 12% manufacturer's rebate on the original price. What fraction of the original price does he pay for the bulbs after he receives the rebate?

57. A census determines that $37\frac{1}{2}\%$ of the residents of a city are age 40 or over and that 45% are age 25 or under. What fraction of the residents are between the ages of 25 and 40?

58. Brooks and Foster form a partnership. If Brook's investment is $62\frac{1}{4}\%$ of the total, what fraction of the total is Foster's share?

59. The spraying for the medfly is found to be 88% successful. What fraction of the medflies is eliminated?

60. The salmon run in an Idaho stream has dropped to 42% of what it was 10 years ago. What fractional part of the run was lost during the 10 years?

STATE YOUR UNDERSTANDING

61. Describe a pair of circumstances that can be described by either a percent or a fraction. Compare the advantages or disadvantages of using percents or fractions.

62. Explain how to change between the fraction and percent forms of a number. Give examples of each.

CHALLENGE

63. Change 0.00025% to a fraction.

64. Change 0.0005% to a fraction.

65. Change 150.005% to a mixed number.

66. Change 180.04% to a mixed number.

GROUP ACTIVITY

67. Have each member of your group read the ads for the local department stores in the weekend paper. Record the "% off" in as many ads as you can find. Convert the percents to fractions. In which form is it easier to estimate the savings because of the sale?

MAINTAIN YOUR SKILLS (Sections 5.9, 6.1, 6.2, 6.3)

Are the following proportions true or false?

68. $\dfrac{\frac{2}{3}}{\frac{4}{5}} = \dfrac{500}{600}$

69. $\dfrac{\frac{1}{2}}{\frac{7}{8}} = \dfrac{50}{87}$

70. $\dfrac{36}{12.5} = \dfrac{81}{35}$

71. $\dfrac{25.75}{12} = \dfrac{103}{48}$

72. Write a ratio to compare \$39,000,000 to 75,000,000 people.

73. Change the ratio in Exercise 72 to a unit rate.

74. Write a ratio to compare 14,765,000 gallons of water to 100,000 people.

75. Change the ratio in Exercise 74 to a unit rate.

76. The taxes on a home valued at \$89,000 are \$2225. At the same rate, what are the taxes on a house valued at \$75,000?

77. Five measurements of the diameter of a wire are taken with a micrometer screw gauge. The five estimated measurements are 2.31 mm, 2.32 mm, 2.30 mm, 2.34 mm, and 2.30 mm. What is the average estimate? (Note that although 2.3 = 2.30, writing 2.30 mm shows greater precision than writing 2.3 mm.)

7.6

Fractions, Decimals, Percents: A Review

OBJECTIVE

Given a decimal, fraction, or percent, rewrite in a related form.

HOW AND WHY

Objective

Given a decimal, fraction, or percent, rewrite in a related form.

Decimals, fractions, and percents can each be expressed in terms of the others. We can:

Write a percent as a decimal and as a fraction.

Write a fraction as a percent and as a decimal.

Write a decimal as a percent and as a fraction.

For example:

$50\% = 50 \cdot \frac{1}{100} = \frac{50}{100} = \frac{1}{2}$ and $50\% = 0.5$

$\frac{3}{4} = 3 \div 4 = 0.75$ and $\frac{3}{4} = 0.75 = 75\%$

$0.65 = 65\%$ and $0.65 = \frac{65}{100} = \frac{13}{20}$

The following table shows some common fractions and their decimal equivalents. Some of the decimals are repeating decimals. Remember that a repeating decimal is shown by the bar over the digits that repeat. These fractions occur often in applications of percents. They should be memorized so that you can recall the patterns when they appear.

TABLE 7.2 COMMON FRACTIONS AND DECIMALS

$\frac{1}{2} = 0.5$					
$\frac{1}{3} = 0.\overline{3}$	$\frac{2}{3} = 0.\overline{6}$				
$\frac{1}{4} = 0.25$		$\frac{3}{4} = 0.75$			
$\frac{1}{5} = 0.2$	$\frac{2}{5} = 0.4$	$\frac{3}{5} = 0.6$	$\frac{4}{5} = 0.8$		
$\frac{1}{6} = 0.1\overline{6}$				$\frac{5}{6} = 0.8\overline{3}$	
$\frac{1}{8} = 0.125$		$\frac{3}{8} = 0.375$		$\frac{5}{8} = 0.625$	$\frac{7}{8} = 0.875$

For example, $2.66\overline{6} = 2\frac{2}{3}$, $8.166\overline{6} = 8\frac{1}{6}$, and $17.125 = 17\frac{1}{8}$.

Warm Ups A–B

A.

Fraction	Decimal	Percent
$\frac{2}{3}$		
		27%
	0.72	
$\frac{73}{100}$		
		160%
	1.3	

B. In Example B, write the percent that is recycled as a decimal and as a fraction.

Examples A–B

Directions: Fill in the empty spaces with the related percent, decimal, or fraction.

Strategy: Use the procedures of the previous sections.

A.

Fraction	Decimal	Percent
		30%
$\frac{7}{8}$		
	0.62	

Fraction	Decimal	Percent
$\frac{3}{10}$	0.30	30%
$\frac{7}{8}$	0.875	87.5% or $87\frac{1}{2}$%
$\frac{31}{50}$	0.62	62%

$$\mathbf{30\% = 0.30 = \frac{30}{100} = \frac{3}{10}}$$

$$\mathbf{\frac{7}{8} = 0.875 = 87.5\% = 87\frac{1}{2}\%}$$

$$\mathbf{0.62 = 62\% = \frac{62}{100} = \frac{31}{50}}$$

B. The average American uses about 200 lb of plastic a year. Approximately 60% of this is used for packaging and about 5% of it is recycled. Write the percent used for packaging as a decimal and a fraction.

$$60\% = 0.60 = \frac{60}{100} = \frac{3}{5}$$

So, 0.6 or $\frac{3}{5}$ of the plastic is used for packaging.

Answers to Warm Ups A.

Fraction	Decimal	Percent
$\frac{2}{3}$	$0.66\overline{6}$	$66\frac{2}{3}$%
$\frac{27}{100}$	0.27	27%
$\frac{18}{25}$	0.72	72%

Fraction	Decimal	Percent
$\frac{73}{100}$	0.73	73%
$1\frac{3}{5}$	1.6	160%
$1\frac{3}{10}$	1.3	130%

B. Of the 200 lb of plastic, 0.05, or $\frac{1}{20}$, is recycled.

Exercises 7.6

OBJECTIVE: *Given a decimal, fraction, or percent, rewrite in a related form.*

Fill in the empty spaces with the related percent, decimal, or fraction.

	Fraction	Decimal	Percent
°	$\frac{1}{10}$		
°			30%
°		0.75	
°	$\frac{9}{10}$		
			145%
°	$\frac{3}{8}$		
		0.001	
°		1	
	$2\frac{1}{4}$		
°		0.8	
			$5\frac{1}{2}\%$
°		0.875	
			$\frac{1}{2}\%$
°		0.6	
°			$62\frac{1}{2}\%$

°If you are going to be working with problems that will involve percent in your job or in your personal finances (loans, savings, insurance, and so on), it is advisable to know (memorize) these special relationships.

Fraction	Decimal	Percent
		50%
	0.86	
$\frac{5}{6}$		
	0.08	
$\frac{2}{3}$		
		25%
	0.20	
		40%
		$33\frac{1}{3}\%$
	0.125	
$\frac{7}{10}$		

1. Louis goes to buy new tires for his truck. He finds them on sale for $\frac{1}{4}$ off. What percent is this?

2. Michael went on a diet. He now weighs $66\frac{2}{3}\%$ of his original weight. What fraction is this?

Exercises 3–6 relate to the chapter application.

3. George buys a new VCR at a 40% off sale. What fraction is this?

4. A local department store is having its red tag sale. All merchandise will now be 20% off the original price. What decimal is this?

NAME ______________________ CLASS __________ DATE __________

5. Melinda is researching the best place to buy a computer. On the same computer, Family Computers offers $\frac{1}{8}$ off, The Computer Store will give a 12% discount, and Machines Etc. will allow a 0.13 discount. Where does Melinda get the best deal?

---

6. Randy is trading in his above-ground swimming pool for a larger model. Prices are the same for Model PS+ and Model PT. PS+ holds 11% more water than his old pool, while PT holds $\frac{1}{9}$ more water. Which should he buy to get the most additional water for his new pool?

7. During the month of August, Super Value Grocery has a special on sweet corn—buy any number of ears and get $\frac{3}{8}$ more for one cent. During the same time period, Hank's Super Market also runs a special on sweet corn—buy any number of ears and get 35% more for one cent. Which store offers the best deal?

8. Three multivitamins contain the following amounts of the RDA (recommended daily allowance) of calcium: Brand A, 15%; Brand B, 0.149; and Brand C, $\frac{1}{7}$. Which brand contains the most calcium?

9. Teresa is purchasing a copy machine. Vender A has offered a bid that is 6% over the target price. Vendor B has offered Teresa a bid that exceeds the target price by $\frac{1}{18}$. Teresa's boss has authorized her to exceed the target price by not more than 0.0575 of the target price. Who has offered the best deal? Do either or both fall within the boss's price range?

10. During the first full month of play in the local high school basketball season, the top three shooters had the following statistics: Shari made 78.3% of her shots, Bev made 0.805 of her shots, and Teri made $\frac{58}{73}$ of her shots. Who was the most accurate shooter?

STATE YOUR UNDERSTANDING

11. Write a short paragraph with examples that illustrate when to use fractions, when to use decimals, and when to use percents to show comparisons.

CHALLENGE

12. Fill in the table.

Fraction	Decimal	Percent
$\frac{14}{3}$		
	1.1875	
		6.3125%

GROUP ACTIVITY

13. Have each member of your group make up a table like the one in Example A. Exchange your table with the other group members and fill in the blanks. Check your answers with the rest of the group.

MAINTAIN YOUR SKILLS (SECTIONS 6.2, 6.3)

Solve the following proportions.

14. $\frac{25}{30} = \frac{x}{45}$

15. $\frac{45}{81} = \frac{16}{y}$

16. $\frac{x}{72} = \frac{38}{9}$

17. $\frac{x}{9.5} = \frac{125}{250}$

18. $\frac{\frac{1}{2}}{100} = \frac{A}{40}$

19. $\frac{50}{100} = \frac{70}{B}$

20. $\frac{R}{100} = \frac{8}{27}$; round R to the nearest tenth.

21. $\frac{17}{100} = \frac{A}{22.3}$; round A to the nearest tenth.

22. Sean drives 413 miles and uses 11.8 gallons of gasoline. At that rate, how many gallons will he need to drive 1032.5 miles?

23. The Bacons' house is worth $78,000 and is insured so that the Bacons will be paid four-fifths of the value for any damage. One third of the house is totally destroyed by fire. How much insurance money should they collect?

7.7

Solving Percent Problems

OBJECTIVES

1. Solve percent problems using the percent formula.
2. Solve percent problems using a proportion.

VOCABULARY

In the statement "R of B is A,"

R is the **rate** of percent.

B is the **base** unit and follows the word "of."

A is the **amount** that is compared to B.

To **solve** a percent problem means to do one of the following:

1. Find A, given R and B.
2. Find B, given R and A.
3. Find R, given A and B.

HOW AND WHY

Objective 1

Solve percent problems using the percent formula.

We show two methods for solving percent problems. We will refer to these as

The percent formula, $R \times B = A$ (see Examples A–E).

The proportion method (see Examples F–H).

In each method we must identify the rate of percent (R), the base (B), and the amount (A). To help determine these, keep in mind that

R, the rate of percent, includes the percent symbol (%).

B, the base, follows the words "of" or "percent of."

A, the amount, sometimes called the *percentage,* is the amount compared to B and follows the word "is."

The method you choose to solve percent problems should depend on

1. The method your instructor recommends.
2. Your major field of study.
3. How you use percent in your day-to-day activities.

What percent of B is A? The word "of" in this context and in other places in mathematics indicates multiplication. The word "is" describes the relationship "is equal to" or " = ." Thus, we may write:

R of B is A

$\downarrow \quad \downarrow$

$R \times B = A$

When solving percent problems identify the rate (%) first, the base (of) next, and the amount (is) last. For example, what percent of 20 is 5?

R of B is A	**The rate R is unknown, the base B (B follows the word "of") is 20, and the amount A (A follows the word "is") is 5.**
$R \times B = A$	
$R(20) = 5$	**Substitute 20 for B and 5 for A.**
$R = 5 \div 20$	**Divide.**
$= 0.25$	
$= 25\%$	**Change to percent.**

So, 5 is 25% of 20.

All percent problems can be solved using $R \times B = A$. However there are two other forms that can speed up the process.

$A = R \times B$

$R = A \div B$

$B = A \div R$

The triangle below is a useful device to help you select the correct form of the formula to use.

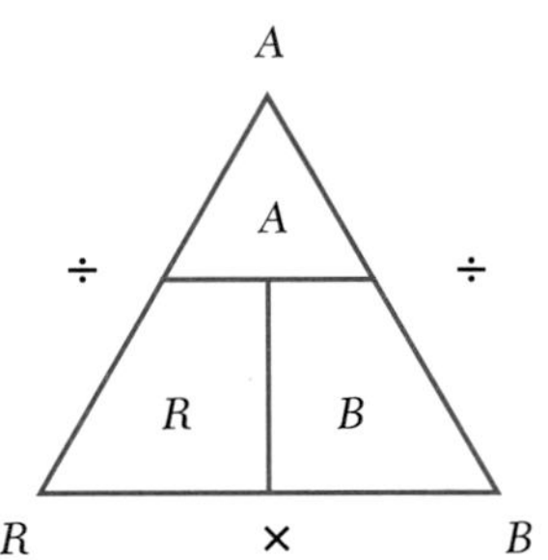

When the unknown value is covered, the positions of the uncovered (known) values help us remember how to see what operation to use:

When A is covered, we see $R \times B$, reading from left to right.

When B is covered, we see $A \div R$, reading from top to bottom.

When R is covered, we see $A \div B$, reading from top to bottom.

For example, 34% of what number is 53.04? The rate R is 34%, B is unknown (B follows the word "of"), and A (A follows the word "is") is 53.04. Fill in the triangle and cover B.

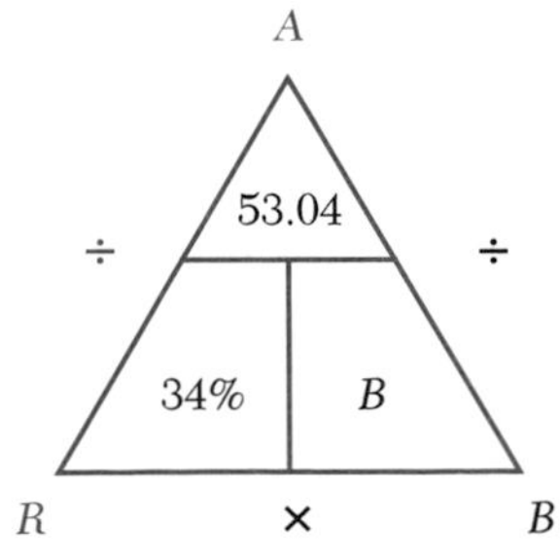

Covering B and reading from the top, we see that A is divided by R. Therefore,

$B = A \div R$

$= 53.04 \div 0.34$ **Substitute 53.04 for *A* and 0.34 for *R*.**

$= 156$

So, 34% of 156 is 53.04.

CAUTION

Remember to use the decimal or fraction form of the rate *R* when solving percent problems.

Examples A–E

Directions: Solve the percent problems using a percent formula.

Strategy: Identify R, B, and A. Select a formula. Substitute the known values and find the unknown value.

A. 32% of what number is 13?

32% of B is 13 **The % symbol follows 32, so $R = 32\%$. The base B (following the word "of") is unknown. A follows the word "is," so $A = 13$.**

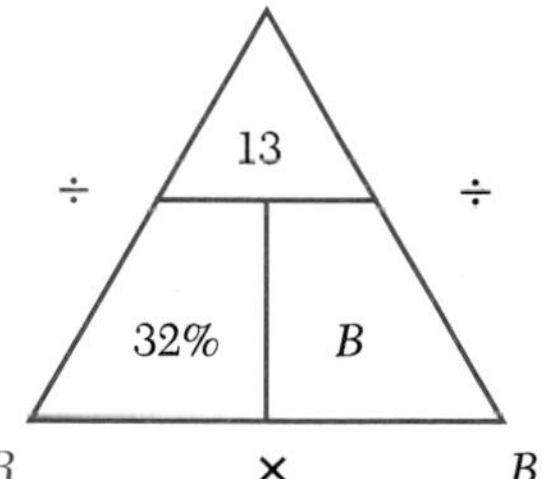

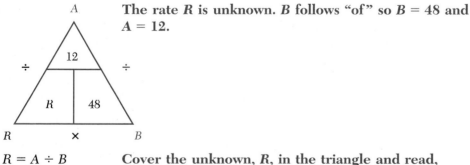

$B = A \div R$ **Cover the unknown, B, in the triangle and read "$A \div R$."**

$B = 13 \div 0.32$ **Use the decimal form of the rate, $32\% = 0.32$.**

$B = 40.625$

So, 32% of 40.625 is 13.

B. 12 is what percent of 48?

The rate R is unknown. B follows "of" so $B = 48$ and $A = 12$.

$R = A \div B$ **Cover the unknown, R, in the triangle and read, "$A \div B$."**

Warm Ups A–E

A. 60% of what number is 51?

B. 8 is what percent of 40?

Answers to Warm Ups A. 85 B. 20%

$R = 12 \div 48$

$R = 0.25 = 25\%$ **Divide. Change the decimal to a percent.**

So, 12 is 25% of 48.

C. What is $66\frac{2}{3}\%$ of 78?

C. What is $33\frac{1}{3}\%$ of 963?

The % symbol follows $33\frac{1}{3}$, so $R = 33\frac{1}{3}\%$.

The base B (following the word "of") is 963. A is unknown.

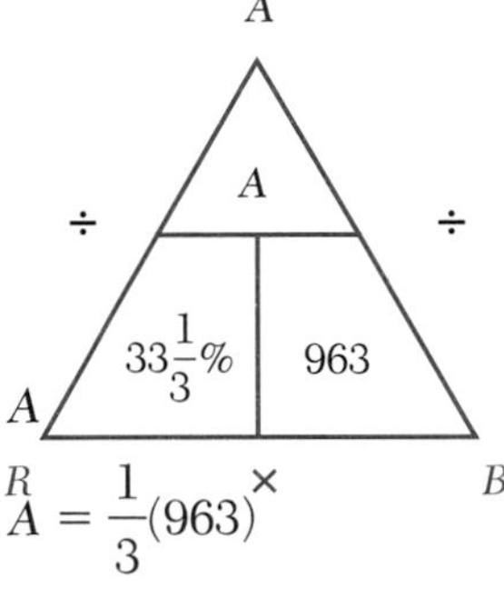

Cover A in the triangle and read "$R \times B$."

$A = \frac{1}{3}(963)$

Change the percent to a fraction.

$$33\frac{1}{3}\% = 33\frac{1}{3}\left(\frac{1}{100}\right)$$
$$= \frac{100}{3}\left(\frac{1}{100}\right)$$
$$= \frac{1}{3}$$

$A = 321$ **Multiply and simplify.**

So, 321 is $33\frac{1}{3}\%$ of 963.

D. 37% of what number is 37.74?

D. 65% of what number is 46.8?

Strategy: We use the formula $R \times B = A$.

$R \times B = A$ **Formula**

$0.65\ B = 46.8$ **$R = 65\% = 0.65$ and $A = 46.8$**

B is unknown.

$B = 46.8 \div 0.65$ **Rewrite as division.**

$B = 72$

So, 65% of 72 is 46.8.

Answers to Warm Ups C. 52 D. 102

E. 10.4 is what percent of 66? Round to the nearest tenth of a percent.

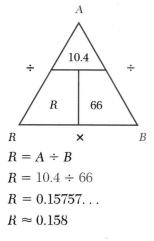

$A = 10.4$ and $B = 66$. R is unknown.

A

÷ 10.4 ÷

R 66

R × B

$R = A \div B$ — **In the percent triangle, cover R and read "$A \div B$."**

$R = 10.4 \div 66$

$R = 0.15757\ldots$

$R \approx 0.158$ — **To round to the nearest tenth of a percent we round the decimal to the nearest thousandth.**

$R \approx 15.8\%$

So, 10.4 is approximately 15.8% of 66.

E. 6.8 is what percent of 138? Round to the nearest tenth of a percent.

HOW AND WHY

Objective 2

Solve percent problems using a proportion.

Since R is a comparison of A to B, and we have seen earlier that this comparison can be written as a ratio, we can write the percent ratio equal to the ratio of A and B. In writing the percent as a ratio we will let $R = X\%$. We can now write the proportion: $\frac{X}{100} = \frac{A}{B}$.

When any one of the values of R, A, and B is unknown it can be found by solving the proportion.

For example, what percent of 85 is 19.55?

$A = 19.55$, $B = 85$, and $R = X\% = ?$

$\frac{X}{100} = \frac{A}{B}$

$\frac{X}{100} = \frac{19.55}{85}$

$85X = 100(19.55)$ — **Cross multiply.**

$85X = 1955$

$X = 1955 \div 85$

$X = 23$

$R = X\% = 23\%$

So, 23% of 85 is 19.55.

Answer to Warm Up E. 4.9%

Warm Ups F–H

F. 45% of 72 is what number?

G. 135% of _____ is 99.9.

H. 92 is what percent of 185, to the nearest tenth of a percent?

Examples F–H

Directions: Solve the percent problem using a proportion.

Strategy: Write the proportion, $R = \frac{X}{100} = \frac{A}{B}$, fill in the known values, and solve.

F. 30% of 80 is what number?

$\frac{X}{100} = \frac{A}{B}$ **Proportion for solving percent problems.**

$\frac{30}{100} = \frac{A}{80}$ **$R = X\% = 30\%$, so $X = 30$, $B = 80$, and A is unknown.**

$30(80) = 100A$ **Cross multiply.**

$2400 = 100A$

$2400 \div 100 = A$

$24 = A$

So, 30% of 80 is 24.

G. 162% of __________ is 145.8.

$\frac{X}{100} = \frac{A}{B}$ **Proportion for solving percent problems.**

$\frac{162}{100} = \frac{145.8}{B}$ **$R = X\% = 162\%$, so $X = 162$, B is unknown, and $A = 145.8$.**

$162(B) = 100(145.8)$ **Cross multiply.**

$162(B) = 14580$

$B = 14580 \div 162$

$B = 90$

So, 162% of 90 is 145.8.

H. 60 is what percent of 172? Round to the nearest tenth of a percent.

$\frac{X}{100} = \frac{A}{B}$

$\frac{X}{100} = \frac{60}{172}$ **$R = X\%$ is unknown, $A = 60$, and $B = 172$.**

$172(X) = 100(60)$ **Cross multiply.**

$172(X) = 6000$

$X = 6000 \div 172$

$X \approx 34.88$ **Carry out the division to two decimal places.**

$X \approx 34.9$ **Round to the nearest tenth.**

$R \approx 34.9\%$ **$R = X\%$**

So, 60 is 34.9% of 172 to the nearest tenth of a percent.

Answers to Warm Ups F. 32.4 G. 74 H. 49.7%

NAME ______________________ CLASS __________ DATE __________

Exercises 7.7

OBJECTIVES: **1.** Solve percent problems using the formula.
2. Solve percent problems using a proportion.

A.

Solve.

1. 12 is 50% of __________.

2. 12 is 20% of __________.

3. What is 80% of 70?

4. What is 30% of 80?

5. 4 is __________% of 2.

6. 6 is __________% of 2.

7. __________% of 80 is 40.

8. __________% of 60 is 15.

9. 60% of __________ is 42.

10. 40% of __________ is 16.

11. 80% of 45 is __________.

12. __________ is 70% of 21.

13. 64 is __________% of 80.

14. __________% of 28 is 7.

15. 23% of __________ is 23.

16. 23 is __________% of 23.

17. $\frac{1}{2}$% of 400 is __________.

18. $\frac{1}{4}$% of 1600 is __________.

19. 125% of 40 is __________.

20. 175% of 80 is __________.

B.

21. 11.5% of 70 is __________.

22. 45.6% of 75 is __________.

23. 0.65 is __________% of 20.

24. 0.28 is __________% of 70.

25. 72 is 48% of __________.

26. 81 is 18% of __________.

27. 34.6% of 52 is __________.

28. 16.8% of 91 is __________.

29. 76% of __________ is 80.18.

30. 47% of __________ is 5.875.

31. 142% of __________ is 443.04.

32. 165% of __________ is 204.6.

33. 96 is __________% of 125.

34. 135 is __________% of 160.

35. 6.73% of 110 is __________.

36. 7.12% of 341 is __________.

37. 0.7 is ___160___% of 1.12.

38. 3.4 is __________% of 13.6.

39. $11\frac{1}{9}$% of 1845 is __________.

40. $16\frac{2}{3}$% of 3522 is __________.

C.

41. What percent of 94 is 12? Round to the nearest tenth of a percent.

42. What percent of 234 is 187? Round to the nearest tenth of a percent.

43. Eighty-four is 35.6% of what number? Round to the nearest hundredth.

44. Fifty-six is 23.8% of what number? Round to the nearest hundredth.

45. Seventeen and six tenths percent of 394 is what number?

46. Eighty-two and sixteen hundredths percent of 42 is what number?

47. Forty-seven is what percent of 29? Round to the nearest tenth of a percent.

48. One hundred sixty-nine is what percent of 54? Round to the nearest tenth of a percent.

Solve.

49. $5\frac{2}{3}\%$ of $8\frac{3}{4}$ is __________. (as a fraction)

50. __________ is $7\frac{3}{8}\%$ of $32\frac{4}{5}$. (as a decimal)

51. $6\frac{4}{15}\%$ of 850 is __________. (as a mixed number)

52. $17\frac{1}{9}\%$ of __________ is 54.95. (as a decimal rounded to the nearest hundredth)

53. __________% of 34.05 is 23.95. (round to the nearest tenth of a percent)

54. __________% of $16\frac{4}{15}$ is $8\frac{2}{3}$. (as a fraction)

STATE YOUR UNDERSTANDING

55. Explain the inaccuracies in this statement: "Starbuck Industries charges 70¢ for a part that cost them 30¢ to make. They're making 40% profit."

56. Explain how to use the percent triangle to solve percent problems.

CHALLENGE

57. $\frac{1}{2}\%$ of $45\frac{1}{3}$ is what fraction?

58. $\frac{3}{7}\%$ of $2\frac{1}{9}$ is what fraction?

GROUP ACTIVITY

59. Divide up the task of computing these percents: 45% of 37; 37% of 45; 18% of 80; 80% of 18; 130% of 22; 22% of 130; 0.6% of 5.5; 5.5% of 0 6. Compare your answers and together write up a statement about the answers.

60. Divide up the task of computing these percents: 30% of the number that is 80% of 250; 80% of the number that is 30% of 250; 60% of the number that is 20% of 340; 20% of the number that is 60% of 230; 150% of the number that is 200% of 40; 200% of the number that is 150% of 40. Compare your answers and together write up a statement about the answers.

MAINTAIN YOUR SKILLS (SECTIONS 6.2, 6.3)

Solve the proportions.

61. $\frac{14}{24} = \frac{x}{30}$

62. $\frac{4.8}{2.5} = \frac{96}{y}$

63. $\frac{a}{\frac{5}{8}} = \frac{1\frac{1}{2}}{3\frac{3}{4}}$

64. $\frac{1\frac{1}{2}}{t} = \frac{5\frac{5}{8}}{1\frac{2}{3}}$

65. $\frac{1.3}{0.07} = \frac{w}{3.01}$

66. $\frac{1.3}{0.07} = \frac{5.59}{t}$

Exercises 67–70: On a certain map, $1\frac{1}{2}$ in. represents 60 miles.

67. How many miles are represented by $2\frac{7}{8}$ in.?

68. How many miles are represented by $3\frac{3}{16}$ in.?

69. How many inches are needed to represent 820 miles?

70. How many inches are needed to represent 22 miles?

7.8

Applications of Percents

OBJECTIVES

1. Solve percent word problems.
2. Solve business-related problems.
3. Read data from a circle graph or construct a circle graph from data.

HOW AND WHY

Objective 1

Solve percent word problems.

When a word problem is translated to the simpler form, "What percent of what is what?" the unknown value can be found using one of the methods of the previous section. For example, "What percent of the loan is interest?" (See Example A.)

Both methods of Section 7.7 are used in the examples. When a value B is increased by an amount A the rate of percent R is called the *percent of increase.* When a value B is decreased by an amount A the rate of percent R is called the *percent of decrease.*

Examples A–E

Directions: Solve the percent word problem.

Strategy: Write the problem in the form, "What percent of what is what?" Fill in the known values and find the unknown value.

A. Rod buys a motorcycle with a 15% one-year loan. The interest payment is $135. How much is his loan?

Strategy: Use a percent formula.

15% of the loan is $135. **The $135 interest is 15% of the loan, so $R = 15\% = 0.15$ and $A = \$135$. Substitute these values in the triangle and solve for B.**

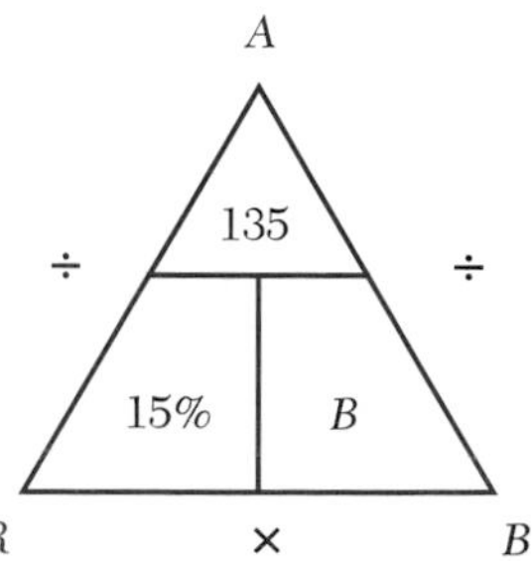

$B = A \div R$

$B = 135 \div 0.15$

$B = 900$

Rod's loan is for $900.

Warm Ups A–E

A. Jamie buys a motorcycle with a 14% one-year loan. If the interest payment is $154, how much is the loan?

Answer to Warm Up A. Jamie's loan is $1100.

B. The cost of a certain model of Oldsmobile is 145% of what it was five years ago. If the cost of the automobile five years ago was $19,400, what is the cost today?

B. This year the population of Dechutes County is 156% of its population 10 years ago. The population 10 years ago was 134,000. What is the population this year?

Strategy: Use the proportion equation.

156% of 134,000 is the current population

$$\frac{156}{100} = \frac{A}{134{,}000}$$

$R = X\% = 156\%$, so $X = 156$, and $B = 134{,}000$. Substitute these values into the proportion $\frac{X}{100} = \frac{A}{B}$, and solve.

$134{,}000(156) = 100A$ **Cross multiply.**

$20{,}904{,}000 = 100A$

$20{,}904{,}000 \div 100 = A$

$209{,}040 = A$

The population this year is 209,040.

C. A bakery has 300 packages of day-old buns to sell. If the price was originally $1.36 per package and they sell them for 99¢ per package, what percent discount, based on the original price, should the bakery advertise?

C. The Davidson Bakery has 750 loaves of day-old bread they want to sell. If the price was originally $1.49 a loaf and they sell it for $1.19 a loaf, what percent discount, based on the original price, should the bakery advertise?

Strategy: Use a percent formula.

$1.49 − $1.19 = $0.30 **First determine the discount amount.**

What % of $1.49 is $0.30? **$A = 0.30$ and $B = 1.49$ because we are asking the percent of discount based on the original price. Fill in the triangle and solve for R.**

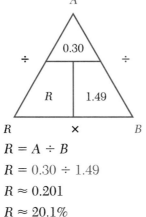

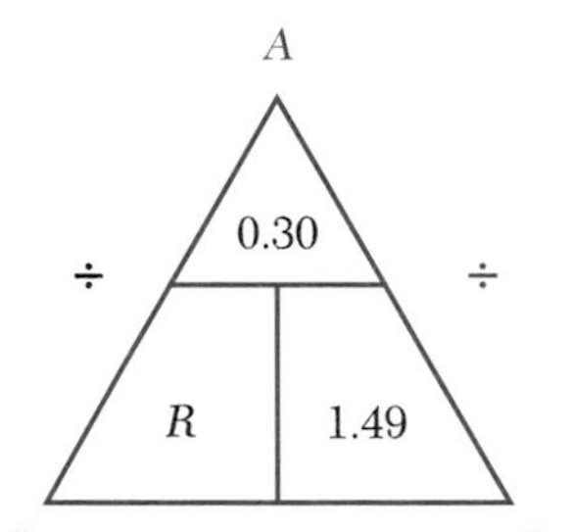

$R = A \div B$

$R = 0.30 \div 1.49$

$R \approx 0.201$ **Round to the nearest thousand.**

$R \approx 20.1\%$ **Change the decimal to percent.**

The discount is about 20.1%, so the bakery will probably advertise "over 20% off."

Answers to Warm Ups B. The cost of the Oldsmobile today is $28,130. C. They should advertise a discount of 27.2%, or over 27%.

D. A student newspaper polls a group of students. Five of them say they walk to school, eleven say they ride the bus, fifteen ride in car pools, and four drive their own cars. What percent of the group ride in car pools? Round to the nearest whole percent.

Strategy: Use the proportion equation.

$5 + 11 + 15 + 4 = 35$	**First find the number in the group.**
What % of 35 is 15?	**There are 35 in the group so $B = 35$, and 15 car poolers so $A = 15$.**

$$\frac{X}{100} = \frac{15}{35}$$

$35(X) = 1500$	**Cross multiply.**
$X = 1500 \div 35$	
$X \approx 43$	**Round to the nearest whole number.**
$R \approx 43\%$	**$R = X\%$.**

Approximately 43% of the students ride in car pools.

D. A list of grades in a math class revealed that 9 students received As, 14 received Bs, 27 received Cs, and 7 received Ds. What percent of the students received a grade of C? Round to the nearest whole percent.

E. In a statistical study of 345 people, 145 said they preferred eating whole wheat bread. What percent of the people surveyed preferred eating whole wheat bread? Round to the nearest whole percent.

Strategy: Use the percent formula.

What percent of 345 is 145?	**$B = 345$ and $A = 145$.**
$R \times B = A$	**Percent formula**
$R(345) = 145$	**Substitute**
$R = 145 \div 345$	
$R \approx 0.42$	**Round to the nearest hundredth.**
$R \approx 42\%$	**Change to a percent.**

So approximately 42% of the people surveyed preferred eating whole wheat bread.

E. In a similar study of 453 people, 115 said they jog for exercise. What percent of those surveyed jog? Round to the nearest whole percent.

Examples F–G

Directions: Find the percent of increase or decrease.

Strategy: Use one of the two methods to solve for R.

Warm Ups F–G

Answers to Warm Ups D. The percent of students receiving a C grade is approximately 47%. E. Of the 453 people, 25% jog.

F. Find the percent of increase from 165 to 231.

F. Find the percent of increase from 380 to 551.

Strategy: Use a percent formula.

$551 - 380 = 171$

The difference, 171, is the amount of increase. The percent of increase is calculated from the amount of increase, 171, based on the original amount, 380.

What percent of 380 is 171? **$B = 380$ and $A = 171$**

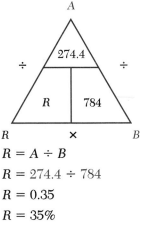

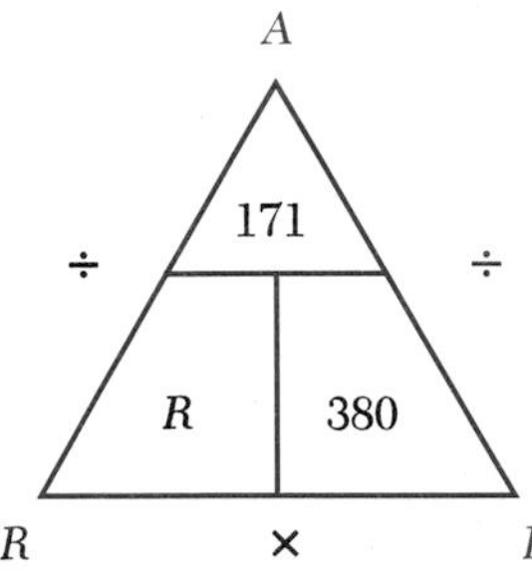

$R = A \div B$

$R = 171 \div 380$

$R = 0.45$

$R = 45\%$

There is a 45% increase from 380 to 551.

G. Find the percent of decrease from 550 to 132.

G. Find the percent of decrease from 784 to 509.6.

Strategy: Use a percent formula.

$784 - 509.6 = 274.4$

The difference, 274.4, is the amount of decrease. The percent of decrease is calculated from the amount of decrease, 274.4, based on the original amount, 784.

What percent of 784 is 274.4? **$B = 784$ and $A = 274.4$**

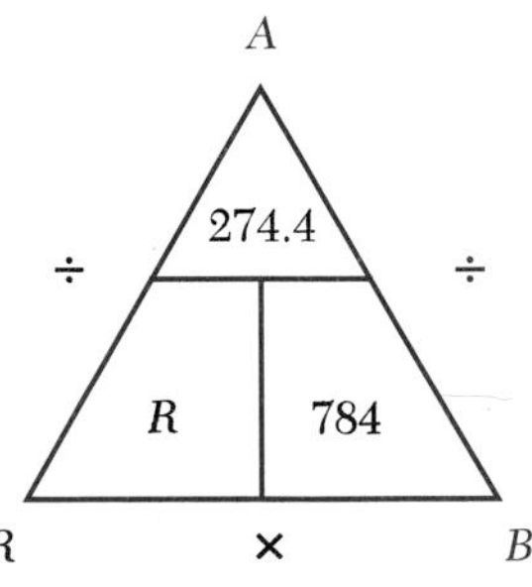

$R = A \div B$

$R = 274.4 \div 784$

$R = 0.35$

$R = 35\%$

There is a 35% decrease from 784 to 509.6.

Answers to Warm Ups F. There is a 40% increase. G. There is a 76% decrease.

HOW AND WHY

Objective 2

Solve business-related percent problems.

Businesses use percent in a variety of ways. Among these are percent of markup, percent of discount, percent of profit, interest rates, taxes, salary increases, and commissions. These words and phrases are explained in the next set of examples.

Examples H–M

Directions: Solve business-related percent problems.

Strategy: Write the problem in the simpler word form, "What percent of what is what?" Fill in the known values and find the unknown value.

H. The cost of an electric iron is $22.50. The markup is 16% of the cost. What is the selling price of the iron?

Strategy: *Markup* is the amount added to the cost, by the store, of an article so the store can pay its expenses and make a profit. Let *M* represent the markup. Use the percent formula.

Simpler Word Form:

What % of the cost is the markup?

16% of $22.50 is *M*.

$R = 16\% = 0.16$
$B = \$22.50$, and
$A = M$

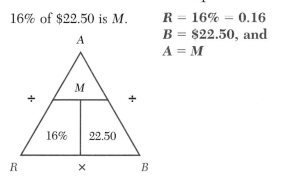

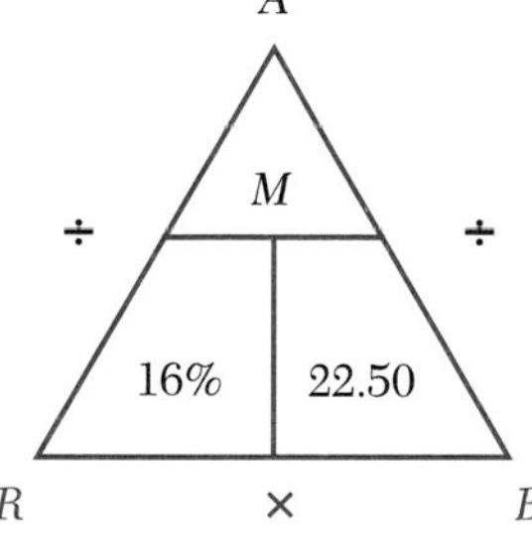

$M = R \times B$

$M = 0.16(22.50)$

$M = 3.60$

So, the markup is $3.60.

S.P. = cost + markup **To find the selling price, add the cost and the markup. Let S.P. represent the selling price.**

= $22.50 + $3.60

= $26.10

The store will sell the iron for $26.10.

Warm Ups H–M

H. The cost of a coffeemaker is $28.50. The markup is 55% of the cost. What is the selling price of the coffeemaker?

Answer to Warm Up H. The coffeemaker sells for $44.18.

I. The selling price of a boombox is \$187.50. What is the sale price of the boombox if the percent of discount is 23%? Round to the nearest cent.

I. The regular selling price of a personal stereo is \$49.95. What is the sale price of the radio if it is discounted 33%? Round to the nearest cent.

Strategy: The *amount of discount* is the amount subtracted from the regular price. The *percent of discount* is 33%. Use the percent formula.

Simpler Word Form:

33% of the selling price is the discount.

$$R \times B = A$$

$R = 33\% = 0.33$ and $B = \$49.95$.

$$0.33(49.95) = A$$

$$16.48 \approx A$$

Round to the nearest cent (hundredth).

The discount is \$16.48.

Strategy: To find the sale price, subtract the discount from the selling price.

$$\text{Sale price} = \$49.95 - \$16.48$$
$$= \$33.47$$

The sale price is \$33.47.

J. A business withholds 7.8% of each employee's gross salary for state taxes. How much is withheld from an employee earning \$45,800?

J. A business pays 0.75% of its net income as a transportation tax to the local transit agency. What is the transit tax on a business with a net income of \$1,975,000?

Strategy: Net income is the amount after expenses have been deducted. Use the proportion equation.

Simpler Word Form:

0.75% of the net income is the transit tax.

$$\frac{X}{100} = \frac{A}{B}$$

Percent proportion.

$$\frac{0.75}{100} = \frac{A}{1{,}975{,}000}$$

$R = X\% = 0.75\%$, so $X = 0.75$ and $B = 1{,}975{,}000$.

$$0.75(1{,}975{,}000) = 100(A)$$

Cross multiply.

$$1{,}481{,}250 = 100(A)$$
$$1{,}481{,}250 \div 100 = A$$
$$14{,}812.50 = A$$

The business pays \$14,812.50 in transit taxes.

K. A CEO's Infinity has depreciated to 65% of its original cost. If the value of the Infinity is now \$23,400, what did it cost originally?

K. A phone company buys a four-wheel drive RV to use while inspecting phone lines. After one year the RV depreciated to 85% of its original value. The RV is now valued at \$20,400. What did it cost originally?

Strategy: *Depreciation* is the name given to the decrease in value caused by age or use. Use the percent formula.

Answers to Warm Ups I. The sale price of the boombox is \$144.38. J. The state taxes withheld are \$3572.40. K. The Infinity originally cost \$36,000.

Simpler Word Form:

85% of the original cost is \$20,400.

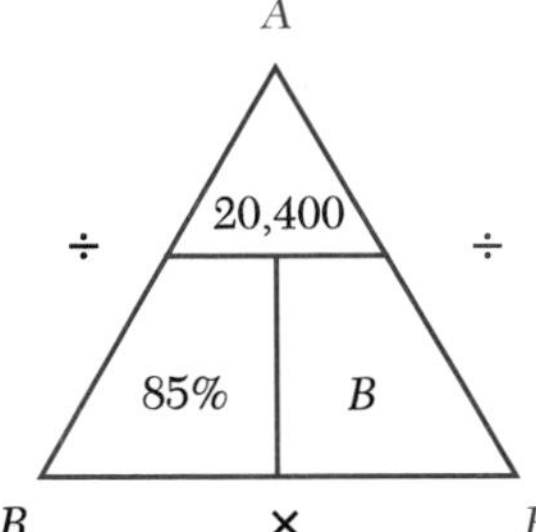

$R = 85\%$ and $A = \$20,400$

$B = A \div R$

$B = 20{,}400 \div 0.85$ **Change 85% to a decimal, 0.85.**

$B = 24{,}000$

The original cost of the RV was \$24,000.

L. Jean's rate of pay is \$13.85 per hour. She gets time and a half for each hour over 40 hr worked in one week. What are her earnings if she worked 48.5 hr in one week?

Strategy: Time and a half means that she will earn 1.5 times, or 150%, of her regular hourly pay. First find her overtime pay rate, then calculate her earnings at each rate and add. To find her overtime rate use the percent formula.

Simpler Word Form:

150% of her pay rate is the overtime pay rate

$R = 150\% = 1.5$ and $B = \$13.85$.

$R \times B = A$

$1.50(13.85) = A$

$20.775 = A$

Jean's overtime pay rate is \$20.775.

Earnings = 40(\$13.85) + 8.5(\$20.775) **Jean worked 40 hr at the regular rate and 8.5 hr at the overtime rate. Round pay to the nearest cent.**

≈ \$554 + \$176.59

≈ \$730.59

Jean earns \$730.59 for the week.

L. What are Jean's earnings for a week in Example L if her hourly rate is \$15.00?

M. Mr. Jordan sells men's clothing. He receives a salary of \$225 per week plus a commission of 9.75% of his total sales. One week he sells \$9865 worth of clothing. What are his total earnings for the week?

Strategy: *Commission* is the money salespeople earn based on goods sold. To find the total earnings add the commission to the base salary. Use the formula.

M. What are Mr. Jordan's total earnings for a week in which his sales total \$5565, if his salary and commission are the same as in Example M?

Answers to Warm Ups L. Jean's earnings for the week are \$791.25. M. Mr. Jordan's earnings for the week are \$767.59.

Simpler Word Form:

9.75% of sales is the commission.

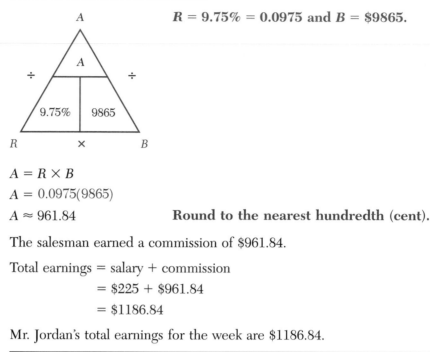

$R = 9.75\% = 0.0975$ and $B = \$9865$.

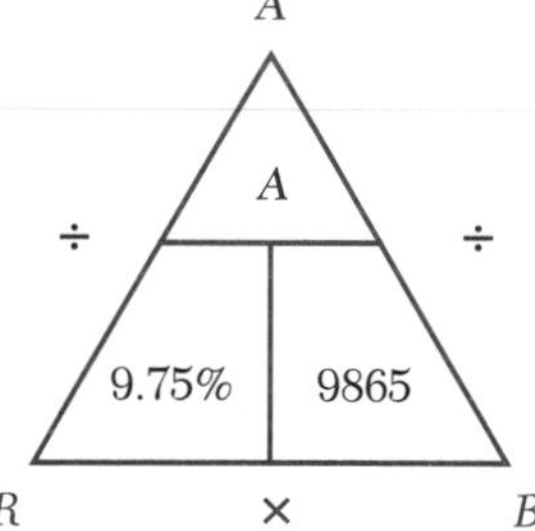

$A = R \times B$

$A = 0.0975(9865)$

$A \approx 961.84$ **Round to the nearest hundredth (cent).**

The salesman earned a commission of \$961.84.

Total earnings = salary + commission
= \$225 + \$961.84
= \$1186.84

Mr. Jordan's total earnings for the week are \$1186.84.

HOW AND WHY

Objective 3

Read data from a circle graph or construct a circle graph from data.

A circle graph or pie chart is used to show how a whole unit is divided into parts. The area of the circle represents the entire unit and each subdivision is represented by a sector. Percents are often used as the unit of measure of the subdivision. Consider the following pie chart.

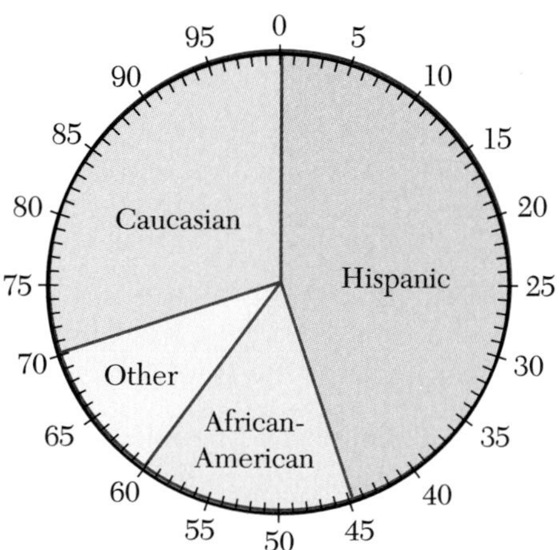

Percent of Population of El Centro by Ethnic Group

From the circle graph we can conclude:

1. The largest ethnic group in El Centro is Hispanic.
2. The Caucasian population is twice the African-American population.
3. The African-American and Hispanic populations are 60% of the total.

If the population of El Centro is 125,000, we can also compute the approximate number in each group. For instance, the number of Hispanics is found by:

$$R \times B = A \qquad \mathbf{R = 45\% = 0.45,\ B = 125{,}000}$$
$$0.45(125{,}000) = A$$
$$56{,}250 = A$$

There are approximately 56,250 Hispanics in El Centro.

To construct a circle graph, determine what fractional part or percent each subdivision is, compared to the total. Then draw a circle and divide it accordingly. We can draw a pie chart of the data in the following table.

	Age Groups		
	0–21	22–50	Over 50
Population	14,560	29,120	14,560

Begin by adding two rows and a column to the data table.

	Age Groups			
	0–21	22–50	Over 50	Total
Population	14,560	29,120	14,560	58,240
Fractional part	$\frac{1}{4}$	$\frac{1}{2}$	$\frac{1}{4}$	1
Percent	25%	50%	25%	100%

The third row is computed by writing each age group as a fraction of the total population and reducing. For example, the 0–21 age group is

$$\frac{14{,}560}{58{,}240} = \frac{1456}{5824} = \frac{364}{1456} = \frac{1}{4} \text{ or } 25\%$$

Now draw the circle graph and label it.

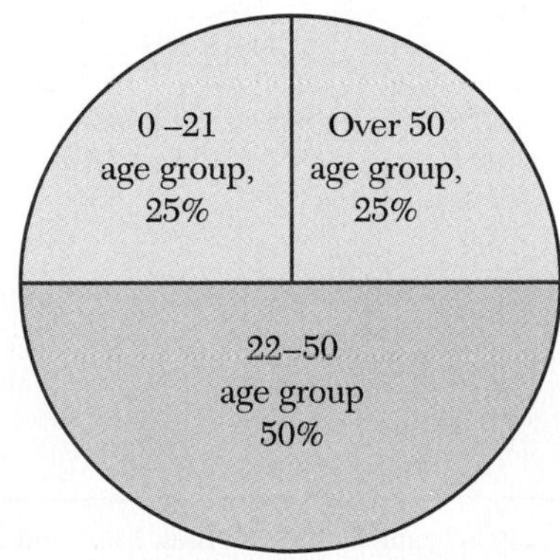

Group Distribution by Age

Sometimes circle graphs are drawn using one degree as the unit of measure for the sectors. This is left for a future course.

Warm Up N

N. The sales of items at Grocery Mart are displayed in the circle graph.

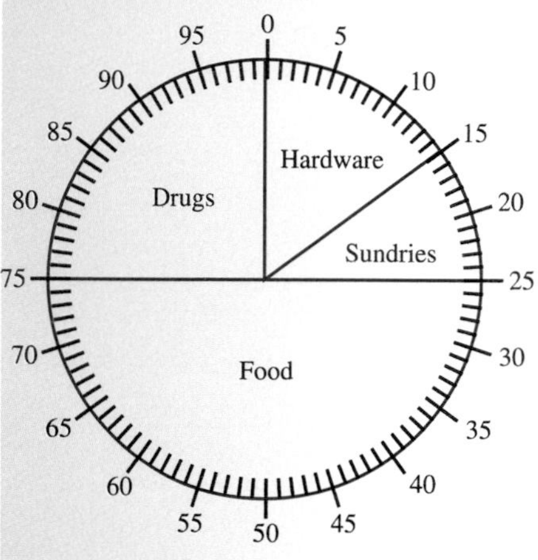

Items Sold

i. What is the area of highest sales?

ii. What percent of total sales is from sundries and drugs?

iii. What percent of total sales is from food and hardware?

Example N

Directions: Answer the questions associated with the graph.

Strategy: Examine the graph to determine the size of the related sector.

N. The sources of City Community College's revenue are displayed in the circle graph.

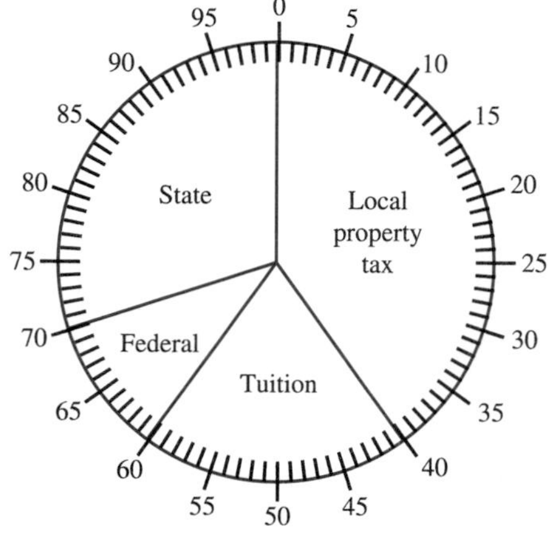

Revenue Sources

i. What percent of the revenue is from the federal government?

ii. What percent of the revenue is from tuition and property taxes?

iii. What percent of the revenue is from the federal and state government?

i. 10% **Read directly from the graph.**

ii. 60% **Add the percents for tuition and taxes.**

iii. 40% **Add the percents for state and federal sources.**

Answer to Warm Up N. The highest sales are in food. Sundries and drugs account for 35% of sales. Food and hardware account for 65% of sales.

Example O

Directions: Make a circle graph which illustrates the information.

Strategy: Use the information to calculate the percents. Divide the circle accordingly and label.

O. Make a circle graph to illustrate that out of 25 students in Frau Heinker's German class, 4 are seniors, 5 are juniors, 14 are sophomores, and 2 are freshmen.

Seniors: $\frac{4}{25} = 16\%$ **Compute the percents.**

Juniors: $\frac{5}{25} = 20\%$

Sophomores: $\frac{14}{25} = 56\%$

Freshmen: $\frac{2}{25} = 8\%$

Construct and label the graph.

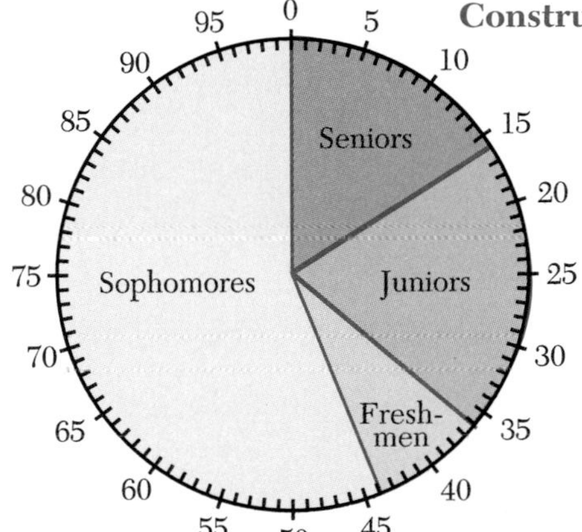

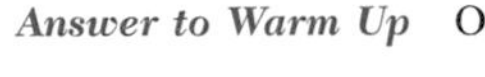

Distribution of Students in Frau Heinker's German Class

Warm Up O

O. Make a circle graph to illustrate that in a survey of 20 people, 7 like football best, 9 like basketball best, and 4 like baseball best.

Answer to Warm Up O.

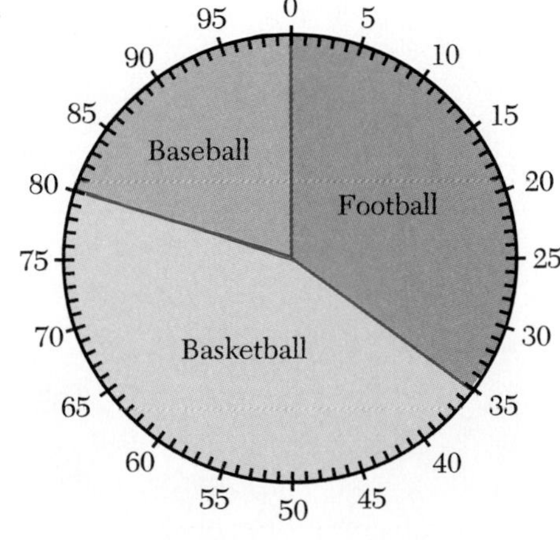

Sport Preference

NAME ____________________ CLASS ____________ DATE ____________

Exercises 7.8

OBJECTIVE 1: *Solve percent word problems.*

1. If there is a 4% sales tax on a television set costing \$129.95, how much is the tax?

2. Dan bought a used motorcycle for \$955. He made a down payment of 18%. How much cash did he pay as a down payment?

3. Last year Joan had 14% of her salary withheld for taxes. If the total amount withheld was \$2193.10, what was Joan's yearly salary?

4. The manager of a fruit stand lost $16\frac{2}{3}\%$ of his bananas to spoilage and sold the rest. He discarded four boxes of bananas in 2 weeks. How many boxes did he have in stock?

5. Maria's base rate of pay is \$7.82 per hour. She receives time and a half for all hours over 40 that she works in 1 week. What were her total earnings if last week she worked a total of 47 hours?

6. Truong's base rate of pay is \$6.72 per hour. He receives time and a half for all hours over 40 that he works in 1 week. What were his earnings if last week he worked a total of 42 hours?

7. John got 25 problems correct on a 30-problem test. What was his percent score to the nearest whole-number percent?

8. To pass a test to qualify for a job interview, Whitney must score at least 70%. If there are 40 questions on the test, how many must she get correct to score 70%?

9. Eddie and his family went to a restaurant for dinner. The dinner check was $33.45. He left the waiter a tip of $5. What percent of the check was the tip, to the nearest whole-number percent?

10. Adams High School's basketball team finished the season with a record of 15 wins and 9 losses. What percent of the games played were won?

11. The town of Verboort has a population of 15,560, which is 45% male. Of the men, 32% are 40 years or older. How many men are there in Verboort who are younger than 40?

12. If you use a 35¢ coupon to buy a $2.89 box of ice cream sandwiches, what percent off is this, to the nearest whole-number percent?

13. A store advertises $20 off every coat in stock. What percent savings is this on a coat that is regularly $72.99, to the nearest tenth of 1%?

14. Good driving habits can increase mileage and save on gas. If good driving causes a car's mileage to go from 31.5 mpg to 35 mpg, what is the percent of increased mileage? Round to the nearest whole-number percent.

15. According to the Bureau of Labor, there were 90,000 physical therapists in the United States in 1992. Physical therapy is one of the fastest-growing occupations, with the Census Bureau predicting an 88% increase over 1992 levels by the year 2005. How many physical therapists are predicted by the year 2005? Round to the nearest thousand.

16. The number of human services workers is growing at an even faster rate than that of physical therapists. The Census Bureau is predicting a 135% increase over 1992 levels by the year 2005. How many human services workers are predicted in the year 2005 if there were 189,000 in the United States in 1992? Round to the nearest thousand.

17. For customers who use a bank's credit card, there is a $1\frac{3}{4}\%$ finance charge on monthly accounts that have a balance of \$400 or less. Merle's finance charge for August was \$2.80. What was the amount of her account for August?

18. A fast-food hamburger has 342 calories from fat. This is approximately 54.3% of the total number of calories in the hamburger. How many calories are in one of the hamburgers? Round to the nearest calorie.

19. In 1980, 64% of all workers drove to work alone. In 1990, 73% drove alone. Find the population of your community and compute the number of single-car drivers for both of these years.

20. Find the percent of the students in your class who drove alone to class today. Compare this percent with the percents in Exercise 19. How does the class compare to the national average for both years?

21. The following label shows the nutrition facts for one serving of Toasted Oatmeal. Use the information on the label to determine the recommended daily intake of (a) total fat, (b) sodium, (c) potassium, and (d) dietary fiber. Use the percentages for cereal alone. Round to the nearest whole number.

Nutrition Facts
Serving Size 1 cup (49g)
Servings Per Container about 9

Amount Per Serving	**Cereal Alone**	**with 1/2 Cup Vitamin A&D Fortified Skim Milk**
Calories	190	230
Calories from Fat	25	25
	% Daily Values**	
Total Fat 2.5g*	**4%**	**4%**
Saturated Fat 0.5g	**3%**	**3%**
Polyunsaturated Fat 0.5g		
Monounsaturated Fat 1g		
Cholesterol 0mg	**0%**	**0%**
Sodium 220mg	**9%**	**12%**
Potassium 180mg	**5%**	**11%**
Total Carbohydrate 39g	**13%**	**15%**
Dietary Fiber 3g	**14%**	**14%**
Soluble Fiber 1g		
Insoluble Fiber 2g		
Sugars 11g		
Other Carbohydrates 24g		
Protein 5g		

22. The following table shows the calories per serving of the item along with the number of fat grams per serving.

Item	Calories per Serving	Fat Grams per Serving
Light mayonnaise	50	4.5
Cocktail peanuts	170	14
Wheat Thin crackers	120	4
Cream sandwich cookies	110	2.5

Assume each fat gram is equivalent to 10 calories. Find the percent of calories that are from fat for each item. Round to the nearest whole percent.

NAME ______________________ CLASS ____________ DATE ____________

OBJECTIVE 2: *Solve business and personal finance percent problems.*

23. An article that costs the store owner $16.80 is marked up $5.04. What is the percent of markup based on the cost?

24. A tool that costs a hardware merchant $8.40 is marked up 30% of the cost. What is the selling price?

25. An article is priced to sell at $39.99. If the markup is 28% of the selling price, how much is the markup to the nearest cent?

26. The Bright TV Store regularly sells a television set for $329.95. An advertisement in the paper shows that it is on sale at a discount of 25%. What is the sale price to the nearest cent?

27. A competitor of the store in Exercise 26 has the same TV set on sale. The competitor normally sells the set for $335.95 and has it advertised at a 27% discount. To the nearest cent, what is the sale price of the TV? Which is the better buy and by how much?

28. A salesman earns a 9% commission on all of his sales. How much did he earn last week if his total sales were $5482?

29. If the salesman in Exercise 28 received a 12% commission on all sales, and his total sales for one week were $4725, what were his earnings for the week?

30. The Top Company offered a 6% rebate on the purchase of their best model of canopy. If the regular price is \$398.98, what is the amount of the rebate to the nearest cent?

31. The taxes on a piece of property are \$2238.30, and the property has an assessed value of \$82,900. What is the percent tax rate in the district where the property is located?

Exercises 32–39 relate to the chapter application.

32. Corduroy overalls that are regularly \$34.99 are on sale for 25% off. What is the sale price of the overalls?

33. A pair of Reebok cross trainers, which regularly sells for \$59.99, goes on sale for \$47.99. What percent off is this? Round to the nearest whole-number percent.

34. A bag of Tootsie Rolls is marked "20% more free—14.5 oz for the price of 12 oz." Assuming there has not been a change in price, is the claim accurate? Explain.

35. A department store puts a blazer, which was originally priced at \$139.95, on sale for 20% off. At the end of the season, the store has an "additional 40% off everything that is already reduced" sale. What is the price of the blazer? What percent saving does this represent over the original price?

36. A shoe store advertises "Buy one pair, get 50% off a second pair of lesser or equal value." The mother of twin boys buys a pair of basketball shoes priced at \$36.99 and a pair of hikers priced at \$27.99. How much did she pay for the two pairs of shoes? What percent savings is this to the nearest tenth of a percent?

37. The Klub House advertises on the radio that all merchandise is on sale at 25% off. When you go in to buy a set of golf clubs that originally sold for \$279.95, you find that the store is giving an additional 10% discount off the original price. What is the price you will pay for the set of clubs?

38. In Exercise 37, if the salesperson says that the 10% discount can only be applied to the sale price, what is the price of the clubs?

39. A store advertises "30% off all clearance items." A boy's knit shirt is on a clearance rack that is marked 20% off. How much is saved on a knit shirt that was originally priced \$14.99?

40. Three people resolve to become business associates. The first associate pays 36% for a share of a franchise, the second associate pays 29%, and the third associate pays for the remaining 35% of the franchise. The profits are to be divided among the three according to each associate's share of the ownership. The profits for last month were \$5274. How much does each associate receive?

41. In 1990, use of office paper resulted in over 11 tons of solid waste. Of this amount, 15% was recycled. How many tons were recycled?

42. A labor union negotiated a contract with the Swift Copier Company. The average wage under the new contract is $4\frac{3}{4}$% higher. The average wage under the old contract was \$9.24 per hour. What is the average wage under the new contract?

OBJECTIVE 3: *Read data from a circle graph or construct a circle graph from data.*

For Exercises 43–44, the circle graph below shows ethnic distribution of children enrolled in Head Start.

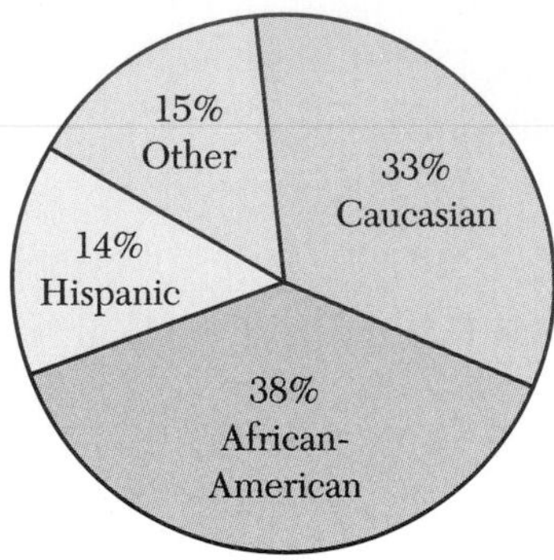

Ethnic Distribution

43. Which group has the smallest number of children in Head Start?

44. What percent of children in Head Start are people of color?

For Exercises 45–48 the circle graph shows how dollars are spent in a particular industry.

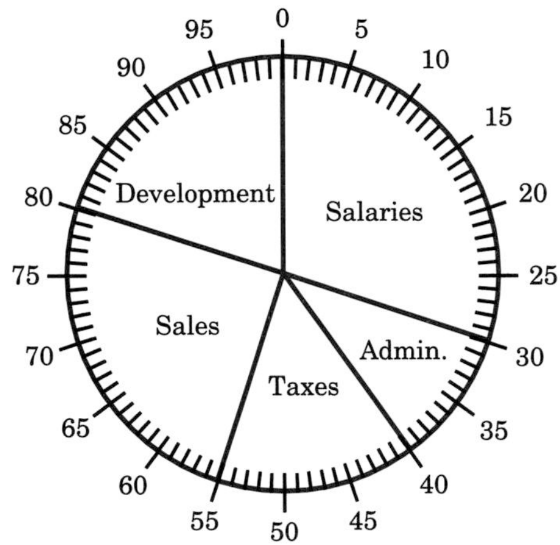

Distribution of Expenditures

45. What item has the greatest expenditure of funds?

46. Which is more costly, development or taxes?

47. If the total expenditures of the industry were $2,500,000, how much was spent on development?

48. Using the total expenditures in Problem 47, how much was spent on salaries?

49. In a family of three children, there are eight possibilities of boys and girls. One possibility is that they are all girls. Another possibility is that they are all boys. There are three ways for the family to have two girls and a boy. There are also three ways for the family to have two boys and a girl. Make a circle graph to illustrate this information.

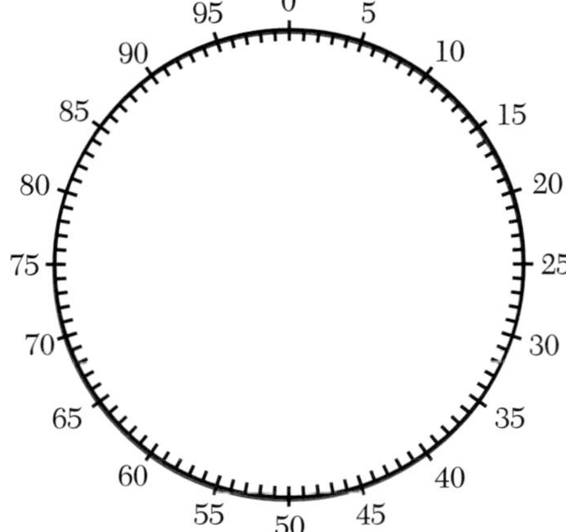

50. According to one state official, 48% of all households have no firearms, 12% of all households store firearms unloaded and ammunition locked up, 10% of all households keep firearms loaded and unlocked, and 30% store firearms another way. Make a circle graph to illustrate this information.

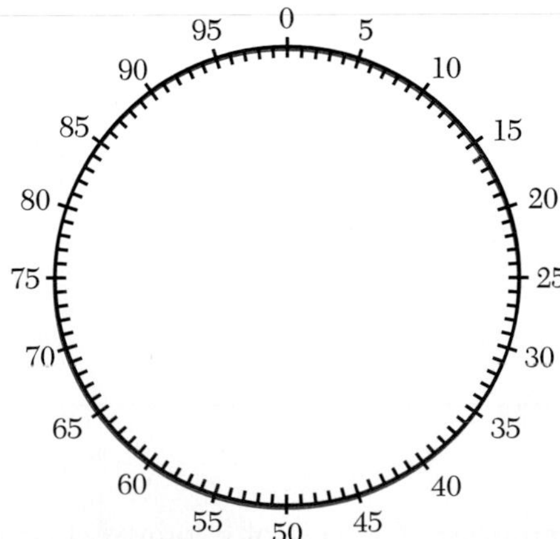

51. The major causes of death worldwide are listed below. Make a circle graph to illustrate this information.

Infectious and parasitic diseases	32%
Heart, circulatory diseases, and stroke	19%
Unknown causes	16%
Cancer	12%
Accidents and violence	8%
Infant death	6%
Chronic lung diseases	6%
Other causes	less than 1%

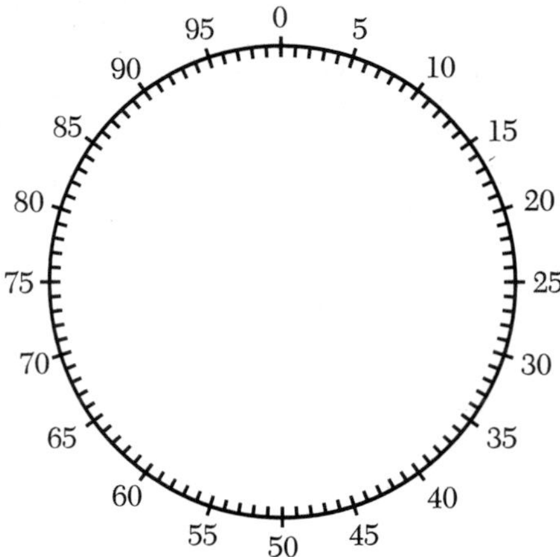

STATE YOUR UNDERSTANDING

52. Explain why the percent of increase from 500 to 750 (50%) is not the same as the percent of decrease from 750 to 500 $\left(33\frac{1}{3}\%\right)$.

53. Explain why a 10% increase in \$100 followed by a 10% decrease does *not* yield \$100.

54. Write a few sentences clarifying the following statements: (a) "a price increase of 100% is double the original price," and (b) "an enlargement of a floor area by 200% of the original is triple the original area."

CHALLENGE

55. The markup on furniture at NuMart is 30% based on the selling price. The cost of a lounge chair is \$58. What is the selling price?

56. A retailer buys a shipment of athletic shoes for \$34.67 a pair and sells them for \$49.99 a pair. On the next shipment the cost is increased 8%, and they, in turn, increase the selling price to \$54.99. Is the percent of markup more or less after the price increase? By how much, rounded to the nearest tenth of a percent?

57. During a six-year period, the cost of maintaining a diesel engine averages \$2650 and the cost of maintaining a gasoline engine averages \$4600. What is the percent of savings of the diesel compared with the gasoline engine? Round to the nearest whole percent.

58. Carol's baby weighed $7\frac{1}{2}$ lb when he was born. On his first birthday he weighed $23\frac{3}{4}$ lb. What was the percent of increase during the year? Round to the nearest whole percent.

GROUP ACTIVITY

59. During one month in 1991, the interest rate on a 30-year home mortgage averaged 9.29%. A 15-year mortgage averaged 8.98%. Two years later, the interest rate on a 30-year mortgage averaged 6.5%. Find the difference in simple interest costs for 1 year on houses that cost \$85,000, \$97,000, \$115,000, and \$166,000. Find the assessed value of your own dwelling and then compute the costs at these different rates. Determine the average interest for the members of your group and for the class. Compare the values with each other, with the other groups, and with the values for 1991 and 1993.

MAINTAIN YOUR SKILLS (SECTIONS 6.3, 7.1, 7.2, 7.3)

60. Peter attended 18 of the 20 GED classes held last month. What percent of the classes did he attend?

61. A family spends \$120 for food out of a budget of \$500. What percent goes for food?

Change to percent.

62. 0.035

63. 2.34

64. 0.018

Write as a decimal.

65. 0.6%

66. 20.5%

67. 400%

68. An engine with a displacement of 400 cubic inches develops 260 horsepower. How much horsepower is developed by an engine with a displacement of 175 cubic inches?

69. Sally and Rita are partners. How much does each receive of the income if they are to share \$8200 in a ratio of 6 to 4 (Sally 6, Rita 4)?

CHAPTER 1

Group Project *(1–2 weeks)*

OPTIONAL

Much research has been conducted in the area of human nutrition, and yet there are still many unanswered questions. Most nutritionists agree, however, that the American diet has too much fat and refined carbohydrates (primarily sugar).

You are about to become a nutritional consultant to John and Jane Doe. John is a 32-year-old male, 5′10″, 165 lb, who consumes 2800 calories per day. Jane is a 31-year-old female, 5′5″, 130 lb, who consumes 2000 calories per day. Both exercise regularly and try to be careful about their diet. However, neither of them likes to cook, so they eat out a lot.

Experts disagree about the exact ratio of carbohydrates, protein, and fat in an optimal diet. Some recommend a 40% carbohydrate, 30% protein, 30% fat ratio. Others would shift the amounts to 50%, 25%, and 25%, respectively. Almost no one recommends more than 30% fat.

All food sold in grocery stores is required by the government to include nutritional labeling. Among other things, the label specifies the total calories per serving and the number of grams of carbohydrates, protein, and fat per serving. Most fast-food chains also provide this information. A gram of carbohydrate and a gram of protein each contain about 4 calories. Assume a gram of fat contains 10 calories.

Your job is to identify exactly what John and Jane will eat for a day so that they meet their nutritional requirements entirely with food purchased from fast-food restaurants. Make a second day's plan for each of them with food purchased from a grocery, bearing in mind that the only cooking they do is heating up in a microwave oven.

Include in your final report:

a. A rationale for the ratio of carbohydrates, protein, and fat selected

b. Two days of menus for John and Jane

c. A table for each person and each day that shows the total calories of carbohydrates, protein, and fat and the final percentages

NAME ______________ CLASS ______________ DATE ______________

CHAPTER 7

True–False Concept Review

Check your understanding of the language of basic mathematics. Tell whether each of the following statements is true (always true) or false (not always true). For each statement you judge to be false, revise it to make a statement that is true.

1. Percent means per one hundred. **1.** ______________

2. The symbol % is read "percent." **2.** ______________

3. To change a fraction to a percent, move the decimal in the numerator two places to the left and add the percent sign. **3.** ______________

4. To change a decimal to a percent, move the decimal point two places to the left. **4.** ______________

5. Percent is a ratio. **5.** ______________

6. In percent the base unit can be more than 100. **6.** ______________

7. To change a percent to a decimal, drop the percent sign and move the decimal point two places to the left. **7.** ______________

8. A percent can be equal to a whole number. **8.** ______________

9. To solve a problem written in the form A is R of B, we can use the proportion $\frac{B}{A} = \frac{X}{100}$, where $R = X\%$. **9.** ______________

10. To solve the problem, "If there is a 5% sales tax on a radio costing \$49.95, how much is the tax?" the simpler word form could be, "5% of \$49.95 is what?" **10.** ______________

11. $1.2 = 120\%$ **11.** ______________

12. $7\frac{1}{2} = 7.5\%$ 12. ______

13. $0.009\% = 0.9$ 13. ______

14. If 0.3% of B is 84, then $B = 280$. 14. ______

15. If some percent of 24 is 12, then the percent is 50%. 15. ______

16. If $2\frac{4}{5}\%$ of 300 is A, then $A = 8.4$. 16. ______

17. Two consecutive decreases of 15% is the same as a decrease of 30%. 17. ______

18. If Selma is given a 10% raise on Monday but has her salary cut 10% on Wednesday, her salary is the same as it was Monday before the raise. 18. ______

19. It is possible to increase a city's population by 110%. 19. ______

20. If the price of a stock increases 100% for each of 3 years, the value of $1 of stock is worth $8 at the end of 3 years. 20. ______

21. A 50% growth in population is the same as 150% of the original population. 21. ______

22. $\frac{1}{2}\% = 0.5$. 22. ______

NAME ______________________ CLASS __________ DATE __________

CHAPTER 7

Test

ANSWERS

1. A computer regularly sells for \$1875. During a sale the dealer discounts the price \$210.50. What is the percent of the discount? Round to the nearest tenth of one percent. **1.** ______

2. Write as a percent: 0.09314 **2.** ______

3. If 12 of every 100 people are left-handed, what percent of the population is left-handed? **3.** ______

4. What percent of $5\frac{3}{8}$ is $3\frac{1}{2}$? (Round to the nearest tenth of one percent.) **4.** ______

5. Change to a percent: $\frac{7}{16}$ **5.** ______

6. One hundred sixteen percent of what number is 65.54? **6.** ______

7. Change to a fraction: $82\frac{1}{7}\%$ **7.** ______

8. Write as a percent: 4.56 **8.** ______

9. What number is 7.8% of 615? **9.** ______

10. Change to a percent (to the nearest tenth of one percent): $5\frac{5}{9}$ **10.** ______

11. Change to a fraction or mixed number: 245% **11.** ______

12. Write as a decimal: 18.93% **12.** ______

13. The Adams family spends 18.9% of their monthly income on food. If their monthly income is \$5234.85, how much money do they spend on food? **13.** ______

14–19. Complete the following table:

Fraction	Decimal	Percent
$\frac{13}{32}$		
	0.624	
		14.5%

20. 12.3 is __________% of 42.9 (to the nearest tenth of a percent). **20.** __________

21. Write as a decimal: 0.67% **21.** __________

22. If a tire costs the dealer $48.52 and the markup is 39% of the cost, what is the selling price of the tire? **22.** __________

23. Nordstrom offers a sale on Tommy Hilfiger jackets at a discount of 23%. What is the sale price of a jacket that originally sold for $89.89? **23.** __________

24. The population of a county in Mississippi grew from 123,456 people to 256,821 people over the past 10 years. What is the percent of increase in the population? (To the nearest tenth of a percent.) **24.** __________

25. The following graph shows the distribution of grades in an American History class. What percent of the class received a B grade? Round to the nearest tenth of a percent. **25.** __________

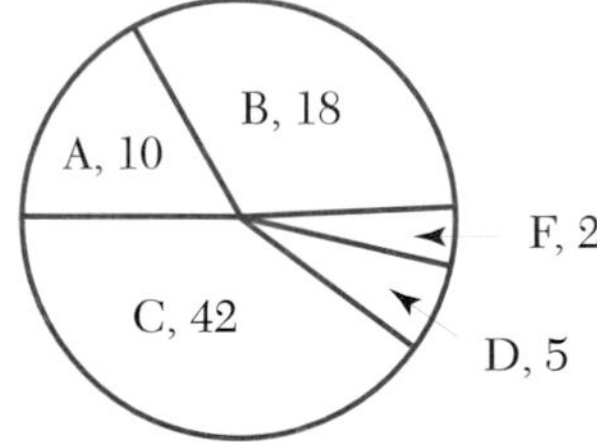

26. Gas prices dropped from summer to winter from $1.189 to $1.059. What is the percent of decrease? Round to the nearest percent. **26.** __________

27. What is the selling price of a suit if the markup is 80% of the cost and the cost is $223.26? **27.** __________

28. If a hamburger contains 780 calories and 37 g of fat, what percent of the calories are from fat? Assume each gram of fat contains 10 calories. Round to the nearest tenth of 1%. **28.** __________

29. In a mathematics class of 145 students, 22.8% withdrew from the course prior to the final exam. How many students took the final exam? **29.** __________

30. The price of a radial tire went from $45.85 to $53.75 over a two-year period. What was the percent of increase in price over the two years? Find to the nearest tenth of a percent. **30.** __________

Good Advice for Studying

Evaluating Your Performance

Pick up your graded test as soon as possible, while the test is still fresh in your mind. No matter what your score, time spent reviewing errors you may have made can help improve future test scores. Don't skip the important step of "evaluating your performance." Begin by categorizing your errors:

1. *Careless errors*. These occur when you know how to do the problem correctly, but don't. Careless errors happen when you read the directions incorrectly, make computational errors, or forget to do the problem.
2. *Concept errors*. These occur because you don't grasp the concept fully or accurately. Even if you did the problem again, you would probably make the same error.
3. *Study errors*. These occur when you do not spend enough time studying the material most pertinent to the test.
4. *Application errors*. These usually occur when you are not sure what concept to use.

If you made a careless error, ask yourself if you followed all the suggestions for better test-taking. If not, vow to do so on the next test. List what you will do differently next time so that you can minimize these careless errors in the future.

Concept errors need more time to correct. You must review what you didn't understand or you will repeat your mistakes. It is important to grasp the concepts because you will use them later. You may need to seek help from your instructor or a tutor for these kinds of errors.

Avoid study errors by asking the teacher before the test what concepts are most important. Also, pay careful attention to the section objectives, because these clarify what you are expected to know.

You can avoid application errors by doing as many of the word problems as you can. Reading the strategies for word problems can help you think about how to start a problem. It also helps to mix up the problems between sections and to do even-numbered problems, for which you do not have answers. It is especially important to try to estimate a reasonable answer before you start calculations.

As a final step in evaluating your performance, take time to think about what you said to yourself during the test. Was this self-talk positive or negative? If the talk was helpful, remember it. Use it again. If not, here are a few suggestions. Be certain that you are keeping a detailed record of your reactions to anxiety. Separate your thoughts from your feelings. Use cue words as signals to begin building coping statements. Your coping statements must challenge your negative belief patterns and they must be believable, not merely pep talk such as "I can do well, I won't worry about it." Make the statements brief and state them in the present tense.

Learn from your mistakes so that you can use what you have learned on the next test. If your are satisfied with the results of the test overall, congratulate yourself. You have earned it.

Algebra Preview: Signed Numbers

APPLICATION

The National Geographic Society was founded in 1888 "for the increase and diffusion of geographic knowledge." The Society has supported more than 5000 explorations and research projects with the intent of "adding to knowledge of earth, sea, and sky." Among the many facts catalogued about our planet, the Society keeps records of the elevations of various geographic features. Table 8.1 lists the highest and lowest points on each continent.

TABLE 8.1 CONTINENTAL ELEVATIONS

Continent	Highest Point	Feet above Sea Level	Lowest Point	Feet below Sea Level
Africa	Kilimanjaro, Tanzania	19,340	Lake Assal	512
Antarctica	Vinson Massif	16,864	Bentley Subglacial Trench	8327
Asia	Mount Everest, Nepal-Tibet	29,028	Dead Sea, Israel-Jordan	1312
Australia	Mount Kosciusko, New South Wales	7,310	Lake Eyre, South Australia	52
Europe	Mount El'brus, Russia	18,510	Caspian Sea, Russia-Azerbaijan	92
North America	Mount McKinley, Alaska	20,320	Death Valley, California	282
South America	Mount Aconcagua, Argentina	22,834	Valdes Peninsula, Argentina	131

When measuring, one has to know where to begin, or the zero point. Notice that when measuring elevations, the zero point is chosen to be sea level. All elevations compare the high or low point to sea level.

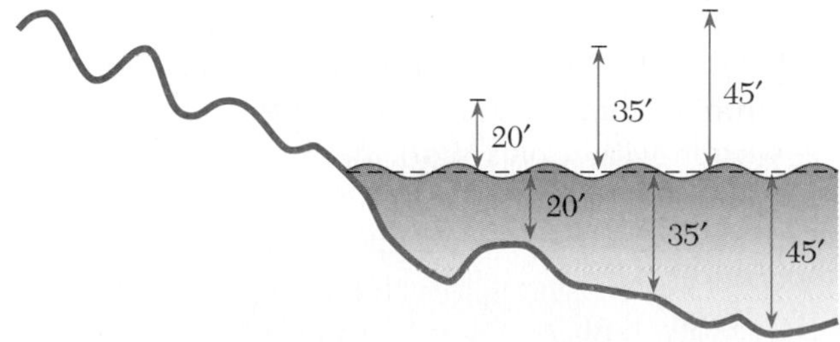

Mathematically we represent quantities under the zero point as negative numbers.

What location has the highest continental altitude on earth? What location has the lowest continental altitude? Is there a location on earth with a lower altitude? Explain.

8.1

Opposites and Absolute Value

OBJECTIVES

1. Find the opposite of a signed number.
2. Find the absolute value of a signed number.

VOCABULARY

Positive numbers are the numbers of arithmetic and are greater than zero. **Negative numbers** are numbers less than zero. Zero is neither positive nor negative. Positive numbers, zero, and negative numbers are called **signed numbers.**

The **opposite** or **additive inverse** of a signed number is the number on the number line that is the same distance from zero but on the opposite side. Zero is its own opposite. The opposite of 5 is written -5. This can be read "the opposite of 5" or "negative 5," since they both name the same number.

The **absolute value** of a signed number is the number of units between the number and zero. The expression $|7|$ is read "the absolute value of 7."

HOW AND WHY

Objective 1

Find the opposite of a signed number.

Exercises such as

$3 - 4$ $8 - 22$ $16 - 17$ and $1 - 561$

do not have answers in the numbers of arithmetic. The answer to each is a signed number. Signed numbers (which include both numbers to the right of zero and to the left of zero) are used to represent quantities with opposite characteristics. For instance,

right and left
up and down
above zero and below zero
gain and loss

A few signed numbers are shown on the following number line:

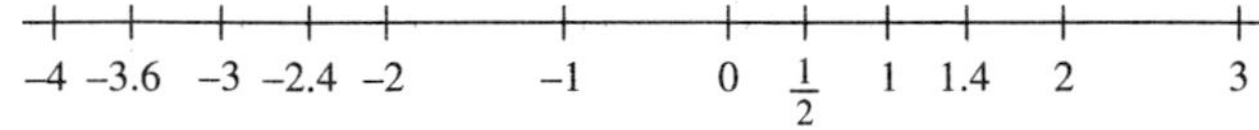

The negative numbers are to the left of zero. The negative numbers have a dash, or negative sign, in front of them. The numbers to the right of zero are called positive (and may be written with a plus sign). Zero is neither positive nor negative. Here are some signed numbers.

7	Seven, or positive seven
-3	Negative three
-0.12	Negative twelve hundredths
0	Zero is neither positive nor negative
$+\frac{1}{2}$	One half, or positive one half

Positive and negative numbers are used many ways in the physical world as shown below:

Positive	*Negative*
Temperatures above zero (72°)	Temperatures below zero (−10°)
Feet above sea level (5000 ft)	Feet below sea level (−50 ft)
Profit ($75)	Loss (−$23)
Right (7)	Left (−4)

In any situation where quantities can be measured in opposite directions, positive and negative numbers can be used to show direction.

The dash in front of a number is read in two different ways:

−19 **The opposite of 19**

−19 **Negative 19**

The opposite of a signed number is the number on the number line that is the same distance from zero but on the opposite side. To find the opposite of a number we refer to a number line.

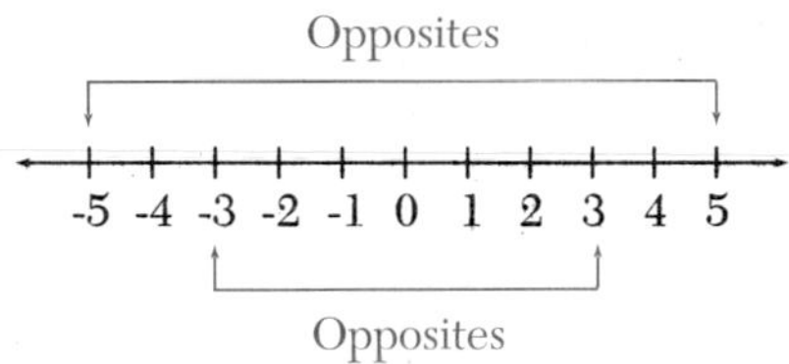

$-(5) = -5$ **The opposite of positive 5 is negative 5.**

$-(-3) = 3$ **The opposite of negative 3 is positive 3.**

$-0 = 0$ **The opposite of 0 is 0.**

To find the opposite of a positive number

1. Locate a positive number on the number line.
2. Count the number of units from zero.
3. Count this many units to the left of zero. Where you stop is the opposite of the positive number.

To find the opposite of a negative number

1. Locate a negative number on the number line.
2. Count the number of units it is from zero.
3. Count this many units to the right of zero. Where you stop is the opposite of the negative number.

The opposite of a positive number is negative and the opposite of a negative number is positive. The opposite of 0 is 0. The expression −5 is sometimes referred to as the "negative of" 5, and 5 is sometimes referred to as the "negative of" −5.

Warm Ups A–E

A. Find the opposite of 22.

B. Find the opposite of −15.

C. Find the opposite of −(−5).

Examples A–E

Directions: Find the opposite.

Strategy: The opposite of a number is found on the number line the same number of units from zero but on the opposite side of zero.

A. Find the opposite of 12.

Strategy: With no sign in front, the number is positive; 12 is read "twelve" or "positive twelve."

$-(12) = ?$ **Since 12 is 12 units to the right of zero, the opposite of 12 is 12 units to the left of zero.**

$-(12) = -12$

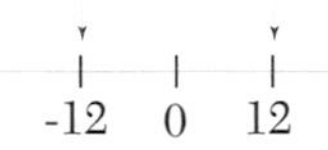

The opposite of 12 is −12.

B. Find the opposite of −16.

Strategy: The number −16 can be thought of in two ways: as a negative number that is 16 units to the left of zero, or as a number that is 16 units on the opposite side of zero from 16; −16 is read "negative 16" or "the opposite of 16."

$-(-16) = ?$ **The opposite of negative 16 is written −(−16) and is found 16 units on the opposite side of zero from −16.**

$-(-16) = 16$

The opposite of −16 is 16.

C. Find the opposite of −(−9).

CAUTION

The dash in front of the parenthesis is *not* read "negative." The dash directly in front of 9 is read "negative" or "the opposite of."

Strategy: First find −(−9). Then find the opposite of that value.

−(−9) is "the opposite of negative 9" **−(−9) can also be read "the opposite of the opposite of 9."**

$-(-9) = ?$ **The opposite of −9 is 9 units to the right of zero, which is 9.**

$-(-9) = 9$

So the opposite of (−9) = 9.

Continuing, we find the opposite of −(−9), which is written −[−(−9)].

$$-[-(-9)] = -[9]$$
$$= -9$$

The opposite of −(−9) is −9.

Answers to Warm Ups A. −22 B. 15 C. −5

Calculator Example

D. Find the opposite of 19, −2.3, and 4.7.

Strategy: The [+/−] key or the [(−)] key on the calculator will give the opposite of the number.

The opposites are −19, 2.3, and −4.7.

D. Find the opposite of 13, −7.8, and 7.1.

E. If 10% of Americans purchased products with no plastic packaging just 10% of the time, approximately 144,000 lb of plastic would be eliminated (taken out of or decreased) from our landfills.

1. Write this decrease as a signed number.
2. Write the opposite of eliminating (decreasing) 225,000 lb of plastic from our landfills as a signed number.

1. Decreases are often represented by negative numbers. Therefore, a decrease of 144,000 lb is −144,000 lb.
2. The opposite of a decrease is an increase, so −(−225,000 pounds) is 225,000 lb.

E. The energy used to produce a pound of rubber is 15,700 BTU. Recycled rubber requires 4100 BTU less.

1. Express this decrease as a signed number.
2. Express the opposite of a decrease of 8500 BTU as a signed number.

HOW AND WHY

Objective 2

Find the absolute value of a signed number.

The absolute value of a signed number is the number of units between the number and zero on the number line. Absolute value is defined as the number of units only, direction is not involved. Therefore, the absolute value is never negative.

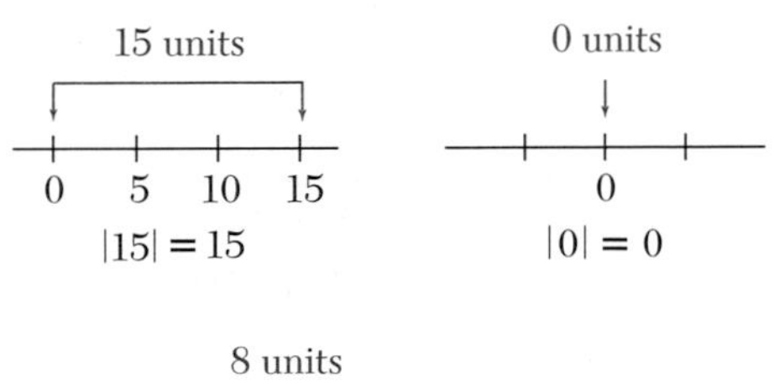

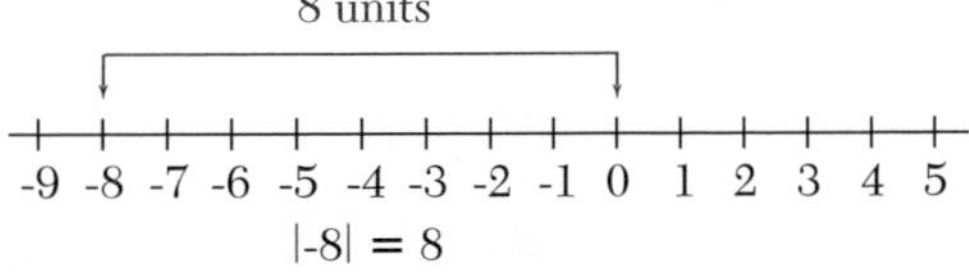

Figure 8.1

Answers to Warm Ups D. −13, 7.8, −7.1 E. As a signed number the decrease is −4100 BTU. As a signed number the opposite of a decrease of 8500 BTU is 8500 BTU.

From Figure 8.1, we can see that the absolute value of a number and the absolute value of the number's opposite are equal. For example

$|19| = |-19|$ since both equal 19

The absolute value of a positive number and zero is the number itself. The absolute value of a negative number is its opposite.

To find the absolute value of a signed number

The value is:

1. 0 if the number is 0.

2. The number if the number is positive.

3. The opposite of the number if the number is negative.

Warm Ups F–I

F. Find the absolute value of 16.

G. Find the absolute value of −16.

H. Find the absolute value of −1.

I. Find the absolute value of $-\frac{3}{5}$.

Examples F–I

Directions: Find the absolute value of the number.

Strategy: If the number is positive or 0, write the number. If the number is negative, write its opposite.

F. Find the absolute value of 9.

$|9| = 9$ **The absolute value of a positive number is the number itself.**

G. Find the absolute value of −22.

$|-22| = -(-22)$ **The absolute value of a negative number is its opposite.**

$= 22$

H. Find the absolute value of 0.

$|0| = 0$ **The absolute value of zero is zero.**

I. Find the absolute value of $-\frac{2}{3}$.

$\left|-\frac{2}{3}\right| = -\left(-\frac{2}{3}\right)$ **The absolute value of a negative number is its opposite.**

$= \frac{2}{3}$

Answers to Warm Ups F. 16 G. 16 H. 1 I. $\frac{3}{5}$

NAME ______________________ CLASS ____________ DATE ____________

Exercises 8.1

OBJECTIVE 1: *Find the opposite of a signed number.*

A.

Find the opposite of the signed number.

1. -4 **2.** -8 **3.** 5 **4.** 13 **5.** -2.3

6. -3.3 **7.** $\frac{3}{5}$ **8.** $\frac{3}{4}$ **9.** $-\frac{6}{7}$ **10.** $-\frac{3}{8}$

11. The opposite of __________ is 61. **12.** The opposite of -23 is __________.

B.

Find the opposite of the signed number.

13. -42 **14.** -57 **15.** -3.78 **16.** -6.27

17. $\frac{17}{5}$ **18.** $\frac{23}{7}$ **19.** 0.55 **20.** 0.732

21. 113.8 **22.** 243.7 **23.** -0.0123 **24.** -0.78

OBJECTIVE 2: *Find the absolute value of a signed number.*

Find the absolute value of the signed number.

25. $|-4|$ **26.** $|-11|$ **27.** $|24|$ **28.** $|61|$

29. $|-3.17|$ **30.** $|-4.6|$ **31.** $\left|\frac{5}{11}\right|$ **32.** $\left|\frac{7}{13}\right|$

33. $\left|-\frac{3}{11}\right|$ **34.** $\left|-\frac{7}{12}\right|$

35. The absolute value of __________ is 32.

36. The absolute value of ________ is 17.

B.

Find the absolute value of a signed number.

37. $|0.0065|$ **38.** $|0.0021|$ **39.** $|-355|$ **40.** $|-922|$

41. $\left|-\frac{11}{3}\right|$ **42.** $\left|-\frac{17}{6}\right|$ **43.** $\left|-\frac{33}{7}\right|$ **44.** $\left|-\frac{18}{5}\right|$

45. $|0|$ **46.** $|-100|$ **47.** $|-0.341|$ **48.** $|-4.96|$

C.

Find the value of the following:

49. Opposite of $\left|-\frac{5}{17}\right|$ **50.** Opposite of $\left|\frac{9}{8}\right|$

51. Opposite of $|78|$ **52.** Opposite of $|-91|$

53. At the New York Stock Exchange, positive and negative numbers are used to record changes in stock prices on the board. What is the opposite of a gain of five eighths $\left(+\frac{5}{8}\right)$?

54. At the American Stock Exchange a stock is shown to have taken a loss of three-eights point $\left(-\frac{3}{8}\right)$. What is the opposite of this loss?

55. On a thermometer temperatures above zero are listed as positive and those below zero as negative. What is the opposite of a reading of 12°C?

56. On a thermometer such as the one in Exercise 55, what is the opposite of a reading of 23°C?

57. The modern calendar counts the years after the birth of Christ as positive numbers (A.D. 1993 or +1993). Years before Christ are listed using negative numbers (2045 B.C. or −2045). What is the opposite of 1875 B.C. or −1875?

58. The empty-weight center of gravity of an airplane is determined. A generator is installed at a moment of −278. At what moment could a weight be placed so that the center of gravity remains the same? (Moment is the product of a quantity such as weight and its distance from a fixed point. In this application the moments must be opposites to keep the same center of gravity.)

59. A cyclist travels up a mountain 1685 ft then turns around and travels down the mountain 1246 ft. Represent each trip as a signed number.

60. How far is the cyclist from her starting point? (See Exercise 59.) Write the distance as a signed number.

61. An energy audit indicates that the Gates family could reduce their average electric bill by $28.76 per month by doing some minor repairs, insulating their attic and crawl space, and caulking around the windows and other cracks in the siding.

a. Express this savings as a signed number.

b. Express the opposite of the savings as a signed number.

62. If 80 miles north is represented by +80, how would you represent 80 miles south?

Exercises 63–65 refer to the chapter application (see page 723).

63. Rewrite the continental altitudes in Table 8.1 using signed numbers.

64. The U.S. Department of Defense has extensive maps of the ocean floors because they are vital information for the country's submarine fleet. The table lists the deepest part and the average depth of the world's major oceans.

Ocean	Deepest Part	Feet	Average Depth (ft)
Pacific	Mariana Trench	35,840	12,925
Atlantic	Puerto Rico Trench	28,232	11,730
Indian	Java Trench	23,376	12,598
Arctic	Eurasia Basin	17,881	3,407
Mediterranean	Ionian Basin	16,896	4,926

Rewrite the table using signed numbers.

65. Is the highest point on Earth further away from sea level than the deepest point in the ocean? Explain. What mathematical concept allows you to answer this question?

Simplify.

66. $-(-24)$ **67.** $-(81)$ **68.** $-(-(-14))$ **69.** $-(-(33))$

70. The Buffalo Bills are playing a football game against the Seattle Seahawks. On the first play the Seahawks lose 8 yd. Represent this as a signed number. What is the opposite of a loss of 8 yd? Represent this as a signed number.

71. The Dallas Cowboys and the New York Giants are having an exhibition game in London, England. The Cowboy offensive team runs a gain of 6 yd, a loss of 8 yd, a gain of 21 yd, and a loss of 15 yd. Represent these yardages as signed numbers.

72. The Golden family is on a vacation in the southwestern United States. Consider north and east as positive directions and south and west as negative directions. On one day they drive north 97 miles then east 152 miles. The next day they drive west 72 miles then 18 miles south. Represent each of these distances as signed numbers.

STATE YOUR UNDERSTANDING

73. Is zero the only number that is its own opposite? Justify your answer.

74. Is there a set of numbers for which the absolute value of each number is the number itself? If yes, identify that set and tell why this is true.

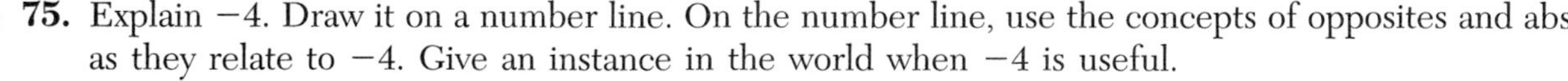

75. Explain -4. Draw it on a number line. On the number line, use the concepts of opposites and absolute value as they relate to -4. Give an instance in the world when -4 is useful.

NAME ______________________ CLASS __________ DATE __________

CHALLENGE

Simplify.

76. $|16 - 10| - |14 - 9| + 6$

77. $8 - |12 - 8| - |10 - 8| + 2$

78. If n is a negative number, what kind a number is $-n$?

79. For what numbers is $|-n| = n$ always true?

GROUP ACTIVITY

80. Develop a rule (procedure) for adding a positive and a negative integer. *Hint:* You might want to visualize that walking forward represents a positive number and walking backward represents a negative integer. See if you can find answers to:

$6 + (-4) = ?$ $7 + (-10) = ?$ $-6 + 11 = ?$

MAINTAIN YOUR SKILLS (SECTIONS 2.1, 5.7)

Multiply.

81. (40 g)30 = ? g

82. (9.01 kℓ)1.5 = ? kℓ

Divide.

83. (84 kg) ÷ 20 = ? kg

84. (100.8 cm) ÷ 5.6 = ? cm

Change units as shown.

85. 8.25 km = ? m

86. 0.82 kg = ? cg

Add.

87. 6 m 250 cm
+7 m 460 cm
= ? m

88. 5 ℓ 78 mℓ
2 ℓ 95 mℓ
= ? mℓ

89. A 12-m coil of wire is to be divided into eight equal parts. How many meters are in each part?

90. Some drug doses are measured in grains. For example, a common aspirin tablet contains 5 grains. If there are 15.43 g in a grain, how many grams do two aspirin tablets contain?

8.2

Adding Signed Numbers

OBJECTIVE

Add signed numbers.

HOW AND WHY

Objective

Add signed numbers.

Positive and negative numbers are used to show opposite quantities:

$+482$ lb may show 482 lb loaded
-577 lb may show 577 lb unloaded
$+27$ dollars may show 27 dollars earned
-19 dollars may show 19 dollars spent

Using this idea we find the sum of signed numbers. We already know how to add two positive numbers, so we concentrate on adding positive and negative numbers.

Think of 27 dollars earned (positive) and 19 dollars spent (negative). The result is 8 dollars left in your pocket (positive). So

$27 + (-19) = 8$

To get this sum we subtract the absolute value of -19 (19) from the absolute value of 27 (27). The sum is positive.

Think of 23 dollars spent (negative) and 15 dollars earned (positive). The result is that you still owe 8 dollars (negative). So

$-23 + 15 = -8$

To get this sum, subtract the absolute value of 15 (15) from the absolute value of -23 (23). The sum is negative.

Think of 5 dollars spent (negative) and another 2 dollars spent (negative). The result is 7 dollars spent (negative). So

$-5 + (-2) = -7$

To get this sum we add 5 and 2 (the absolute value of each). The sum is negative.

The results of the examples lead us to the procedure for adding signed numbers.

To add signed numbers

1. If the signs are alike, add their absolute values and use the common sign.
2. If the signs are not alike, subtract the smaller absolute value from the larger absolute value. The sum will have the sign of the number with the larger absolute value.

As a result of this definition, the sum of a number and its opposite is zero. Thus, $5 + (-5) = 0$, $8 + (-8) = 0$, $-12 + 12 = 0$, etc.

Warm Ups A–J

A. Add: $48 + (-29)$

B. Find the sum: $-51 + 34$

C. Add: $29 + (-75)$

D. Add: $-1.3 + (-2.5)$

E. Find the sum: $\frac{7}{8} + \left(-\frac{5}{6}\right)$

Examples A–J

Directions: Add.

Strategy: If the signs are alike, add the absolute values and use the common sign. If the signs are unlike, subtract their absolute values and use the sign of the number with larger absolute value.

A. Add: $37 + (-12)$

Strategy: Since the signs are unlike, subtract their absolute values.

$$|37| - |-12| = 37 - 12$$
$$= 25$$

Because the positive number has the larger absolute value, the sum is positive.

$37 + (-12) = 25$

B. Find the sum: $-52 + 28$

$$|-52| - |28| = 52 - 28$$
$$= 24$$

$-52 + 28 = -24$ **The sum is negative since −52 has the larger absolute value.**

C. Add: $41 + (-63)$

$$|-63| - |41| = 63 - 41$$
$$= 22$$

$41 + (-63) = -22$ **The sum is negative since −63 has the larger absolute value.**

D. Add: $-0.33 + (-1.7)$

Strategy: The signs are the same, so add their absolute values.

$$|-0.33| + |-1.7| = 0.33 + 1.7$$
$$= 2.03$$

$-0.33 + (-1.7) = -2.03$ **The numbers are negative, therefore the sum is negative.**

E. Find the sum: $\frac{4}{5} + \left(-\frac{3}{2}\right)$

$$\left|-\frac{3}{2}\right| - \left|\frac{4}{5}\right| = \frac{15}{10} - \frac{8}{10}$$
$$= \frac{7}{10}$$

$\frac{4}{5} + \left(-\frac{3}{2}\right) = -\frac{7}{10}$ **Since the negative number has the larger absolute value, the sum is negative.**

Answers to Warm Ups A. 19 B. −17 C. −46 D. −3.8 E. $\frac{1}{24}$

F. Find the sum of -13, 53, -27, and -21.

Strategy: Where there are more than two numbers to add, it may be easier to add the numbers with the same sign first.

$$\begin{array}{r} -13 \\ -27 \\ +\underline{-21} \\ -61 \end{array}$$

Add the negative numbers. Add their sum to 53. Since the signs are different, find the difference in their absolute values.

$$\begin{aligned} |-61| - |53| &= 61 - 53 \\ &= 8 \end{aligned}$$

$-13 + 53 + (-27) + (-21) = -8$ **The sum is negative since -61 has the larger absolute value.**

F. Find the sum of -35, 76, -101, and 37.

G. Add: $-0.1 + 0.5 + (-3.4) + 0.8$

Strategy: Add the negative numbers and add the positive numbers.

$$\begin{array}{r} -0.1 \\ +\underline{-3.4} \\ -3.5 \end{array} \qquad \begin{array}{r} 0.5 \\ +\underline{0.8} \\ 1.3 \end{array}$$

The signs are different so subtract their absolute values.

$$\begin{aligned} |-3.5| - |1.3| &= 3.5 - 1.3 \\ &= 2.2 \end{aligned}$$

$-0.1 + 0.5 + (-3.4) + 0.8 = -2.2$ **Since -3.5 has the larger absolute value, the sum is negative.**

G. Add: $0.32 + (-0.54) + (-0.73) + 1.2$

H. Add: $-\frac{3}{4} + \frac{7}{8} + \left(-\frac{1}{2}\right) + \left(-\frac{7}{8}\right)$

Strategy: First, add the negative numbers.

$$\begin{aligned} -\frac{3}{4} + \left(-\frac{1}{2}\right) + \left(-\frac{7}{8}\right) &= \left(-\frac{6}{8}\right) + \left(-\frac{4}{8}\right) + \left(-\frac{7}{8}\right) \\ &= -\frac{17}{8} \end{aligned}$$

Add the sum of the negative numbers and the positive number. To do this find the difference of their absolute values.

$$\frac{17}{8} - \frac{7}{8} = \frac{10}{8} = \frac{5}{4}$$

$-\frac{3}{4} + \frac{7}{8} + \left(-\frac{1}{2}\right) + \left(-\frac{7}{8}\right) = -\frac{5}{4}$ **The sum is negative since the negative number has the larger absolute value.**

H. Add: $-\frac{5}{6} + \frac{3}{2} + \left(-\frac{8}{9}\right) + \left(-\frac{2}{3}\right)$

Answers to Warm Ups F. -23 G. 0.25 H. $-\frac{8}{9}$

Calculator Example

I. Find the sum: $-55 + 47 + (-77) + (-55)$

I. Find the sum: $-68 + 43 + (-64) + (-11)$

Use the [+/−] key or the [(−)] key on the calculator to display negative numbers.

The sum is -100.

J. A second stock that John owns has the following changes for a week: Monday, gains \$1.25; Tuesday, gains \$0.75; Wednesday, loses \$2.625; Thursday, loses \$1.50; Friday, gains \$1.375. What is the net change in the price of the stock for the week?

J. John owns stock that is traded on the American Stock Exchange. On Monday the stock gains \$2, on Tuesday it loses \$3, on Wednesday it loses \$2, on Thursday it gains \$5, and on Friday it gains \$2. What is the net change in the price of the stock for the week?

Strategy: To find the net change in the price of the stock, write the daily changes as signed numbers and find the sum of these numbers.

Monday	gains \$2	2
Tuesday	loses \$3	−3
Wednesday	loses \$2	−2
Thursday	gains \$5	5
Friday	gains \$2	2

$$2 + (-3) + (-2) + 5 + 2 = 9 + (-5)$$
$$= 4$$

Add the positive numbers and add the negative numbers.

The stock gains \$4 during the week.

Answers to Warm Ups I. -140 J. The stock lost \$0.75 ($-\$0.75$) for the week.

NAME ______________________ CLASS ______________ DATE ______________

Exercises 8.2

OBJECTIVE: *Add signed numbers.*

A.

Add.

1. $-6 + 8$ **2.** $-8 + 2$ **3.** $6 + (-7)$ **4.** $9 + (-4)$

5. $-8 + (-6)$ **6.** $-5 + (-9)$ **7.** $-10 + (-6)$ **8.** $-12 + (-7)$

9. $-7 + 7$ **10.** $11 + (-11)$ **11.** $0 + (-13)$ **12.** $-15 + 0$

13. $-17 + (-17)$ **14.** $-21 + (-21)$ **15.** $7 + (-15)$ **16.** $9 + (-18)$

17. $-15 + (-12)$ **18.** $-10 + (-17)$ **19.** $23 + (-18)$ **20.** $-13 + 21$

B.

21. $-3 + (-6) + 5$ **22.** $-2 + (-5) + (-6)$ **23.** $-72 + (-72)$

24. $-48 + (-48)$ **25.** $-40 + (-40)$ **26.** $56 + (-56)$

27. $-72 + 72$ **28.** $48 + (-39)$ **29.** $-18 + 21 + (-3)$

30. $-22 + 25 + (-4)$ **31.** $47 + (-32) + (-14)$ **32.** $-17 + (-12) + 32$

33. $-2.3 + (-4.3)$ **34.** $-8.2 + (-3.2)$ **35.** $6.3 + (-3.7)$

36. $-7.4 + 4.1$ **37.** $-\frac{2}{3} + \frac{1}{6}$ **38.** $\frac{5}{6} + \left(-\frac{3}{4}\right)$

39. $-\frac{3}{4} + \left(-\frac{7}{8}\right)$

40. $-\frac{5}{6} + \left(-\frac{1}{3}\right)$

C.

Simplify.

41. $135 + (-256)$

42. $233 + (-332)$

43. $-81 + (-32) + (-76)$

44. $-75 + (-82) + (-71)$

45. $-31 + 28 + (-63) + 36$

46. $-44 + 37 + (-59) + 45$

47. $49 + (-67) + 27 + 72$

48. $81 + (-72) + 33 + 49$

49. $356 + (-762) + (-892) + 541$

50. $-923 + 672 + (-823) + (-247)$

51. Find the sum of 542, -481, and -175.

52. Find the sum of 293, -122, and -211.

NAME ______________________ CLASS __________ DATE __________

Exercises 53–56. *Viking II* records the following temperatures for a five-day period on one point on the surface of Mars.

	Day 1	Day 2	Day 3	Day 4	Day 5
5:00 A.M.	−92°C	−88°C	−115°C	−103°C	−74°C
9:00 A.M.	−57°C	−49°C	−86°C	−93°C	−64°C
1:00 P.M.	−52°C	−33°C	−46°C	−48°C	−10°C
6:00 P.M.	−45°C	−90°C	−102°C	−36°C	−42°C
11:00 P.M.	−107°C	−105°C	−105°C	−98°C	−90°C

53. What is the sum of the temperatures recorded at 9:00 A.M.?

54. What is the sum of the temperatures recorded on day 1?

55. What is the sum of the temperatures recorded at 11:00 P.M.?

56. What is the sum of the temperatures recorded on day 4?

57. An airplane is being reloaded; 877 lb of baggage and mail are removed (−877 lb) and 764 lb of baggage and mail are loaded on (+764 lb). What net change in weight should the cargo master report?

58. At another stop, the plane in Exercise 57 unloads 1842 lb of baggage and mail and takes on 1974 lb. What net change should the cargo master report?

59. The change in altitude of a plane in flight is measured every 10 min. The figures between 3:00 P.M. and 4:00 P.M. are as follows:

3:00 P.M.	30,000 ft initially	(+30,000)
3:10 P.M.	increase of 220 ft	(+220)
3:20 P.M.	decrease of 200 ft	(−200)
3:30 P.M.	increase of 55 ft	(+55)
3:40 P.M.	decrease of 110 ft	(−110)
3:50 P.M.	decrease of 55 ft	(−55)
4:00 P.M.	decrease of 40 ft	(−40)

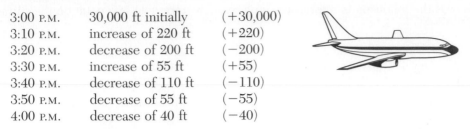

What is the altitude of the plane at 4 P.M.? (*Hint:* Find the sum of the initial altitude and the six measured changes between 3 and 4 P.M.)

60. What is the final altitude of the airplane in Exercise 59 if it is initially flying at 23,000 ft with the following changes in altitude?

3:00 P.M.	23,000 ft initially	(+23,000)
3:10 P.M.	increase of 315 ft	(+315)
3:20 P.M.	decrease of 825 ft	(−825)
3:30 P.M.	increase of 75 ft	(+75)
3:40 P.M.	decrease of 250 ft	(−250)
3:50 P.M.	decrease of 85 ft	(−85)
4:00 P.M.	decrease of 70 ft	(−70)

61. The Pacific Northwest Book Depository handles most textbooks for the local schools. On September 1 the inventory is 28,945 volumes. During the month the company makes the following transactions (positive numbers represent volumes received, negative numbers represent shipments): 2386, −497, −924, 475, −997. What is the inventory at the end of the month?

62. The Pacific Northwest Book Depository has 12,895 volumes on November 1. During the month the depository has the following transactions: −2478, 514, −877, −213, 97, −482. What is the inventory at the end of the month?

63. The Buffalo Bills made the following consecutive plays during a recent Monday night football game: 8-yd loss, 10-yd gain, and a 7-yd gain. A first down requires a gain of 10 yd. Did they get a first down?

64. The Seattle Seahawks have these consecutive plays one Sunday: 12-yd loss, 19-yd gain, and a 4-yd gain. Do they get a first down?

65. Nordstrom stock has the following changes in one week: up $\frac{5}{8}$, down $\frac{1}{2}$, down $1\frac{3}{4}$, up $2\frac{1}{4}$, and up $\frac{5}{8}$. What is the net change for the week?

66. If the Nordstrom stock in Exercise 65 starts at $72\frac{7}{8}$ at the beginning of that week, what is the closing price?

67. On a January morning in a small town in upstate New York, the lowest temperature is recorded as 19 degrees below zero. During the following week the daily lowest temperature readings are up 6 degrees, up 8 degrees, down 2 degrees, up 3 degrees, no change, up 1 degree, and down 5 degrees. What is the low temperature reading for the last day?

68. A new company has the following weekly balances after the first month of business: a loss of $56, a gain of $8, a loss of $95, and a gain of $27. What is their net gain or loss for the month?

69. Marie decided to play the state lottery for one month. This meant that she played every Wednesday and Saturday for a total of nine times. This is her record: lost $4, won $5, lost $3, lost $8, won $9, lost $6, lost $2, lost $4, won $15. What is the net result of her playing?

STATE YOUR UNDERSTANDING

70. When adding a positive and a negative number, explain how to determine whether the sum is positive or negative.

CHALLENGE

71. What is the sum of 25 and (−31) increased by 17?

72. What is the sum of (-41) and (-18) added to (-32)?

73. What number added to 25 equals 18?

74. What number added to (-32) equals (-17)?

75. What number added to (-48) equals (-72)?

GROUP ACTIVITY

76. Develop a rule for subtracting two signed numbers. (*Hint:* Subtraction is the inverse of addition—it undoes addition. So if adding $+6$ can be thought of as walking forward six then subtracting $+6$ may be thought of as walking backward 6 paces.) See if you can find answers to

$4 - (+6) = ?$ $\quad -5 - (+6) = ?$ $\quad$ and $\quad 8 - (-4) = ?$

MAINTAIN YOUR SKILLS (SECTIONS 2.2, 2.3, 4.5, 5.7, 5.9, 6.1)

Convert the units as shown.

77. $\dfrac{450 \text{ miles}}{1 \text{ hr}} = \dfrac{? \text{ ft}}{1 \text{ sec}}$

78. $\dfrac{6 \text{ lb}}{1 \text{ ft}} = \dfrac{? \text{ oz}}{1 \text{ in.}}$

79. 8136 in. = ? feet

80. $\dfrac{\$8.40}{1 \text{ hr}} = \dfrac{? \text{ cents}}{1 \text{ min}}$

81. Find the perimeter of a rectangle that is 42 m long and 29 m wide.

82. Find the circumference of a circle with a radius of 24 cm. The formula is $C = \pi d$, where d is the diameter of the circle. Let $\pi \approx 3.14$.

83. Find the perimeter of a square that is 18 in. on each side.

84. Find the perimeter of a triangle with sides 48 cm, 36 cm, and 18 cm.

85. Bars of soap are sold at three bars for \$1.95. What is the cost of one bar?

86. The cost of pouring a 3-ft-wide cement sidewalk is estimated to be \$18 per square foot. If a walk is to be placed around a rectangular plot of ground that is 24 ft along the width and 48 ft along the length, what is the cost of pouring the walk?

8.3

Subtracting Signed Numbers

OBJECTIVE:

Subtract signed numbers.

HOW AND WHY

Objective

Subtract signed numbers.

The expression $11 - 8 = ?$, in terms of addition, asks $8 + (?) = 11$. We know $8 + 3 = 11$, so $11 - 8 = 3$. The expression $-3 - 5 = ?$, in terms of addition, asks $5 + (?) = -3$. We know $5 + (-8) = -3$, so $-3 - 5 = -8$. The expression $-4 - (-7) = ?$, in terms of addition, asks $-7 + (?) = -4$. We know $-7 + 3 = -4$, so $-4 - (-7) = 3$.

Compare

$11 - 8$	and	$11 + (-8)$	Both equal 3.
$-3 - 5$	and	$-3 + (-5)$	Both equal -8.
$-4 - (-7)$	and	$-4 + 7$	Both equal 3.

Every subtraction problem can be worked by asking what number added to the subtrahend will yield the minuend. However, when we look at the second column we see an addition problem that gives the same answer as the original subtraction problem. In each case the opposite (additive inverse) of the number being subtracted is added. Let's look at three more examples.

Answer Obtained by Adding to the Subtrahend	*Answer Obtained by Adding the Opposite*	
$12 - 7 = 5$	$12 + (-7) = 5$	**-7 is the opposite of 7, the number to be subtracted.**
$-2 - 4 = -6$	$-2 + (-4) = -6$	**-4 is the opposite of 4, the number to be subtracted.**
$-5 - (-8) = 3$	$-5 + 8 = 3$	**8 is the opposite of -8, the number to be subtracted.**

This leads us to the rule for subtracting signed numbers.

To subtract signed numbers

1. Rewrite as an addition problem by adding the opposite of the number to be subtracted.
2. Find the sum.

Warm Ups A–J

Examples A–J

Directions: Subtract.

Strategy: Add the opposite of the number to be subtracted.

A. Subtract: $82 - 76$

A. Subtract: $45 - 33$

Strategy: Rewrite as an addition problem by adding -33, which is the opposite of 33.

$45 - 33 = 45 + (-33)$

$= 12$ **Add. Since the signs are different, subtract their absolute values and use the sign of the number with the larger absolute value, which is 45.**

Since both numbers are positive we can also do the subtraction in the usual manner: $45 - 33 = 12$.

B. Subtract: $-41 - 36$

B. Subtract: $-38 - 41$

$-38 - 41 = -38 + (-41)$

$= -79$ **Add. Since both numbers are negative, add their absolute values and keep the common sign.**

C. Find the difference of -71 and -41.

C. Find the difference of -49 and -24.

$-49 - (-24) = -49 + 24$

$= -25$ **Add.**

D. Subtract: $67 - 94$

D. Subtract: $88 - 107$

$88 - 107 = 88 + (-107)$

$= -19$ **Add.**

E. Find the difference: $\frac{3}{4} - \left(-\frac{5}{6}\right)$

E. Find the difference: $\frac{2}{3} - \left(-\frac{1}{2}\right)$

$\frac{2}{3} - \left(-\frac{1}{2}\right) = \frac{2}{3} + \frac{1}{2}$

$= \frac{4}{6} + \frac{3}{6}$ **Write each fraction with a common denominator and add.**

$= \frac{7}{6}$

F. Subtract: $21 - (-32) - 19$

F. Subtract: $16 - (-24) - 17$

Strategy: Change both subtractions to add the opposite.

$16 - (-24) - 17 = 16 + 24 + (-17)$

$= 40 + (-17)$ **Add 16 and 24.**

$= 23$ **Add.**

Answers to Warm Ups A. 6 B. -77 C. -30 D. -27 E. $\frac{19}{12}$ F. 34

G. Subtract: $-0.21 - (-4.2) - (-0.18) - 0.75$

Strategy: Change all subtractions to add the opposite and add.

$$-0.21 - (-4.2) - (-0.18) - 0.75 = -0.21 + 4.2 + 0.18 + (-0.75)$$
$$= 3.42$$

G. Subtract: $-0.67 - 0.76 - (-0.45) - (-1.23)$

H. Subtract: $-\frac{3}{4} - \left(\frac{7}{8}\right) - \left(-\frac{1}{2}\right) - \left(-\frac{1}{8}\right)$

$$-\frac{3}{4} - \left(\frac{7}{8}\right) - \left(-\frac{1}{2}\right) - \left(-\frac{1}{8}\right) = -\frac{3}{4} + \left(-\frac{7}{8}\right) + \frac{1}{2} + \frac{1}{8}$$
$$= -\frac{6}{8} + \left(-\frac{7}{8}\right) + \frac{4}{8} + \frac{1}{8}$$
$$= -\frac{8}{8} = -1$$

H. Subtract: $-\frac{3}{5} - \left(-\frac{3}{8}\right) - \left(-\frac{7}{10}\right) - \left(\frac{7}{20}\right)$

Calculator Example

I. Subtract: $-481.92 - (-284.7)$

Strategy: The calculator does not require you to change subtraction to add the opposite.

The difference is -197.22.

I. Subtract: $-346.98 - (-245.82)$

J. The highest point in North America is Mount McKinley, a peak in central Alaska, which is approximately 20,320 ft above sea level. The lowest point in North America is Death Valley, a deep desert basin in southeastern California, which is approximately 282 ft below sea level. What is the difference in height between Mount McKinley and Death Valley? (Above sea level is positive and below sea level is negative.)

Strategy: To find the difference in height, write each height as a signed number and subtract the lower height from the higher height.

Mount McKinley	20,320 above	20,320
Death Valley	282 below	−282

$20{,}320 - (-282) = 20{,}320 + 282$ **Change subtraction to add the opposite and add.**
$= 20{,}602$

The difference in height is approximately 20,602 ft.

J. One night last winter the temperature dropped from 18°F to −12°F. What was the difference between the high and the low temperatures?

Answers to Warm Ups G. 0.25 H. $\frac{1}{8}$ I. −101.16 J. The difference in temperatures is 30°F.

NAME ______________________________ CLASS ______________ DATE ______________

Exercises 8.3

OBJECTIVE: *Subtract signed numbers.*

A.

Subtract.

1. $6 - 4$ **2.** $8 - 3$ **3.** $-6 - 4$ **4.** $-8 - 3$

5. $-6 - (-4)$ **6.** $-8 - (-3)$ **7.** $10 - 7$ **8.** $12 - 5$

9. $-10 - 7$ **10.** $-12 - 5$ **11.** $-15 - (-3)$ **12.** $-19 - (-7)$

13. $17 - (-15)$ **14.** $21 - (-12)$ **15.** $-16 - 15$ **16.** $-18 - 11$

17. $-14 - 12$ **18.** $-22 - 19$ **19.** $-16 - (-17)$ **20.** $-27 - (-28)$

21. $-9 - (-9)$ **22.** $-5 - (-5)$ **23.** $-12 - 12$ **24.** $-25 - 25$

25. $-30 - 30$ **26.** $-29 - 29$

B.

27. $56 - (-31)$ **28.** $48 - (-43)$ **29.** $-65 - 73$ **30.** $-72 - 87$

31. $-65 - (-69)$ **32.** $-49 - (-73)$ **33.** $-89 - 89$ **34.** $-72 - 72$

35. $91 - (-91)$ **36.** $83 - (-83)$ **37.** $-74 - (-74)$ **38.** $-99 - (-99)$

39. $101 - 101$ **40.** $121 - 121$ **41.** $145 - (-32)$ **42.** $136 - (-29)$

43. $-9.56 - 8.32$ **44.** $-8.12 - 5.71$ **45.** $-7.45 - (-2.11)$

46. $-5.74 - (-3.22)$ **47.** $-11.14 - 12.89$ **48.** $-17.98 - (-9.99)$

C.

49. Find the difference between 43 and -73.

50. Find the difference between -88 and -97.

51. Subtract 328 from -349.

52. Subtract 145 from -251.

53. *Viking II* records high and low temperatures of $-22°C$ and $-107°C$ for one day on the surface of Mars. What is the change in temperature for that day?

54. The surface temperature of one of Jupiter's satellites is measured for one week. The highest temperature recorded is $-75°C$ and the lowest is $-139°C$. What is the difference in the extreme temperatures for the week?

55. At the beginning of the month, Joe's bank account had a balance of \$487.52. At the end of the month, the account was overdrawn by \$63.34 (−\$63.34). If there were no deposits during the month, what was the total amount of checks Joe wrote? (*Hint:* Subtract the ending balance from the original balance.)

56. At the beginning of the month Jack's bank account had a balance of \$295.72. At the end of the month the balance was −\$8.73. If there were no deposits, find the amount of checks Jack wrote. (Refer to Exercise 55.)

Exercises 57–59 refer to the chapter application (see pages 723 and 731).

57. The range of a set of numbers is defined as the difference between the largest and the smallest numbers in the set. Calculate the range of altitude for each continent. Which continent has the smallest range and what does this mean in physical terms?

58. What is the difference between the lowest point in the Mediterranean and the lowest point in the Atlantic?

59. Some people consider Mauna Kea, Hawaii to be the tallest mountain in the world. It rises 33,476 ft from the ocean floor, but is only 13,796 ft above sea level. What is the depth of the ocean floor at this location?

Exercises 60 to 63. A Martian probe records the following temperatures for a five-day period at one point on the surface of Mars.

	Day 1	Day 2	Day 3	Day 4	Day 5
5:00 A.M.	−92°C	−88°C	−115°C	−103°C	−74°C
9:00 A.M.	−57°C	−49°C	−86°C	−93°C	−64°C
1:00 P.M.	−52°C	−33°C	−46°C	−48°C	−10°C
6:00 P.M.	−45°C	−90°C	−102°C	−36°C	−42°C
11:00 P.M.	−107°C	−105°C	−105°C	−98°C	−90°C

60. What is the difference between the high and low temperatures recorded on day 3?

61. What is the difference between the temperature recorded at 11:00 P.M. on day 3 and day 5?

62. What is the difference between the temperatures recorded at 5:00 A.M. on day 2 and 6:00 P.M. on day 4?

63. What is the difference between the highest and lowest temperatures recorded during the five days?

64. Al's bank account had a balance of \$235. He writes a check for \$310. What is his account balance now?

65. Thomas started with \$117 in his account. He writes a check for \$188. What is his account balance now?

66. Carol started school owing her mother \$12; by school's end she borrowed \$85 more from her mother. How does her account with her mother stand now?

67. At the beginning of the month Janna's bank account had a balance of \$249.78. At the end of the month the account was overdrawn \$2.09. If there were no deposits during the month, what was the total amount of the checks Janna wrote? (*Hint:* Subtract the ending balance from the original balance.)

68. What is the difference in altitude between the highest point in the world and the lowest point in the United States?

Highest point: Mount Everest is 29,028 ft above sea level (+29,028)

Lowest point: Death Valley is 282 ft below sea level (−282)

69. Marie's bank account had a balance of \$195.84. She writes a check for \$212.69. What is her account balance now?

70. Thomas started with \$75.32 in his account. He writes a check for \$92.17. What is his account balance now?

71. On the first sale of the day Gina makes a profit of \$125.45. However, on the next sale, Gina loses \$245.36 because of another employee's misquote on the price of an item. After these two sales, what is the status of Gina's sales?

72. The temperature at 2 A.M. was 5° below zero. At 6 A.M. it was 12° below zero. What is the difference between the 6 A.M. and the 2 A.M. temperatures?

NAME ________________ CLASS ________ DATE ________

73. The New England Patriots started on their 40-yd line. After three plays they were on their 12-yd line. Did they lose or gain yards? Represent this loss or gain with a signed number.

74. During 1970 the net import of coal was -1.93 quadrillion BTU. In 1980 the net import of coal was -2.39 quadrillion BTU. In 1990 the net import of coal was -2.70 quadrillion BTU. Find the difference in coal imports for each decade and also for the 20-yr span.

STATE YOUR UNDERSTANDING

75. Explain the difference between adding and subtracting signed numbers.

76. How would you explain to a 10-yr-old how to subtract -8 from 12?

77. Explain why the order in which two numbers are subtracted is important, but the order in which they are added is not.

78. Explain the difference between the problems $5 + (-8)$ and $5 - 8$.

CHALLENGE

79. $-12 - (-4.5) - |-16|$

80. $-8.72 - [-(-5)]$

81. $-34 - (-42) - |-(-32)|$

82. $48 - |-42 - (-43)| - 64$

83. $|-12.7 - 6.2| - |-19.3 - (-8.7)|$

GROUP ACTIVITY

84. Determine the normal daily mean (average) temperature for your city for each month of the year. Find the differences from month to month. Chart this result.

MAINTAIN YOUR SKILLS (SECTIONS 2.3, 4.5, 5.7, 5.9)

85. Find the area of a circle that has a radius of 12 in. (Let $\pi \approx 3.14$.)

86. Find the area of a square that is 16 m on each side.

87. Find the area of a rectangle that is 36 cm long and 24 cm wide.

88. Find the area of a triangle that has a base of 3 m and a height of 3 m.

89. A secretary can type 90 words per minute. How many words can she type in one second?

90. Find the area of a trapezoid that has a height of 1 meter and bases of 34 cm and 42 cm.

91. Find the area of a circle that has a diameter of 13.5 km. (Let $\pi \approx 3.14$.)

92. Find the area of a triangle that has a base of 2 ft and a height of 18 in.

93. How many square yards of wall-to-wall carpeting are needed to carpet a rectangular floor that measures 36 ft by 45 ft?

94. How many square tiles, which are 9 in. on a side, are needed to cover a floor that is 27 ft by 36 ft?

8.4 Multiplying Signed Numbers

OBJECTIVES

1. Multiply a positive number and a negative number.
2. Multiply two negative numbers.

HOW AND WHY

Objective 1

Multiply a positive number and a negative number.

Consider the following multiplications:

$3(4) = 12$
$3(3) = 9$
$3(2) = 6$
$3(1) = 3$
$3(0) = 0$
$3(-1) = ?$
$3(-2) = ?$

Each product is 3 smaller than the one before it. Continuing this pattern,

$3(-1) = -3$ **3(1) = 3 and the product is negative.**

and

$3(-2) = -6$ **3(2) = 6 and the product is negative.**

The pattern indicates that the product of a positive and a negative number is negative; that is, the opposite of the product of their absolute values.

To find the product of a positive and a negative number

1. Find the product of the absolute values.
2. Make this product negative.

This is sometimes stated, "The product of two unlike signs is negative."

The commutative property of multiplication dictates that no matter the order in which the positive and negative numbers appear, their product is always negative. Thus,

$3(-4) = -12$ and $-4(3) = -12$

The product of two numbers with unlike signs is negative.

Warm Ups A–F

A. Find the product: $-11(4)$

B. Find the product: $24(-6)$

C. Find the product: $5(-1.7)$

D. Multiply: $\left(-\frac{5}{6}\right)\left(\frac{3}{4}\right)$

E. Multiply: $8(-3)(4)$

F. James wants to lose weight. His goal is to lose an average of 4.5 lb (-4.5 lb) per week. At this rate, how much will he lose in eight weeks? Express this weight loss as a signed number.

Examples A–F

Directions: Multiply.

Strategy: Multiply the absolute values and write the opposite of that product.

A. Find the product: $-6(5)$

$-6(5) = -30$ The product of a positive and a negative number is negative.

B. Find the product: $17(-8)$

$17(-8) = -136$ The product of two factors with unlike signs is negative.

C. Find the product: $4(-2.4)$

$4(-2.4) = -9.6$ The product of unlike signs is negative.

D. Multiply: $\left(-\frac{2}{3}\right)\left(\frac{3}{5}\right)$

$\left(-\frac{2}{\cancel{3}_1}\right)\left(\frac{\cancel{3}^1}{5}\right) = -\frac{2}{5}$ **Simplify. Find the product of the numerators and the denominators. The product is negative.**

E. Multiply: $7(-6)(5)$

$7(-6)(5) = (-42)5$ **Multiply the first two factors.**

$= -210$ **Multiply again.**

F. In order to attract business, the Family grocery store ran a "loss leader" sale last week. The store sold eggs at a loss of \$0.20 ($-\$0.20$) per dozen. If 248 dozen eggs were sold last weekend, what was the total loss from the sale of the eggs? Express this loss as a signed number.

$248(-0.20) = -49.60$ **To find the total loss, multiply the loss per dozen by the number of dozens sold.**

Therefore, the loss, written as a signed number, is $-\$49.60$.

HOW AND WHY

Objective 2

Multiply two negative numbers.

We use the product of a positive and a negative number to develop a pattern for multiplying two negative numbers.

$-3(4) = -12$
$-3(3) = -9$
$-3(2) = -6$
$-3(1) = -3$
$-3(0) = 0$
$-3(-1) = ?$
$-3(-2) = ?$

Answers to Warm Ups A. -44 B. -144 C. -8.5 D. $-\frac{5}{8}$ E. -96 F. James's weight loss will be -36 lb.

Each product is three larger than the one before it. Continuing this pattern,

$-3(-1) = 3$ **3(1) = 3 and the product is positive.**

$-3(-2) = 6$ **3(2) = 6 and the product is positive.**

In each case the product is positive.

To multiply two negative numbers

1. Find the product of the absolute values.

2. Make this product positive.

The product of two like signs is positive. When multiplying more than two signed numbers, if there are an even number of negative factors, the product is positive.

Examples G–M

Directions: Multiply.

Strategy: Multiply the absolute values; make the answer positive.

G. Multiply: $-6(-7)$

$-6(-7) = 42$ **The product of two negative numbers is positive.**

H. Multiply: $-3.2(-0.7)$

$-3.2(-0.7) = 2.24$ **The product of two like signs is positive.**

I. Find the product: $-121(-3)$

$-121(-3) = 363$

J. Find the product of $-\frac{2}{7}$ and $-\frac{3}{8}$.

$$\left(-\frac{\overset{1}{\cancel{2}}}{7}\right)\left(-\frac{3}{\underset{4}{\cancel{8}}}\right) = \frac{3}{28}$$

So, the product is $\frac{3}{28}$.

K. Multiply: $-6(-3)(-4)$

Strategy: There are an odd number of negative factors, therefore the product is negative.

$-6(-3)(-4) = 18(-4)$ **Multiply the first two factors.**

$= -72$ **Multiply again.**

Warm Ups G–M

G. Multiply: $-8(-9)$

H. Multiply: $-1.5(-0.8)$

I. Find the product: $-124(-4)$

J. Find the product of $-\frac{7}{12}$ and $-\frac{8}{25}$.

K. Multiply: $-8(-6)(-5)$

Answers to Warm Ups G. 72 H. 1.2 I. 496 J. $\frac{14}{75}$ K. −240

L. Find the product of -2, 11, 4, -2 and -3.

L. Find the product of -3, 12, 3, -1, and -5.

$(-3)(12)(3)(-1)(-5) = -540$

The product is -540.

M. Multiply: $-36(-1.9)(3.5)$

Calculator Example

M. Multiply: $-28(-4.6)(-2.9)$

The product is -373.52.

Answers to Warm Ups L. -528 M. 239.4

NAME ____________ CLASS ________ DATE ________

Exercises 8.4

OBJECTIVE 1: *Multiply a positive number and a negative number.*

A.

Multiply.

1. $-2(4)$ **2.** $4(-5)$ **3.** $-5(2)$ **4.** $(-6)(8)$

5. $10(-8)$ **6.** $-13(5)$ **7.** $-11(7)$ **8.** $12(-3)$

9. The product of -5 and ________ is -55.

10. The product of 8 and ________ is -72.

B.

Multiply.

11. $-9(34)$ **12.** $11(-23)$ **13.** $-17(15)$ **14.** $23(-18)$

15. $2.5(-3.6)$ **16.** $-3.4(2.7)$ **17.** $-0.35(1000)$

18. $2.57(-10000)$ **19.** $-\frac{2}{3} \cdot \frac{3}{8}$ **20.** $\left(\frac{3}{8}\right)\left(-\frac{4}{5}\right)$

OBJECTIVE 2: *Multiply two negative numbers.*

A.

Multiply.

21. $(-1)(-3)$ **22.** $(-2)(-4)$ **23.** $-7(-4)$ **24.** $-6(-5)$

25. $-11(-9)$ **26.** $(-4)(-3)$ **27.** $-12(-5)$ **28.** $-6(-16)$

29. The product of -6 and __________ is 66.

30. The product of -13 and __________ is 39.

B.

Multiply.

31. $-14(-15)$

32. $-23(-17)$

33. $-1.2(-4.5)$

34. $-0.9(-0.72)$

35. $(-5.5)(-4.4)$

36. $(-6.3)(-2.3)$

37. $\left(-\frac{3}{14}\right)\left(-\frac{7}{9}\right)$

38. $\left(-\frac{8}{15}\right)\left(-\frac{5}{24}\right)$

39. $(-0.35)(-4.7)$

40. $(-7.2)(-2.1)$

C.

Multiply.

41. $(-56)(45)$

42. $(16)(-32)$

43. $(15)(31)$

44. $(-23)(71)$

45. $(-1.4)(-5.1)$

46. $(-2.4)(6.1)$

47. $\left(-\frac{9}{16}\right)\left(\frac{8}{15}\right)$

48. $\left(-\frac{8}{21}\right)\left(-\frac{7}{16}\right)$

49. $(-4.01)(3.5)$

50. $(-6.7)(-0.45)$

51. $(-3.19)(-1.7)(0.1)$

52. $-1.3(4.6)(-0.2)$

53. $-2(4)(-1)(0)(-5)$

54. $(-4)(-7)(3)(-8)(0)$

55. $-0.07(0.3)(-10)(100)$

56. $(0.3)(-0.05)(-10)(-10)$

57. $-2(-5)(-6)(-4)(-1)$

58. $9(-2)(-3)(-5)(-4)$

59. $\left(-\frac{2}{3}\right)\left(-\frac{3}{4}\right)\left(-\frac{4}{5}\right)\left(-\frac{5}{6}\right)$

60. $\left(-\frac{5}{12}\right)\left(\frac{7}{8}\right)\left(\frac{3}{14}\right)\left(-\frac{8}{15}\right)$

61. The formula for converting a temperature measurement from Fahrenheit to Celsius is $C = \frac{5}{9}(F - 32)$. What Celsius measure is equal to $-4°F$?

62. Use the formula in Exercise 61 to find the Celsius measure that is equal to 68°F.

63. While on a diet for eight consecutive weeks Ms. Riles averages a weight loss of 3.5 lb each week. If each loss is represented by -3.5 lb, what is her total weight loss for the eight weeks, expressed as a signed number?

64. Mr. Riles goes on a diet for eight consecutive weeks. He averages a loss of 2.5 lb each week. If each loss is represented by -2.5 lb, what is his total weight loss for the eight weeks, expressed as a signed number?

65. The Dow Jones Industrial Average sustains 12 straight days of a 2.83-point decline. What is the total decline during the 12-day period, expressed as a signed number?

66. The Dow Jones Industrial Average sustains eight straight days of a loss of 1.75 points. What is the total decline in this period, expressed as a signed number?

Simplify.

67. $(15 - 8)(5 - 12)$

68. $(17 - 20)(5 - 9)$

69. $(15 - 21)(13 - 6)$

70. $(25 - 36)(5 - 9)$

71. $(-12 + 30)(-4 - 10)$

72. $(11 - 18)(-13 + 5)$

73. Safeway Inc. offers as a loss leader 10 lb of sugar at a loss of 12¢ per bag (−12¢). If 560 bags are sold during the sale, what is the total loss, expressed as a signed number?

74. Albertsons offers a loss leader of coffee at a loss of 18¢ per can (−18¢). If they sell 235 cans of coffee, find the total loss, expressed as a signed number.

75. Winn Dixie's loss leader is a soft drink that loses 8¢ per six-pack. They sell 251 of these six-packs. What is Winn Dixie's total loss expressed as a signed number?

76. Kroger's loss leader is soap powder that loses 15¢ per carton. They sell 326 cartons. What is Kroger's total loss expressed as a signed number?

77. Safeway's loss leader is 1 dozen eggs that lose 14¢ per dozen. The store sells 712 dozen eggs that week. Express Safeway's total loss as a signed number.

78. A scientist is studying the movement, within its web, of a certain spider. Any movement up is considered to be positive, while any movement down is negative. Determine the net movement of a spider that goes up 2 cm five times and down 3 cm twice.

79. A certain junk bond trader purchased 850 shares of stock at $8\frac{3}{8}$ $\left(\$8\frac{3}{8}\right)$. When she sold her shares the stock sold for $7\frac{7}{8}$ $\left(\$7\frac{7}{8}\right)$. What did she pay for the stock? How much money did she receive when she sold this stock? How much did she lose or gain? Represent the loss or gain with a signed number.

80. A company bought 300 items at \$0.89 each. They tried to sell them for \$1.19 and sold only 26. They lowered the price to \$1.06 and sold 34 more. The price was lowered a second time to \$0.89 and 125 items were sold. Finally they advertised a close-out price of \$0.84 and sold the remaining items. Determine the net profit or loss for each price. Did they make a profit or lose money on this item overall?

Exercises 81–82 refer to the chapter application (see page 723).

81. Which continent has a low point that is approximately ten times the low point of South America?

NAME ______________________ CLASS ____________ DATE ____________

82. Which continent has a high point that is approximately twice the absolute value of its lowest point?

STATE YOUR UNDERSTANDING

83. Explain the difference between -3^2 and $(-3)^2$.

84. Explain the procedure for multiplying two signed numbers.

CHALLENGE

Simplify.

85. $|-(-5)|(-9 - [-(-5)])$

86. $|-(-8)|(-8 - [-(-9)])$

87. Find the product of -8 and the opposite of 7.

88. Find the product of the opposite of 12 and the absolute value of -9.

GROUP ACTIVITY

89. Throughout the years, mathematicians have used a variety of examples to explain to students why the product of two negative numbers is positive. Talk to science and math instructors and record their favorite explanation. Discuss your findings in class.

MAINTAIN YOUR SKILLS (SECTIONS 2.3, 5.7, 5.9, 6.1)

90. What is the equivalent piecework wage (dollars per piece) if the hourly wage is $15.86 and the average number of articles completed in 1 hr is 6.1?

91. A wheel on Ellie's automobile is 16 in. in diameter. To the nearest whole number, how many revolutions will the wheel turn if the automobile travels 2 miles? Let $\pi \approx 3.14$.

92. How far does the tip of the hour hand of a clock travel in 6 hr if the length of the hand is 3 in.? Let $\pi \approx 3.14$.

93. The following diagram shows the Smiths' yard with respect to their house. How much grass seed is needed to sow the lawn if 1 lb of seed will sow 1000 ft^2? Find the weight to the nearest pound.

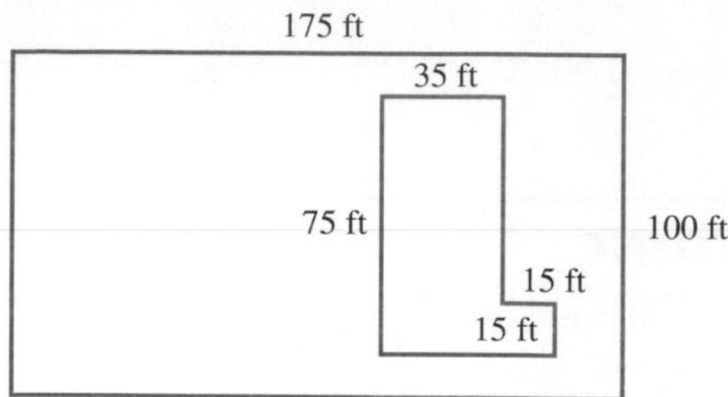

94. If carpeting costs $17.95 per square yard, what is the cost of wall-to-wall carpeting needed to cover the floor in a rectangular room that is 20 ft wide and 27 ft long?

95. How many square feet of sheet metal are needed to make a box without a top that has measurements of 5 ft 6 in. by 4 ft 6 in. by 9 in.?

96. A cylindrical tank designed to hold acidic liquids must be coated on the inside to prevent corrosion. The tank has an inside diameter of 7 ft 6 in. and a height of 4 ft. If one gallon of coating will cover 3.5 ft^2, how many gallons are needed to coat the tank? (Include the top and bottom.)

97. A mini-storage complex has one unit that is 40 ft by 80 ft and rents for $1800 per year. What is the cost of a square foot of storage for a year?

8.5
Dividing Signed Numbers

OBJECTIVES

1. Divide a positive number and a negative number.
2. Divide two negative numbers.

HOW AND WHY

Objective 1

Divide a positive and a negative number.

To divide two signed numbers, we find the number that when multiplied times the divisor equals the dividend. The expression $-8 \div 4 = ?$ asks $4(?) = -8$; we know $4(-2) = -8$ so $-8 \div 4 = -2$. The expression $18 \div (-2) = ?$ asks $-2(?) = 18$; we know $-2(-9) = 18$ so $18 \div (-2) = -9$.

When dividing unlike signs, we see that the quotient is negative. We use these examples to state how to divide a negative and a positive number.

To divide a positive and a negative number

1. Find the quotient of the absolute values.
2. Find the opposite of that quotient.

Examples A–D

Directions: Divide.

Strategy: Divide the absolute values and find the opposite of that quotient.

A. Divide: $28 \div (-4)$

$28 \div (-4) = -7$ **The quotient of two numbers with unlike signs is negative.**

B. Divide: $(-9.6) \div 1.6$

$(-9.6) \div 1.6 = -6$ **When dividing unlike signs, the quotient is negative.**

C. Divide: $\left(\frac{5}{9}\right) \div \left(-\frac{5}{3}\right)$

$\left(\frac{5}{9}\right) \div \left(-\frac{5}{3}\right) = \left(\frac{\overset{1}{\cancel{5}}}{\underset{3}{\cancel{9}}}\right)\left(-\frac{\overset{1}{\cancel{3}}}{\underset{1}{\cancel{5}}}\right)$ **Multiply by the reciprocal and simplify.**

$= -\frac{1}{3}$ **The product is negative.**

Warm Ups A–D

A. Divide: $78 \div (-6)$

B. Divide: $(-1.44) \div 1.2$

C. Divide: $\left(\frac{7}{8}\right) \div \left(-\frac{7}{6}\right)$

Answers to Warm Ups A. −13 B. −1.2 C. $-\frac{3}{4}$

D. Ms. Rich loses \$1152 in her stock market account over a period of 24 consecutive weeks. What is her average loss per week, expressed as a signed number?

D. Over a period of 24 weeks, Mr. Rich loses a total of \$8736 (−\$8736) in his stock market account. What is his average loss per week, expressed as a signed number?

Strategy: To find the average loss per week, divide the total loss by the number of weeks.

$-8736 \div 24 = -364$

Mr. Rich has an average loss of \$364 (−\$364) per week.

HOW AND WHY

Objective 2

Divide two negative numbers.

To determine how to divide two negative numbers, we again use the relationship to multiplication.

The expression $-21 \div (-3) = ?$ asks $(-3)(?) = -21$; we know that $-3(7) = -21$, so $-21 \div (-3) = 7$. The expression $-15 \div (-5) = ?$ asks $(-5)(?) = -15$; we know that $-5(3) = -15$, so $-15 \div (-5) = 3$. We see that in each case when dividing a negative number by a negative number, the quotient is positive. These examples lead us to the following rule:

To divide two negative numbers

1. Find the quotient of the absolute values.
2. Leave the answer positive.

Warm Ups E–G

Examples E–G

Directions: Divide.

Strategy: Find the quotient of the absolute values.

E. Find the quotient: $(-36) \div (-4)$

E. Find the quotient: $-24 \div (-3)$

$-24 \div (-3) = 8$ **The quotient of two negative numbers is positive.**

Calculator Example

F. Find the quotient: $(-18.6) \div (-0.6)$

F. Find the quotient: $-12.5 \div (-0.5)$

$-12.5 \div (-0.5) = 25$

G. Divide: $\left(-\frac{5}{6}\right)$ by $\left(-\frac{3}{4}\right)$

G. Divide: $\left(-\frac{3}{4}\right)$ by $\left(-\frac{1}{2}\right)$

$$\left(-\frac{3}{4}\right) \div \left(-\frac{1}{2}\right) = \left(-\frac{3}{\underset{2}{\cancel{4}}}\right)\left(-\frac{\overset{1}{\cancel{2}}}{1}\right) = \frac{3}{2}$$ **Invert and multiply. Divide out like factors.**

Answers to Warm Ups D. Ms. Rich lost \$48 (−\$48) per week. E. 9 F. 31 G. $\frac{10}{9}$

Exercises 8.5

OBJECTIVE 1: *Divide a positive number by a negative number.*

A.

Divide.

1. $-10 \div 5$ **2.** $10 \div (-2)$ **3.** $-16 \div 4$ **4.** $15 \div (-3)$

5. $18 \div (-6)$ **6.** $-18 \div 3$ **7.** $24 \div (-3)$ **8.** $-33 \div 11$

9. The quotient of -48 and __________ is -6.

10. The quotient of 70 and __________ is -14.

B.

11. $72 \div (-12)$ **12.** $84 \div (-12)$ **13.** $6.06 \div (-3)$

14. $3.05 \div (-5)$ **15.** $-210 \div 6$ **16.** $-315 \div 9$

17. $\left(-\frac{6}{7}\right) \div \frac{2}{7}$ **18.** $\left(-\frac{4}{3}\right) \div \frac{8}{3}$

19. $0.75 \div (-0.625)$ **20.** $0.125 \div (-0.625)$

OBJECTIVE 2: *Divide two negative numbers.*

A.

Divide.

21. $-10 \div (-5)$ **22.** $-10 \div (-2)$ **23.** $-12 \div (-4)$ **24.** $-14 \div (-2)$

25. $-28 \div (-4)$ **26.** $-32 \div (-4)$ **27.** $-54 \div (-9)$ **28.** $-63 \div (-7)$

29. The quotient of -105 and __________ is 21.

30. The quotient of -75 and __________ is 15.

B.

31. $-98 \div (-14)$ **32.** $-88 \div (-11)$ **33.** $-96 \div (-12)$

34. $-210 \div (-10)$ **35.** $-12.12 \div (-3)$ **36.** $-18.16 \div (-4)$

37. $\left(-\frac{3}{8}\right) \div \left(-\frac{3}{4}\right)$ **38.** $\left(-\frac{1}{2}\right) \div \left(-\frac{5}{8}\right)$

39. $-0.65 \div (-0.13)$ **40.** $-0.056 \div (-0.4)$

C.

Divide.

41. $-540 \div 12$ **42.** $-1071 \div 17$ **43.** $-3364 \div (-29)$

44. $-4872 \div (-48)$ **45.** $0.75 \div (-0.625)$ **46.** $-0.125 \div (-0.625)$

47. $0 \div (-35)$ **48.** $-85 \div 0$ **49.** $-0.26 \div 100$

50. $-0.56 \div (-100)$ **51.** $\frac{-16{,}272}{36}$ **52.** $\frac{-34{,}083}{-63}$

53. Find the quotient of -384 and -24.

54. Find the quotient of -357 and 21.

55. The membership of the Burlap Baggers Investment Club takes a loss of \$284.22 (-284.22) on the sale of stock. If there are six co-equal members in the club, what is each member's share of the loss, expressed as a signed number?

56. The temperature in Fairbanks, Alaska, drops from 10° above zero ($+10°$) to 22° below zero ($-22°$) in an eight-hour period. What is the average drop in temperature per hour, expressed as a signed number?

57. Mr. Harkness loses a total of 108 lb in 24 weeks. Express the average weekly loss as a signed number.

58. Ms. Harkness loses a total of 60 lb in 24 weeks. Express the average weekly loss as a signed number.

59. A certain stock loses $31\frac{1}{2}$ points in 12 days. Express the average daily loss as a signed number.

60. A certain stock loses $30\frac{3}{8}$ points in 9 days. Express the average daily loss as a signed number.

61. Determine the population of Los Angeles in 1970, in 1980, and in 1990. Determine the population of Detroit in 1970, in 1980, and in 1990. Find the average yearly loss or gain for each decade and also for the 20-year period for each city (written as a signed number). List the possible reasons for these changes.

62. A certain company loses $862,200 during one 20-month period. Determine the average monthly loss (written as a signed number). If there are 30 stockholders in this company, determine the total loss per stockholder (written as a signed number).

Exercises 63–64 refer to the chapter application (see page 723).

63. Which continent has a high point that is approximately one-fourth the height of Mount Everest?

64. Which continent has a low point that is approximately one-fourth the low point of Africa?

STATE YOUR UNDERSTANDING

65. The sign rules for multiplication and division of signed numbers may be summarized as follows:

If the numbers have the same sign, the answer is positive.
If the numbers have different signs, the answer is negative.

Explain why the rules for division are the same as the rules for multiplication.

66. When dividing signed numbers, care must be taken not to divide by zero. Why?

CHALLENGE

Simplify.

67. $[-|-9|(8-12)] \div [(9-13)(8-7)]$

68. $[(14-20)(-5-9)] \div [-(-12)(-8+7)]$

69. $\left(-\frac{5}{6}-\frac{1}{2}\right)\left(-\frac{2}{3}+\frac{1}{6}\right) \div \left(\frac{1}{3}-\frac{3}{4}\right)$

70. $\left(-\frac{1}{3}-\frac{1}{4}\right)\left(-\frac{1}{3}+\frac{1}{6}\right) \div \left(\frac{1}{3}-\frac{3}{4}\right)$

71. $(-0.82-1.28)(1.84-2.12) \div [3.14+(-3.56)]$

GROUP ACTIVITY

72. Determine the temperature on the first of each month over a 12-month period in one city in Alaska, one city in Canada, and one city in Hawaii. Find the average of the monthly changes in temperature for each city. Find the averages for the temperatures on the first of the month for each city. Make a chart or graph from these data. Be sure no two groups choose the same cities. Compare the results with your classmates.

MAINTAIN YOUR SKILLS (SECTIONS 2.3, 2.4, 5.7, 5.9, 7.8)

73. Find the volume of a cylinder that has a radius of 8 in. and a height of 24 in. (Let $\pi \approx 3.14$.)

74. Find the volume of a sphere with a diameter of 18 cm. (Let $\pi \approx 3.14$.)

75. Find the volume of a cone that has a radius of 12 in. and a height of 9 in. (Let $\pi \approx 3.14$.)

76. Find the volume of a pyramid with a square base of 12 cm on each side and a height of 12 cm, to the nearest cubic centimeter. Use the formula $V = \frac{1}{3}Bh$.

77. An underground gasoline storage tank is a cylinder that is 72 in. in diameter and 18 ft long. If there are 231 in^3 in a gallon, how many gallons of gasoline will the tank hold? Round the answer to the nearest gallon. (Let $\pi \approx 3.14$.)

78. A swimming pool is to be dug and the dirt hauled away. The pool is to be 27 ft long, 16 ft wide, and 6 ft deep. How many cubic *yards* of dirt must be removed?

79. To remove the dirt for the swimming pool in Exercise 78, trucks that can haul 8 yd^3 per load are used. How many truckloads will there be?

80. A real estate broker sells a lot that measures 88.75 ft by 180 ft. The sale price is $2 per square foot. If the broker's commission is 8%, how much does she make?

8.6

Order of Operations: A Review

OBJECTIVE

Do any combination of operations with signed numbers.

HOW AND WHY

Objective

Do any combination of operations with signed numbers.

The order of operations for signed numbers is the same as that for whole numbers, fractions, and decimals.

To evaluate expressions with more than one operation

Step 1. Parentheses—Do the operations within grouping symbols first (parentheses, fraction bar, etc.), in the order given in steps 2, 3, and 4.

Step 2. Exponents—Do the operations indicated by exponents.

Step 3. Multiply and **divide**—Do only multiplication and division as they appear from left to right.

Step 4. Add and **subtract**—Do addition and subtraction as they appear from left to right.

Examples A–G

Directions: Perform the indicated operations.

Strategy: Follow the order of operations.

A. Perform the indicated operations: $-64 + (-22) \div 2$

$-64 + (-22) \div 2 = -64 + (-11)$

$= -75$

B. Perform the indicated operations: $(-12)(5) - 54 \div (-3)$

$(-12)(5) - 54 \div (-3) = -60 - (-18)$

$= -60 + 18$ **Add the opposite of −18.**

$= -42$

C. Perform the indicated operations: $4 \div (-0.8) + 2(-2.4)$

$4 \div (-0.8) + 2(-2.4) = -5 + (-4.8)$

$= -9.8$

Warm Ups A–G

A. Perform the indicated operations: $-36 + (-48) \div 6$

B. Perform the indicated operations: $(-9)(4) - 24 \div (-6)$

C. Perform the indicated operations: $6 \div (-0.5) + 3(-1.7)$

Answers to Warm-Ups A. −44 B. −32 C. −17.1

D. Perform the indicated operations:
$9 - \left(\frac{2}{3}\right)(-6)$

D. Perform the indicated operations: $8 - \left(\frac{3}{4}\right)(-4)$

$$8 - \left(\frac{3}{4}\right)(-4) = 8 - (-3)$$
$$= 8 + 3 \qquad \textbf{Add the opposite of } -3.$$
$$= 11$$

E. Perform the indicated operations:
$(-5)(-3)^2 + 32 - (-4)^2$

E. Perform the indicated operations: $2(-4)^2 - 5^2 + 3(-2)^2$

$$2(-4)^2 - 5^2 + 3(-2)^2 = 2(16) - 25 + 3(4)$$
$$= 32 - 25 + 12$$
$$= 19$$

Calculator Example

F. Perform the indicated operations:
$(-18)(11) - 75 \div (-3)$

F. Perform the indicated operations:

$(-13)(12) - (-42) \div (-7)$

The result is -162.

G. How many degrees Celsius is $-40°$F?

G. Hilda keeps the thermostat on her furnace set at 68°F. Her pen pal in Germany says that her thermostat is set at 20°C. They wonder whether the two temperatures are equal. Hilda finds the formula, given below, for changing degrees Celsius to degrees Fahrenheit. Use this formula to find out whether 68°F is equal to 20°C.

$$C = \frac{5}{9}(F - 32)$$

Strategy: To find out whether 68°F = 20°C, substitute 68 for F in the formula.

$$C = \frac{5}{9}(F - 32)$$

$$C = \frac{5}{9}(68 - 32)$$

$$C = \frac{5}{9}(36)$$

$$C = 20$$

Therefore, 68°F equals 20°C.

Answers to Warm-Ups D. 13 E. -29 F. -173 G. The Celsius temperature is $-40°$C.

Exercises 8.6

OBJECTIVE: *Do any combination of operations with signed numbers.*

A.

Perform the indicated operations.

1. $2(-6) - 8$

2. $12 + 3(-4)$

3. $(-3)(-2) + 14$

4. $12 + (-3)(-6)$

5. $2(-8) + 10$

6. $(-3)4 + 9$

7. $-7 + 2(-3)$

8. $-12 + (-3)4$

9. $(-2)10 \div (5)$

10. $(-7)6 \div 3$

11. $(-4)8 \div (-4)$

12. $(-9)12 \div (-6)$

13. $(-6) \div 4(2)$

14. $(-18) \div 3(2)$

15. $3^2 + 2^2$

16. $5^2 - 3^2$

17. $(10 - 2) + (8 - 5)$

18. $(9 - 3) + (12 - 4)$

19. $(4 - 7)(8 - 11)$

20. $(6 - 4)(9 - 14)$

21. $(-2)^2 + 4(2)$

22. $(-3)^2 + 3(5)$

23. $-3 + (3 - 5) - 4(2)$

24. $-5 + (4 - 8) - 3(4)$

B.

25. $(-12)(-3) + (-15)2$

26. $(-15)(-4) + (-12)4$

27. $(9 - 7)(-2 - 5) + (15 - 9)(2 + 7)$

28. $(10 - 15)(-4 - 3) + (12 - 7)(3 + 2)$

29. $6(-10 + 4) - 33 \div (-11)$

30. $7(-9 + 4) - 45 \div (-9)$

31. $16(-2) \div (-4) - 12$

32. $(-3)(-8) \div (-6) + 10$

33. $-120 \div (-20) - (9 - 11)$

34. $-135 \div (-15) - (12 - 17)$

35. $-2^3 - (-2)^3$

36. $-4^3 - (-4)^3$

37. $-35 \div (-5)7 - 7^2$

38. $-28 \div (-4)7 - 7^2$

39. $2^2(5 - 4)(7 - 3)^2$

40. $3^2(8 - 6)(6 - 8)^2$

41. $(8 - 12) + (-4)(-2) - (-3)4 - 2^2$

42. $(9 - 14) - (7)(-2) + (-5)(2) - 3^2$

43. $(-2)(-3)(-4) - (-4)(3) - (-2)(-5)$

44. $(-4)(-5)(-1) - (-2)(5) - (-3)(2)$

45. $(-1)(-6)^2(-1) - (-2)^2(-3)^2$

46. $(-1)(-2)^2(-9) - (-3)^2(-2)^2$

C.

47. Find the sum of the product of 12 and -4 and the product of -3 and -12.

NAME ______________________ CLASS ____________ DATE ____________

48. Find the difference of the product of 3 and 9 and the product of −8 and 3.

Exercises 49–52. A satellite records the following temperatures for a five day period on one point on the surface of Mars.

	Day 1	Day 2	Day 3	Day 4	Day 5
5:00 A.M.	−92°C	−88°C	−115°C	−103°C	−74°C
9:00 A.M.	−57°C	−49°C	−86°C	−93°C	−64°C
1:00 P.M.	−52°C	−33°C	−46°C	−48°C	−10°C
6:00 P.M.	−45°C	−90°C	102°C	36°C	−42°C
11:00 P.M.	−107°C	−105°C	−105°C	−98°C	−90°C

49. What is the average temperature recorded during day 5?

50. What is the average temperature recorded at 6:00 P.M.?

51. What was the average high temperature recorded for the five days?

52. What was the average low temperature recorded for the five days?

Simplify.

53. $[-3 + (-6)]^2 - [-8 - 2(-3)]^2$

54. $[-5(-9) - (-6)^2]^2 + [(-8)(-1)^3 + 2]^2$

55. $[46 - 3(-4)^2]^3 - [-7(1)^3 + (-5)(-8)]$

56. $[30 - (-5)^2]^2 - [-8(-2) - (-2)(-4)]^2$

57. $-15 - \dfrac{8^2 - (-4)}{3^2 + 3}$

58. $-22 + \dfrac{9^2 - 6}{6^2 - 11}$

59. $\frac{12(8-24)}{5^2-3^2} \div (-12)$

60. $\frac{15(12-45)}{6^2-5^2} \div (-9)$

61. $-8|125-321| - 21^2 + 8(-7)$

62. $-9|482-632| - 17^2 + 9(-9)$

63. $-6(8^2-9^2)^2 - (-7)20$

64. $-5(6^2-7^2)^2 - (-8)19$

65. Find the difference of the quotient of 28 and −7 and the product of −4 and −3.

66. Find the sum of the product of −3 and 7 and the quotient of −15 and −5.

67. Keshia buys a TV for $95 down and $47 per month for 15 months. What is the total price she pays for the TV? (*Hint:* When a deferred payment plan is used, the total cost of the article is the down payment plus the total of the monthly payments.)

68. The E-Z Chair Company advertises recliners for $40 down and $17 per month for 24 months. What is the total cost of a recliner?

69. During a "blue light" special K-Mart sold 24 fishing poles at a loss of $3 per pole. During the remainder of the day they sold 9 poles at a profit of $7 per pole. Express the profit or loss on the sale of fishing poles as a signed number.

70. Fly America sells 40 seats on Flight 402 at a loss of \$52 per seat (−\$52). Fly America also sells 67 seats at a profit of \$78 per seat. Express the profit or loss on the sale of the seats as a signed number.

Exercises 71–74 refer to the chapter application (see page 723).

71. For each continent calculate the average of the highest and lowest point. Which continent has the largest average and which has the smallest average?

The following table gives the altitudes of selected cities around the world.

City	Altitude (ft)	City	Altitude (ft)
Athens, Greece	300	Mexico City, Mexico	7347
Bangkok, Thailand	0	New Delhi, India	770
Berlin, Germany	110	Quito, Ecuador	9222
Bogota, Columbia	8660	Rome, Italy	95
Jakarta, Indonesia	26	Tehran, Iran	5937
Jerusalem, Israel	2500	Tokyo, Japan	30

72. Find five cities with an average altitude of less than 100 ft.

73. Find three cities with an average altitude of approximately 350 ft.

74. Find four cities with an average altitude of approximately 7000 ft.

75. The treasurer of a local club records the following transactions during one month:

Opening balance	\$4756
Deposit	\$345
Check #34	\$212
Check #35	\$1218
New check cost	\$15
Deposit	\$98
National dues paid	\$450
Electric bill	\$78

What is the balance at the end of the month?

76. The Chicago Bears made the following plays during a quarter of a game:

3 plays lost 8 yd each
8 plays lost 5 yd each
1 quarterback sack lost 23 yd
1 pass for 85 yd
5 plays gained 3 yd each
2 plays gained 12 yd each
1 fumble lost 7 yd
2 passes for 10 yd each

Determine the average movement per play during this quarter. Round to the nearest tenth.

77. Consider traveling north and east as positive values and traveling south and west as negative values. A certain trip requires 81 miles north followed by 67 miles west. The next day the trip requires 213 miles south followed by 107 miles west. The last day of the trip takes 210 miles north and 83 miles east. What is the net result of this trip? Determine your position at the end of your trip in relation to your starting point.

78. Try this game on your friends. Have them pick a number. Tell them to double it, then add 20 to that number, divide the sum by 4, subtract 5 from that quotient, square the difference, and multiply the square by 4. They should now have the square of the original number. Write a mathematical representation of this riddle.

STATE YOUR UNDERSTANDING

Locate the error in Exercises 79 and 80. Indicate why each is not correct. Determine the correct answer.

79.
$$\begin{aligned} 2[3 + 5(-4)] &= 2[8(-4)] \\ &= 2[-32] \\ &= -64 \end{aligned}$$

80.
$$\begin{aligned} 3 - [5 - 2(6 - 4^2)^3] &= 3 - [5 - 2(6 - 16)^3] \\ &= 3 - [5 - 2(-10)^3] \\ &= 3 - [5 - (-20)^3] \\ &= 3 - [5 - (-8000)] \\ &= 3 - [5 + 8000] \\ &= 3 - 8005 \\ &= -8002 \end{aligned}$$

81. Is there ever a case in which exponents are not computed first? If so, give an example.

CHALLENGE

Simplify.

82. $\dfrac{3^2 - 5(-2)^2 + 8 + [4 - 3(-3)]}{4 - 3(-2)^3 - 18}$

83. $\dfrac{(5 - 9)^2 + (-6 + 8)^2 - (14 - 6)^2}{[3 - 4(7) + 3^3]^2}$

84. $\dfrac{3(4 - 7)^2 + 2(5 - 8)^3 - 18}{(6 - 9)^2 + 6}$

GROUP ACTIVITY

85. Engage the entire class in a game of KRYPTO. This card game consists of 41 cards numbered from −20 to 20. Each group gets four cards. A card is chosen at random from the remaining cards and the number is put on the board. Each group must find a way using addition, subtraction, multiplication, and/or division to combine their given cards to equal the number on the board. Operations may be used more than once.

MAINTAIN YOUR SKILLS (SECTIONS 8.2, 8.3, 8.4, 8.5)

Add.

86. $(-17.2) + (-18.6) + (-2.7) + 9.1$

87. $(28.31) + (-8.14) + (-21.26) + (-16)$

Subtract.

88. $48 - (-136)$

89. $-62.7 - (-78.8)$

Multiply.

90. $(-36)(84)(-21)$

91. $(-62)(-22)(-30)$

Divide.

92. $(-800) \div (-32)$

93. $(-25.781) \div (3.5)$

94. The four Zapple brothers form a company. The first year the company loses $5832 (−5832). The brothers share equally in the loss. Represent each brother's loss as a signed number.

95. The WOW-Smith stock average records the following gains and losses for the week:

Monday	loss 2.25
Tuesday	loss 3.125
Wednesday	gain 4.5
Thursday	gain 2
Friday	loss 1.125

Use signed numbers to find out whether the stock average gains or loses for the week.

8.7

Solving Equations

OBJECTIVE

Solve equations of the form $ax + b = c$ or $ax - b = c$, where a, b, and c, are signed numbers.

VOCABULARY

Recall that the **coefficient** of the variable is the number that is multiplied times the variable.

HOW AND WHY

Objective

Solve equations of the form $ax + b = c$ or $ax - b = c$, where a, b, and c, are signed numbers.

The solutions of equations that are of the form $ax + b = c$ and $ax - b = c$, using signed numbers, involve two operations to isolate the variable. To isolate the variable is to get an equation in which the variable is the only symbol on a particular side of the equation.

 ***To find the solution of an equation of the form, $ax + b = c$ or $ax - b = c$***

1. Add (subtract) the constant to (from) each side of the equation.
2. Divide both sides by the coefficient of the variable.

Examples A–D

Directions: Solve.

Strategy: First, add or subtract the constant to or from both sides of the equation. Second, divide both sides of the equation by the coefficient of the variable.

A. Solve: $-5x + 17 = -8$

$$-5x + 17 = -8 \quad \textbf{Original equation.}$$
$$-5x + 17 - 17 = -8 - 17$$
$$-5x = -25 \quad \textbf{Subtract.}$$
$$\frac{-5x}{-5} = \frac{-25}{-5} \quad \textbf{Divide.}$$
$$x = 5$$

CHECK: Substitute 5 for x in the original equation.

$$-5(5) + 17 = -8$$
$$-25 + 17 = -8$$
$$-8 = -8$$

The solution is $x = 5$.

Warm Ups A–D

A. Solve: $-6x + 18 = -12$

Answer to Warm Up A. $x = 5$

B. Solve: $9 = -3x - 27$

B. Solve: $-18 = -13x - 44$

$$-18 = -13x - 44 \quad \textbf{Original equation.}$$
$$-18 + 44 = -13x - 44 + 44$$
$$26 = -13x \quad \textbf{Add.}$$
$$\frac{26}{-13} = \frac{-13x}{-13} \quad \textbf{Divide.}$$
$$-2 = x$$

CHECK: Substitute -2 in the original equation.

$$-18 = -13(-2) - 44$$
$$-18 = 26 - 44$$
$$-18 = -18$$

The solution is $x = -2$.

C. Solve: $4x - 5 = 11$

C. Solve: $2x - 4 = 32$

Strategy: An alternate format is used to add 4 to both sides of the equation.

$$2x - 4 = 32 \quad \textbf{Original equation.}$$
$$\underline{+4 \;\; +4} \quad \textbf{Add 4 to both sides.}$$
$$2x \qquad = 36$$
$$\frac{2x}{2} = \frac{36}{2} \quad \textbf{Divide both sides by 2.}$$
$$x = 18 \quad \textbf{The check is left for the student.}$$

The solution is $x = 18$.

D. Using the formula $s = v + gt$, find the initial velocity (v) in feet per second of a sky diver if after a time (t) of 5 sec she reaches a speed (s) of 180 feet per second and $g = 32$ ft per second squared.

D. Using the formula $s = v + gt$, find the initial velocity (v) in feet per second of a sky diver if after a time (t) of 3.5 sec she reaches a speed (s) of 125 ft per second and $g = 32$ feet per second squared.

Strategy: Substitute into the formula and solve the resulting equation.

Formula:
$s = v + gt$

Substitute:

$$\frac{125 \text{ ft}}{1 \text{ sec}} = v + \left(\frac{32 \text{ ft}}{\text{sec}^2}\right)(3.5 \text{ sec})$$
$$\frac{125 \text{ ft}}{1 \text{ sec}} = v + \frac{112 \text{ ft}}{1 \text{ sec}} \quad \textbf{One factor of seconds divides out.}$$
$$\frac{125 \text{ ft}}{1 \text{ sec}} - \frac{112 \text{ ft}}{1 \text{ sec}} = v + \frac{112 \text{ ft}}{1 \text{ sec}} - \frac{112 \text{ ft}}{1 \text{ sec}} \quad \textbf{Subtract 112 ft per second from both sides.}$$
$$\frac{13 \text{ ft}}{1 \text{ sec}} = v \quad \textbf{The check is left for the student.}$$

The sky diver's initial velocity was 13 ft per second.

Answers to Warm Ups B. $x = -12$ C. $x = 4$ D. The initial velocity is 20 ft per second.

Exercises 8.7

OBJECTIVE: *Solve equations of the form $ax + b = c$ or $ax - b = c$, where a, b, and c are signed numbers.*

A.

Solve.

1. $-3x + 25 = 4$

2. $-4y + 11 = -9$

3. $-6 + 3x = 9$

4. $-11 + 5y = 14$

5. $4y - 9 = -29$

6. $3x - 13 = -43$

7. $2a - 11 = 3$

8. $5a + 17 = 17$

9. $-5x + 12 = -23$

10. $-11y - 32 = -65$

11. $4x - 12 = 28$

12. $9y - 14 = 4$

B.

13. $-14 = 2x - 8$

14. $26 = 3x - 4$

15. $-40 = 5x - 10$

16. $-30 = -5x - 10$

17. $-6 = -8x - 6$

18. $8 = -5x + 8$

19. $-10 = -4x + 2$

20. $20 = -8x + 4$

21. $-14y - 1 = -99$

22. $-16x + 5 = -27$

23. $-3 = -8a - 3$

24. $-12 = 5b - 12$

C.

25. $-0.6x - 0.15 = 0.15$

26. $-1.05y + 5.08 = 1.72$

27. $0.03x + 2.3 = 1.55$

28. $0.02x - 2.4 = 1.22$

29. $-135x - 674 = 1486$

30. $94y + 307 = -257$

31. $-102y + 6 = 414$

32. $-63c + 22 = 400$

33. Find the time (t) it takes a free-falling skydiver to reach a falling speed (s) of 147 ft per second (147 fps) if $v = 19$ fps and $g = 32$ ft per second squared. The formula is $s = v + gt$.

34. Find the time it takes a free-falling skydiver to reach a falling speed (s) of 211 ft per second (211 fps) if $v = 19$ fps and $g = 32$ ft per second squared. The formula is $s = v + gt$.

35. If 98 is added to six times some number, the sum is 266. What is the number?

36. If 73 is added to eleven times a number, the sum is -158. What is the number?

37. The difference of 15 times a number and 181 is -61. What is the number?

38. The difference of 24 times a number and 32 is -248. What is the number?

39. A formula for distance traveled is $2d = t^2a + 2v$, where d represents distance, v represents initial velocity, t represents time, and a represents acceleration. Find a if $d = 244$, $v = -20$, and $t = 4$.

40. Use the formula in Exercise 39 to find a if $d = 240$, $v = -35$, and $t = 5$.

41. The formula for the balance of a loan (D) is $D = B - NP$, where P represents the monthly payment, N represents the number of payments, and B represents the money borrowed including interest. Find N when $D = \$575$, $B = \$925$, and $P = \$25$.

42. Use the formula in Exercise 41 to determine the monthly payment (P) if $D = \$820$, $B = \$1020$, and $N = 5$.

Exercises 43–45 refer to the chapter application (see page 723 and 731). Use negative numbers to represent feet below sea level.

43. The high point of Australia is 12,558 ft more than 4 times the lowest point of one of the continents. Write an algebraic equation that describes this relationship. Which continent's lowest point fits the description?

44. The lowest point of Antarctica is 2183 ft less than 12 times the lowest point of one of the continents. Write an algebraic equation that describes this relationship. Which continent's lowest point fits this description?

45. The Mariana Trench in the Pacific Ocean is about 2000 ft deeper than twice one of the other oceans' deepest parts. Write an algebraic equation that describes this relationship. Which ocean's deepest part fits this description?

Solve.

46. $5x + 12 + (-9) = 18$

47. $8z - 12 + (-6) = 38$

48. $-3b - 12 + (-4) = 11 + (-6)$

49. $-5z - 15 + 6 = -21 - 18$

NAME ______________________ CLASS ____________ DATE ____________

STATE YOUR UNDERSTANDING

50. Explain what it means to solve an equation.

51. Explain how to solve the equation $-3x + 10 = 4$.

CHALLENGE

Solve.

52. $8x - 9 = 3x + 6$

53. $7x + 14 = 3x - 2$

54. $9x + 16 = 7x - 12$

55. $10x + 16 = 5x + 6$

GROUP ACTIVITY

56. Bring in your last two electricity bills. Develop a formula for determining how your bills are computed. Share your group's synopsis with the class. Try to predict your next month's bill.

NAME ______________________ CLASS ______________ DATE ______________

CHAPTER 8

Group Project *(4 weeks)*

OPTIONAL

For three consecutive weeks, on Monday, locate the final scores for each of the three major professional golf tours in the United States, the Professional Golf Association, PGA; the Ladies Professional Golf Association, LPGA; and the Senior Professional Golf Association, Senior PGA. These scores can usually be found on the summary page in the sports section of the daily newspaper.

a. Record the scores, against par, for the 30 top finishers and ties on each tour. Display the data using bar graphs for week one, line graphs for week two, and pictorial graphs for week three. Which type of graph best displays the data? Why?

b. Calculate the average score, against par, for each tour for each week. When finding the average if there is a remainder and if it is half of or more than the divisor, round up, otherwise round down. Now average the average scores for each tour. Which tour scored the best? Why?

c. What is the difference between the best and worst scores on each tour for each week?

d. What is the average amount of money earned by the players whose scores were recorded on each tour for each week? Which tour pays the best?

e. How much did the winner on each tour earn per stroke under par in the second week of your data? Compare the results. Is this a good way to compare the earnings on the tour? If not, why not?

NAME ______________________ CLASS ____________ DATE ____________

CHAPTER 8

True–False Concept Review

Check your understanding of the language of basic mathematics. Tell whether each of the following statements is True (always true) or False (not always true). For those statements that you judge to be false, revise them to make them true.

ANSWERS

1. Negative numbers are found to the left of zero on the number line. **1.** ____________

2. The opposite of a signed number is always negative. **2.** ____________

3. The absolute value of a number is always positive. **3.** ____________

4. The opposite of a signed number is the same distance from zero as the number on the number line but in the opposite direction. **4.** ____________

5. The sum of two signed numbers is always positive or negative. **5.** ____________

6. The sum of a positive signed number and a negative signed number is always positive. **6.** ____________

7. To find the sum of a positive signed number and a negative signed number, subtract their absolute values and use the sign of the number with the larger absolute value. **7.** ____________

8. To subtract two signed numbers, add their absolute values. **8.** ____________

9. If a negative number is subtracted from a positive number, the difference is always positive. **9.** ____________

10. The product of two negative numbers is never negative. **10.** ____________

11. The sign of the product of a positive number and a negative number depends on which number has the larger absolute value. **11.** ____________

12. The sign of the quotient, when dividing two signed numbers, is the same as the sign obtained when multiplying the two numbers. **12.** ____________

13. The order of operations for signed numbers is the same as the order of operations for positive numbers. **13.** ______

14. Subtracting a number from both sides of an equation results in an equation that has the same solution as the original equation. **14.** ______

NAME ______________________ CLASS ____________ DATE ____________

CHAPTER 8

Test

Perform the indicated operation.

1. $-21 + (-17) + 42 + (-18)$ — **1.** ______

2. $(36 - 42)(-18 + 6)$ — **2.** ______

3. $\left(-\frac{3}{8}\right) \div \left(\frac{3}{10}\right)$ — **3.** ______

4. $\left(-\frac{7}{15}\right) - \left(-\frac{3}{5}\right)$ — **4.** ______

5. $(-8 - 3) - (7 - 21) + (-4)$ — **5.** ______

6. $-3.65 + 4.72$ — **6.** ______

7. a. $-(-21)$
b. $|-21|$ — **7.** ______

8. $-52 - (-17)$ — **8.** ______

9. $(-16 + 4) \div 3 \cdot 5 - (-6)(-3)(-1)$ — **9.** ______

10. $-88 \div (-22)$ — **10.** ______

11. $(-7)(-9)(2)$ — **11** ______

12. $(|-6|)(-4)(-1)(-1)$ — **12.** ______

13. $-57.9 - 32.5$ — **13.** ______

14. $(-2)^2(-2)^2 + 4^2 \div (2)(3)$ — **14.** ______

15. $30.66 \div (-0.6)$ — **15.** ______

16. $\left(-\frac{1}{3}\right)+\frac{5}{6}+\left(-\frac{1}{2}\right)+\left(-\frac{1}{6}\right)$ 16. ______

17. $-96 \div (-12)$ 17. ______

18. $-37 - 6$ 18. ______

19. $(-7)(3 - 11)(-2) - 4(-3 - 5)$ 19. ______

20. $\left(-\frac{1}{3}\right)\left(\frac{6}{7}\right)$ 20. ______

21. $-26 + (-27)$ 21. ______

22. $5(-8) + 44$ 22. ______

23. $-13 = 4x + 7$ 23. ______

24. $6x - 25 = 17$ 24. ______

25. $16 - 4a = 64$ 25. ______

26. Ms. Rosier loses an average of 1.03 lb per week (−1.03 lb) during her 16-wk diet. Express Ms. Rosier's total weight loss during the 16 wk as a signed number. 26. ______

27. The temperature in Chicago ranges from a high of 12°F to a low of −9°F within a 24-hr period. What is the drop in temperature, expressed as a signed number? 27. ______

28. A stock on the New York Stock Exchange opens at $6\frac{5}{8}$ on Monday. It records the following changes during the week: Monday, $+\frac{1}{8}$; Tuesday, $-\frac{3}{8}$; Wednesday, $+1\frac{1}{4}$; Thursday, $-\frac{7}{8}$; Friday $+\frac{1}{4}$. What is its closing price on Friday? 28. ______

29. What Fahrenheit temperature is equal to a reading of −10°C? Use the formula $F = \frac{9}{5}C + 32$. 29. ______

30. Find the average of −8, −11, 6, −14, 9, and −12. 30. ______

Appendix A

Calculators

The wide availability and economical price of current hand-held calculators make them ideal for doing time-consuming arithmetic operations. Even people who are very good at math use calculators under certain circumstances (for instance, when balancing their checkbooks). You are encouraged to use a calculator as you work through this text. Learning the proper and appropriate use of a calculator is a vital skill for today's math students.

As with all new skills, your instructor will give you guidance as to where and when to use it. Calculators are especially useful in the following instances.

1. For doing the fundamental operations of arithmetic (addition, subtraction, multiplication, and division);
2. For finding powers or square roots of numbers;
3. For evaluating complicated arithmetic expressions; and
4. For checking solutions to equations.

There are several different kinds of calculators available.

- A basic 4-function calculator will add, subtract, multiply and divide. Sometimes these calculators also have a square root key. These calculators are not powerful enough to do all of the math in this text, and they are not recommended for math students at this level.
- A scientific calculator generally has about eight rows of keys on it, and is usually labeled "scientific." Look for keys labeled "sin," "tan," and "log." Scientific calculators also have power keys and parentheses keys, and the order of operations is built into them. These calculators are recommended for math students at this level.
- A graphing calculator also has about eight rows of keys, but it has a large, nearly square display screen. These calculators are very powerful, and you may be required to purchase them in later math courses. However, you will not need all that power to be successful in this course, and they are significantly more expensive than scientific calculators.

We will assume that you are operating a scientific calculator. (Some of the keystrokes are different on graphing calculators, so if you are using one of these calculators, please consult your owner's manual.) Study the following table to discover how the basic keys are used.

Expression	Key Strokes	Display
$144 \div 3 - 7$	144 ÷ 3 − 7 =	41.
$3(2) + 4(5)$	3 × 2 + 4 × 5 =	26.
$13^2 - 2(12 + 10)$	13 x^2 − 2 × (12 + 10) =	125.
$\frac{28 + 42}{10}$	(28 + 42) ÷ 10 =	7.
	or	
	28 + 42 = ÷ 10 =	7.
$\frac{288}{6 + 12}$	288 ÷ (6 + 12) =	16.
$19^2 - 3^5$	19 x^2 − 3 x^y 5 =	118.
$2\frac{1}{3} + \frac{5}{6}$	2 a^b/c 1 a^b/c 3 + 5 a^b/c 6 =	3 ⌟1 ⌟6

Notice that the calculator does calculations when you hit the = key. The calculator automatically uses the order of operations when you enter more than one operation before hitting =. Notice that if you begin a sequence with an operation sign, the calculator automatically uses the number currently displayed as part of the calculation. There are three operations that require only one number: squaring a number, square rooting a number, and taking the opposite of a number. In each case, enter the number first and then hit the appropriate operation key. Be especially careful with fractions. Remember that when there is addition or subtraction inside a fraction, the fraction bar acts as a grouping symbol. But the only way to convey this to your calculator is by using the grouping symbols (and). Notice that the fraction key is used between the numerator and denominator of a fraction and also between the whole number and fractional part of a mixed number. It automatically calculates the common denominator when necessary.

Model Problem Solving

Practice the following problems until you can get the results shown.

	Answers
a. $47 + \frac{525}{105}$	52
b. $\frac{45 + 525}{38}$	15
c. $\frac{648}{17 + 15}$	20.25
d. $\frac{140 - 5(6)}{11}$	10
e. $\frac{3870}{9(7) + 23}$	45
f. $\frac{5(73) + 130}{33}$	15
g. $100 - 2^5$	68
h. $100 - (-2)^5$	132
i. $4\frac{2}{7} - 3\frac{3}{5}$	$\frac{24}{35}$

Prime Factors of Numbers 1 through 100

	Prime Factors		Prime Factors		Prime Factors		Prime Factors
1	none	26	$2 \cdot 13$	51	$3 \cdot 17$	76	$2^2 \cdot 19$
2	2	27	3^3	52	$2^2 \cdot 13$	77	$7 \cdot 11$
3	3	28	$2^2 \cdot 7$	53	53	78	$2 \cdot 3 \cdot 13$
4	2^2	29	29	54	$2 \cdot 3^3$	79	79
5	5	30	$2 \cdot 3 \cdot 5$	55	$5 \cdot 11$	80	$2^4 \cdot 5$
6	$2 \cdot 3$	31	31	56	$2^3 \cdot 7$	81	3^4
7	7	32	2^5	57	$3 \cdot 19$	82	$2 \cdot 41$
8	2^3	33	$3 \cdot 11$	58	$2 \cdot 29$	83	83
9	3^2	34	$2 \cdot 17$	59	59	84	$2^2 \cdot 3 \cdot 7$
10	$2 \cdot 5$	35	$5 \cdot 7$	60	$2^2 \cdot 3 \cdot 5$	85	$5 \cdot 17$
11	11	36	$2^2 \cdot 3^2$	61	61	86	$2 \cdot 43$
12	$2^2 \cdot 3$	37	37	62	$2 \cdot 31$	87	$3 \cdot 29$
13	13	38	$2 \cdot 19$	63	$3^2 \cdot 7$	88	$2^3 \cdot 11$
14	$2 \cdot 7$	39	$3 \cdot 13$	64	2^6	89	89
15	$3 \cdot 5$	40	$2^3 \cdot 5$	65	$5 \cdot 13$	90	$2 \cdot 3^2 \cdot 5$
16	2^4	41	41	66	$2 \cdot 3 \cdot 11$	91	$7 \cdot 13$
17	17	42	$2 \cdot 3 \cdot 7$	67	67	92	$2^2 \cdot 23$
18	$2 \cdot 3^2$	43	43	68	$2^2 \cdot 17$	93	$3 \cdot 31$
19	19	44	$2^2 \cdot 11$	69	$3 \cdot 23$	94	$2 \cdot 47$
20	$2^2 \cdot 5$	45	$3^2 \cdot 5$	70	$2 \cdot 5 \cdot 7$	95	$5 \cdot 19$
21	$3 \cdot 7$	46	$2 \cdot 23$	71	71	96	$2^5 \cdot 3$
22	$2 \cdot 11$	47	47	72	$2^3 \cdot 3^2$	97	97
23	23	48	$2^4 \cdot 3$	73	73	98	$2 \cdot 7^2$
24	$2^3 \cdot 3$	49	7^2	74	$2 \cdot 37$	99	$3^2 \cdot 11$
25	5^2	50	$2 \cdot 5^2$	75	$3 \cdot 5^2$	100	$2^2 \cdot 5^2$

NAME ______________________ CLASS ______________ DATE ______________

CHAPTERS 1 to 4

Midterm Examination

ANSWERS

1. Write the place value of the digit 6 in 569,234. — 1. ________
2. Add: — 2. ________

$$\begin{array}{r} 3{,}678 \\ 91 \\ 341 \\ 89{,}612 \\ +\ \ 297 \\ \hline \end{array}$$

3. Write the word name for 345,804. — 3. ________
4. Subtract: — 4. ________

$$\begin{array}{r} 8934 \\ -\ 5877 \\ \hline \end{array}$$

5. Add: $843 + 16{,}867 + 45 + 8$ — 5. ________
6. Multiply: (532)(78) — 6. ________
7. Estimate the product and multiply: — 7. ________

$$\begin{array}{r} 803 \\ \times\ 906 \\ \hline \end{array}$$

8. Divide: $565\overline{)94{,}355}$ — 8. ________
9. Divide: $23\overline{)83{,}672}$ — 9. ________
10. Perform the indicated operations: $8 - 3 \cdot 2 + 15 \div 3$ — 10. ________
11. Find the sum of the quotient of 72 and 9 and the product of 33 and 2. — 11 ________
12. Find the average of 452, 149, 86, and 101. — 12. ________
13. Write the least common multiple (LCM) of 12, 15, and 18. — 13. ________
14. Is 107 a prime number or a composite number? — 14. ________
15. Is 396 a multiple of 9? — 15. ________
16. List the first five multiples of 23. — 16. ________
17. List all the factors of 304. — 17. ________
18. Write the prime factorization of 304. — 18. ________
19. Change to a mixed number: $\frac{67}{8}$ — 19. ________
20. Change to an improper fraction: $17\frac{4}{7}$ — 20. ________
21. Which of these fractions are proper? $\frac{3}{4}, \frac{7}{6}, \frac{8}{8}, \frac{7}{8}, \frac{9}{8}, \frac{5}{4}, \frac{4}{4}$ — 21. ________
22. List these fractions from the smallest to the largest: $\frac{5}{9}, \frac{5}{8}, \frac{7}{12}, \frac{2}{3}$ — 22. ________

23. Simplify: $\frac{144}{216}$ 23. ____

24. Multiply and simplify: $\frac{2}{15} \cdot \frac{5}{8} \cdot \frac{6}{7}$ 24. ____

25. Multiply. Write the answer as a mixed number. $\left(4\frac{3}{4}\right)\left(11\frac{5}{8}\right)$ 25. ____

26. Divide and simplify: $\frac{21}{15} \div \frac{14}{25}$ 26. ____

27. What is the reciprocal of $4\frac{1}{3}$? 27. ____

28. Add: $\frac{4}{15} + \frac{7}{12}$ 28. ____

29. Add: 29. ____

$$\begin{array}{r} 6\frac{2}{3} \\ +\ 11\frac{7}{8} \\ \hline \end{array}$$

30. Subtract: $11 - 7\frac{3}{5}$ 30. ____

31. Subtract: 31. ____

$$\begin{array}{r} 34\frac{4}{9} \\ -\ 26\frac{4}{5} \\ \hline \end{array}$$

32. Find the average of $4\frac{5}{6}$, $6\frac{2}{3}$, and $11\frac{3}{4}$. 32. ____

33. Perform the indicated operations: $\frac{4}{5} - \frac{1}{2} \cdot \frac{5}{6} \div \frac{5}{6}$ 33. ____

34. Write the place value name for "Seventy-one thousand, three hundred two." 34. ____

35. Add: 35. ____

$$\begin{array}{r} 3 \text{ yd } 1 \text{ ft } 9 \text{ in.} \\ +\ 7 \text{ yd } 2 \text{ ft } 7 \text{ in.} \\ \hline \end{array}$$

36. Find the perimeter of a trapezoid with bases of 34 ft and 29 ft and sides of 7 ft and 5 ft. 36. ____

37. Find the area of a rectangle that is $3\frac{3}{4}$ ft wide and $7\frac{2}{5}$ ft long. 37. ____

38. Find the volume of a box with a length 4 ft, height 6 in., and width 5 ft. Find the volume in cubic feet. 38. ____

39. Find the area of a square that is 34 cm on a side. 39. ____

40. Find the perimeter of the following figure. 40. ____

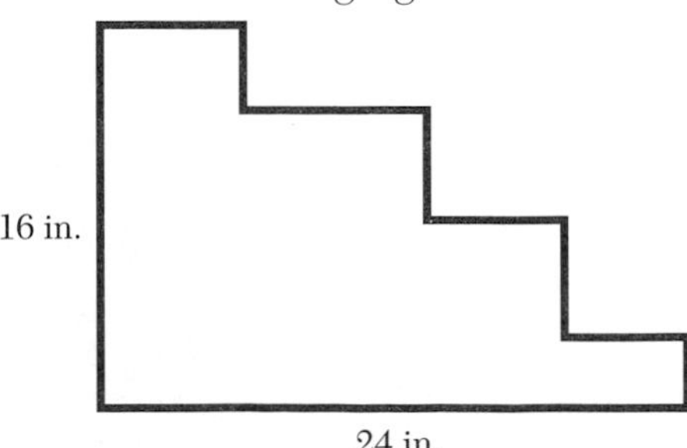

NAME ______________________ CLASS ______________ DATE ______________

CHAPTER 1 to 8

Final Examination

ANSWERS

1. Add: $\frac{7}{8} + \frac{9}{16}$ **1.** ______
2. Write the LCM (least common multiple) of 14, 18, and 21. **2.** ______
3. Subtract: $\begin{array}{r} 73.56 \\ -\ \underline{39.67} \end{array}$ **3.** ______
4. Add: 19.04 + 0.073 + 0.74 **4.** ______
5. Divide: $\frac{5}{6} \div \frac{5}{7}$ **5.** ______
6. Multiply: $2\frac{4}{9} \cdot 21$ **6.** ______
7. Which of these numbers is a prime number? 99, 199, 299, 699 **7.** ______
8. Divide: Round the answer to the nearest hundredth: $0.58\overline{)56.75}$ **8.** ______
9. Subtract: $14 - 7\frac{7}{10}$ **9.** ______
10. Multiply: (0.00721)(100,000) **10.** ______
11. Round to the nearest hundredth: 82.04499 **11.** ______
12. Multiply: (8.3)(7.02) **12.** ______
13. Write as a fraction and simplify: 86% **13.** ______
14. Add: $\begin{array}{r} 9\frac{11}{15} \\ +\ \underline{\frac{5}{6}} \end{array}$ **14.** ______
15. Solve the proportion: $\frac{7}{12} = \frac{x}{4}$ **15.** ______
16. What is the place value of the 6 in 23.0067? **16.** ______
17. List these fractions from the smallest to the largest: $\frac{3}{4}, \frac{4}{5}, \frac{14}{20}$ **17.** ______
18. Change to percent: $\frac{17}{20}$ **18.** ______
19. Divide: $27\overline{)659}$ **19.** ______
20. Write as an approximate decimal to the nearest thousandth: $\frac{11}{23}$ **20.** ______
21. Write the place value name for "Six thousand and fifteen thousandths." **21.** ______
22. Divide: $0.016\overline{)61}$ **22.** ______

23. Write as a decimal: $11\frac{3}{8}\%$ 23. ______

24. Forty-six percent of what number is 16.1? 24. ______

25. Write as a percent: 4.52 25. ______

26. Write as a decimal: $\frac{53}{200}$ 26. ______

27. Eighty-one percent of 45 is what number? 27. ______

28. Multiply: (0.17)(1.2)(0.15) 28. ______

29. If 6 books cost $117.48, how much would 15 books cost? 29. ______

30. A CD player is priced at $480. It is on sale for $345. What is the percent of discount based on the original price? 30. ______

31. List the first five multiples of 41. 31. ______

32. Write the word name for 9045.045 32. ______

33. Is the following proportion true or false? $\frac{1.9}{22} = \frac{5.8}{59}$ 33. ______

34. Write the prime factorization of 420. 34. ______

35. Simplify: $\frac{105}{180}$ 35. ______

36. Change to a fraction and simplify: 0.935 36. ______

37. Change to a mixed number: $\frac{255}{7}$ 37. ______

38. Multiply and simplify: $\frac{5}{28} \cdot \frac{35}{20}$ 38. ______

39. Change to a fraction: 0.075 39. ______

40. At a service station, 29 out of 50 drivers asked for a "fill-up." What percent of the drivers wanted a full tank of gas? 40. ______

41. Is 733 a multiple of 3? 41. ______

42. Subtract: $14\frac{3}{5} - \frac{11}{15}$ 42. ______

43. Change to an improper fraction: $13\frac{7}{9}$ 43. ______

44. List the following decimals from smallest to largest: 2.32, 2.332, 2.299, 2.322 44. ______

45. Divide: $674.81 \div 10{,}000$ 45. ______

46. Divide: $6\frac{3}{8} \div 4\frac{5}{16}$ 46. ______

47. Write a ratio to compare 5 in. to 5 ft (using common units) and simplify. 47. ______

48. Mildred calculates that she pays $1.21 for gas and oil to drive 5 miles. In addition, she pays 30¢ for maintenance for each 5 miles she travels. How much will it cost her to drive 6000 miles? 48. ______

49. Twenty percent of a family's income is spent on food, 8 percent on transportation, 35 percent on housing, 10 percent on heat and utilities, 7 percent on insurance, and the rest on miscellaneous expenses. If the family's income is $2100 per month, how much is spent on miscellaneous expenses? 49. ______

50. The sales tax on a $55 purchase is $4.75. What is the sales tax rate, to the nearest tenth of a percent? 50. ______

ANSWERS

51. Perform the indicated operations: $32 - 5 \cdot 6 + 45 \div 9$ **51.** ______

52. Find the average of $\frac{7}{15}$, $3\frac{3}{4}$, and $5\frac{5}{6}$. **52.** ______

53. Perform the indicated operations: $7.8 - 2.1(3.6) \div 4$ **53.** ______

54. Add: 4 hr 35 min 16 sec + 2 hr 42 min 55 sec **54.** ______

55. Subtract: 7 m 27 cm − 4 m 35 cm **55.** ______

56. Convert 5¢ per gram to dollars per kilogram. **56.** ______

57. Find the perimeter of a trapezoid with bases of 45.7 ft and 52.8 ft and sides of 13.7 ft and 15.3 ft. **57.** ______

58. Find the area of a triangle with base 5.7 m and height 4.7 m **58.** ______

59. Find the area of the following geometric figure (let $\pi \approx 3.14$): **59.** ______

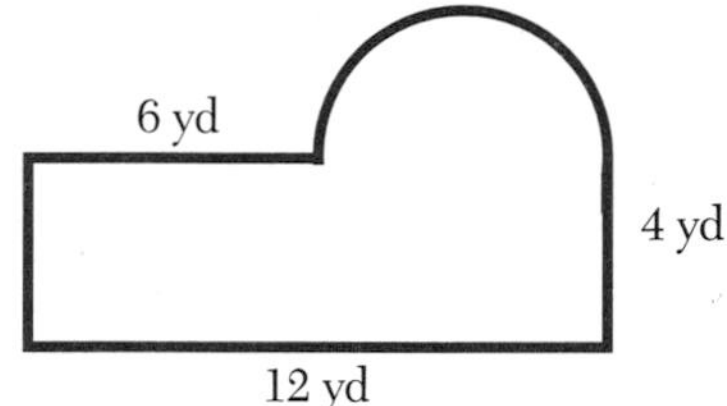

60. Find the volume of a box with length $2\frac{1}{4}$ ft, width 18 in., and height 8 in. (in cubic inches). **60.** ______

61. Add: $(-34) + (-23) + 41$ **61.** ______

62. Subtract: $(-45) - (-72)$ **62.** ______

63. Multiply: $(3)(-4)(-1)(-6)$ **63.** ______

64. Divide: $(-125) \div (-25)$ **64.** ______

65. Perform the indicated operations: $(-6 - 4)(-4) \div (-5) - (-10)$ **65.** ______

66. Solve: $12a + 98 = 2$ **66.** ______

Answers to Odd-Numbered Questions

CHAPTER 1

Exercises 1.1

1. five hundred forty-two **3.** eight hundred ninety **5.** seven thousand, fifteen **7.** 57 **9.** 7500 **11.** 10,000,000 **13.** twenty-five thousand, three hundred ten **15.** two hundred five thousand, three hundred ten **17.** forty-five million **19.** 243,700 **21.** 23,470 **23.** 17,000,000 **25.** < **27.** > **29.** < **31.** < **33.** 690 **35.** 1700 **37.** 102,390, 102,400, 102,000, 100,000 **39.** 7,250,980, 7,251,000, 7,251,000, 7,250,000 **41.** 560,353,730 **43.** Thirteen thousand, one hundred forty-eight dollars **45.** < **47.** 1000 **49.** 43,780,000 **51.** 63,700: 63,800: Yes. Rounding the second time changes the tens digit to 5, so rounding to the hundreds place is inaccurate. The first method is correct. **53.** Kimo wrote "Eighteen thousand, four hundred sixty-five" on the check. **55.** Ducks Unlimited estimated that three hundred eighty-nine thousand, five hundred ducks spent the winter at the refuge. **57.** The place value name for the bid is \$36,407. **59.** To the nearest thousand dollars the value of the sale of Intel shares is \$847,000. **61.** The metal industries released 327,000,000 pounds of toxic material. **63.** The per capita personal income in Maine is twenty thousand, five hundred twenty-seven dollars. **65.** Maine has the smallest per capita income. **67.** To the nearest million miles, the distance from the earth to the moon is 93,000,000 miles. **69.** The word name for the Yankee payroll is sixty-three million, seven hundred thousand dollars. **71.** There are two ways to write an inequality involving the payrolls: \$56,700,000 > \$53,300,000 or \$53,300,000 < \$56,700,000. **73.** The land area of the earth to the nearest million square miles is 52,000,000 square miles. **75.** The word name for Scottie Pippen's salary is two million, two hundred fifty thousand dollars. **77.** Jif has fewer of the following nutrients: fat, sodium, carbohydrates, and sugars. **79.** Base-ten is a good name for our number system because each place value in the system is 10 times the previous place and one-tenth the succeeding place. **81.** Rounding a number is a method of calculating an approximation of that number. The purpose is to get an idea of the value of the number without listing digits that do not add to our understanding. The number 87,452 rounds to 87,000, to the nearest thousand, because 452 is less than halfway between 0 and 1000. Rounded to the nearest hundred it is 87,500, because 52 is over halfway between 0 and 100. **83.** Three trillion, four hundred fifty-six billion, seven hundred nine million, two hundred thirty thousand. **85.** 5 **87.** 0

Exercises 1.2

1. 97 **3.** 979 **5.** 509 **7.** 1 **9.** 4399 **11.** 6571 **13.** 17,500 **15.** 1100 **17.** 10,000 **19.** 19,000; 19,918 **21.** 160,000; 160,588 **23.** 201 **25.** 526 **27.** 451 **29.** 10 **31.** 158 **33.** 422 **35.** 4700 **37.** 300 **39.** 20,000 **41.** 500: 469 **43.** 4000; 3363 **45.** 40,000; 40,883 **47.** 0; 8,233 **49.** The total number of Fords, Toyotas, and Lexuses sold is 3561. **51.** The Hondas sold were 163 more than the Fords sold. **53.** The total of the best three selling cars is 5320. **55.** The estimate of the total cars sold is 8000. The exact number sold is 9106. **57.** The sum of the payrolls is \$229,900,000. **59.** For the week, 3589 salmon went through the ladder. Tuesday's count was 363 more than Saturday's count. **61.** The gross sales for the week is \$767,847. **63.** The estimated number of arrests for violent crimes is 600,313. **65.** The number of aggravated assaults is 264,271 more than the number of robberies. **67.** Sasha consumes a total of 1070 calories. **69.** The total attendance at the three games is 268,181. The largest attendance is 24,513 more than the smallest attendance. **71.** The new line will handle 167,210 gallons per minute more than the old line. **73.** To the nearest million gallons, 36,000,000 gallons were pumped to the second tanker. **75.** The median family income

was probably rounded to the nearest hundred. The San Francisco median income is $9300 higher than Seattle's median income. **77.** Dennis Rodman earned $6,750,000 more than Scottie Pippen. **79.** The total number of HIV infections in Asia is 3,100,000.
81. Explanations for subtracting which are aimed at 8-year-olds, are usually based on physical objects. So $15 - 9 = 6$ because when 9 circles (or pencils or apples or whatever) are removed from 15 circles, there are 6 circles remaining.

Count the circles that are empty.

83. A sum is the result of adding numbers. The sum of 8, 8, and 2 is 18. Mathematically we write $8 + 8 + 2 = 18$.
85. Seven hundred thousand, nine hundred
87. The total dollar sales for the nine cars is $165,991. The Accord's sales were $26,585 more than the Civic's sales. **89.** $A = 7, B = 2, C = 3, D = 1$

Getting Ready for Algebra, page 39

1. $x = 9$ **3.** $x = 11$ **5.** $z = 7$ **7.** $c = 29$
9. $a = 183$ **11.** $x = 34$ **13.** $y = 47$
15. $k = 246$ **17.** $37 = x$ **19.** $130 = w$
21. The markup is $393. **23.** The length of the garage is 9 meters. **25.** Let S represent the EPA city rating of the Saturn and I represent the EPA city rating of the Impreza. $I + 4 = S$; The Impreza has a rating of 24 mpg. **27.** Let B represent the total dollars budgeted in a category, S represent the dollars spent in a category, and R represent the dollars not yet spent in a category. $S + R = B$.

Exercises 1.3

1. 210 **3.** 132 **5.** 288 **7.** 483
9. 328 **11.** 0 **13.** 2640 **15.** thousands
17. 2178 **19.** 4030 **21.** 1596 **23.** 2848
25. 153,900 **27.** 47,101 **29.** 29,070
31. 13,300 **33.** 1600 **35.** 16,000
37. 8000 **39.** 35,000 **41.** 15,000; 15,600
43. 32,000; 34,608 **45.** 35,000; 32,148
47. 50,000; 66,636 **49.** 251,490 **51.** 1,200,000; 1,494,513 **53.** The Rotary Club estimates it will sell 4230 dozen roses. **55.** The estimated gross receipts from the sale of Cougars is $600,000. **57.** The estimated gross receipts from the sale of Villagers is $400,000 and the actual gross receipts is $370,515. **59.** 40,000; 38,880 **61.** During the 17-day period a total of 2278 salmon is counted. **63.** The CEO realized $77,070 from the sale of the shares. **65.** The bacteria count at 2 P.M. will be 30,375 bacteria. **67.** Eight times Ron Harper's salary is larger than Michael Jordan's salary.
69. To the nearest thousand gallons, in a 31-day month the water usage is 9,934,000 gallons. **71.** The radios cost Ms. Muzos $35,100. The net income from the sale of the radios is $50,400. The profit on the radios is $15,300.
73. The place value name for 36 billion is 36,000,000,000. You would spend about $1,007,400,000.
75. If the Astros doubled their payroll they would rank 1st. **77.** Carol burned the following calories: in a week, 3136 calories; in a month, 13,440 calories: in a year, 163,520 calories. **79.** When 74 is multiplied by 8, we first multiply 8 times 4 to get 32. The "3" in 32 represents 3 tens or 30, not 3 ones. Therefore we "carry" the "3" to the tens column to be added to the 8 times 70.
81. Five billion, one hundred thirty million, three hundred fifty-four thousand, one hundred eighty
83. $A = 6, B = 3, C = 8, E = 1$

Exercises 1.4

1. 12 **3.** 5 **5.** 107 **7.** 61 **9.** 15
11. 40 **13.** 71 **15.** 2 R 3 **17.** Divisor
19. 3052 **21.** 24 **23.** 26 **25.** 96
27. 51 **29.** 174 R 32 **31.** 15 R 30
33. 58 **35.** 810 **37.** 10 **39.** 100
41. 60 **43.** 50 **45.** 300; 271 **47.** 500; 439 **49.** 600; 578 R 68 **51.** 2000; 1790 R 414 **53.** 20,000; 16,039 R 324 **55.** The estimated taxes paid per return in week 2 is $5000.
57. The actual taxes paid per return in week 3 is $4100, rounded to the nearest hundred. **59.** 200; 188 R 759
61. The survey finds that 128 trees per acre are ready to harvest. **63.** Each nephew will receive $69,575.
65. A total of 2305 radios can be assembled with 8 resistors left over. **67.** It takes the company 520 hours to process the beans. **69.** You would need to spend $1,430,000 per day. **71.** To the nearest hundred thousand, the Braves payroll per game is $300,000. **73.** To the nearest ten, the cost per mile per passenger is $8060. **75.** Los Angeles has the better deal. The cost per mile per passenger is about half that of Portland.
77. To the nearest ten thousand, Michael Jordan is paid $370,000 per game. **79.** Explanations for division which are aimed at 8-year-olds are usually based on physical objects. So $45 \div 9 = 5$ because 45 circles (or pencils or apples or whatever) can be separated into 5 groups with 9 circles in each group.

Divide into 5 groups of 9.

81. A quotient is the result of dividing numbers. The quotient of 30 and 10 is 3. Mathematically we write

$30 \div 10 = 3$ or $10\overline{)30}^{\,3}$.

83. $A = 5, B = 1, C = 6$

Getting Ready for Algebra, page 69

1. $x = 5$ **3.** $c = 18$ **5.** $x = 4$ **7.** $b = 120$ **9.** $x = 12$ **11.** $y = 312$ **13.** $x = 24$ **15.** $b = 45{,}414$ **17.** $5 = x$ **19.** $1278 = w$ **21.** The width of the garden plot is 17 ft. **23.** He sells 2340 lb of crab. **25.** The wholesale cost of one set is \$310. **27.** Let L represent the low temperature in July and H represent the high temperature in January. $2H - L$; The average high temperature is 30° F.

Exercises 1.5

1. 12^6 **3.** 81 **5.** 8 **7.** 1 **9.** base; exponent; power or value **11.** 216 **13.** 361 **15.** 10,000 **17.** 512 **19.** 6561 **21.** 4500 **23.** 70,000 **25.** 12 **27.** 340 **29.** exponent **31.** 4,350,000 **33.** 1200 **35.** 35,910,000 **37.** 302 **39.** 70,500,000,000 **41.** 9700 **43.** 10^{11} **45.** 38,416 **47.** 387,420,489 **49.** 3,350,000,000,000 **51.** 4380 **53.** The college's operating budget is \$73,000,000. **55.** During the week, 32,000,000 shares of Microsoft were traded. **57.** The WHO estimates 10^7 cases of HIV infection in sub-Saharan Africa, and 10^5 cases in Northern Africa/Middle East. **59.** Five hundred thousand can be written as 5×10^5, 50×10^4, 500×10^3, 5000×10^2, or $50{,}000 \times 10$. **61.** The player bets 5^9 dollars on the ninth bet, or \$1,953,125. **63.** The distance, 6 trillion miles, can also be written as 6×10^{12} and as 6,000,000,000,000. **65.** The Cardinals' payroll is $\$477 \times 10^5$. **67.** 25^6 is the first power of 25 that is larger than the Astro's payroll. **69.** A salary of $\$5^{11}$ would pay \$18,688,125 more than Michael Jordan's salary. **71.** To multiply a whole number by a power of ten, write the number followed by as many zeros as the exponent in the power of 10. For example, $32 \times 10^3 = 32{,}000$ (the exponent is 3, so write 3 zeros after 32) and $5 \times 10^6 = 5{,}000{,}000$ (write 5 followed by 6 zeros). **73.** 2025 **75.** The sun is estimated to weigh 2,000,000,000,000,000,000,000,000,000 tons or 2×10^{27} tons.

Exercises 1.6

1. 53 **3.** 0 **5.** 27 **7.** 15 **9.** 25 **11.** 39 **13.** 10 **15.** 43 **17.** 31 **19.** 7 **21.** 35 **23.** 217 **25.** 24 **27.** 28 **29.** 145 **31.** 102 **33.** 45 **35.** There were 595 more mallards and canvasbacks than teal and woodducks. **37.** Four times the number of woodducks added to the number of mallards would be 165 more than the number of teal and canvasbacks. **39.** 18 **41.** 436 **43.** The store sold \$40,275 in guitars and pianos. **45.** Pete's son earns \$85. **47.** The Hair Barn used supplies costing \$1130 for the month. **49.** The trucker's average weekly income is \$1225. His yearly income is \$61,250. **51.** The first golfer scored the most points. **53.** Marla consumes 1150 calories for breakfast. **55.** To simplify $2(1 + 36 \div 3^2) - 3$, begin by doing the operations inside the parentheses. Since there is more than one operation inside, work them according to the order of operations. In this case square first, then divide, and finally add.

$$\begin{aligned} 2(1 + 36 \div 3^2) - 3 &= 2(1 + 36 \div 9) - 3 \\ &= 2(1 + 4) - 3 \\ &= 2(5) - 3 \end{aligned}$$

At this point we have two operations remaining, so multiply first and then subtract.

$$\begin{aligned} &= 10 - 3 \\ &= 7 \end{aligned}$$

57. USA Video must still raise \$154.

Getting Ready for Algebra, page 95

1. $x = 5$ **3.** $y = 24$ **5.** $x = 4$ **7.** $c = 88$ **9.** $x = 6$ **11.** $c = 12$ **13.** $a = 800$ **15.** $b = 22$ **17.** Remy bought 7 tickets. **19.** Rana made 6 arrangements. **21.** $15m = C$; Jessica can purchase 200 minutes from AT&T. **23.** $9m + 696 = C$; Jessica can purchase 256 minutes from Pace.

Exercises 1.7

1. 5 **3.** 9 **5.** 13 **7.** 6 **9.** 4 **11.** 5 **13.** 25 **15.** 12 **17.** 26 **19.** 40 **21.** 18 **23.** 31 **25.** 110 **27.** 145 **29.** 101 **31.** 54 **33.** 540 **35.** The average number of ducks per species is 675. **37.** 2104 **39.** 27,458 **41.** The average score per round of golf is 84. **43.** Mr. Adams' average caloric intake is 4025 calories. **45.** The average payroll for the four division winners was \$57,500,000. **47.** The average number of HIV cases is 5,050,000. No—The actual cases in the two regions are too far apart for the average to be meaningful. **49.** The average price per can of fruit is 76¢.

51. The average score at the Rock Creek tournament is 73. **53.** The average cost in the districts is $4925 and $5540. The average cost in the combined districts is $5300. **55.** The average salary paid the Bulls is $4,480,000. No—Jordan's salary throws the average off and distorts what is being paid. **57.** The average, or mean, of 2, 4, 5, 5, and 9 is $\frac{2+4+5+5+9}{5} = \frac{25}{5} = 5$. The average gives one possible measure of the center of the group.

Exercises 1.8

1. The broadcast rights to the Summer Olympics were least expensive in 1992. **3.** The total cost of broadcasting the Olympics in the 1990s is $857 million. **5.** The difference in the projected costs is $248 million. **7.** The estimated cost to NBC is $4700 million. The actual cost will be $4707 million. **9.** Fish cakes have the highest level of cholesterol per serving. **11.** You consume 48 mg less when ordering chicken dijon. **13.** There are 72 g of fat in 3 servings of veal chops. **15.** Dan consumes 908 calories. **17.** Jerry can eat fish cakes, chicken dijon, and pepper steak. **19.** The gain in value is $106,930. **21.** If cashed in at 59, the loss is $14,291. **23.** At age 61 the difference between the death benefit and the account value is $210,478. **25.** The largest increase in account value occurred between 63 and 65. **27.** The most picnics are held in August. **29.** The total number of overnight campers is 3793. **31.** The income from picnics in July was $11,855. **33.** There were 144 more hikers/climbers in August than May. **35.** Fisher Zoo had the greatest increase in attendance. **37.** The 5-year estimate of attendance at Fisher Zoo is 13,000,000. **39.** The average attendance at Delaney Zoo is 1,251,036. **41.** To the nearest thousand, the revenue at Fisher Zoo for 1995 is $68,628,000. **43.** The two bottom rows of the table are: Total animals; 262, 51, 494; Total; 640, 61, 495. **45.** Income tax tables, nutritional value tables on food labels, mileage tables on maps, postage tables, sports league records, to name a few. No. **47.** The 1997 attendance at Utaki Park must be 598,345.

Exercises 1.9

1. 10 − 11 **3.** 250 **5.** The total number of calls made is 1425. **7.** 40 **9.** Full size **11.** The number of cars in for repair is 400. **13.** 1996 **15.** The increase in production is 12,500 units. **17.** The average production is 20,500 units. **19.** The total paid for paint and lumber is $45,000. **21.** Steel castings cost $15,000 less than plastics. **23.** They will pay $40,000 for doubling production.

25.

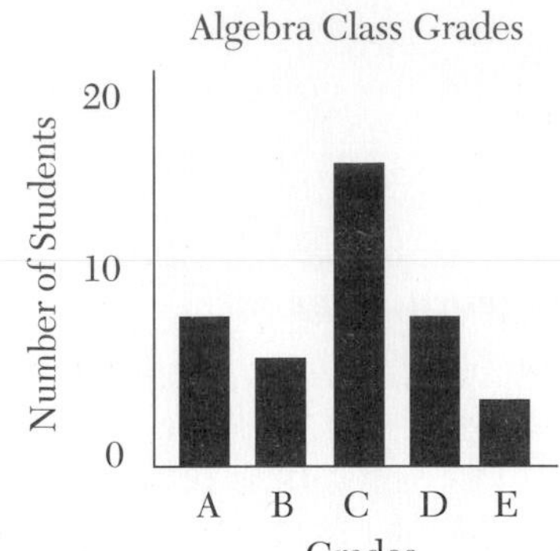

27.

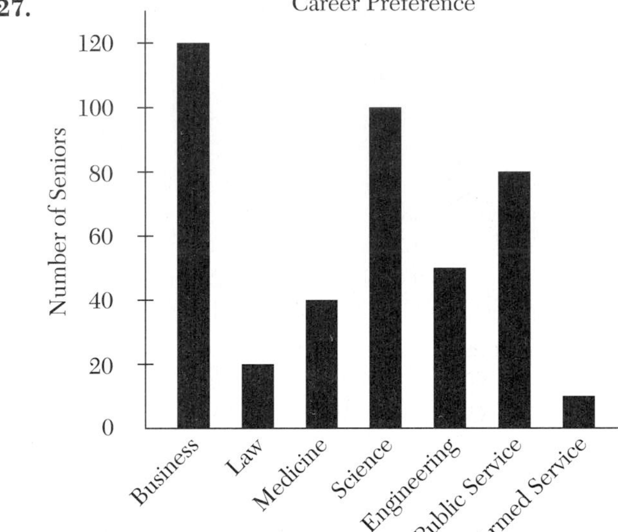

29.

31.

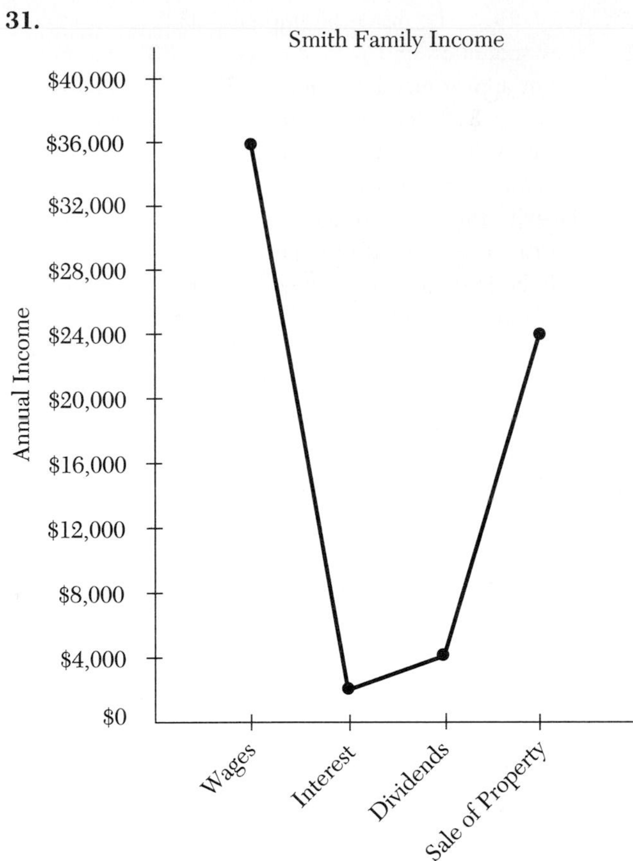

33. AVERAGE COST OF A THREE-BEDROOM HOUSE

Year	= $10,000 = $5000
1970	
1975	
1980	
1985	
1990	
1995	

35. OIL PRODUCTION IN BARRELS

Year	= 5000 = 2500
1991	
1992	
1993	
1994	
1995	

37. General Motors; Chrysler

39. The total number of U.S. workers is 675,000.

41.

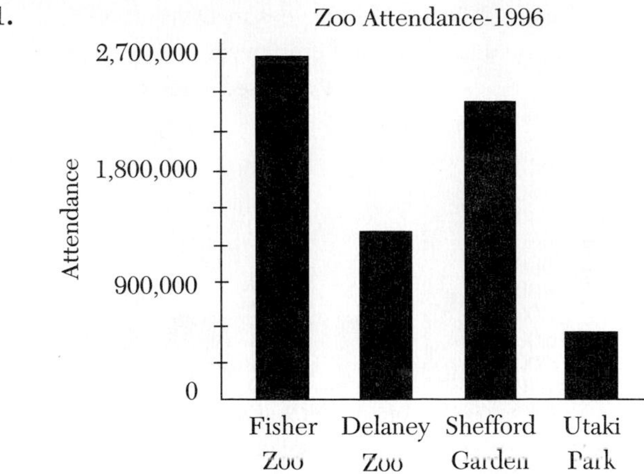

43. A pictograph gives instant recognition of what you are graphing, a line graph shows changes as the line rises or falls, and the bar graph shows quantity comparisons by the height of the bars. The line graph, as it shows a continuous flow of the data.

Chapter 1 True-False Concept Review

1. False; To write a billion takes 10 digits.
2. True **3.** True **4.** False; Five is less than eighteen. **5.** False; 1345 > 1344 **6.** True
7. True **8.** True **9.** False; The sum is 64.
10. True **11.** True **12.** False; The product is 28. **13.** True **14.** True **15.** True
16. False; A number multiplied by 0 is 0.
17. True **18.** False; It represents 160 people.
19. False; The quotient is 23. **20.** True
21. True **22.** False; Division by 0 is undefined.
23. False; The value is 64. **24.** True
25. True **26.** False; The product is 34,000.
27. True **28.** True **29.** False; In $(2 + 3)^2$ the addition is done first. **30.** False; In $(9 - 4) \cdot 5$, subtraction is done first. **31.** True **32.** True
33. True **34.** True

Chapter 1 Test

1. 109 **2.** 2279 **3.** 19 **4.** 40,704
5. > **6.** 760,000 **7.** 273,674 **8.** 450,082
9. 907 **10.** 73,300 **11.** 17,900
12. 160,000 **13.** 729 **14.** 5882
15. 10,000 **16.** 6453 **17.** Six thousand, seven
18. 30 **19.** 236,047 **20.** 60,500
21. 524,940,000 **22.** 600; 620 R 5 **23.** 17
24. 611 **25.** 666 **26.** It will take the secretary 105 minutes or 1 hour 45 minutes to type 12 pages.
27. Each person won $788,000. Each will receive $39,400 per year. **28.** **a.** Chevrolet **b.** 700

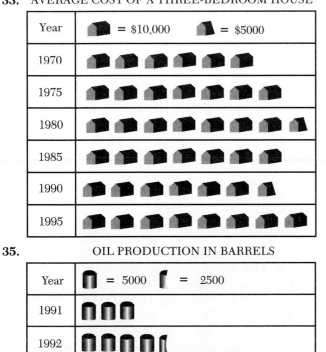

c. 100 **29. a.** Divisions A & B have the same number of and the most employees.
b. There are 225 more employees in Division A.
c. There are a total of 1200 employees.
30.

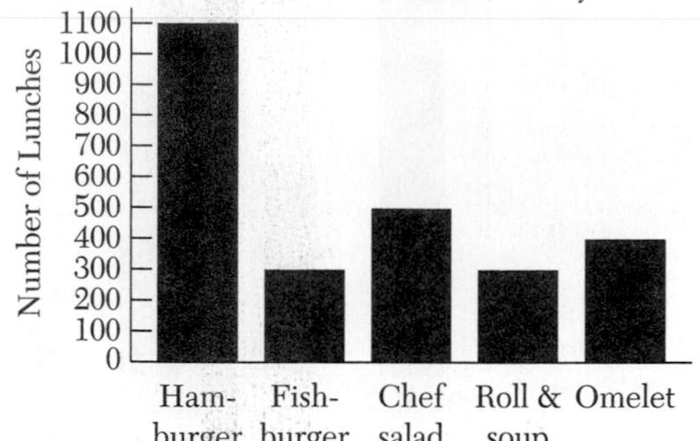

48 min. **29.** The inside perimeter is 92 in. **31.** They should order 117 ft of fence. **33.** He ran 19,376 yd or approximately 11 mi. **35.** The patio is an octagon. **37.** The perimeter of the right side flowerbed is 60 ft. **39.** Rectangles and squares are symmetrical. **41.** Perimeter is the distance around a flat surface. To find the perimeter of the figure, add the lengths of the four sides. If the figure were a park, the sides could be measured in miles or kilometers. If the figure were a room, the sides could be measured in feet or meters. If it were a piece of paper, the sides could be measured in inches or centimeters. **43.** A square has all four sides equal. This does not have to be true for a rectangle. **47.** 950,000 **49.** 385 **51.** 3004 **53.** 5 ft 5 in. **55.** The total cost is $51.

CHAPTER 2

Exercises 2.1

1. 24 ft **3.** 8 mℓ **5.** 4 gal **7.** 951 oz **9.** 50 hours **11.** 208 lb **13.** 1120 sec **15.** 20 lb **17.** 17 yd **19.** 20 g **21.** 205 kℓ **23.** 90 mm **25.** 104 yd **27.** 553 gal **29.** 24 cm **31.** Answers vary **33.** 19 ft 7 in. **35.** 6 min 52 sec **37.** 3 hr 31 min 12 sec **39.** 6 yd 1 ft 2 in. **41.** The total of the putts was 44 ft 6 in. **43.** Seven bags of potato chips contain 2086 g. **45.** The average low temperature was 12° C. **47.** The patient will receive 840 mg in a week. **49.** Nine lanes will require 8 lengths of dividers which is a total of 400 m. **51. a.** Figures for the Amazon and the Congo Rivers appear to be rounded, which implies they are probably estimates. **b.** The total length is 18,389 miles. **c.** The Amazon River is just over twice as long as the São Francisco River. **d.** The Nile River is 1260 miles longer than the Congo River. **53.** The patio is 14 ft wide and 20 ft long. **55.** Equivalent measures measure the same amount using different units. 12 inches and 1 foot are equivalent measures. So are 1 mi and 5280 ft. **57.** There are 16 oz in one pound but only 12 in. in one foot. So 8 in. + 10 in. = 18 in. = 1 ft 6 in., but 8 oz + 10 oz = 1 lb 2 oz. **59.** The second ring must weigh 12 dwt. **63.** 832 **65.** They took in $3920. **67.** 98,000 **69.** 3 **71.** 65

Exercises 2.2

1. 19 in. **3.** 28 m **5.** 34 mm **7.** 26 km **9.** 83 mm **11.** 160 m **13.** 90 ft **15.** 100 cm **17.** 78 yd **19.** 168 m **21.** 54 cm **23.** 348 mm **25.** They will need 13 ft 4 in. of molding. **27.** It will take Hazel 1 hr

Exercises 2.3

1. 16 km^2 **3.** 20 yd^2 **5.** 60 ft^2 **7.** 52 m^2 **9.** 132 km^2 **11.** 238 yd^2 **13.** 396 cm^2 **15.** 135 in^2 **17.** 612 m^2 **19.** 101 ft^2 **21.** 6550 m^2 **23.** 5655 in^2 **25.** 9 ft^2 = 1 yd^2 **27.** No, she needs more than 3 gal. **29.** Debbie will need 240 oz of weed killer. **31.** A total of 36 ft^2 of glass is needed. **33.** A total of 162 ft^2 of sheathing is needed. **35.** It will take 1332 in^2 of padding. **37.** A total of 6 bags of seed will cover the region. **39.** The patio requires 2282 bricks. **41.** To find the area of the figure, calculate the area of the triangle and then subtract the area of the rectangle. For the triangle, use the formula $A = bh \div 2$, and for the rectangle use the formula $A = \ell w$. **43.** Joe needs 72 6-in. black and 64 6-in. white squares. He needs 28 black and 28 white 12-in. squares. **45.** It will cost Ingrid $3348. **49.** 3050 **51.** It costs $29. **53.** 4425 **55.** 56 in. **57.** 2134

Exercises 2.4

1. 1000 m^3 **3.** 250 ft^3 **5.** 120 cm^3 **7.** It will take 21,840 mℓ of water. **9.** The can holds 48 ft^3. **11.** 3060 in^3 **13.** 43,310 cm^3 **15.** The can holds 5880 in^3. **17.** 3 ft^3 **19.** 10 yd^3 **21.** The concrete needed is 93,312 in^3 or 2 yd^3. **23.** 1200 in^3 **25.** 1320 yd^3 **27.** 177,120 cm^3 **29.** It will take 16 truckloads. **31.** The bricklayer needs 21 ft^3 of mortar. **33.** It is necessary to round up in this case because we cannot allow ourselves to run short of needed materials. Numbers rounded according to the rule in Section 1.1 could be less than the actual amount necessary. **35.** The garden is approximately 33,696 in^2, so they will need 269,568 in^3 = 156 ft^3 ≈ 6 yd^3 of top soil. **37.** Consider the pointed end as a solid on its side with a triangular base. The remainder of the figure is a rectangular solid. Calculate the volume of each piece and add the volumes together. **39.** The pool holds 2400 ft^3. Seven loads of dirt were hauled away. **43.** 41 **45.** 65 **47.** 132 **49.** 10 **51.** He needs 4 jars.

Chapter 2 True-False Concept Review

1. False; The metric system is most commonly used. **2.** True **3.** False; A liter is a measure of volume. **4.** True **5.** False; Volume is the measure of the inside of a solid. **6.** False; The volume of a cube is $V = s^3$. **7.** True **8.** True **9.** True **10.** True **11.** True **12.** False; A parallelogram has 4 sides. **13.** False; Area can be thought of as the number of squares in an object. **14.** True **15.** True **16.** True **17.** False; Volume can also be measured in cups or liters. **18.** True **19.** False; 1 ft^2 = 144 in^2 **20.** False, "Kilo" means 1000.

Chapter 2 Test

1. 7454 mm **2.** The perimeter is 136 cm. **3.** The volume is 1728 in^3. **4.** The flooring needed is 80 ft^2. **5.** Each will receive 27 lb. **6.** The perimeter is 22 ft. **7.** The area is 30 in^2. **8.** 1 gal 3 qt 1 pt **9.** Possible answers are: oz, lb, ton, mg, cg, g, kg. **10.** 288 in^2 **11.** A total of 46 ft of molding is needed. **12.** The volume is 264 m^3. **13.** The area is 48 cm^2. **14.** The volume of the tank is 20 ft^3. **15.** Each silver bar weighs 114 g. **16.** The fencing will cost $132. **17.** Possible answers are: cup, pint, qt, gal, mℓ, ℓ, kℓ, in^3, ft^3, yd^3, mm^3, cm^3, m^3. **18.** The area is 2736 mm^2. **19.** Each player runs 1920 yd. **20.** Area measures the inside of a two dimensional figure. Volume measures the inside of a three dimensional object. **21.** His average bench press weight is 193 lb. **22.** She needs 336 in. She must buy 10 yd. **23.** Yes, the dog food will fit. **24.** The total cost is $100,100. **25.** The minimum gift wrap needed is 400 in^2.

CHAPTER 3

Exercises 3.1

1. yes **3.** yes **5.** yes **7.** yes **9.** yes **11.** no **13.** yes **15.** yes **17.** no **19.** 2, 3, 5 **21.** 2 **23.** 3, 5 **25.** 2, 3, 5 **27.** 3, 5 **29.** yes **31.** no **33.** yes **35.** yes **37.** no **39.** yes **41.** no **43.** yes **45.** yes **47.** 6, 10 **49.** 6, 9 **51.** none **53.** 6, 10 **55.** 6, 10 **57.** It is possible for each one to contribute the same whole number amount because 1221 is divisible by 3. **59.** Yes, the profit can be evenly divided because 2060 is divisible by 5. **61.** Huang can choose bin sizes of 2 lb, 3 lb, 5 lb, 6 lb, or 10 lb since 4560 is divisible by 2, 3, 5, 6, and 10. **63.** Rows of 10 seats each are possible because 250 is divisible by 10. Rows of 15 seats each are not possible because 15 is divisible by 3 but 250 is not. **65.** ORYH PDNHV WKH ZRUOG JR URXQG **67.** Love is a many splendored thing **69.** If a number is divisible by 2 then the number is twice some number. If the number is also divisible by 3 then the number is three times some number. This means that the number is 2 × 3 × another number or 6 × another number. Therefore, the number is divisible by 6. **71.** If the sum of the digits of a number is divisible by 9, then the sum is also divisible by 3. So the number is divisible by 3. **73.** Yes. A number divisible by 3 and 5 is divisible by 15. **75.** **a.** No, the ones-place digit is odd. **b.** Yes, the sum of the digits is divisible by 3. **c.** Yes, the ones-place digit is 5. **d.** No, the number is *not* divisible by 2. **e.** Yes, the sum of the digits is 45. **f.** No, the ones-place digit is not 0. **79.** 87,750 **81.** 346,000,000 **83.** False **85.** 4 yd 2 ft 3 in. **87.** 169 cm^2

Exercises 3.2

1. 3, 6, 9, 12, 15 **3.** 17, 34, 51, 68, 85 **5.** 12, 24, 36, 48, 60 **7.** 25, 50, 75, 100, 125 **9.** 50, 100, 150, 200, 250 **11.** 48, 96, 144, 192, 240 **13.** 68, 136, 204, 272, 340 **15.** 83, 166, 249, 332, 415 **17.** 123, 246, 369, 492, 615 **19.** 345, 690, 1035, 1380, 1725 **21.** yes **23.** yes **25.** no **27.** no **29.** no **31.** yes **33.** yes **35.** no **37.** yes **39.** Multiple of 6 and 9 **41.** Multiple of 9 and 15 **43.** Multiple of 6 and 15 **45.** Multiple of 13 **47.** Multiple of 13 and 19 **49.** Jean's car will not be selected because 16 is not divisible by 5. **51.** The students should work problems 4, 8, 12, 16, 20, 24, 28, 32, 36, 40, 44, 48, and 52. **53.** Yes, the prediction is accurate because 605 is the eleventh multiple of 55. **55.** The team will check bottles with the numbers 810, 825, 840, 855, 870, 885, 900, 915, 930, 945, 960, 975, and 990. **57.** The first year in the 21st century that is divisible by 35 is 2030. **59.** Yes. There were approximately three times as many injuries from in-line skating as from skateboarding in 1996. **61.** They are all 0 mod 5. **63.** The equivalents are: 18 = 6 mod 12; 27 = 3 mod 12; 34 = 10 mod 12; 48 = 0 mod 12; and 62 = 2 mod 12. **65.** When a number is a multiple of 12, the number is

12 × some whole number. Therefore, the number is also 6 × 2 × some whole number. So the number is a multiple of 6. **67.** First divide 77 by 7 and write the product: 7 × 11 = 77. Therefore the next multiple is 7 × 12 = 84 followed by 7 × 13 = 91, and so on. Stop at 7 × 18 = 126.
69. yes **71.** 1333 **75.** 44,534
77. 11,128 **79.** 6336 **81.** 29,106
83. The twelve cases will last for 13 days. There will be 30 cans of soup left.

Exercises 3.3

1. 1 · 16, 2 · 8, 4 · 4 **3.** 1 · 23 **5.** 1 · 20, 2 · 10, 4 · 5 **7.** 1 · 46, 2 · 23 **9.** 1 · 47 **11.** 1 · 98, 2 · 49, 7 · 14 **13.** 1 · 100, 2 · 50, 4 · 25, 5 · 20, 10 · 10 **15.** 1 · 114, 2 · 57, 3 · 38, 6 · 19
17. 1 · 115, 5 · 23 **19.** 1 · 444, 2 · 222, 3 · 148, 4 · 111, 6 · 74, 12 · 37 **21.** 1, 2, 4, 8, 16 **23.** 1, 23 **25.** 1, 2, 3, 5, 6, 10, 15, 30 **27.** 1, 5, 11, 55 **29.** 1, 5, 13, 65 **31.** 1, 2, 3, 4, 6, 8, 9, 12, 18, 24, 36, 72 **33.** 1, 2, 4, 23, 46, 92 **35.** 1, 2, 3, 6, 17, 34, 51, 102 **37.** 1, 2, 61, 122 **39.** 1, 2, 71, 142 **41.** 1 · 485, 5 · 97 **43.** 1 · 650, 2 · 325, 5 · 130, 10 · 65, 13 · 50, 25 · 26 **45.** 1 · 660, 2 · 330, 3 · 220, 4 · 165, 5 · 132, 6 · 110, 10 · 66, 11 · 60, 12 · 55, 15 · 44, 20 · 33, 22 · 30

47.

Number of Programs	Length of Each
1	120 min
2	60 min
3	40 min
4	30 min
5	24 min
6	20 min
8	15 min
10	12 min
12	10 min
15	8 min
20	6 min
24	5 min
30	4 min
40	3 min
60	2 min
120	1 min

49. To provide 1 adult for every 4 infants, they need 6 adults. To provide 1 adult for every 3 infants, they need 8 adults. **51.** The possible energy costs are $1230 for 1 person; $615 for 2 people; $410 for 3 people; $246 for 5 people; $205 for 6 people; and $123 for 10 people.
53. The band members can march in 6 ways: 1 by 72; 2 by 36; 3 by 24; 4 by 18; 6 by 12; and 8 by 9.
55. Addicted to love **57.** The difference is the process used to discover the relationship. If the process is multiplication, we generally use the word "factor." If the process is division, we generally use the word "divisor." **59.** 269 **63.** 2499 **65.** 38,016
67. 31 **69.** 24 R 40 **71.** There is enough wire to wire 41 speakers with 16 feet left.

Exercises 3.4

1. Composite **3.** Prime **5.** Composite
7. Prime **9.** Composite **11.** Prime
13. Composite **15.** Prime **17.** Composite
19. Prime **21.** Composite **23.** Composite
25. Prime **27.** Composite **29.** Prime
31. Composite **33.** Prime **35.** Composite
37. Composite **39.** Composite **41.** Composite
43. Composite **45.** Composite **47.** Prime
49. Prime **51.** Composite **53.** Composite
55. The prime number year after 1999 is 2003.
57. No, 1903 is not a prime number because it is divisible by 11. **59.** The next prime number after 47 is 53.
61. Only one arrangement is possible because 17 is a prime number. **63.** BIOOO EUNFV UTGAE LDMLH OAOLO VRSIW ELTLY M **65.** A prime number has only two factors, itself and 1. The number 7 is a prime number. A composite number has three or more factors, that is, at least itself, 1, and another factor. The number 8 is composite because the factors of 8 are 1, 2, 4, and 8. Composite numbers are composed of numbers other than itself and 1. **67.** There is no limit to the number of factors a composite number can have. If you started multiplying counting numbers together, no matter when you stopped, the product would be a composite number (with a lot of factors!). **69.** Composite
73. True **75.** 28,561 **77.** 2,900,000
79. 645 **81.** 7524

Exercises 3.5

1. $2^2 \cdot 3$ **3.** 2^4 **5.** $2^2 \cdot 5$ **7.** $2^3 \cdot 3$
9. $2^2 \cdot 7$ **11.** 2^5 **13.** $2^4 \cdot 3$ **15.** $2^3 \cdot 3^2$
17. 3^4 **19.** $2 \cdot 3^2 \cdot 5$ **21.** $7 \cdot 13$ **23.** 2^6
25. $2 \cdot 3 \cdot 17$ **27.** $2^2 \cdot 3 \cdot 11$ **29.** $3^2 \cdot 17$
31. $2^2 \cdot 3^2 \cdot 5$ **33.** $2 \cdot 7 \cdot 13$ **35.** $3^2 \cdot 5^2$
37. $2 \cdot 3^2 \cdot 17$ **39.** $2^2 \cdot 3^2 \cdot 11$ **41.** $3 \cdot 101$
43. $17 \cdot 19$ **45.** $11 \cdot 19$ **47.** Prime **49.** 5^4
51. $2 \cdot 3 \cdot 7 \cdot 29$ **53.** Answers will vary
55. Answers will vary **57.** I'm everything I am because you love me. **59.** A number is written in prime factored form when it is the product of prime numbers. No composite factors are allowed. For example $18 = 2 \cdot 3^2$ is in prime factored form whereas $18 = 2 \cdot 9$ and $18 = 3 \cdot 6$ are not. **61.** $2 \cdot 3 \cdot 7 \cdot 71$ **65.** Yes
67. Yes **69.** Yes **71.** The perimeter is 56 in. **73.** The volume is 1331 cm^3.

Exercises 3.6

1. 8 **3.** 10 **5.** 10 **7.** 14 **9.** 12
11. 24 **13.** 12 **15.** 18 **17.** 24
19. 30 **21.** 30 **23.** 48 **25.** 40
27. 48 **29.** 72 **31.** 120 **33.** 48
35. 24 **37.** 200 **39.** 72 **41.** 72
43. 204 **45.** 1400 **47.** 24 **49.** 240
51. Answers will vary **53.** The least price they could pay is \$100, which is the LCM of 20, 50, and 5.
55. Light of the world, shine on me, love is the answer.
57. The LCM must have a factor of 2^2 because 2^2 is a factor of 100. The correct LCM is $2^2 \cdot 3 \cdot 5^2 = 300$
59. 768 **61.** 18,000 **63.** 1, 3, 5, 15, 25, 75, 125, 375 **65.** 1, 2, 4, 8, 61, 122, 244, 488
67. Yes **69.** No **71.** They must sell at least 9 more appliances.

Chapter 3 True-False Concept Review

1. False. Not all multiples of 6 end with the digit 6. For example, 12 is a multiple of 6. **2.** True
3. True **4.** True **5.** True **6.** True
7. False. Only one multiple of 200 is also a factor of 200, —itself. **8.** False. The square of 200 is 40,000. One half of 200 is 100. **9.** False. Not all natural numbers ending in 4 are divisible by 4; for example, 54 is not divisible by 4. **10.** True **11.** False. Not all natural numbers ending in 9 are divisible by 3. For example, 19 is not divisible by 3. **12.** True **13.** False. The number 123,321,234 is *not* divisible by 4. It is divisible by 2, 3, and 6. **14.** True **15.** False. Two is the only prime number that is not odd. **16.** False. Not all composite numbers end in 1, 3, 7, or 9. All even numbers larger than 2 are composite as are all numbers larger than 5 that end in 5. **17.** False. Every composite number has three or more factors. **18.** False. Every prime number has exactly two factors. **19.** True
20. False. All of the prime factors of a composite number are smaller than the number. **21.** True **22.** True
23. False. The largest divisor of the least common multiple (LCM) of three numbers is not necessarily one of the three numbers. For example, the LCM of 4, 6, and 8 is 24, whose largest divisor is 24. **24.** False. It is not possible for a group of numbers to have two different LCM's.
25. True

Chapter 3 Test

1. Yes **2.** 1, 2, 5, 10, 11, 22, 55, 110
3. Yes **4.** Yes **5.** 160 **6.** $1 \cdot 75, 3 \cdot 25, 5 \cdot 15$ **7.** $2^2 \cdot 5 \cdot 13$ **8.** 252 **9.** $2 \cdot 3^2 \cdot 47$
10. 104, 117, 130, 143 **11.** No **12.** Prime
13. Composite **14.** $7 \cdot 11^2$ **15.** 504 **16.** 2
17. 198 **18.** 360 **19.** No, a set of prime factors has only one product. **20.** Any two of the following sets of numbers: 2, 3, 5 or 2, 3, 10 or 2, 3, 15 or 2, 3, 30 or 2, 5, 30 or 3, 5, 10 or 3, 5, 30 or 5, 6, 30 or 10, 15, 30.

CHAPTER 4

Exercises 4.1

1. $\frac{4}{7}$ **3.** $\frac{4}{7}$ **5.** $\frac{4}{5}$ **7.** $\frac{4}{3}$ **9.** $\frac{5}{3}$

11. $\frac{13}{9}$ **13.** Proper fractions: $\frac{3}{7}, \frac{4}{7}, \frac{5}{7}, \frac{6}{7}$; Improper fractions: $\frac{7}{7}, \frac{8}{7}, \frac{9}{7}$ **15.** Proper fractions: $\frac{7}{13}, \frac{8}{15}, \frac{10}{13}, \frac{11}{15}, \frac{12}{23}$; Improper fractions: none

17. Proper fractions: $\frac{10}{11}, \frac{3}{5}$; Improper fractions: $\frac{7}{4}, \frac{13}{13}, \frac{20}{19}$ **19.** Proper fractions: $\frac{9}{10}, \frac{102}{103}$; Improper fractions: $\frac{5}{5}, \frac{18}{18}, \frac{147}{147}$ **21.** $42\frac{3}{5}$ **23.** $26\frac{5}{8}$

25. $12\frac{4}{9}$ **27.** $7\frac{7}{13}$ **29.** $14\frac{21}{22}$ **31.** $22\frac{13}{17}$

33. $\frac{39}{7}$ **35.** $\frac{12}{1}$ **37.** $\frac{31}{4}$ **39.** $\frac{89}{3}$

41. $\frac{77}{6}$ **43.** $\frac{155}{8}$

45. The fraction $\frac{0}{\boxed{1}}$ is a proper fraction. Any whole number except 0 can be put in the denominator.
47. The error is that the whole number and numerator are added. $15\frac{2}{7} = \frac{107}{7}$ **49.** $\frac{7}{10}$ **51.** $\frac{11}{8}$

53.

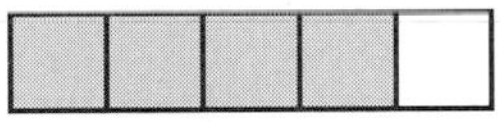

55. Of the class, $\frac{13}{21}$ are women.

57. Only $\frac{5}{6}$ of the cylinders are firing.

59. The tank is $\frac{3}{8}$ full.

61. The number 3 is closest to the mark.
63. The company needs 2495 sections.

65. The processor can make $2722\frac{17}{24}$ cases.

67. She can use five $\frac{1}{2}$-cup measures.

69. She can use seven $\frac{1}{4}$-tsp measures.

71. If you have 55 eggs, for example, a mixed number will help to identify how many dozen eggs there are.

Divide 55 by 12 to see that $\frac{55}{12} = 4\frac{7}{12}$. There are $4\frac{7}{12}$ dozen eggs. On the other hand, the number of eggs in $7\frac{5}{12}$ eggs is the numerator of the improper fraction. So $7\frac{5}{12} = \frac{89}{12}$ tells us that there are 89 eggs.

73. $\frac{117}{9}$; $\frac{1521}{117}$

75. They can make $2263\frac{2}{30}$ special boxes and $90\frac{13}{25}$ cartons. The shipping cost for full cartons is $4050.

81. 9070 **83.** 65,536 **85.** 4600

87. Bonnie went 935 miles.

89. The motor makes 8125 revolutions per minute.

Exercises 4.2

1. $\frac{1}{2}$ **3.** $\frac{2}{3}$ **5.** $\frac{2}{5}$ **7.** $\frac{3}{5}$ **9.** $\frac{3}{4}$

11. $\frac{3}{5}$ **13.** $\frac{4}{5}$ **15.** $\frac{5}{3}$ **17.** $\frac{7}{9}$ **19.** $\frac{3}{5}$

21. 4 **23.** $\frac{7}{3}$ **25.** $\frac{1}{3}$ **27.** $\frac{1}{3}$ **29.** $\frac{3}{4}$

31. $\frac{29}{36}$ **33.** $\frac{2}{3}$ **35.** $\frac{11}{15}$ **37.** $\frac{3}{4}$

39. $\frac{9}{16}$ **41.** $\frac{3}{4}$ **43.** 6 **45.** $\frac{3}{5}$ **47.** $\frac{16}{21}$

49. $\frac{3}{4}$ **51.** $\frac{34}{49}$ **53.** $\frac{5}{6}$ **55.** $\frac{2}{3}$ **57.** $\frac{3}{5}$

59. There are $\frac{2}{3}$ of the plugs fouled.

61. She has $\frac{3}{10}$ of her shift left.

63. Of the salmon, $\frac{11}{216}$ are cohos.

65. The student answered $\frac{7}{10}$ correctly.

67. To produce a pound of recycled rubber $\frac{23}{79}$ of the BTU's are needed.

69. The profit is $\frac{7}{36}$ of the selling price.

71. Prime factor both numerator and denominator. Then eliminate the common factors. The product of the remaining factors form the simplified fraction.

$$\frac{(\cancel{3})(\cancel{5})(\cancel{5})(7)}{(\cancel{3})(3)(\cancel{5})(\cancel{5})(5)} = \frac{7}{15}$$

73. Yes, all simplify to $\frac{5}{12}$. **77.** 1906 **79.** 2212

81. 2015 **83.** $2 \cdot 2 \cdot 2 \cdot 2 \cdot 2 \cdot 2 \cdot 5$ or $2^6 \cdot 5$

85. Trudy must average 14 pages a day.

Exercises 4.3

1. $\frac{2}{25}$ **3.** $\frac{15}{28}$ **5.** $\frac{7}{8}$ **7.** $\frac{1}{4}$ **9.** $\frac{1}{6}$

11. $\frac{1}{3}$ **13.** $\frac{2}{3}$ **15.** 1 **17.** $\frac{5}{9}$ **19.** 0

21. $\frac{8}{3}$ **23.** $\frac{1}{5}$ **25.** $\frac{2}{9}$ **27.** $\frac{14}{17}$ **29.** 8

31. $\frac{27}{28}$ **33.** $\frac{3}{8}$ **35.** $\frac{1}{3}$ **37.** $\frac{2}{3}$ **39.** 1

41. $\frac{64}{15}$ **43.** $\frac{3}{4}$ **45.** $\frac{21}{50}$ **47.** $\frac{10}{11}$

49. 1 **51.** $\frac{7}{12}$ **53.** $\frac{1}{2}$ **55.** $\frac{2}{33}$ **57.** $\frac{8}{3}$

59. $\frac{2}{3} \cdot \frac{\boxed{5}}{16} = \frac{5}{24}$; $\frac{2}{3} \cdot \frac{\boxed{5}}{16}$ can be simplified to $\frac{1}{3} \cdot \frac{\boxed{5}}{8}$, then we multiply 1 by 5 to get 5.

61. The error comes from multiplying the numerators, 3 and 4, times the denominators, 5 and 8. The numerators and denominators are multiplied by each other. $\frac{3}{8} \cdot \frac{4}{5} = \frac{12}{40}$ or $\frac{3}{10}$.

63. Three fourths full, it holds $\frac{21}{20}$ or $1\frac{1}{20}$ gallons

65. The sale price is $64.

67. One package will feed 10 gerbils.

69. Melvin might save 41 gallons.

71. **a.** Sandy needs 140 cookies.

b. Six recipes will make 144 cookies.

c. She needs 3 tsp of salt.

d. She needs $4\frac{1}{2}$ cups of brown sugar.

73. **a.** Sally would have to make $\frac{3}{8}$ of the recipe.

b. Sally needs $1\frac{1}{8}$ lb of ricotta cheese.

75. Multiply $\frac{35}{24}$ by $\frac{14}{40}$ $\left(\text{the reciprocal of } \frac{40}{14}.\right)$ Then simplify. $\frac{35}{24} \cdot \frac{14}{40} = \frac{49}{96}$. **77.** $\frac{9}{10}$ **83.** 120

85. 238,000 **87.** $6\frac{1}{13}$ **89.** No, she will not.

91. Only $\frac{5}{9}$ of the class bring their books.

Exercises 4.4

1. $1\frac{5}{16}$ **3.** 12 **5.** 8 **7.** $22\frac{1}{2}$ **9.** 24

11. $3\frac{1}{3}$ **13.** $3\frac{1}{16}$ **15.** 0 **17.** 75

19. $18\frac{3}{4}$ **21.** $35\frac{1}{5}$ **23.** 105 **25.** 2

27. $\frac{3}{4}$ **29.** $2\frac{1}{12}$ **31.** $\frac{13}{18}$ **33.** $1\frac{1}{3}$

35. $\frac{2}{3}$ **37.** $\frac{7}{30}$ **39.** $\frac{2}{3}$ **41.** $18\frac{1}{3}$

43. $8\frac{1}{28}$ **45.** $28\frac{1}{5}$ **47.** $4\frac{2}{3}$ **49.** 186

51. $5\frac{5}{6}$

53. Mixed numbers cannot be multiplied by multiplying the whole number parts and the fraction parts separately. Change them to fractions first. $\frac{5}{3} \cdot \frac{3}{2} = \frac{5}{2} = 2\frac{1}{2}$

55. The iron content is 22 parts per million.
57. Karla can make 192 glasses of juice.
59. The reduced pressure is 39 pounds per square inch.
61. The less-efficient car emits 3000 pounds of CO_2.

63. The area is $10\frac{5}{16}$ in^2.

65. **a.** Sandy needs $22\frac{1}{2}$ c flour. **b.** Sandy needs 15 c sugar.

67. **a.** Quan can make $\frac{2}{3}$ of the recipe.
b. Quan should use 4 tbs butter.
c. Quan should use 1 tsp vanilla.

69. When we divide by $1\frac{1}{2}$ we are dividing by $\frac{3}{2}$. Dividing by $\frac{3}{2}$ is the same as multiplying by $\frac{2}{3}$. Since $\frac{2}{3}$ is less than one, the multiplication gives us a smaller number.

71. 1230 **75.** 15,153 **77.** 41,580 **79.** 360

81. $36\frac{13}{15}$

83. No, because 234,572 is not a multiple of 3.

Getting Ready for Algebra, page 331

1. $x = \frac{3}{4}$ **3.** $y = \frac{16}{15}$ or $y = 1\frac{1}{15}$ **5.** $z = \frac{15}{16}$

7. $\frac{17}{8} = x$ or $2\frac{1}{8} = x$ **9.** $a = \frac{10}{7}$ or $a = 1\frac{3}{7}$

11. $\frac{3}{2} = b$ or $1\frac{1}{2} = b$ **13.** $z = 2$

15. $a = \frac{165}{8}$ or $a = 20\frac{5}{8}$

17. The distance is $\frac{3}{2}$ miles or $1\frac{1}{2}$ miles.

19. There were 180 lb of tin recycled.

Exercises 4.5

1. 14 days **3.** 24 months **5.** 24 in.
7. 5280 ft **9.** 200 cm **11.** 9 ft^2 **13.** 5 lb
15. 150 mm **17.** 5 gal **19.** 6 kg

21. 10,000 g **23.** 3 yd **25.** 500 cm

27. 12 lb **29.** 1 mm **31.** $2\frac{2}{3}$ ft

33. 120 in. **35.** $1\frac{1}{2}$ tons **37.** 7 days

39. 13,000 lb **41.** 66 ft/sec **43.** 1500 lb/in.
45. 50 m/sec **47.** 24,000 kg/km^2

49. He can type $1\frac{1}{2}$ words per second.

51. The order requires 150 cc.
53. They donate $240,000.

55. It will take $4\frac{4}{5}$ minutes.

57. One cubic inch of the alloy weighs about 1 oz.
59. There are 30 g per bottle for a cost of $6.
61. Multiply by conversion factors that will eliminate the ft^2/sec, and result in mi^2/hr. These factors are $\frac{1 \text{ mi}}{5280 \text{ ft}}$ (used twice), $\frac{60 \text{ sec}}{1 \text{ min}}$, and $\frac{60 \text{ min}}{1 \text{ hr}}$. $\frac{x \text{ ft}^2}{\text{sec}} \cdot \frac{1 \text{ mi}}{5280 \text{ ft}} \cdot$
$\frac{1 \text{ mi}}{5280 \text{ ft}} \cdot \frac{60 \text{ sec}}{1 \text{ min}} \cdot \frac{60 \text{ min}}{1 \text{ hr}} = \frac{x\,(60)(60) \text{ mi}^2}{(5280)(5280) \text{ hr}} = \frac{x \text{ mi}^2}{7744 \text{ hr}}$

63. The profit is $2500 per kilogram. **67.** $\frac{2}{5}$

69. $\frac{5}{4}$ or $1\frac{1}{4}$ **71.** $2 \cdot 3^2 \cdot 7 \cdot 17$

73. The area is 695 cm^2.

75. The volume is 18,144 in^3 or $10\frac{1}{2}$ ft^3.

Exercises 4.6

1. $\frac{4}{6}, \frac{6}{9}, \frac{8}{12}, \frac{10}{15}$ **3.** $\frac{14}{16}, \frac{21}{24}, \frac{28}{32}, \frac{35}{40}$

5. $\frac{8}{18}, \frac{12}{27}, \frac{16}{36}, \frac{20}{45}$ **7.** $\frac{8}{22}, \frac{12}{33}, \frac{16}{44}, \frac{20}{55}$

9. $\frac{14}{6}, \frac{21}{9}, \frac{28}{12}, \frac{35}{15}$ **11.** 5 **13.** 10

15. 16 **17.** 27 **19.** 15 **21.** 8 **23.** 46

25. 140 **27.** 57 **29.** 120 **31.** $\frac{3}{7}, \frac{4}{7}, \frac{5}{7}$

33. $\frac{1}{4}, \frac{3}{8}, \frac{1}{2}$ **35.** $\frac{1}{3}, \frac{3}{8}, \frac{1}{2}$ **37.** True

39. False **41.** True **43.** $\frac{2}{3}, \frac{3}{4}, \frac{4}{5}$

45. $\frac{4}{5}, \frac{5}{6}, \frac{13}{15}, \frac{9}{10}$ **47.** $\frac{11}{24}, \frac{17}{36}, \frac{35}{72}$

49. $\frac{6}{14}, \frac{13}{28}, \frac{17}{35}$ **51.** $2\frac{3}{4}, 2\frac{5}{6}, 2\frac{7}{8}$

53. False **55.** True **57.** True

59. LCM = 24; $\frac{12}{24}, \frac{16}{24}, \frac{4}{24}, \frac{15}{24}$

61. Janie must get 32 problems correct.

63. From smallest to largest the diameters are $\frac{11}{16}$, $\frac{3}{4}$, $\frac{7}{8}$, $1\frac{1}{16}$, $1\frac{3}{32}$, and $1\frac{1}{8}$ inches.
65. Chang's measure is heaviest.
67. There are 90 oz.
69. **a.** Pecan chocolate chip cookies gives them the bigger profit.
b. Three-fourths of a cup is more than she needs.
71. To simplify a fraction is to find a fraction with smaller numerator and smaller denominator that is equivalent to the original fraction. To build a fraction is to find a fraction with a larger numerator and larger denominator that is equivalent to the original.
73. $\frac{50}{70}$, $\frac{65}{91}$, $\frac{115}{161}$, $\frac{560}{784}$, $\frac{2905}{4067}$ **77.** Composite
79. 65,460,000 **81.** $\frac{7}{8}$ **83.** $18\frac{1}{12}$
85. The pharmacist will give her 56 capsules.

Exercises 4.7

1. $\frac{9}{11}$ **3.** $\frac{2}{3}$ **5.** 2 **7.** $\frac{4}{5}$ **9.** $\frac{10}{13}$
11. $1\frac{1}{4}$ **13.** $\frac{1}{2}$ **15.** $\frac{5}{16}$ **17.** $\frac{3}{5}$
19. $\frac{7}{8}$ **21.** $\frac{19}{24}$ **23.** $\frac{2}{3}$ **25.** $\frac{13}{16}$
27. $\frac{19}{30}$ **29.** $\frac{9}{10}$ **31.** $\frac{7}{15}$ **33.** $1\frac{7}{60}$
35. $1\frac{2}{5}$ **37.** $1\frac{5}{16}$ **39.** $2\frac{5}{8}$ **41.** $1\frac{77}{144}$
43. $2\frac{19}{90}$ **45.** $\frac{3}{5}$ **47.** $\frac{7}{15}$ **49.** $1\frac{13}{144}$
51. $1\frac{3}{10}$ **53.** $\frac{57}{80}$
55. When adding fractions we need a common denominator for the addends as well as the sum. $\frac{1}{2}+\frac{4}{7}=\frac{7}{14}+\frac{8}{14}=\frac{15}{14}$.
57. She advises a total of $2\frac{3}{16}$ miles.
59. Jonnie needs a bolt that is $1\frac{13}{16}$ inches long.
61. The perimeter is $1\frac{11}{12}$ yd.
63. The bamboo grew $1\frac{1}{8}$ inches.
65. The length of the rod is $3\frac{3}{8}$ inches.
67. **a.** The three recipes require $1\frac{1}{4}$ tsp baking soda.
b. The order requires $3\frac{3}{4}$ tsp baking soda.
c. The three recipes require 2 tsp salt.
d. The order requires 3 tsp salt.
69. Each denominator indicates how many parts a unit has been divided into. Unless the units are the same size it would be like adding apples and oranges. You cannot add units of different sizes. **71.** $\frac{4}{7}$
73. Jim consumes 4 g of fat, which is $\frac{6}{55}$ of the calories.
75. $2 \cdot 2 \cdot 2 \cdot 31$ or $2^3 \cdot 31$ **77.** 678,224,000,000
79. 42 **81.** 48 **83.** Each should sell for $461.

Exercises 4.8

1. $3\frac{6}{7}$ **3.** $14\frac{2}{5}$ **5.** $8\frac{1}{6}$ **7.** $6\frac{2}{9}$
9. $6\frac{5}{14}$ **11.** $15\frac{13}{15}$ **13.** $15\frac{1}{6}$ **15.** 11
17. $26\frac{5}{8}$ **19.** $13\frac{5}{24}$ **21.** $10\frac{1}{6}$ **23.** $25\frac{43}{80}$
25. $49\frac{5}{6}$ **27.** $719\frac{31}{36}$ **29.** $86\frac{11}{30}$
31. $60\frac{7}{12}$ **33.** $118\frac{1}{45}$ **35.** $103\frac{11}{70}$
37. $62\frac{3}{20}$ **39.** $12\frac{14}{15}$ **41.** $48\frac{7}{60}$
43. $106\frac{5}{72}$ **45.** $64\frac{3}{8}$
47. Juanita worked $76\frac{3}{4}$ hr.
49. The perimeter is $84\frac{1}{2}$ feet.
51. The bolt is $2\frac{1}{8}$ in. long.
53. The total rainfall for the six months was $21\frac{1}{24}$ inches, which is 50,500,000 gallons.
55. **a.** The three recipes will require $6\frac{1}{4}$ c flour.
b. The order requires $31\frac{1}{4}$ c flour.
c. The three recipes require $3\frac{1}{2}$ c sugar.
d. The order requires $17\frac{1}{2}$ c sugar.
57. The sum of two mixed numbers sometimes results in an improper fraction. When this occurs we must rename the sum. For example, $3\frac{3}{4}+5\frac{3}{4}=8\frac{6}{4}=8+1\frac{2}{4}=9\frac{1}{2}$.
59. No, the statement is not true.

61. The snowfall total is $39\frac{3}{20}$ in., which exceeds the average. **63.** $\frac{1}{3}$ **65.** $\frac{8}{45}$ **67.** $3 \cdot 37$ **69.** $\frac{10}{21}$ **71.** Her score is 35, which is under par.

Exercises 4.9

1. $\frac{1}{4}$ **3.** $\frac{1}{4}$ **5.** $\frac{1}{3}$ **7.** $\frac{1}{2}$ **9.** $\frac{7}{16}$

11. $\frac{7}{45}$ **13.** $\frac{1}{2}$ **15.** $\frac{5}{18}$ **17.** $\frac{13}{20}$

19. $\frac{9}{20}$ **21.** $\frac{1}{24}$ **23.** $\frac{17}{48}$ **25.** $\frac{4}{21}$

27. $\frac{19}{48}$ **29.** $\frac{1}{18}$ **31.** $\frac{13}{24}$ **33.** $\frac{9}{20}$

35. $\frac{5}{36}$ **37.** $\frac{19}{75}$ **39.** $\frac{17}{48}$ **41.** $\frac{1}{36}$

43. $\frac{7}{30}$ **45.** $\frac{23}{60}$

47. The thickness will be $1\frac{1}{8}$ in.

49. She has $\frac{5}{12}$ oz left.

51. Lake Tuscumba has the greatest content by one part per million $\left(\frac{1}{1{,}000{,}000}\right)$.

53. John has grown the most, by $\frac{1}{24}$ in.

55. The carpenter removes $\frac{3}{16}$ inch.

57. The project will require 102 bricks but four of them will be cut to fit. One brick $3\frac{3}{4}$ in. wide, plus the mortar on one end, $\frac{3}{8}$ in., has a width of $4\frac{1}{8}$ in. The length of the garden is 10 ft or 120 in. We must add 16 in. to the length to account for the border on the ends. The width of the garden, is 72 in. We do not need to add on to the width. Each length requires $136 \div 4\frac{1}{8}$ bricks or about 33 bricks. Each width requires $72 \div 4\frac{1}{8}$ bricks or about 18 bricks.

59. I would show the child how fractions can be represented using unit regions with physical objects. Perhaps use a cardboard circle or rectangle that is cut into pieces. If the pieces or units are not the same size, the child could see that we cannot subtract (as we can with whole numbers.) By using the same size pieces (units), that is, common denominators, we can then remove (take away) the smaller number from the larger number of pieces and count the pieces left over.

61. $\frac{835}{4816}$ **63.** Yes, Skola outdistances Sheila by $\frac{1}{120}$ mi. The donor contributes $44,000.

65. 22,654 **67.** 2080 R 102 **69.** $\frac{5}{7}$

71. $\frac{9}{28}$ **73.** They can lay 11,925 bricks.

Exercises 4.10

1. $7\frac{2}{7}$ **3.** $103\frac{1}{5}$ **5.** $5\frac{1}{8}$ **7.** $6\frac{4}{7}$

9. $118\frac{1}{6}$ **11.** $2\frac{1}{2}$ **13.** $15\frac{1}{5}$ **15.** $155\frac{1}{9}$

17. $1\frac{5}{8}$ **19.** $9\frac{1}{2}$ **21.** $56\frac{1}{3}$ **23.** $26\frac{23}{60}$

25. $21\frac{3}{4}$ **27.** $7\frac{29}{48}$ **29.** $28\frac{1}{3}$ **31.** $5\frac{25}{32}$

33. $7\frac{11}{36}$ **35.** $22\frac{5}{6}$ **37.** $36\frac{17}{120}$

39. $27\frac{31}{36}$ **41.** $7\frac{2}{15}$ **43.** $1\frac{3}{20}$ **45.** $\frac{5}{6}$

47. If "1" is borrowed from 16 in order to subtract the fraction part, $\frac{1}{4}$, then the whole number part is $15 - 13$ or 2. So $16 - 13\frac{1}{4} = \left(15 + \frac{4}{4}\right) - \left(13 + \frac{1}{4}\right) = 2\frac{3}{4}$.

49. $1\frac{74}{75}$ **51.** She trims $1\frac{1}{2}$ lb.

53. He has $18\frac{9}{20}$ tons left.

55. They have $17\frac{7}{10}$ miles to go.

57. The Williams recycle $381\frac{3}{10}$ more pounds.

59. **a.** During the year, $3\frac{5}{8}$ in. more rain falls in Westport. **b.** In a 10-yr period, $277\frac{1}{12}$ in. more rain falls in Salem. **c.** Westview would get $28\frac{1}{6}$ in. more rain. **61.** This recipe requires $\frac{15}{16}$ c milk.

63. When you "borrow" 1 to subtract mixed numbers, the fraction for 1 is in the form: $\frac{\text{denominator}}{\text{denominator}}$, where we use the common denominator of the fractions. We do this so the fractions can be subtracted.

65. Yes, they are equal. **69.** $\frac{2}{3}$ **71.** $\frac{255}{8}$

73. $\frac{8}{15}$ **75.** 2

77. The shipping weight is 1160 oz or $72\frac{1}{2}$ lb.

Getting Ready for Algebra, page 411

1. $a = \frac{1}{2}$ **3.** $c = \frac{5}{8}$ **5.** $x = \frac{11}{72}$

7. $y = 1\frac{38}{63}$ **9.** $a = 1\frac{11}{40}$ **11.** $c = 4\frac{1}{6}$

13. $x = 3\frac{5}{36}$ **15.** $3\frac{1}{6} = w$ **17.** $a = 36\frac{4}{9}$

19. $c = 21\frac{2}{21}$ **21.** The share price was $91\frac{5}{8}$ points.

23. She bought 46 lb of nails.

Exercises 4.11

1. $\frac{2}{9}$ **3.** $\frac{1}{7}$ **5.** $\frac{1}{2}$ **7.** 1 **9.** 0

11. $1\frac{7}{12}$ **13.** $\frac{11}{12}$ **15.** $\frac{3}{8}$ **17.** $\frac{3}{5}$

19. $\frac{5}{6}$ **21.** $\frac{2}{3}$ **23.** 0 **25.** $1\frac{13}{64}$

27. $\frac{1}{24}$ **29.** $\frac{4}{9}$ **31.** $\frac{3}{7}$ **33.** $\frac{10}{21}$

35. $\frac{35}{36}$ **37.** 4 **39.** $1\frac{1}{16}$ **41.** $\frac{15}{32}$

43. $3\frac{37}{54}$ **45.** $7\frac{1}{5}$ **47.** $C \approx 44$ cm

49. $C \approx 88$ ft **51.** $C \approx \frac{110}{7}$ m or $15\frac{5}{7}$ m

53. $V \approx 88 \text{ in}^3$ **55.** $P \approx \frac{288}{7}$ ft or $41\frac{1}{7}$ ft

57. $A \approx \frac{1782}{7} \text{ in}^2$ or $254\frac{4}{7} \text{ in}^2$ **59.** $A \approx 121\frac{23}{28} \text{ in}^2$

61. $V \approx 6373\frac{5}{7} \text{ in}^3$ or $3\frac{347}{504} \text{ ft}^3$ **63.** $1\frac{5}{18}$

65. 1 **67.** $4\frac{19}{54}$ **69.** $3\frac{8}{15}$

71. $A \approx \frac{14}{11} \text{ in}^2$ or $1\frac{3}{11} \text{ in}^2$

73. The average length is $33\frac{2}{3}$ in.

75. The class average is $\frac{7}{10}$ correct.

77. The average is $\frac{22}{25}$ correct.

79. There are $108\frac{1}{4}$ oz of seafood. The average cost is 48¢ per ounce. **81.** $V \approx 10{,}899\frac{3}{7} \text{ mm}^3$

83. **a.** The average amount is $2\frac{1}{12}$ c of flour.

b. The average amount is $1\frac{1}{6}$ c sugar.

85. 1. Do operations in parentheses first, following steps 2, 3, and 4. 2. Do exponents next. 3. Do multiplication and division as they occur. 4. Do addition and subtraction as they occur. The order is the same as for whole numbers.

87. $2\frac{57}{1000}$ **89.** The total amount paid was $19,444.

91. $\frac{7}{48}$ **93.** $1\frac{7}{8}$ **95.** $\frac{1}{7}$ **97.** $2 \cdot 5^2 \cdot 13$

99. The table top is $1\frac{1}{4}$ in. thick.

Chapter 4 True-False Concept Review

1. False. Use more than one unit with one partially shaded and the rest totally shaded. **2.** True **3.** False. The numerator always equals the denominator, so the fraction is improper. **4.** True **5.** True **6.** False. Every improper fraction can be changed to a mixed number or a whole number. **7.** True **8.** True **9.** True **10.** True **11.** True **12.** True **13.** False. Unlike fractions have different denominators. **14.** False. Add the whole numbers and add the fractions. **15.** True **16.** True **17.** True **18.** False. The product of two fractions is sometimes smaller than either of the fractions.

Chapter 4 Test

1. $3\frac{13}{16}$ **2.** $1\frac{7}{24}$ **3.** $\frac{79}{9}$ **4.** $\frac{3}{10}, \frac{3}{8}, \frac{2}{5}$

5. $\frac{17}{1}$ **6.** 35 **7.** $9\frac{2}{15}$ **8.** $17\frac{23}{40}$ **9.** $\frac{3}{4}$

10. $\frac{2}{3}$ **11.** $6\frac{4}{5}$ **12.** $\frac{5}{8}$ **13.** $\frac{1}{2}$ **14.** $\frac{3}{8}$

15. $\frac{12}{35}$ **16.** $6\frac{13}{20}$ **17.** $\frac{3}{8}$ **18.** $\frac{2}{5}$

19. $\frac{8}{21}$ **20.** $3\frac{1}{5}$ **21.** $\frac{7}{8}, \frac{7}{9}, \frac{8}{9}$ **22.** $3\frac{3}{4}$

23. $8\frac{13}{40}$ **24.** $\frac{7}{12}$ **25.** $5\frac{5}{12}$ **26.** $\frac{5}{7}$

27. $1\frac{3}{8}$ **28.** $\frac{9}{50}$ **29.** True **30.** $\frac{5}{5}, \frac{6}{6}, \frac{7}{7}$

31. There are 22 truckloads of hay in the rail car.
32. She needs to make 25 lb of candy.

CHAPTER 5

Exercises 5.1

1. twelve hundredths
3. two hundred sixty-seven thousandths
5. six and four ten-thousandths
7. four and sixty-seven hundredths **9.** 0.11
11. 0.111 **13.** 2.019 **15.** 0.0021
17. five hundred four thousandths
19. fifty and four hundredths
21. eighteen and two hundred five ten-thousandths
23. thirty and eight thousandths **25.** 0.012
27. 700.096 **29.** 505.005 **31.** 14.0014

	Unit	Tenth	Hundredth
33. 15.888	16	15.9	15.89
35. 477.774	478	477.8	477.77
37. 0.7392	1	0.7	0.74

39. \$33.54 **41.** \$246.49

	Ten	Hundredth	Thousandth
43. 12.5532	10	12.55	12.553
45. 245.2454	250	245.25	245.245
47. 0.5536	0	0.55	0.554

49. \$11 **51.** \$1129
53. The word name is "one and three tenths inches."
55. To the nearest tenth of an inch the precipitation in week four is 1.1 inches.
57. Dan writes "Sixty-four and seventy-nine hundredths dollars."
59. The position of the arrow to the nearest hundredth is 1.62.
61. The position of the arrow to the nearest tenth is 1.6.
63. five hundred sixty-seven and nine thousand twenty-three ten-thousandths
65. The account value to the nearest cent is \$1617.37.
67. 213.1101

	Hundred	Hundredth	Ten-thousandth
69. 982.678456	1000	982.69	982.6785
71. 86025.47782	86000	86025.48	86025.4778

73. To the nearest tenth of a mph the Stanley car was going 127.7 mph.
75. The digit after the round-off place is zero so we choose the smaller number.
77. In 43.29, the numeral 4 is in the tens place, which is two places to the left of the decimal point. This means there are four tens in the number. In 18.64, the 4 is in the hundredths place, which is two places to the right of the decimal point. This means that there are four hundredths in the number.
79. The rounded value, 8.283, is greater than the original value. $8.283 > 8.28282828$ or $8.28282828 < 8.283$
81. The place value name is 0.0000215. **85.** $8\frac{4}{15}$
87. $\frac{138}{5}$ **89.** $\frac{16}{15}$ or $1\frac{1}{15}$ **91.** $\frac{7}{11}$
93. From 200 inches of wire you can make $6\frac{2}{33}$ springs.

Exercises 5.2

1. $\frac{33}{100}$ **3.** $\frac{3}{4}$ **5.** $\frac{111}{1000}$ **7.** $\frac{17}{50}$ **9.** $\frac{12}{25}$
11. $2\frac{73}{100}$ **13.** $\frac{7}{8}$ **15.** $3\frac{16}{25}$ **17.** $3\frac{141}{250}$
19. $\frac{1}{5}$ **21.** 0.1, 0.6, 0.7 **23.** 0.05, 0.07, 0.6
25. 4.159, 4.16, 4.161 **27.** False **29.** True
31. 0.072, 0.0729, 0.073, 0.073001, 0.073015
33. 0.88579, 0.88799, 0.888, 0.8881
35. 20.004, 20.039, 20.04, 20.093 **37.** False
39. True **41.** As a reduced fraction the probability is $\frac{1}{8}$. **43.** As a simplified fraction the number of inches of precipitation in the first week is $\frac{3}{50}$ inch.
45. The least precipitation is in week one. The most precipitation is in week two.
47. Mitchell loses the most weight this week.
49. $\frac{821}{2000}$ **51.** $65\frac{3}{400}$
53. Hoa needs less soap.
55. 5.00088, 5.0009, 5.00091, 5.001, 5.00101
57. 22.01, 22.81, 22.89, 23.67, 23.76, 23.86, 23.98, 24.99
59. Maria should choose the $\frac{6}{7}$ yard.
61. Tim Hardaway made $\frac{167}{200}$ of his free throws. He missed $\frac{33}{200}$ of his free throws.
63. The fat grams are $8\frac{19}{20}$, $8\frac{9}{10}$, and $10\frac{17}{20}$. The least amount of fat is in the frozen hash browns with butter sauce.
65. $\frac{11}{25}, \frac{101}{250}, \frac{1011}{25000}$ **69.** $\frac{2}{9}$ **71.** $12\frac{8}{25}$
73. $23\frac{5}{8}$ **75.** 4 **77.** 10

Exercises 5.3

1. 0.7 **3.** 3.8 **5.** 7.7 **7.** 27.43 **9.** 3
11. 13.841 **13.** 23.40 **15.** 2.755
17. 9.0797 **19.** 82.593 **21.** 55.5363
23. 831.267 **25.** 26.981 **27.** 0.3 **29.** 3.4
31. 5.1 **33.** 6.41 **35.** 8.09 **37.** 0.457
39. 1.564 **41.** 2.609 **43.** 67.044 **45.** 2.76
47. 0.069 **49.** 0.17 **51.** 0.3 **53.** 6
55. 0.08 **57.** 1.6 **59.** 0.04; 0.04019
61. 0.11; 0.106794 **63.** 10; 11.6793
65. 1.3; 1.2635 **67.** 0.5; 0.4686
69. Manuel bought 49.1 gallons of gas on his trip.
71. 471.9
73. Sera is 0.778 seconds faster than Muthoni.
75. The relay race was completed in 38.95 sec.
77. The fourth sprinter must have a time of 13.16 sec.
79. $18.7 billion more was invested in 1992 than in 1989.
81. Doris' take home pay is $1945.90.
83. The drop in interest rate was 0.35 percent.
85. The smoke stack is 75 ft high.
87. The distance is 6.59375 in.
89. Align the decimal points to add decimals. This gives us columns with like place values. We then add the columns, carrying when necessary. So

$$\begin{array}{r} 2.005 \\ 8.2 \\ 0.00004 \\ +\ 3. \\ \hline 13.20504 \end{array}$$

91.

Operation on Decimals	Procedure	Example
Addition	Write the decimals in columns, aligning the decimal points. Now add the columns, carrying when necessary.	$\begin{array}{r} 3.45 \\ 12.06 \\ +\ 7.012 \\ \hline 22.522 \end{array}$
Subtraction	Write the decimals in columns, aligning the decimal points. Now, subtract the columns, borrowing when necessary.	$\begin{array}{r} {}^{6\,16} \\ 24.7\not{6} \\ -13.49 \\ \hline 11.27 \end{array}$

93. The missing number is 1.9.
95. The number is 978.889.
97. The rounded sum is 33.7.
99. $\frac{11}{8}$ or $1\frac{3}{8}$ **101.** $17\frac{1}{2}$ **103.** $\frac{3}{32}$
105. $8\frac{3}{4}$ **107.** The board will be $4\frac{3}{4}$ in. thick.

Getting Ready for Algebra, page 485

1. $11.5 = x$ **3.** $y = 14.98$ **5.** $t = 0.073$
7. $x = 12$ **9.** $2.62 = w$ **11.** $t = 7.23$
13. $a = 0.78$ **15.** $x = 6.56$ **17.** $a = 21.6$
19. $s = 6.289$ **21.** $c = 467.02$
23. Two years ago the price of the water heater was $406.97. **25.** The markup is $140.66.
27. Let C represent the cost of the groceries, then $\$19 + C = \$43$. The shopper can spend $24 on groceries.

Exercises 5.4

1. 3.2 **3.** 9 **5.** 0.27 **7.** 0.2 **9.** 0.015
11. 0.064 **13.** five **15.** 0.0149 **17.** 6.67
19. 4.1552 **21.** 2.6988 **23.** 30.1608
25. 0.117 **27.** 24 **29.** 0.0035 **31.** 1.6
33. 0.0018 **35.** 10; 11.045
37. 0.028; 0.023852 **39.** 0.4; 0.3822
41. 0.004; 0.004455
43. Grant purchased a total of 97.7 gallons of gas.
45. To the nearest cent, Grant paid $28.02 for the fifth fill-up.
47. Grant paid the least for his fill-up at $1.393 per gallon.
49. 400; 347.3184 **51.** 8000; 9576.252
53. 21,392.542 **55.** 835.15
57. Joanne pays $558.88 for 14.75 yards.
59. It costs $219 to rent a mid-size car.
61. It costs less to rent the compact car. It costs $13.56 less.
63. Store 3 is selling the freezer-refrigerator for the least total cost.
65. The weight of the bar is 71.725 lb.
67. The dive was awarded 66.99 points.
69. The total weight of red meat consumed is 528 lb in 1970, 505.6 lb in 1980, and 449.2 lb in 1990. It could be due to greater awareness of potential heart problems associated with the consumption of large quantities of red meat.
71. The number of gallons of water used by each type of toilet is as follows: pre-1970, 956,505 gallons; 1970s, 608,685 gallons; low-flow, 269,560.5 gallons. A total of 686,944.5 gallons of water is saved by using the low-flow model.
73. The Gregory Estate owes $23,413 in taxes.
75. The number of decimal places in a product is the sum of the number of decimal places in the individual factors. So the product (0.006)(3.2)(68) should have four decimal places, but 13.056 only had three places. This makes the answer suspect. This error could be due to careless entry. Beyond that, we could and should round and estimate the product as a check on the reasonableness of our answer. $(0.006)(3.2)(68) \approx (0.006)(3)(70) = 1.26$. Our estimate, 1.26, is not particularly close to 13.056, but it is close to 1.3056, which is the actual answer.

77. The largest whole number is 75.
79. The missing number is 0.0031. **81.** 44,500
83. 783 **85.** 12,000,000 **87.** 670
89. 230,000,000

Exercises 5.5

1. 0.425 **3.** 367 **5.** 628.33 **7.** 0.5692
9. 56.45 **11.** 8700 **13.** right **15.** 6274
17. 0.185 **19.** 36,900 **21.** 0.68953
23. 1,478,000 **25.** 13.6794 **27.** 0.000458
29. 2.3×10^5 **31.** 3.5×10^{-4}
33. 4.6795×10^2 **35.** 60,000 **37.** 0.008
39. 4780 **41.** 7.8×10^5 **43.** 3.45×10^{-5}
45. 8.21×10^{-11} **47.** 3.567003×10^3
49. 0.000001345 **51.** 7,110,000,000
53. 0.000000444 **55.** 567.23
57. The total cost of the shoes is $2229.
59. The total cost of Ms. James' land is $98,500.
61. The land area of the earth is 5.2×10^7 square miles.
63. They are multiplying the actual number by 1000.
65. The mathematically calculated average is 0.424.
67. It takes 3.3×10^{-6} sec for light to travel 1 km.
69. The earth is approximately 150,000,000 km from the sun.
71. The thickness of a sheet of paper is 0.0013 in.
73. The amount consumed by 100,000 people is 1.15×10^6 lb of fish, 6.36×10^6 lb of poultry, and 1.123×10^7 lb of red meat.
75. In 1990, $\$1.582 \times 10^9$ more was spent than in 1980. The average increase per year is $\$1.582 \times 10^8$.
77. The approximate length of one parsec is 1.92×10^{13} miles.
79. 6.5×10^6 **83.** 108 **85.** 28 R 4
87. 2416 **89.** 4820 **91.** The area is 648 in^2.

Exercises 5.6

1. 0.5 **3.** 9.8 **5.** 183.1 **7.** 2020
9. whole number **11.** 12.6 **13.** 0.52
15. 1.3 **17.** 9.1 **19.** 15.555 **21.** 0.02
23. 10 **25.** 1 **27.** 400 **29.** 100; 97.06
31. 0.07; 0.076 **33.** 10; 17.143 **35.** 83.55
37. To the nearest tenth of a cent, the unit price of oranges is $1.473 per pound.
39. To the nearest tenth of a cent, the unit price of rib steak is $2.969 per pound.
41. To the nearest cent, the cost of 7 lb of ham is $24.42.
43. The total cost is $27.08. **45.** 200; 212.64
47. 5; 6.21
49. Vern paid $4.15 for the pair of socks.
51. To the nearest day, the solvent will last 34 days.
53. The average cost per connection is $1588.87.
55. To the nearest tenth, the length of the beam is 19.3 ft.
57. The cost of the flue is $11.16.
59. To the nearest hundredth, the ERA is 2.93.
61. The stolen base average is 0.621.
63. The decimal point in a quotient is positioned directly above the decimal point in the dividend when dividing by a whole number. When dividing by a decimal, move the decimal point in the divisor so as to make it a whole number. Move the decimal point in the dividend the same number of places. Now the decimal point in the quotient is positioned above the relocated decimal in the dividend. This works because a division problem may be thought of as a fraction $\left(\frac{\text{dividend}}{\text{divisor}}\right)$, and moving the decimals in both the same number of places to the right is the same as multiplying a fraction by 1 in the form of some power of 10.
65.

Operation on Decimals	Procedure	Example
Division	If the divisor is not a whole number, move the decimal point to the right until it is a whole number. Move the decimal point in the dividend the same number of places to the right. Now place the decimal point in the quotient directly above the decimal point in the dividend. Now divide.	$0.25\overline{)3.5678}$ 14.27 $25\overline{)356.75}$ 25 106 100 67 50 175 175 0

69. $3\frac{1}{8}$ **71.** $11\frac{2}{9}$
73. Mr. Lewis gives away 140 books. He makes a profit of $32.50. **75.** 31 lb
77. Each piece is 11 in. long.

Getting Ready for Algebra, page 531

1. $x = 6$ **3.** $y = 204$ **5.** $0.06 = t$
7. $m = 0.04$ **9.** $q = 437.5$ **11.** $500 = h$
13. $y = 2.673$ **15.** $0.1032 = c$ **17.** $0.9775 = x$
19. $w = 0.0141$ **21.** $z = 30.16$
23. There are 25 servings.
25. The current is 8.4 amp.
27. The length of the rectangle is 21.5 ft.
29. Let S represent the number of students, then $20S = 3500$. The instructor has 175 students in her classes.

Exercises 5.7

1. 0.5 lb **3.** 1.5 cm **5.** 0.55 ℓ **7.** 2.3 yd
9. 36.24 in. **11.** 1830 mm **13.** 24,076.8 ft

15. 40.64 cm **17.** 4.76 qt **19.** 14.92 in.
21. $0.07\ \text{ft}^2$ **23.** $5.34\ \text{m}^2$ **25.** 4.5 m
27. 7.0 lb **29.** 5.1 mi **31.** 177.4 cℓ

33. $1.35\ \text{m}^2$ **35.** $\dfrac{2232.23\ \text{g}}{\text{m}}$ **37.** $\dfrac{2.54\ \text{g}}{\text{cm}^2}$

39. $\dfrac{1.10\ \text{lb}}{\text{qt}}$

41. The box of Wheat Bran Flakes weighs 535.8 g.
43. The company should pay 19 cents per kilometer.
45. The baby elephant weighs 133.8 kg.
47. A typical pole measures 29.0 m.
49. The Canadian pack is the better buy.
51. The average rainfall in Freeport is 68.1 cm.
53. The rate is $\dfrac{0.7\ \text{sec}}{1\ \text{yd}}$ or $\dfrac{0.7655\ \text{sec}}{1\ \text{m}}$.
55. Doug's expected time is 1 min 18.74 sec.
57. The snow pack should be reported as 9.0 ft.
59. It takes 121 tablets to have at least an ounce of aspirin.
61. Each system was defined to be consistent with itself. Conversions within each system are exact because the definitions are exact. For example, 1 yd = 3 ft = 36 in. On the other hand, the metric system was not intended to be consistent with the English system; it was intended to replace it. It is certainly consistent with itself, but no effort was made to be consistent with the English system. Therefore, we have conversions like 1 in. ≈ 2.54 cm and 1 liter ≈ 1.057 qt.
63. The general strategy is to multiply by one in the form of unit conversion factors as many times as is necessary in order to convert all the units. Then simplify the results. Choose unit conversion factors so that the existing units are canceled and the new units remain.

$$40\ \text{mph} \approx \left(\frac{40\ \cancel{\text{mi}}}{1\ \cancel{\text{hr}}}\right)\left(\frac{1.609\ \text{km}}{1\ \cancel{\text{mi}}}\right)\left(\frac{1\ \cancel{\text{hr}}}{60\ \text{min}}\right) = \frac{64.36\ \text{km}}{60\ \text{min}} \approx 1.07\ \text{km/min}$$

65. Mary has 26.0 ℓ, to the nearest tenth of a liter.
69. $\dfrac{7}{9}$ **71.** $\dfrac{131}{7}$ **73.** $18\dfrac{9}{25}$ **75.** $\dfrac{8}{9}$
77. The area is $210\ \text{cm}^2$.

Exercises 5.8

1. 0.75 **3.** 0.375 **5.** 0.6875 **7.** 3.55
9. 11.024 **11.** 0.4; 0.43 **13.** 0.5; 0.45
15. 0.2; 0.15 **17.** 0.5; 0.53 **19.** 5.9; 5.94
21. $0.45\overline{45}$ **23.** $0.416\overline{6}$ **25.** 0.35; 0.349
27. 0.49; 0.491 **29.** $0.\overline{384615}$ **31.** $0.\overline{857142}$
33. The micrometer reading will be 1.625 in.
35. The measurements are 0.375 in., 1.25 in., and 0.5 in.
37. 0.94; 0.939; 0.9394 **39.** 7.61; 7.609; 7.6090
41. The cost of the fabric is \$4.864 per yard. Decimals are easier because you can avoid using fractions like $\dfrac{608}{100}$.
43. Texas has the winning jump.
45. The larger is 0.0036.
47. The fraction is less than 0.1. Yes, the decimal equivalent is 0.015375.
51. 2362.8041 **53.** 4670.644 **55.** 13.721
57. 0.34
59. The balance of Katya's account is \$544.87.

Exercises 5.9

1. 0.9 **3.** 0.02 **5.** 0.3 **7.** 8.7 **9.** 0.2
11. 0.52 **13.** 3.24 **15.** 130 **17.** 35.86
19. 26.264 **21.** 5.5 **23.** 8.9 **25.** 3
27. 4.2 **29.** 3.1 **31.** 6.255 **33.** 9.9625
35. 12.12 **37.** 0.5705 **39.** 0.64774
41. Elmer spent \$13.31.
43. The average closing price of Oracle is \$50.65.
45. 37.27 **47.** 0.361 **49.** 37.492
51. The Adams family will make monthly payments of \$58.23.
53. Tim has an average of 6.7 yd per carry.
55. The total cost is \$14.15.
57. The area of the stage is approximately $24{,}531.25\ \text{ft}^2$.
59. The volume of the tank is approximately $105.975\ \text{ft}^3$. It will hold approximately 793 gal of water.
61. In the problem $0.3(5.1)^2 + 8.3 \div 5$, we will divide 8.3 by 5 and add the result to $0.3(5.1)^2$. This is because the order of operations requires that we divide before adding. In the problem $[0.3(5.1)^2 + 8.3] \div 5$, the entire quantity $[0.3(5.1)^2 + 8.3]$ is divided by 5. The insertion of the brackets has changed the order of operations so that we add before dividing because the addition is inside grouping symbols.
63. $3.62 \div (0.02 + 72.3 \cdot 0.2) = 0.25$
65. $(1.4^2 - 0.8)^2 = 1.3456$
69. 0.78125 **71.** $\dfrac{23}{25}$ **73.** $\dfrac{9}{200}$
75. The clerk should price the VCR at \$289.45.
77. The volume is $1331\ \text{cm}^3$.

Getting Ready for Algebra, page 571

1. $x = 9$ **3.** $x = 0.6$ **5.** $0.1 = t$ **7.** $x = 524$
9. $x = 0.32$ **11.** $m = 67$ **13.** $16.5 = y$
15. $p = 3.002$ **17.** $40 = x$ **19.** $7.28 = h$
21. $17.25 = c$ **23.** The Celsius temperature is 2° C.
25. The monthly payment is \$118.33.
27. Let M represent the number of miles driven, then $0.22M + 3(23) = 344$. The customer drove 1250 miles.

Chapter 5 True-False Concept Review

1. False: The word name is "five hundred two thousandths." **2.** True

3. False: The expanded form is $\frac{7}{10} + \frac{5}{100}$. **4.** True
5. False: The number at the left is smaller because 821,597 is less than 840,000. **6.** True **7.** True
8. True **9.** False: The numeral in the tenths place is 3 followed by 4, so it rounds to 123.3. **10.** True
11. False: $8.5 - 0.2 = 8.3$, as we can see when the decimal points are aligned.
12. False: Not if we leave out the extra zeros in a product such as $0.5(3.02) = 1.51$. **13.** True
14. False: The decimal moves to the left, making the dividend smaller.
15. False: Move the decimal left when the exponent is negative. **16.** True
17. False: Most fractions have no exact terminating decimal form. For example, the fractions $\frac{1}{3}$, $\frac{1}{6}$, $\frac{1}{7}$, and $\frac{1}{9}$ have no exact terminating decimal equivalents.
18. False: The rounded value is 0.56. **19.** True
20. True

Chapter 5 Test

1. 2.837 **2.** 0.7279, 0.728, 0.7299, 0.7308, 0.731
3. Thirty-two and thirty-two thousandths **4.** 56.274
5. 0.056 **6.** 46.00 **7.** 20.264 **8.** $16\frac{37}{40}$
9. 5.2×10^{-4} **10.** 0.62 **11.** 56,900
12. 15.328 **13.** 2.908 **14.** 0.00266
15. 312.058 **16.** 342 **17.** 5.581×10^{4}
18. 23.143 **19.** 0.29248 **20.** 0.068
21. 90 kpm **22.** The average Sunday offering was $2284.75. **23.** 1739.814 **24.** Grant pays $6.27.
25. Michael Jordan had played in 683 games.
26. The player's slugging percentage is 672.
27. The perimeter of the rectangle is 47.68 m.
28. Harold lost 0.66 lb more than Jerry lost.

CHAPTER 6

Exercises 6.1

1. $\frac{1}{5}$ **3.** $\frac{6}{5}$ **5.** $\frac{8}{9}$ **7.** $\frac{1}{2}$ **9.** $\frac{2}{3}$
11. $\frac{1}{1500}$ **13.** $\frac{8 \text{ people}}{11 \text{ chairs}}$ **15.** $\frac{55 \text{ miles}}{1 \text{ hr}}$
17. $\frac{21 \text{ miles}}{1 \text{ gallon}}$ **19.** $\frac{8 \text{ lb}}{3 \text{ ft}}$ **21.** $\frac{2 \text{ trees}}{7 \text{ ft}}$
23. $\frac{2 \text{ books}}{5 \text{ students}}$ **25.** $\frac{85 \text{ people}}{3 \text{ rooms}}$ **27.** $\frac{15 \text{ pies}}{2 \text{ sales}}$
29. 25 mph **31.** 4 ft per second
33. 66 yd per billboard
35. 9¢ per pound of potatoes **37.** 37.5 mpg
39. 83.3 ft per second **41.** 268.8 lb per square mile
43. 16.1 gal per minute
45. **a.** The ratio of compact spaces to larger spaces is $\frac{3}{4}$.
b. The ratio of compact spaces to total spaces is $\frac{3}{7}$.
47. No, the first part of the country has a rate of 3.5 TVs per house and the second part of the country has a rate of about 3.3 TVs per house.
49. The ratio of cost to selling price is $\frac{7}{12}$.
51. The ratio of sale price to regular price is $\frac{2}{3}$.
53. The population density of Dryton is about 97.6 people per square mile. **55.** Answers will vary.
57. **a.** The ratio of laundry use to toilet use is $\frac{7}{20}$.
b. The ratio of showering to dishwashing is $\frac{16}{3}$.
59. A total of 1.25 mg of lead is enough to pollute 25 ℓ of water.
61. The 14-oz can is 6.36¢ per ounce and the 1-lb can is 6.25¢ per ounce, so the 1-lb can is the better buy.
63. The 5-lb sale bag is the better buy, at a cost of 99¢ per pound. **65.** The 16-oz can has a smaller unit price.
67. The ratio of reduced size to regular size is $\frac{0.25}{66}$ or $\frac{1}{264}$.
69. Ratios are useful ways to compare like measurements because we can more quickly understand the relative values of the two comparisons. For example, let's compare how many people live in different places. It is easier to grasp the ratios 65.4 people per square mile and 13.2 people per square mile than it is to relate the pairs of numbers 150,420 people per 2300 square miles and 24,684 people per 1870 square miles.
71. All rates are ratios, so $4.56 per pound is a rate that is also a ratio. However, rates must compare unlike measurements, so the ratio $\frac{5}{3}$ is technically not a rate.
73. The ratios of fat calories to total calories are as follows:
a. $\frac{21}{92}$ **b.** $\frac{24}{89}$ **c.** $\frac{117}{370}$ **d.** $\frac{18}{65}$
e. $\frac{243}{482}$ **f.** $\frac{171}{415}$ **g.** $\frac{30}{71}$ **h.** $\frac{63}{290}$
77. $\frac{7}{9}$ **79.** 5.14 **81.** $5\frac{27}{40}$ **83.** 0.0312
85. $\frac{18}{25}$

Exercises 6.2

1. True **3.** True **5.** False **7.** True
9. False **11.** False **13.** True **15.** $a = 9$

17. $c = 6$ **19.** $a = 4$ **21.** $c = 10$
23. $x = 18$ **25.** $w = 0.5$ **27.** $x = 10.5$
29. $z = 4.4$ **31.** $y = 10\frac{2}{3}$ **33.** $a = 12.8$
35. $c = 0.6$ **37.** $b = \frac{3}{8}$ **39.** $x = 0.25$
41. $w = 0.5$ **43.** $y = 216$ **45.** $t = 8$
47. $w \approx 1.4$ **49.** $y \approx 54.4$ **51.** $a \approx 0.22$
53. $c \approx 0.95$
55. The value in the box must be 10 so that the cross products are the same.
57. The error is multiplying straight across instead of finding cross products. Multiplying straight across goes with fraction multiplication, but cross multiplication goes with proportions.
59. The store brand must be cheaper than \$5.07 for a 25-load box.
61. **a.** The number of migrane sufferers is 245. **b.** The number of men who suffer from migranes in the group is 102. **c.** Out of 350 migrane sufferers, 70 could be related to diet.
63. The proportion is true if the butterfly has a wingspan of about 1.9 in. This is realistic.
65. To solve $\frac{3.5}{\frac{1}{4}} = \frac{7}{y}$, first cross multiply.

$$(3.5)(y) = \left(\frac{1}{4}\right)(7)$$

$$(3.5)(y) = \frac{7}{4}$$

$$y = \frac{7}{4} \div 3.5$$ Rewrite as division. At this point we can simplify using either fractions or decimals.

If we choose fractions, then we get $y = \frac{7}{4} \div \frac{7}{2}$ or $y = \frac{1}{2}$.
If we choose decimals, then we get $y = 1.75 \div 3.5$ or $y = 0.5$.
67. $a = 10$ **69.** $w \approx 8.809$ **73.** 6.8265
75. 48,350 **77.** 25.375 **79.** 2.001
81. Quan pays \$15.99.

Exercises 6.3

1. 6 **3.** 15 **5.** $\frac{6}{4} = \frac{15}{h}$ **7.** 30
9. h **11.** $\frac{30}{18} = \frac{h}{48}$ **13.** 8 **15.** t
17. $\frac{8}{48} = \frac{t}{288}$ **19.** T **21.** $\frac{3}{65} = \frac{T}{910}$
23. The school will need 4 additional teachers.
25. $\frac{1.5}{30} = \frac{14}{x}$ **27.** Merle must knit 100 rows.
29. 2500 ft^2 of lawn can be covered with $26\frac{2}{3}$ lb of fertilizer.
31.

	Case I	Case II
Games won	12	x
Games played	15	30

33. The team should win 24 games.
35. John must work 4 weeks to pay for his tuition.
37. Hazel will be paid \$145.
39. The cost of 20 oz of soap powder is \$2.20.
41. The cost of 10 lb of tomatoes is \$9.48.
43. The brine solution requires 6 gallons of water.
45. The second room will cost \$350.90 to be carpeted.
47. Ida needs 0.4 cm^3 of the drug.
49. A pound of gumdrops should cost \$4.80.
51. Betty needs 12 lb of cashews and 28 lb of peanuts.
53. Debra needs 42 qt of blue paint.
55. Mario needs 70 lb of ground round for the meatballs.
57. The computer costs £1495.
59. A 60-month battery should cost \$59.75.
61. 12 m^3 of air must have less than 2820 mg of ozone.
63. The store must reduce the price of each box by 10¢.
65. Assuming a Cheerio is about 0.5 in., the set designer must make them 11 ft in diameter.
67. A proportion is composed of two equal ratios or rates. The speed of a vehicle can be measured in different units: $\frac{60 \text{ mi}}{1 \text{ hr}} = \frac{88 \text{ ft}}{1 \text{ sec}}$. Food recipes can be cut or expanded by keeping the ingredient measures in proportion. Prices for many items, especially food and clothing, are proportional to the number purchased.
69. From the consumer's viewpoint it is better if there is a price break when buying larger quantities. Many products have a lower unit price when bought in bulk or larger packages.
71. The drive shaft speed is 1120 revolutions per minute.
75. 37.415 **77.** $0.00872 < 0.011$
79. The gas costs \$16.41. **81.** $\frac{173}{200}$ **83.** 0.547

Chapter 6 True-False Concept Review

1. True **2.** False: A ratio is a comparison of two numbers or measures written as a fraction.
3. False: $\frac{7 \text{ miles}}{1 \text{ gallon}} = \frac{14 \text{ miles}}{2 \text{ gallons}}$ or $\frac{7 \text{ miles}}{1 \text{ hour}} = \frac{14 \text{ miles}}{2 \text{ hours}}$
4. True **5.** False: If $\frac{7}{3} = \frac{t}{4}$, then $t = \frac{28}{3}$.
6. False: In a proportion, two rates are *stated* to be equal but they may not be equal.
7. False: Like measures have the same units. Twelve inches and one foot are *equivalent* measures.

8. False: Ratios called rates compare unlike units.
9. True
10. False: The table should look like this:

	First Tree	Second Tree
Height	18	x
Shadow	17	25

Chapter 6 Test

1. $\frac{4}{5}$ **2.** Ken would answer 80 questions correctly.
3. $w = 1.2$ **4.** True **5.** $y = 6.5$ **6.** True
7. Mary should be paid \$84.24. **8.** $\frac{1}{12}$
9. The price of 10 cans of peas is \$8.20.
10. One would expect 17.5 lb of bones. **11.** $x = 2.4$
12. They will catch 24 fish.
13. Jennie will need 10 gallons. **14.** $a = 5.35$
15. Yes **16.** The tree will cast a 10.5 ft shadow.
17. The population density is 131.75 people per square mile. **18.** $y \approx 1.65$
19. The crew could do 26 jobs.
20. There are 35 females in the class.

CHAPTER 7

Exercises 7.1

1. 15% **3.** 63% **5.** 140% **7.** 19%
9. 7% **11.** 22% **13.** 44% **15.** 35%
17. 170% **19.** 100% **21.** 150%
23. 37.5% **25.** 60% **27.** 93.75%
29. $66\frac{2}{3}\%$ **31.** $83\frac{1}{3}\%$ **33.** 12
35. The percent of eligible voters who exercised their right to vote was 76%.
37. $87\frac{1}{2}\%$ **39.** 420% **41.** $53\frac{1}{6}\%$
43. The telephone tax is 8%.
45. The annual interest rate is 5.65%.
47. Carol spends 67% of her money on the outfit.
49. The percent paid in interest is 15%.
51. The tax rate is 2.75%.
53. 25% can be pictured as 125% can be pictured as

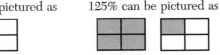

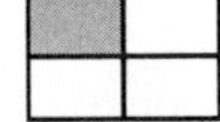

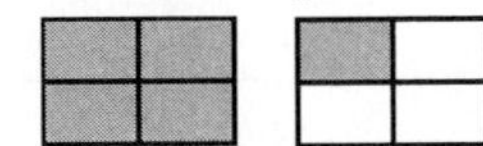

Both quantities involve a unit with $\frac{1}{4}$ shaded. But in 125%, the 1 in the hundreds place indicates that there is also 1 entire region shaded.
55. $\frac{503}{800}$; 62.875% **57.** 0.53125 **59.** 0.72
61. 16.3 **63.** 0.00567002
65. Yes, Bill can make the purchases.

Exercises 7.2

1. 47% **3.** 651% **5.** 7% **7.** 117%
9. 1700% **11.** 0.6% **13.** 73.2% **15.** 45%
17. 300% **19.** 41.6% **21.** 9.56%
23. 8000% **25.** 1475% **27.** 0.034%
29. 330% **31.** 3.45% **33.** 0.011%
35. 20.54% **37.** 74.166% or $74\frac{1}{6}\%$
39. 10.05% **41.** The income tax rate is 22%.
43. The percent of the items sold is 52%.
45. The percent of contestants who withdrew is 13.7%.
47. The price of a snow blower increased by 173%.
49. The swim team won 68.5% of their meets.
51. The sale price is 89% of the original price. The store will advertise "11% off."
53. The percent of income paid for both Social Security and Medicare is 7.65%.
55. Percent means parts out of 100, so a number written as a percent represents a part of 100. For example, 36% means 36 parts out of 100. Another way to represent 36 out of 100 is to write 0.36, or 36 hundredths. The two forms differ by the position of the decimal point and the presence of the % symbol. When a number is in its percent form, the % symbol is used and the decimal point is two places farther to the right than in the decimal form of the number.
57. 1,800,000% **59.** 42.5%; $42\frac{5}{9}\%$
61. 3.873 **63.** 2.6472 **65.** 0.0185
67. 0.6088
69. The total cost of Marilyn's books is \$167. Marilyn pays an average of \$41.75 per book.

Exercises 7.3

1. 0.18 **3.** 0.91 **5.** 0.73 **7.** 0.0346
9. 4.16 **11.** 2.165 **13.** 0.0003 **15.** 2.13
17. 0.0013 **19.** 23.4 **21.** 0.00058 **23.** 1
25. 1.25 **27.** 0.005 **29.** 0.00625
31. 0.2975 **33.** 0.0000074 **35.** 0.00875
37. 0.67443 **39.** 4.755
41. The employees will get a raise of 0.0975.
43. The Credit Union will use 0.1075 to compute the interest.
45. Unemployment is down by a factor of 0.0035.
47. 0.002 **49.** 0.351 **51.** 0.018
53. The river pollution due to agricultural runoff is 0.6 of the total pollution.
55. In 1993, Americans had cholesterol levels a factor of 0.04 less than in 1981.

57. Carol will use 0.45 to calculate the savings. **59.** Answers will vary. **61.** Randy Johnson won 0.833 of the games in which he was involved in the decision. **63.** The decimal will be greater than one when the percent is greater than 100%. For example, 235% written as a decimal is 2.35 and 3.4% written as a decimal is 0.034. **65.** 0.6; 0.569 **67.** 0.875

69. 1.1875 **71.** 1.8 **73.** $\frac{143}{200}$

75. Greg's take-home pay is $206.55.

Exercises 7.4

1. 63% **3.** 78% **5.** 85% **7.** 50%
9. 44% **11.** 15% **13.** 130% **15.** 137.5%
17. 0.9% **19.** 52% **21.** 460% **23.** $66\frac{2}{3}\%$
25. $183\frac{1}{3}\%$ **27.** $283\frac{1}{3}\%$
29. $5\frac{1}{4}\%$ or 5.25% **31.** 73.3% **33.** 55.6%
35. 63.6% **37.** 261.5% **39.** 9.5%
41. Mike Bibby made 85% of his free throws.
43. The voter turnout was 71.4%.
45. The soccer club won 52.9% of their games.
47. 11.75% **49.** 1.86% **51.** 484.62%
53. The cost of food is 108.3% of what it was a year ago.
55. Miguel is taking 162.5% of the recommended allowance.
57. In 1995, 57.8% of the days were smoggy. In 1998, 45.8% of the days were smoggy. Possible reasons include better emission controls on cars and trucks and more pollution controls on factories.
59. 2.3% **63.** 0.2185 **65.** 8.2215
67. 0.0326 **69.** 0.129
71. Spencer's average loss per day is 0.83 lb.

Exercises 7.5

1. $\frac{3}{50}$ **3.** $\frac{9}{20}$ **5.** $\frac{5}{4}$ or $1\frac{1}{4}$ **7.** 3
9. $\frac{4}{5}$ **11.** $\frac{16}{25}$ **13.** $\frac{3}{4}$ **15.** $\frac{3}{20}$
17. $\frac{3}{2}$ or $1\frac{1}{2}$ **19.** $\frac{83}{100}$ **21.** $\frac{57}{200}$
23. $\frac{17}{250}$ **25.** $\frac{61}{200}$ **27.** $\frac{1}{300}$ **29.** $\frac{3}{40}$
31. $\frac{3}{400}$ **33.** $\frac{21}{40}$ **35.** $\frac{13}{70}$ **37.** $\frac{3}{80}$
39. $\frac{7}{6}$ or $1\frac{1}{6}$
41. George and Ethel paid $\frac{4}{25}$ of their income in taxes.

43. The monthly mortgage payment is $\frac{9}{25}$ of their monthly income.
45. Grant Hill made $\frac{19}{25}$ of his shots. **47.** $\frac{14}{125}$
49. $\frac{2}{45}$ **51.** $\frac{21}{4000}$
53. It represents $\frac{29}{25}$ or $1\frac{4}{25}$ of last year's enrollment.
55. Juanita pays $\frac{18}{25}$ of the original price.
57. The fraction of residents between the ages of 25 and 40 is $\frac{7}{40}$.
59. The fraction of medflies that is eliminated is $\frac{22}{25}$.
61. Sales can be described either by fractions or percents. For example, you might hear about a sale of $\frac{1}{3}$ off or of 33% off. Common statistics are also given in either form. You might hear that $\frac{1}{5}$ of the residents of a town are over 65 years old or that 20% of the residents are over 65 years old. It is important to note that in such circumstances, fractions are generally approximations of the actual statistics. Percent forms of statistics may be approximations but are generally considered more accurate than fraction forms.
63. $\frac{1}{400{,}000}$ **65.** $1\frac{10{,}001}{20{,}000}$ **69.** False
71. True **73.** $\frac{\$0.52}{1\text{ person}}$ **75.** $\frac{147.65\text{ gallons}}{1\text{ person}}$
77. The average estimate is 2.314 mm.

Exercises 7.6

Fraction	Decimal	Percent
$\frac{1}{10}$	0.1	10%
$\frac{3}{10}$	0.3	30%
$\frac{3}{4}$	0.75	75%
$\frac{9}{10}$	0.9	90%
$1\frac{9}{20}$	1.45	145%

Fraction	Decimal	Percent
$\frac{3}{8}$	0.375	37.5% or $37\frac{1}{2}$%
$\frac{1}{1000}$	0.001	0.1%
$\frac{1}{1}$ or 1	1	100%
$2\frac{1}{4}$	2.25	225%
$\frac{4}{5}$	0.8	80%
$\frac{11}{200}$	0.055	$5\frac{1}{2}$%
$\frac{7}{8}$	0.875	87.5% or $87\frac{1}{2}$%
$\frac{1}{200}$	0.005	$\frac{1}{2}$%
$\frac{3}{5}$	0.6	60%
$\frac{5}{8}$	0.625	$62\frac{1}{2}$%
$\frac{1}{2}$	0.50	50%
$\frac{43}{50}$	0.86	86%
$\frac{5}{6}$	$0.833\overline{3}$	$83\frac{1}{3}$%
$\frac{2}{25}$	0.08	8%
$\frac{2}{3}$	$0.66\overline{6}$	$66\frac{2}{3}$%
$\frac{1}{4}$	0.25	25%
$\frac{1}{5}$	0.20	20%
$\frac{2}{5}$	0.4	40%
$\frac{1}{3}$	$0.33\overline{3}$	$33\frac{1}{3}$%
$\frac{1}{8}$	0.125	12.5% or $12\frac{1}{2}$%
$\frac{7}{10}$	0.7	70%

1. The tires are 25% off the regular price.
3. The VCR sells for $\frac{2}{5}$ off the regular price.
5. Melinda gets the best deal from Machines Etc.
7. Super Value Grocery offers the best deal.
9. Vendor B offers the best deal and meets the boss's authorized amount.
11. Fractions are commonly used to describe units with subdivisions that are not powers of 10. For example, inches are subdivided into fourths and eighths so it is common to have measures of $14\frac{3}{8}$ in. Rates are usually stated using decimal forms. For example, unit pricing is generally given in tenths of a cent as in Cheerios costing 19.4¢ per ounce. Statistics are most often given using percent forms. For example, 52% of the total orchestra expenses are salaries of the musicians.
15. $y = 28.8$ **17.** $x = 4.75$ **19.** $B = 140$
21. $A = 3.8$
23. The Bacons should collect $20,800 in insurance money.

Exercises 7.7

1. 24 **3.** 56 **5.** 200% **7.** 50% **9.** 70
11. 36 **13.** 80% **15.** 100 **17.** 2
19. 50 **21.** 8.05 **23.** 3.25% **25.** 150
27. 17.992 **29.** 105.5 **31.** 312 **33.** 76.8%
35. 7.403 **37.** 62.5% **39.** 205 **41.** 12.8%
43. 235.96 **45.** 69.344 **47.** 162.1%
49. $\frac{119}{240}$ **51.** $53\frac{4}{15}$ **53.** 70.3%
55. The problem is in equating a 40¢ profit with a 40% profit. A 40¢ profit means that the company makes 40¢ on each part sold. A 40% profit means that the company makes 40% of the cost on each part. But 40% of 30¢ is only 12¢. So a 40% profit would mean selling the parts for 42¢. The company is making a much higher profit than that. The 40¢ profit is actually 133% of the cost.
57. $\frac{17}{75}$ **61.** $x = 17.5$ **63.** $a = 0.25$ or $a = \frac{1}{4}$
65. $w = 55.9$ **67.** The number of miles is 115.
69. The number of inches needed is $20\frac{1}{2}$.

Exercises 7.8

1. The tax is $5.20.
3. Joan's yearly salary was $15,665.
5. Maria's total salary was $394.91.
7. John's score was 83%.
9. The tip was 15% of the check.
11. The number of men younger than 40 in Verboort is 4761.
13. The percent of savings is 27.4%.

15. By 2005 it is predicted there will be 169,000 physical therapists. **17.** Merle had \$160 in her account in August. **19.** Answers will vary.
21. The recommended daily values are as follows: fat, 63 g; sodium, 2444 mg; potassium, 3600 mg; dietary fiber, 21 g. **23.** The percent of markup is 30%.
25. The markup is \$11.20.
27. The sale price of the TV is \$245.24. The competitor of the store in Exercise 26 has the better buy by \$2.22.
29. The salesman's earnings were \$567.
31. The property tax rate is 2.7%.
33. The Reeboks are sold at 20% off.
35. The final price of the blazer is \$67.18. The savings is 52% off the original price.
37. You pay \$181.97 for the golf clubs.
39. You will save \$6.60 on the purchase of the knit shirt.
41. In 1990, 1.65 tons of paper were recycled.
43. Hispanics have the smallest number of children in Head Start.
45. The greatest expenditure of funds is for salaries.
47. The amount spent on development was \$500,000.

49.

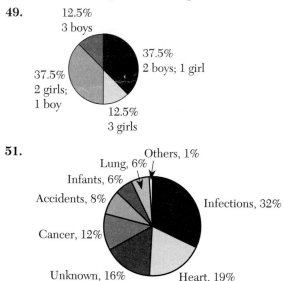

12.5%
3 boys
37.5%
2 boys; 1 girl
37.5%
2 girls;
1 boy
12.5%
3 girls

51.

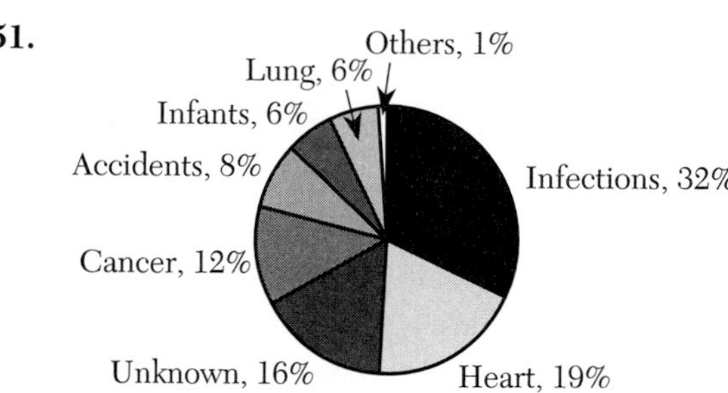

53. A 10% increase in \$100 is a \$10 increase and results in \$110. Now a 10% decrease from \$110 is a decrease of \$11 and results in \$99. The second 10% is taken on a larger base number.
55. The selling price of the lounge chair is \$82.86.
57. The percent of savings with the diesel engine is 42%.
61. The family spends 24% of the budget for food.
63. 234% **65.** 0.006 **67.** 4
69. Rita receives \$3280 and Sally receives \$4920.

Chapter 7 True-False Concept Review

1. True **2.** True
3. False, rewrite the fraction with a denominator of 100 and then write the numerator and add a percent symbol.
4. False, move the decimal point two places to the right and add a percent symbol.
5. True **6.** False, the base unit is always 100.
7. True **8.** True
9. False, the proportion is $\frac{A}{B} = \frac{X}{100}$ **10.** True
11. True **12.** False, $7\frac{1}{2} = 750\%$.
13. False, 0.009% = 0.00009.
14. False, B = 28,000. **15.** True **16.** True
17. False, it is the same as a decrease of 27.75%.
18. False, her salary is 99% of what it was Monday.
19. True **20.** True **21.** True
22. False, $\frac{1}{2}\% = 0.005$.

Chapter 7 Test

1. The percent of discount on the computer is 11.2%.
2. 9.314% **3.** The population is 12% left-handed.
4. 65.1% **5.** 43.75% **6.** 56.5 **7.** $\frac{23}{28}$
8. 456% **9.** 47.97 **10.** 555.6%
11. $\frac{49}{20}$ or $2\frac{9}{20}$ **12.** 0.1893
13. The Adams family spends \$989.39 per month on food.
14–19.

Fraction	Decimal	Percent
$\frac{13}{32}$	0.40625	40.625%
$\frac{78}{125}$	0.624	62.4%
$\frac{29}{200}$	0.145	14.5%

20. 28.7% **21.** 0.0067
22. The selling price of the tire is \$67.44.
23. The sale price of the Hilfiger jacket is \$69.22.
24. The percent of increase in population is 108.0%.
25. The percent of B grades is 23.4%.
26. Gas prices decreased 11%.
27. The suit sells for \$401.87.
28. The percent of calories from fat is 47.4%.
29. The number of students who took the final exam is 112. **30.** The percent of increase in the price of the radial tire was 17.2%.

CHAPTER 8

Exercises 8.1

1. 4 **3.** −5 **5.** 2.3 **7.** $-\frac{3}{5}$ **9.** $\frac{6}{7}$
11. −61 **13.** 42 **15.** 3.78 **17.** $-\frac{17}{5}$

19. −0.55 **21.** −113.8 **23.** 0.0123 **25.** 4
27. 24 **29.** 3.17 **31.** $\frac{5}{11}$ **33.** $\frac{3}{11}$
35. 32 or −32 **37.** 0.0065 **39.** 355
41. $\frac{11}{3}$ **43.** $\frac{33}{7}$ **45.** 0 **47.** 0.341
49. $-\frac{5}{17}$ **51.** −78
53. The opposite of a gain of five eighths is $-\frac{5}{8}$.
55. The opposite of a reading of 12° C is −12° C.
57. The opposite of 1875 B.C. is 1875 A.D.
59. As a signed number the distance up the mountain is +1685 ft and the distance down the mountain is −1246 ft.
61. **a.** As a signed number the savings is −\$28.76.
b. As a signed number the opposite of the savings is +\$28.76.
63. The continental altitudes are as follows: Africa—Kilimanjaro, +19,340 ft; Lake Assal, −512 ft; Antarctica—Vinson Massif, +16,864 ft; Bently Trench, −8327 ft; Asia—Mt. Everest, +29,028 ft; Dead Sea, −1312 ft; Australia—Mount Kosciusko, +7310 ft; Lake Eyre, −52 ft; Europe—Mount El'brus, +18,510 ft; Caspian Sea, −92 ft; North America—Mount McKinley, +20,320 ft, Death Valley, −282 ft; South America—Mount Aconcagua, +22,834 ft; Valdes Peninsula, −131 ft.
65. The highest point on the earth (Mount Everest) is nearer to sea level than the deepest point in the ocean (Mariana Trench) by 6812 ft. The mathematical concept is absolute value. **67.** −81 **69.** 33
71. As signed numbers the Dallas Cowboys made the following yards, 6 yd, −8 yd, 21 yd, and −15 yd.
73. Zero is the only number that is its own opposite. This is because opposites must be the same distance from zero on the number line but in opposite directions. Every other number is some distance from zero and so has an opposite on the other side of zero that is not itself.
75.

```
        |<------ -4 ------|----- -(-4) = 4 -->|
---|----|----|----|----|----|----|----|----|----|----|----|---
  -5   -4   -3   -2   -1    0    1    2    3    4    5    6
        |<--- |-4| = 4 ---|
```

Four units to the left of 0 is −4. The opposite of −4 is 4 units to the right of 0, or 4. The absolute value of −4 is its distance from 0, $|-4| = 4$. A practical use of −4 is a golfer's having a round of 4 under par.
77. 4
79. For all numbers greater than or equal to zero.
81. 1200 g **83.** 4.2 kg **85.** 8250 m
87. 20.1 m **89.** Each part of the coil is 1.5 m.

Exercises 8.2

1. 2 **3.** −1 **5.** −14 **7.** −16 **9.** 0
11. −13 **13.** −34 **15.** −8 **17.** −27
19. 5 **21.** −4 **23.** −144 **25.** −80
27. 0 **29.** 0 **31.** 1 **33.** −6.6 **35.** 2.6
37. $-\frac{1}{2}$ **39.** $-\frac{13}{8}$ **41.** −121 **43.** −189
45. −30 **45.** −30 **47.** 81 **49.** −757
51. −114
53. The sum of the temperatures recorded at 9:00 A.M. is −349° C.
55. The sum of the temperatures recorded at 11:00 P.M. is −505° C.
57. The net change in weight the cargo master should report is −133 lb.
59. The altitude of the airplane at 4 P.M. is 29,870 ft.
61. The inventory at the end of the month is 29,388 volumes.
63. No, the Bills were one yard short of a first down.
65. The net change in the Nordstrom stock for the week was up $1\frac{1}{4}$.
67. The low temperature reading for the last day is −8°.
69. Marie won \$2. **71.** 11 **73.** −7
75. −24 **77.** 660 fps **79.** 678 ft
81. The perimeter is 142 m.
83. The perimeter is 72 in.
85. One bar of soap costs \$0.65.

Exercises 8.3

1. 2 **3.** −10 **5.** −2 **7.** 3 **9.** −17
11. −12 **13.** 32 **15.** −31 **17.** −26
19. 1 **21.** 0 **23.** −24 **25.** −60
27. 87 **29.** −138 **31.** 4 **33.** −178
35. 182 **37.** 0 **39.** 0 **41.** 177
43. −17.88 **45.** −5.34 **47.** −24.03
49. 116 **51.** −677
53. The change in temperature is −85° C.
55. Joe wrote checks totaling \$550.86.
57. The range of altitude for each continent is Africa, 19,852 ft; Antarctica, 25,191 ft; Asia, 30,340 ft; Australia, 7362 ft; Europe, 18,602 ft; North America, 20,602 ft; and South America, 22,965 ft. The smallest range is in Australia. It means it is a rather flat continent.
59. The depth of the ocean floor at Mauna Kea is −19,680 ft or 19,680 ft below sea level.
61. The difference between the temperature recorded at 11:00 P.M. on day 3 and day 5 is −15° C.
63. The difference between the highest and lowest temperatures recorded during the five days is 105° C.
65. Thomas' bank balance is −\$71 or \$71 overdrawn.
67. Janna wrote checks totaling \$251.87.
69. Marie's bank balance is now −\$16.85.
71. Gina's sales total −\$119.91.
73. The New England Patriots lost yards. The loss as a signed number is −28 yd.

75. Adding signed numbers means to move on the number line, starting at zero, the specified number of units in the direction indicated, + to the right and − to the left. Subtraction requires that the direction of the number being subtracted be reversed. This is why we say to subtract by adding the opposite of the second number.
77. Since adding involves moving the indicated direction on the number line, any order will result in the same outcome. But subtraction is defined as reversing the direction of one of the numbers and then adding. So it stands to reason that reversing the direction of one of the numbers will not be the same as reversing the direction of the other before adding.
79. -23.5 **81.** -24 **83.** 8.3
85. The area of the circle is 452.16 in^2.
87. The area of the rectangle is 864 cm^2.
89. The secretary can type 1.5 words per second.
91. The area of the circle is 143.06625 km^2.
93. To carpet the floor you need 180 yd^2.

Exercises 8.4

1. -8 **3.** -10 **5.** -80 **7.** -77 **9.** 11
11. -306 **13.** -255 **15.** -9 **17.** -350
19. $-\frac{1}{4}$ **21.** 3 **23.** 28 **25.** 99
27. 60 **29.** -11 **31.** 210 **33.** 5.4
35. 24.2 **37.** $\frac{1}{6}$ **39.** 1.645 **41.** -2520
43. 465 **45.** 7.14 **47.** $-\frac{3}{10}$ **49.** -14.035
51. 0.5423 **53.** 0 **55.** 21 **57.** -240
59. $\frac{1}{3}$ **61.** The Celsius measure is $-20°$ C.
63. Ms. Riles' loss expressed as a signed number is -28 lb.
65. The Dow Jones decline over the 12 days is -33.96.
67. -49 **69.** -42 **71.** -252
73. Safeway's loss expressed as a signed number is $-\$67.20$.
75. Winn Dixie's loss expressed as a signed number is $-\$20.08$.
77. Safeway's loss expressed as a signed number is $-\$99.68$.
79. The trader paid $7118.75 for the stock. She received $6693.75 when she sold the stock. She lost $425 or $-\$425$ on the sale of the stock.
81. The continent is Asia.
83. One of the ways to interpret the opposite of a number is to consider it as -1 times the number. So in the expression -3^2 we have two operations, multiplying by -1 and squaring. The rules for order of operations require that we do the exponent before multiplying. So, $-3^2 = -1(3)^2 = -1(9) = -9$. In the expression $(-3)^2$ we have the same two operations but the parentheses require that we multiply first, so -3 is squared or $(-3)^2 = 9$.

85. -70 **87.** 56
91. The wheel will make 2522 revolutions in 2 miles.
93. The Smiths will need 15 pounds of grass seed.
95. It takes 39.75 ft^2 of sheet metal to make the box.
97. The cost per square foot for the rental is $0.5625 per square foot per year.

Exercises 8.5

1. -2 **3.** -4 **5.** -3 **7.** -8 **9.** 8
11. -6 **13.** -2.02 **15.** -35 **17.** -3
19. -1.2 **21.** 2 **23.** 3 **25.** 7 **27.** 6
29. -5 **31.** 7 **33.** 8 **35.** 4.04 **37.** $\frac{1}{2}$
39. 5 **41.** -45 **43.** 116 **45.** -1.2
47. 0 **49.** -0.0026 **51.** -452 **53.** 16
55. Each member's share of the loss is $-\$47.37$.
57. Mr. Harkness lost an average of -4.5 lb per week.
59. The average daily loss for the stock is $-\left(2\frac{5}{8}\right)$ points.
61. Answers will vary.
63. The continent is Australia.
65. The division rules are the same as a consequence of the fact that division is the inverse of multiplication. Every division fact can be rewritten as a multiplication fact. So we have: $a \div b = a$(reciprocal of b).
67. -9 **69.** $-\frac{8}{5}$ **71.** -1.4
73. The volume of the cylinder is 4823.04 in^3.
75. The volume of the cone is 1356.48 in^3.
77. The tank will hold 3805 gallons.
79. There will be 12 truckloads.

Exercises 8.6

1. -20 **3.** 20 **5.** -6 **7.** -13 **9.** -4
11. 8 **13.** -3 **15.** 13 **17.** 11 **19.** 9
21. 12 **23.** -13 **25.** 6 **27.** 40
29. -33 **31.** -4 **33.** 8 **35.** 0
37. 0 **39.** 64 **41.** 12 **43.** -22
45. 0 **47.** -12
49. The average temperature recorded during day 5 is $-56°$ C.
51. The average high temperature recorded for the five days is $-34°$ C.
53. 77 **55.** -41 **57.** $-\frac{62}{3}$ **59.** 1
61. -2065 **63.** -1594 **65.** -16
67. Keshia pays a total of $800 for the TV set.
69. K-Mart has a loss on the fishing poles of $-\$9$.
71. The average of the highest and lowest points are: Africa, 9414 ft; Antarctica, 4269 ft; Asia, 13,858 ft; Australia, 3629 ft; Europe, 9209 ft; North America, 10,019 ft; and South America, 11,352 ft. The largest average is in Asia. The smallest average is in Australia.

73. The three cities are Athens, Bangkok, and New Delhi.
75. The balance at the end of the month is $3226.
77. The net result of the trip is −13 miles. At the end of the trip the traveler is 78 miles north and 91 miles west of the starting point.
79. The problem is worked according to the order of operations. So we work inside the brackets first, $3 + 5(-4)$. This is where the error has been made. We must multiply first, so

$$\begin{aligned} 2[3 + 5(-4)] &= 2[3 + (-20)] \\ &= 2[-17] \\ &= -34 \end{aligned}$$

81. Yes, if a different operation is inside parentheses. For example in $(3 - 6)^2$ the subtraction is done first. So, $(3 - 6)^2 = (-3)^2 = 9$.
83. −11 **87.** −17.09 **89.** 16.1
91. −40,920 **93.** −7.366
95. There was no change in the stock for the week.

Exercises 8.7

1. $x = 7$ **3.** $x = 5$ **5.** $y = -5$ **7.** $a = 7$
9. $x = 7$ **11.** $x = 10$ **13.** $x = -3$
15. $x = -6$ **17.** $x = 0$ **19.** $x = 3$
21. $y = 7$ **23.** $a = 0$ **25.** $x = -0.5$
27. $x = -25$ **29.** $x = -16$ **31.** $y = -4$
33. It will take the skydiver 4 sec to reach a falling speed of 147 ft per second.
35. The number is 28. **37.** The number is 8.
39. The acceleration is 33.
41. There will be 14 payments.
43. The equation is $7310 - 4L = 12{,}558$, where L represents the lowest point. Asia's lowest point fits the description.
45. The equation is $-35{,}840 - 2D = -2000$, where D represents the deepest part. The Ionian Basin's deepest part fits this description.
47. $z = 7$ **49.** $z = 6$
51.

$$\begin{aligned} -3x + 10 &= \quad 4 \\ \underline{-10} &\quad \underline{-10} && \text{Add } -10 \text{ to both sides.} \\ -3x &= \quad -6 && \text{Simplify.} \\ \frac{-3x}{-3} &= \frac{-6}{-3} && \text{Divide both sides by } -3. \\ x &= 2 && \text{Simplify.} \end{aligned}$$

53. $x = -4$ **55.** $x = -2$

Chapter 8 True-False Concept Review

1. True **2.** False: The opposite of a positive number is negative.
3. False: The absolute value of a nonzero number is always positive.
4. True **5.** False: The sum of two signed numbers may be positive, negative, or zero.
6. False: The sum of a positive signed number and a negative signed number may be positive, negative, or zero.
7. True
8. False: To subtract two signed numbers, add the opposite of the number to be subtracted.
9. True **10.** True
11. False: The sign of the product of a positive number and a negative number is negative.
12. True **13.** True **14.** True

Chapter 8 Test

1. −14 **2.** 72 **3.** $-\frac{5}{4}$ **4.** $\frac{2}{15}$ **5.** −1
6. 1.07 **7. a.** 21 **b.** 21 **8.** −35
9. −2 **10.** 4 **11.** 126 **12.** −24
13. −90.4 **14.** 40 **15.** −51.1 **16.** $-\frac{1}{6}$
17. 8 **18.** −43 **19.** −80 **20.** $-\frac{2}{7}$
21. −53 **22.** 4 **23.** $x = -5$ **24.** $x = 7$
25. $a = -12$ **26.** Ms. Rosier lost −16.48 lb.
27. The drop in temperature is −21° F.
28. The closing price on Friday is 7.
29. The Fahrenheit temperature is 14° F.
30. The average is −5.

Midterm Examination

1. ten thousand **2.** 94,019
3. Three hundred forty-five thousand, eight hundred four
4. 3057 **5.** 17,763 **6.** 41,496
7. 720,000: 727,518 **8.** 167 **9.** 3637 R 21
10. 7 **11.** 74 **12.** 197 **13.** 180
14. Prime Number **15.** Yes
16. 23, 46, 69, 92, 115
17. 1, 2, 4, 8, 16, 19, 38, 76, 152, 304
18. $2 \cdot 2 \cdot 2 \cdot 2 \cdot 19$ or $2^4 \cdot 19$ **19.** $8\frac{3}{8}$
20. $\frac{123}{7}$ **21.** $\frac{3}{4}, \frac{7}{8}$ **22.** $\frac{5}{9}, \frac{7}{12}, \frac{5}{8}, \frac{2}{3}$
23. $\frac{2}{3}$ **24.** $\frac{1}{14}$ **25.** $55\frac{7}{32}$ **26.** $2\frac{1}{2}$
27. $\frac{3}{13}$ **28.** $\frac{17}{20}$ **29.** $18\frac{13}{24}$ **30.** $3\frac{2}{5}$
31. $7\frac{29}{45}$ **32.** $7\frac{3}{4}$ **33.** $\frac{3}{10}$ **34.** 71,302
35. 11 yd 1 ft 4 in. **36.** The perimeter is 75 ft.
37. The area is $27\frac{3}{4}$ ft². **38.** The volume is 10 ft³.
39. The area is 1156 cm².
40. The perimeter is 80 in.

Final Examination

1. $\frac{23}{16}$ or $1\frac{7}{16}$ **2.** 126 **3.** 33.89 **4.** 19.853

5. $\frac{7}{6}$ or $1\frac{1}{6}$ **6.** $51\frac{1}{3}$ **7.** 199 **8.** 97.84

9. $6\frac{3}{10}$ **10.** 721 **11.** 82.04 **12.** 58.266

13. $\frac{43}{50}$ **14.** $10\frac{17}{30}$ **15.** $x = \frac{7}{3}$ or $x = 2\frac{1}{3}$

16. thousandths **17.** $\frac{14}{20}, \frac{3}{4}, \frac{4}{5}$ **18.** 85%

19. 24 R 11 **20.** 0.478 **21.** 6000.015

22. 3812.5 **23.** 0.11375 **24.** 35 **25.** 452%

26. 0.265 **27.** 36.45 **28.** 0.0306

29. $239.70 **30.** 28.125%

31. 41, 82, 123, 164, 205

32. Nine thousand, forty-five and forty-five thousandths

33. False **34.** $2 \cdot 2 \cdot 3 \cdot 5 \cdot 7$ or $2^2 \cdot 3 \cdot 5 \cdot 7$

35. $\frac{7}{12}$ **36.** $\frac{187}{200}$ **37.** $36\frac{3}{7}$ **38.** $\frac{5}{16}$

39. $\frac{3}{40}$ **40.** 58% **41.** No **42.** $13\frac{13}{15}$

43. $\frac{124}{9}$ **44.** 2.299, 2.32, 2.322, 2.332

45. 0.067481 **46.** $1\frac{11}{23}$ **47.** $\frac{1}{12}$

48. $1812 **49.** $420 **50.** 8.6% **51.** 7

52. $3\frac{7}{20}$ **53.** 5.91 **54.** 7 hr 18 min 11 sec

55. 2 m 92 cm **56.** $50 per kilogram

57. The perimeter is 127.5 ft. **58.** 13.395 m

59. 62.13 yd^2 **60.** 3888 in^3 **61.** -16

62. 27 **63.** -72 **64.** 5 **65.** 2

66. $a = -8$

Index

B

C

D

E

N

O

P

Q